Synthetic Receptors for Biomolecules
Design Principles and Applications

Monographs in Supramolecular Chemistry

Series Editors:
Professor Philip Gale, *University of Southampton, UK*
Professor Jonathan Steed, *Durham University, UK*

Titles in this Series:
1: Cyclophanes
2: Calixarenes
3: Crown Ethers and Cryptands
4: Container Molecules and Their Guests
5: Membranes and Molecular Assemblies: The Synkinetic Approach
6: Calixarenes Revisited
7: Self-assembly in Supramolecular Systems
8: Anion Receptor Chemistry
9: Boronic Acids in Saccharide Recognition
10: Calixarenes: An Introduction, 2nd Edition
11: Polymeric and Self Assembled Hydrogels: From Fundamental Understanding to Applications
12: Molecular Logic-based Computation
13: Supramolecular Systems in Biomedical Fields
14: Synthetic Receptors for Biomolecules: Design Principles and Applications

How to obtain future titles on publication:
A standing order plan is available for this series. A standing order will bring delivery of each new volume immediately on publication.

For further information please contact:
Book Sales Department, Royal Society of Chemistry, Thomas Graham House, Science Park, Milton Road, Cambridge, CB4 0WF, UK
Telephone: +44 (0)1223 420066, Fax: +44 (0)1223 420247
Email: booksales@rsc.org
Visit our website at http://www.rsc.org/Shop/Books/

Synthetic Receptors for Biomolecules
Design Principles and Applications

Edited by

Bradley D. Smith
University of Notre Dame, Indiana, USA
Email: smith.115@nd.edu

ROYAL SOCIETY OF CHEMISTRY

THE QUEEN'S AWARDS FOR ENTERPRISE: INTERNATIONAL TRADE 2013

Monographs in Supramolecular Chemistry No. 14

Print ISBN: 978-1-84973-971-9
PDF eISBN: 978-1-78262-206-2
ISSN: 1368-8642

A catalogue record for this book is available from the British Library

Published by The Royal Society of Chemistry,
Thomas Graham House, Science Park, Milton Road,
Cambridge CB4 0WF, UK

Registered Charity Number 207890

Visit our website at www.rsc.org/books

Preface

The field of supramolecular chemistry continues to grow at an accelerated pace, and it is especially active at the interface of chemistry and biology. Pioneering research conducted during the last quarter of the 20th century focused initially on synthetic receptors for structurally simple guests, such as metal cations. As the field expanded, researchers pursued more elaborate synthetic receptors for binding anions and small neutral molecules. The early pioneers recognized the technical advantages that would be gained by developing synthetic mimics of large protein-based receptor systems, such as antibodies, enzymes, and membrane transporters. But time was needed for the field to discover and conceptualize the fundamental physical-organic factors that produce strong and selective binding of biological molecules in aqueous solution. Also required were new synthetic methods and nanoscale fabrication techniques for reliable construction of suitably functionalized synthetic receptors with appropriate geometric arrangement of molecular recognition motifs. These intellectual and technical attributes have now been acquired and merged to create effective receptor design and discovery platforms.

The classic approach to synthetic receptor development is a step-wise process that involves *de novo* molecular design followed by receptor synthesis and evaluation of supramolecular performance. For some types of research projects, this approach is quite suitable, but for others it is unacceptably slow and inefficient. In response, researchers have developed alternative accelerated discovery processes. One strategy is to prepare libraries of multiple receptor candidates and rapidly screen them for library members that exhibit appropriate binding affinity. A conceptually powerful refinement of this screening approach is to create dynamic combinatorial libraries of receptor building blocks that have been structurally programmed to self-assemble in the presence of a suitable guest template and create a

Monographs in Supramolecular Chemistry No. 14
Synthetic Receptors for Biomolecules: Design Principles and Applications
Edited by Bradley D. Smith

Published by the Royal Society of Chemistry, www.rsc.org

high-affinity receptor. Using these complementary methods, the field has produced biomolecule-binding systems with functional properties for various applications in complicated biological matrices, including living subjects. These applications are the basis of emerging new nanoscale technologies that will have broad future impact in modern society.

This volume describes the receptor design principles and discovery algorithms, and shows how they have been used to produce synthetic receptors for each major class of biomolecules. A unique feature of the book is the presentation style. To date, virtually all review articles or books on synthetic receptors have been presented in a format that is primarily a list of receptor chemical structures with an ensuing discussion of the various biomolecule guests they bind. This book reverses the format and devotes a chapter to each specific class of biomolecule, with a methodical summary of the different biological and synthetic receptors. The systematic layout allows readers to quickly identify the sections that are relevant to their information needs. The target audience is a broad group of scientists and engineers (academic and industrial) who have a biomolecule targeting problem in mind and want to know what synthetic receptors will likely lead to a solution. What are the unique molecular recognition challenges for each type of biomolecule? What is the best synthetic receptor system for a specific type of biomolecule and a specific type of application? Another large group of readers is the regular community of supramolecular chemists who are developing next-generation synthetic receptors with improved performance. They want an integrated picture of the current state of the art and a sense of future directions and expected challenges.

The opening chapter provides a concise summary of the technical applications that utilize synthetic receptors for biomolecules. The applications are classified broadly into four groups: separations, imaging and sensing, catalysis, and pharmaceutical activity. The various receptor classes are separated into six categories: organic molecules, inorganic molecules, synthetic oligomers, molecular imprinted polymers, dendrimers, and nanoparticles. The chemical and supramolecular attributes of each receptor category are summarized, along with illustrative examples of how the synthetic receptors are employed for each type of application. The second chapter is a tutorial on the fundamental structural and thermodynamic factors that affect biomolecule recognition, and includes the following topics: non-covalent interactions, receptor preorganization and shape complementarity, cooperative binding, solvent effects, and enthalpy–entropy compensation. There is also an informative presentation of the various ways to design and discover new classes of synthetic receptors.

The next eight chapters provide a systematic summary of the best-known receptors for alkali metal cations, organic and inorganic anions, carbohydrates, nucleosides and nucleotides, oligonucleotides and nucleic acids, amino acids and peptides, protein surfaces, and polar lipids. Each chapter follows a similar structural format of: (a) chemical structure and physical properties of the biomolecule; (b) biological recognition of the biomolecule;

(c) synthetic receptors for the biomolecule; and (d) future directions and challenges.

The order of chapters is well suited for instructors who wish to use the book for a graduate-level class on supramolecular chemistry. Alternatively, the book is an excellent source of supplementary information for a range of undergraduate chemistry, biochemistry, health science, and engineering classes. A valuable pedagogical addition is an accompanying series of PowerPoint slides containing graphics from each of the 10 chapters. The PowerPoint files are provided as electronic supplementary information on the Royal Society of Chemistry website (DOI: 10.1039/9781782622062).

I am grateful to the chapter authors for completing their writing tasks in a timely fashion, and for agreeing to follow a consistent presentation format. I warmly acknowledge the wonderful technical assistance of Theresa Bollinger, Kasey Clear, Evan Peck, and the publication staff at the Royal Society of Chemistry. I am grateful for funding and resources from the University of Notre Dame, and the US National Science Foundation (grant: CHE1401783).

Bradley D. Smith
Notre Dame, IN, USA

Contents

Monographs in Supramolecular Chemistry No. 14
Synthetic Receptors for Biomolecules: Design Principles and Applications
Edited by Bradley D. Smith

Published by the Royal Society of Chemistry, www.rsc.org

CHAPTER 1

Applications of Synthetic Receptors for Biomolecules

EVAN M. PECK AND BRADLEY D. SMITH*

Department of Chemistry and Biochemistry, 236 Nieuwland Science Hall, University of Notre Dame, Notre Dame, 46556 IN, USA
*Email: smith.115@nd.edu

1.1 Biomolecule Recognition

This chapter describes the major classes of synthetic receptors for biomolecules and how they are employed for four different types of practically useful applications: separations, imaging and sensing, catalysis, and pharmaceutical activity. These applications play crucial roles within the frontier technology areas of health care, environmental remediation, nanotechnology, and advanced materials.[1] Each application requires a synthetic receptor that is endowed with a specific set of functional properties. In all cases, the functions are triggered by a molecular recognition event that involves association of the receptor (sometimes called a host) with the target biomolecule (sometimes called a guest, substrate, ligand, or analyte) (Scheme 1.1).[2] The associated complex is held together by a collection of weak and reversible bonds, and typically the association alters the chemical and physical properties of the binding partners. It is this change in molecular properties that enables the subsequent functions to occur.

The association event can be characterized in terms of thermodynamics (the extent to which association occurs) and kinetics (the rate at which it occurs). The energy diagram in Scheme 1.1 defines the free energy of association (ΔG_a), which determines the association constant (K_a). Another

Monographs in Supramolecular Chemistry No. 14
Synthetic Receptors for Biomolecules: Design Principles and Applications
Edited by Bradley D. Smith

Published by the Royal Society of Chemistry, www.rsc.org

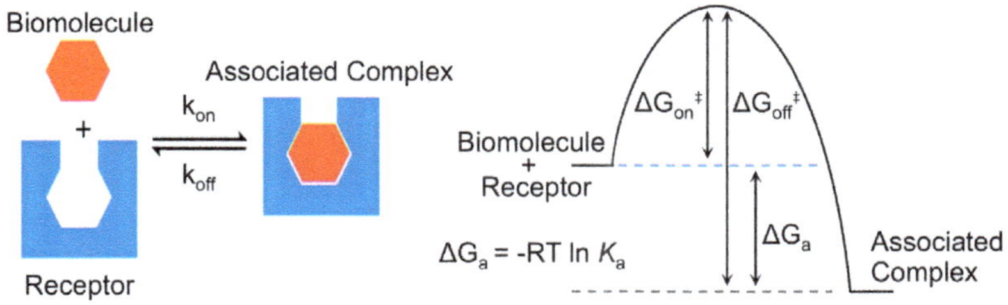

Scheme 1.1 Receptor biomolecule association with 1 : 1 stoichiometry.

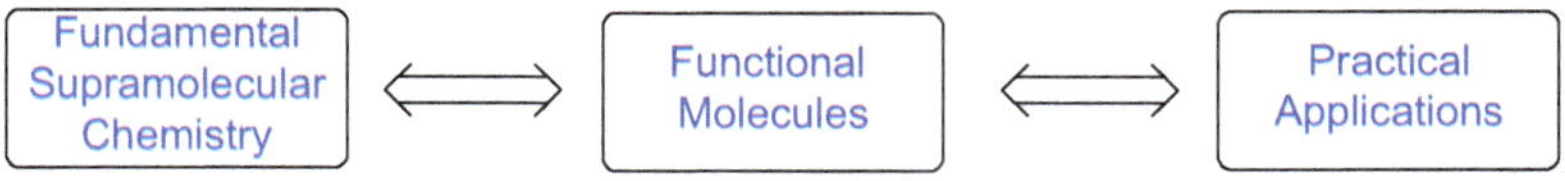

Scheme 1.2 Workflow diagram for supramolecular chemistry research.

important receptor binding property is thermodynamic selectivity, which is equal to the ratio of association constants for two separate guests. The energy diagram also defines the free energies of activation for complex association ($\Delta G_{on}^{\ddagger}$) and dissociation ($\Delta G_{off}^{\ddagger}$), which dictate the corresponding rate constants for these steps (k_{on} and k_{off}, respectively). Different applications have distinctive thermodynamic and kinetic requirements for optimal performance and the kinetic and thermodynamic properties of biological receptors are often informative benchmark values.

Scheme 1.2 is a simple workflow diagram that connects three important stages in supramolecular chemistry research. In the case of biomolecule recognition, the vast majority of research projects move through these stages from left to right. Researchers typically start out by conducting fundamental studies of model receptor/biomolecule association systems and work to develop systematic rules and generalizable concepts that rationalize experimental observations. Based on the new insight, a typical project progresses to the second stage by developing prototype examples of novel functional molecules. Only occasionally does a biomolecule recognition project progress all the way and generate robust new technology with useful applications.

Putting aside discussion concerning the mission of academic research, there are several scientific reasons for the modest number of current industrial and commercial processes that use synthetic receptors for biomolecules. The most obvious point is the relative newness of supramolecular chemistry as a sub-discipline within molecular science. There is strong consensus that the discovery of crown ethers in the late 1960s was the seminal event that coalesced supramolecular chemistry as a field of systematic study.[3] Within few a years supramolecular chemists were pursuing synthetic receptors for small biomolecules. But molecular recognition in water is a complicated physical process, and it has taken some time for the field to elucidate the fundamental thermodynamic factors that produce selective non-covalent association. As discussed further in Chapter 2, relatively intuitive supramolecular concepts such as the hydrophobic effect, binding cooperativity, and enthalpy–entropy compensation are still under intellectual refinement.[4–6] Computational algorithms for rational design of effective receptors are in early-stage development, and so are alternative statistical discovery approaches that aim to uncover effective receptors by combinatorial synthesis and screening methods.

Nonetheless, examples of practically useful applications based on synthetic receptors are emerging at an accelerated pace. It is worth noting that many applications mimic biological processes that evolved by natural selection over billions of years. In comparison, the remarkable progress of supramolecular chemistry over the last 50 years is quite impressive. Continued expansion is expected for several reasons. One is the growing synergistic overlap of supramolecular chemistry with advances in nanoscience. The development of versatile nanoscale fabrication methods has led to new synthetic nanoparticle platforms for constructing multivalent biomolecule recognition systems with relatively large binding surface areas. In addition, major advances in biomolecular engineering now allow rapid construction of biopolymers with unnatural sequences, and new bioconjugation techniques enable ready modification of these biopolymers with synthetic components such as indicator groups, catalytic residues, and secondary binding sites, to produce natural/synthetic hybrid receptor molecules.

Indeed, the field of synthetic receptors for biomolecules is sufficiently advanced that researchers can realistically have the ambition to progress all the way through the workflow diagram in Scheme 1.2. It is quite feasible for a scientist or engineer to use the modern principles of receptor design and discovery to produce supramolecular systems that have practical applications. One of the first major decisions in this workflow process is whether the receptor system should be based on a small molecule, polymer, or nanoparticle platform and the next section in this chapter provides an overview of the attributes of each major choice.

1.2 Types of Synthetic Receptors

The pool of protein-based receptors is a logical resource for researchers who wish to develop a biomolecule-binding receptor that will perform a specific function. Protein-based molecular recognition systems include antibodies, enzymes, lectins, signaling partners, transcription factors, and membrane transport proteins. Antibodies are the most widely utilized protein systems, with a wide range of binding affinities (K_a typically 10^5–10^{12} M^{-1}) and target selectivities. But for certain applications, antibodies and related binding proteins are not feasible. They may be too expensive to produce, or they cannot be stored for extended periods. It may be difficult or impossible to chemically convert them into reporting agents for certain types of imaging or sensing applications. A concern with pharmaceutical applications is the possibility of an undesirable immune response, or the challenge of optimizing the pharmacokinetic properties of a high molecular weight structure.

In principle, all of these drawbacks can be circumvented by developing synthetic receptors.

This chapter classifies synthetic receptors into six categories. The majority of synthetic receptors have a binding pocket that is formed by a convergent arrangement of functional groups appended to either a macrocyclic or cleft-shaped scaffold (Scheme 1.3). The receptor structural factors that control guest affinity, binding selectivity, and binding kinetics are described in Chapter 2. As a rule, preorganized macrocyclic structures tend to bind their complementary guests with higher affinity and selectivity than flexible acyclic receptors. However, the higher affinity usually means that the rates of association and dissociation are slower, which can be an undesirable property for certain applications.[4]

Receptors can also be constructed by attaching multiple functional groups to the exterior of nanoscale scaffolds such as dendrimers or nanoparticles. This divergent arrangement of functional groups enables the receptors to form multiple interactions with the exposed surfaces of relatively large binding partners such as proteins or nucleic acids. But selective binding of biopolymer surfaces in aqueous solution is quite challenging. Exposed polar functional groups are heavily solvated, and thus association processes must overcome a high desolvation penalty. The presence of some nonpolar groups on the receptor surface is often needed to drive association, but too many hydrophobic groups can lead to undesired receptor self-aggregation.

1.2.1 Small Organic Molecules

The largest numerical group of synthetic receptors is comprised of organic molecules. There is a tendency to work with privileged organic molecular scaffolds that already have an inherent molecular recognition capability. Many of these scaffolds are macrocycles, such as crown ethers, calixarenes, cucurbiturils, cyclodextrins, or alternatively acyclic structures with cleft or tweezer shapes (Scheme 1.4, and a more detailed description is provided in Chapter 2). Some organic structures are relatively easy to covalently modify, such as crown ethers, while others are technically harder to transform and purify, such as cyclodextrins and cucurbiturils. The structural modifications include linkers to solubilize or immobilize the receptor, chemical

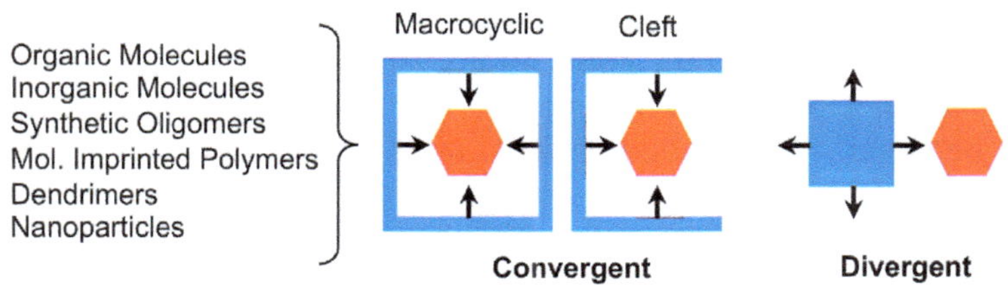

Scheme 1.3 Major categories of synthetic receptors.

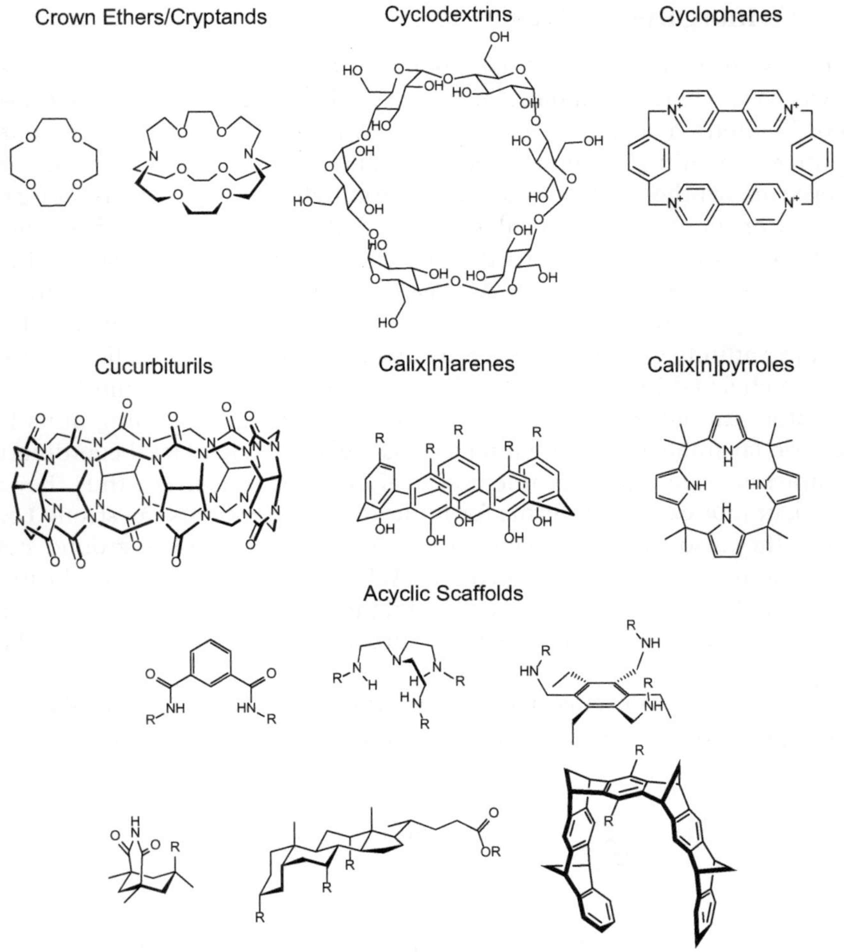

Scheme 1.4 Common scaffolds for organic molecular receptors.

functionality for enhanced binding affinity or recognition selectivity, catalytic residues, or reporter groups for imaging or sensing applications.

Pharmaceutical applications typically require a receptor structure that is biocompatible, and it is relatively easy to modify the structure of a small organic molecule and improve its pharmacokinetic or ADMET (absorption, distribution, metabolism, excretion, and toxicity) properties. Organic receptors are excellent scaffolds for binding ions or small biomolecule guests. But organic receptors have a small number of non-covalent interaction sites and low surface area for hydrophobic contact. Thus, it is difficult to produce organic receptors with high affinity for larger biomolecules such as proteins or cell membranes. One potential solution is to devise receptor self-aggregation methods that assemble a large number of small molecule receptors to produce a larger multivalent conglomerate with increased biomolecule affinity.[7]

1.2.2 Inorganic Molecules

While water is a strong hydrogen bonding molecule, it is a relatively weak Lewis base. Therefore, molecular recognition systems that utilize Lewis acid coordination in aqueous solvent are usually more effective than purely hydrogen bonding systems. Metal coordination complexes are often employed as synthetic receptors for electron-rich species such as inorganic anions or anionic biomolecules. The receptor typically comprises one or more transition metal cations that are chelated by an organic scaffold containing nitrogen donor atoms. At least one of the metal cation coordination sites is open and able to accept the electron-rich guest (Scheme 1.5a). Binding affinity can be enhanced by using multiple Lewis acidic centers to form a chelated binding site for the target biomolecule (Scheme 1.5b). The receptor scaffold can also be decorated with hydrogen bonding units that provide favorable secondary interactions (Scheme 1.5c). Metal coordination complexes often have prominent optical or redox properties, thus they are excellent fabrication platforms for indicators and reporters. As strong Lewis acids, they also can catalyze chemical reactions by polarizing ground states and stabilizing reactive intermediates. While some transition metal cations are potentially toxic, it is usually straightforward to alter the molecular structure of a coordination complex in order to improve solubility or ADMET properties.

In addition to molecular recognition strategies that employ metal centers with open coordination sites for direct contact with a Lewis basic guest, new

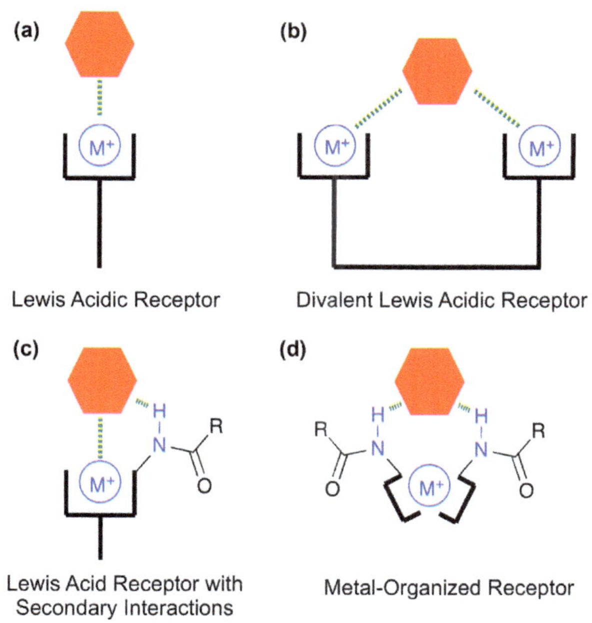

Scheme 1.5 Receptor scaffolds that incorporate Lewis acids.

tactics are emerging that use metal coordination as a structure organizing element. One approach employs metal cation coordination as a construction method to assemble multiple hydrogen bonding units (Scheme 1.5d) and create a convergent binding pocket. An alternative strategy is to combine metal centers with appropriately designed rigid organic building blocks and generate self-assembled soluble cages with large enough cavities to encapsulate molecular guests. This metal organic self-assembly approach has many attractive features, including rapid, often single-step, production of cage structures, which can be highly charged and water soluble. In many cases the cage interior is lined with aromatic surfaces and the hydrophobic effect can be used to drive guest association. The solid-state versions of these self-assembled cages are metal organic frameworks (MOFs). In many cases, MOFs are porous networks and they have typically been used for small molecule capture, as their large surface areas allow for an excellent storage capacity of gases and volatile solvents.[8] Researchers are beginning to pursue additional applications by incorporating fluorescent reporters or catalytic sites within a MOF.[9] It seems likely that porous MOFs will soon be used to selectively trap small biomolecules.

1.2.3 Synthetic Oligomers

Proteins and nucleic acids have excellent molecular recognition abilities, and they can be synthesized using reliable and efficient chemical or biosynthetic procedures. Thus, it is logical to design or discover synthetic receptors that are based on biopolymers with unnatural sequences. The biopolymers can fold to produce structural features such as hairpins, bulges, loops, and binding pockets. Often there is concomitant coordination of a proximal metal cation that can enhance guest binding or reaction catalysis. Polypeptides have the benefit of a wider choice of naturally occurring building blocks (22 naturally occurring amino acids *versus* four naturally occurring ribonucleotides) to construct a wider and more diverse set of receptor candidates. But oligonucleotide systems have a significant advantage in that the binding properties of the receptor structures can be refined by *in vitro* selection methods using biotechnology.

Aptamers are single strands of oligonucleotides with 15–50 building blocks that can fold to form a tertiary structure with the capability to bind a guest molecule (Scheme 1.6).[10] The sequences are usually ribonucleotides rather than deoxyribonucleotides, since the extra hydroxyl provides additional interactions. RNA aptamers are synthetic versions of riboswitches, which are segments within naturally occurring mRNA that can bind a small molecule (usually a metabolite) and change the production of proteins encoded by the mRNA. Aptamers can exhibit extremely high binding constants and high guest selectivities.[11] For example, an aptamer has been reported with 12 000 times higher affinity for L-arginine over its enantiomer, D-arginine.[12] While aptamers can be generated for virtually any molecular target, there are some biomolecules with few functional groups, such as sugars, that

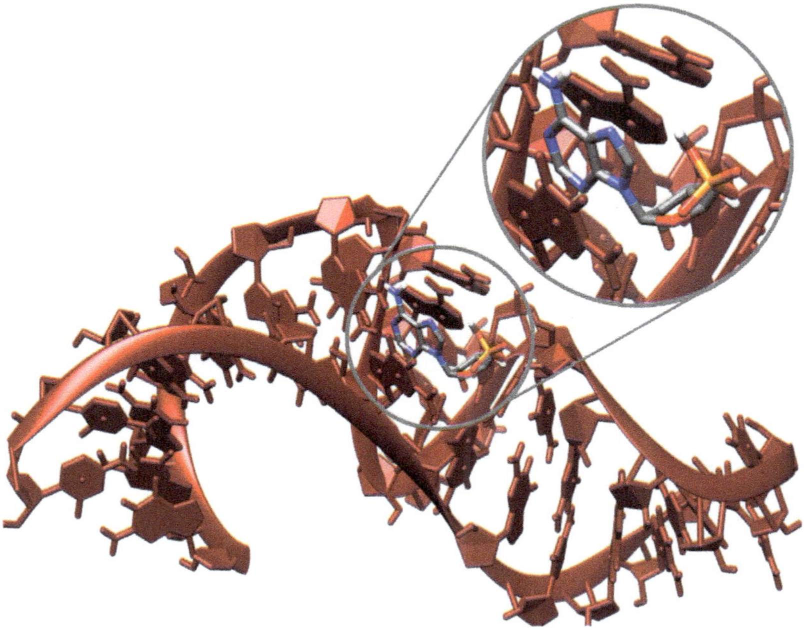

Scheme 1.6 X-ray structure of RNA aptamer complex with 5′-adenosine monophosphate (AMP) guest.

cannot be recognized with high affinity. One way to circumvent this problem is to develop aptamers that form a high-affinity ternary complex with the biomolecule bound to a small organic receptor. An example of this strategy is an aptamer that can recognize glucose bound to a small boronic acid receptor.[13]

Most aptamers are produced using the SELEX (systematic evolution of ligands by exponential enrichment) method or a variant, which involves iterative cycles of oligonucleotide selection, amplification and mutation (see Chapter 2). The technology can be automated to produce an effective binding sequence within a few days. The fact that aptamers are well defined chemical entities with little or no immunogenicity makes them attractive for pharmaceutical development. But aptamers have limited stability in physiological solution due to nuclease degradation. In some cases the rapid degradation is an advantage, but often aptamer stability needs to be improved. Avoidance of the degradation enzymes can be achieved by either: (a) attaching long polyethylene glycol (PEG) chains to sterically protect the aptamer; (b) constructing a Spiegelmer, a sequence composed of non-natural L-ribonucleotides that are not recognized by enzymes; or (c) selective incorporation of a non-cleavable nucleoside, such as a 2′-fluoro-2′-deoxyribonucleoside, at specific sequence locations that are susceptible to enzyme attack.

Aptamers have been investigated extensively as pharmaceutical agents.[14,15] In many cases, the pharmaceutical action is due to selective recognition of a biological target, usually a small section of protein surface (epitope). At present, the only clinically approved aptamer-based therapeutic agent is pegaptanib (brand name MACUGEN®), which is used for treating age-related macular degeneration. Clinical trials have investigated other aptamers for various molecular targets and they also show promise as delivery vehicles for therapeutic small interfering (si)RNA. A related pharmaceutical application is prodrug delivery. A recent study incorporated 30 copies of the cytotoxic nucleoside analogue 5-fluoro-2′-deoxyuridine (5-FUdr) into the sequence of an aptamer (comprised of 106 nucleotides) that binds strongly and specifically to a cell surface receptor.[16] Receptor binding led to endocytosis and aptamer entry into the cell, where it was degraded by intracellular nucleases to release the 5-FUdr drug.

Aptamers are broadly useful in a range of other recognition-based technologies, such as separations, environmental purification, or diagnostics.[11,17,18] In most of these cases, the aptamers are immobilized by covalent attachment to a surface. Considerable effort has focused on developing aptamers as the recognition unit within chemical detection systems. Optical sensing with aptamers is used often for quantifying specific biomolecules in environmental samples, biomedical tissue, or cell culture. Molecular beacons are a broad family of fluorescent oligonucleotide sensors that exploit the conformational switching caused by guest binding to modulate the distance between a fluorophore and an energy accepting partner. Other aptamer design configurations use biomolecule binding to trigger association and enhanced emission of a reporter dye. Although highly effective and very popular, the sensitivity of molecular beacons is limited because of the 1 : 1 signaling stoichiometry. Approaches to produce amplified signaling have investigated ribozymes (or DNAzymes), which are aptamers having catalytic activity.[19] Suitable combination of an aptamer with a ribozyme creates a sensing ensemble that can bind a guest molecule and turn on a catalytic process producing an amplified sensing signal.

Polypeptide versions of aptamers are known, but the combined process of synthesis and screening is much slower and there is no peptide equivalent of SELEX for *in vitro* selection and refinement of binding ability.[20] Peptide bonds are kinetically stable at physiological pH but they are susceptible to catalytic cleavage by protease enzymes. This has led researchers to develop alternative polyamide systems that do not have side chains connected to the α-carbon and thus resist protease action. These structures include peptoids with the side chain connected to the peptide backbone nitrogen,[21] and β-peptides with the amino group bonded to the β-carbon.[22]

1.2.4 Molecular Imprinted Polymers

The general idea of molecular imprinted polymers (MIPs) is to generate cross-linked polymeric materials with template imprinted cavities that act as selective molecular recognition sites. The picture in Scheme 1.7 shows the method of non-covalent imprinting, where a polymerization reaction is

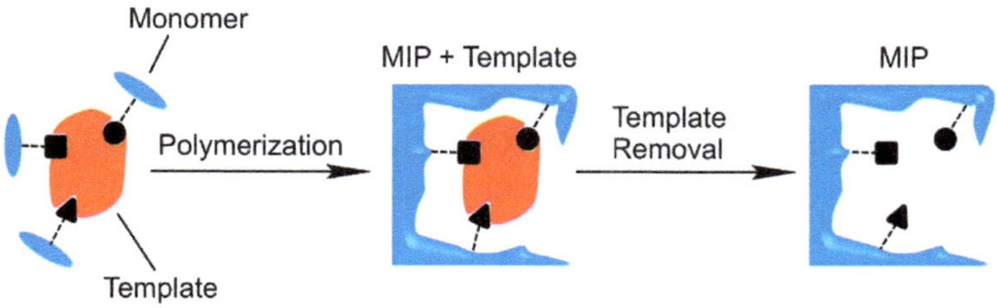

Scheme 1.7 Molecular imprinted polymer created by non-covalent imprinting.

conducted in the presence of a molecule template. The template is incorporated into the cross-linked polymer matrix by means of non-covalent interactions with the monomers and the growing polymer chains. An alternative approach is covalent imprinting, which uses a functional monomer that is comprised of the molecular template with polymerizable groups attached by cleavable covalent bonds. In both cases, the template is removed after the polymerization reaction is complete to reveal the MIP. Optimization of the template structure and the polymer cross-linking conditions produces MIPs with strong and selective affinity for many types of analytes, including various biomolecules. There are reports of MIPs for proteins and even complete micro-organisms. The most reproducible results are obtained when the biological template is a rigid structure like a virus. MIPs that use templates based on flexible protein chains or multiple membrane diffusible epitopes are often more problematic.

MIPs are attractive materials for a broad range of applications in separations, molecular sensors, and catalysis.[23] Although the basic concept of an MIP is easy to visualize, there are several technical challenges that have to be overcome for successful implementation. An obvious requirement is efficient removal of the template from the cross-linked polymer matrix. Sometimes complete removal of the template is not possible without damaging the imprinted cavity, leading to a lower binding activity. One way to overcome this limitation is to change from a bulk imprinting method to a surface imprinting approach.[24] A related problem is the inherent heterogeneity of the imprinting process, which produces a range of binding sites with different degrees of guest affinity and selectivity. This problem can be circumvented by using blocking methods that selectively eliminate the lower binding sites. Regardless of the fabrication method, the performances of MIPs are usually assessed by comparisons to control, non-imprinted polymers according to three quantitative binding factors: capacity, affinity, and selectivity.

Most early-generation MIPs were prepared in organic solvents, which were not compatible with most biological templates. But newer polymerization methods have been developed to make water-based structures such as sol–gel or hydrogel MIPs.[25] The latter are made of water compatible polymers like polyacrylate or polyacrylamide, and they can be prepared by copolymerizing mixtures of monomers containing different polar functional groups. In addition, methods to make imprinted inorganic surfaces, imprinted membranes, and imprinted nanoparticles are emerging.[26] The recent progress with imprinted polymer nanoparticles as "plastic antibodies" is quite exciting. For example, peptide imprinted nanoparticles have been prepared and shown to extract toxic peptides from the bloodstream of living animal models.[27]

1.2.5 Dendrimers

Dendrimers are branched polymers with symmetrical structures and monodisperse molecular weights. The tree-like architecture is usually visualized as a focal core with surrounding layers, or generations, of attached monomers. As the structures get larger (later generations) they begin to adopt a spherical shape with nanoscale dimensions, a sterically crowded surface, and internal pores. While dendrimers are aesthetically pleasing structures, the synthetic effort to make customized systems can be time-consuming and expensive. Therefore many researchers focus on commercially available dendrimers and dendron building blocks. Alternatively, there is increasing interest in hyperbranched polymers, which can be prepared on a large scale using one-pot synthetic methods. Although hyperbranched polymers are not monodisperse structures, they often exhibit many of the same supramolecular properties as dendrimers.[28]

Dendrimers and hyperbranched polymers are investigated for multifarious purposes such as advanced materials, supports for inorganic catalysts, and drug delivery agents.[29,30] In terms of biomolecule recognition, dendrimers and hyperbranched polymers are often converted into multivalent targeting systems by covalent attachment of multiple affinity ligands to the exterior surface of the polymer. This type of coated nanoparticle will be discussed further in the following section. Here, the focus is on the ability of high-generation dendrimers and hyperbranched polymers to encapsulate guest biomolecules within their internal cavities (Scheme 1.8). They act like

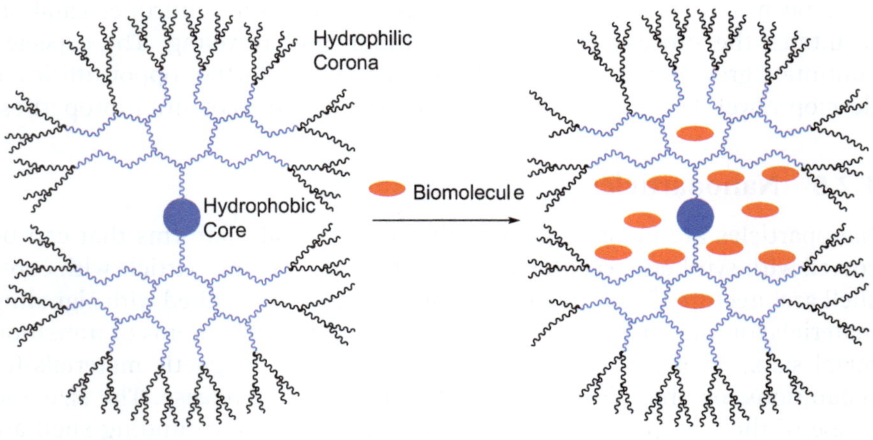

Scheme 1.8 Guest encapsulation within the internal cavities of a dendrimer or hyperbranched polymer.

"unimolecular micelles" with potential applications in drug delivery and supramolecular catalysis.

Most designs of unimolecular micelles have an amphiphilic core–shell architecture, with either a hydrophobic core and hydrophilic corona for extraction of nonpolar molecules into water or *vice versa*.[31] Early work with dendrimer systems demonstrated encapsulation of hydrophobic dyes or drugs, and in most cases the basis for the molecular recognition was a complementary match of guest size with the hydrophobic cavities inside the dendritic structure. Attempts to increase guest selectivity have investigated two related synthetic strategies to create a well-defined binding cavity. One approach is to append multiple dendritic branches to the periphery of a container molecule like a cyclophane or cyclodextrin.[32] The other approach is to use a templated imprinting process to produce a cross-linked dendrimer with a single binding site at the core.[33]

More recent work has investigated new methods of triggering guest release from the dendrimer core. One approach uses "facially amphiphilic dendrimers".[34] The dendritic structures incorporate biphenyl building blocks, each equipped with a polar arm and a nonpolar arm. The aryl groups respond to solvophobic forces by folding and minimizing interfaces with mismatched polarities. In water, the pores in the dendrimer core are lipophilic and can encapsulate hydrophobic guests, which can be subsequently released by different types of stimuli, including protein binding to the dendrimer surface. This causes a change in the dendrimer folding and concomitant polarity inversion within the pores.

The binding pockets within high generation dendrimers and hyperbranched polymers are reminiscent of enzyme active sites and researchers have worked to convert these synthetic receptors into enzyme mimetics (see below). The recent developments in organocatalysis have broadened the number of reactions that can be promoted by dendrimer catalysts.[35] In addition to bond forming and cleavage reactions, oxidations and reductions have been studied. The large size of high-generation dendrimer catalysts facilitates the process of catalyst separation and recycling. The expected continued growth in organocatalysis should lead to further opportunities to develop novel dendrimer catalysts with enzyme-like recognition properties.

1.2.6 Nanoparticles

Nanoparticles are increasingly valuable as spherical platforms that can be coated with synthetic receptors. Scheme 1.9 shows a nanoparticle with core–shell architecture.[36] The core of the nanoparticle can be filled with signaling materials for imaging (*e.g.* inorganic lattices such as gold atoms or transition metal salts, or organic polymers doped with dyes), magnetic materials for separations, or pharmaceutical agents for controlled release. The size and shape of the nanoparticles and the thickness of the surrounding shell are parameters that affect their physical, chemical, and biological performance.[37,38] An important point for molecular recognition is the loading

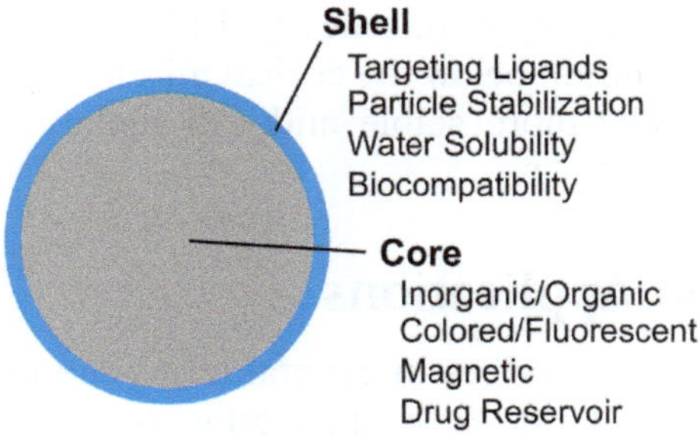

Scheme 1.9 Nanoparticle with core–shell architecture.

density of affinity ligands on the particle surface and their surface mobilities.

Inorganic core–shell nanoparticles are usually prepared by stepwise processes that coat a preformed inorganic core with an organic shell. Organic particles can be prepared by microemulsion polymerization methods and surface functional groups are added either by copolymerization of monomer mixtures or covalent modification of the particle exterior. A wide variety of organic particles are commercially available with diameters ranging from microns to hundreds of nanometers. An alternative way to fabricate organic nanoparticles is through self-assembly of appropriately designed amphiphilic building blocks. Vesicles and micelles are the best known colloidal structures that are formed by amphiphile self-aggregation. Hollow vesicles composed of polar lipids are called liposomes and they are used widely as biocompatible delivery vehicles for drugs and imaging contrast agents. Liposome assembly methods are straightforward and it is easy to control particle size and loading capacity. In addition, modern conjugation methods allow the liposome surface to be functionalized with targeting ligands or reporter groups. The past decade has seen great expansion in the types of amphiphilic building blocks that can be assembled into well-defined nanoparticles. A wide range of amphiphilic block copolymers have been created with various chain architectures, including linear, graft, cyclic, and star polymers. This work has been greatly facilitated by the development of controlled polymerization techniques such as reversible addition–fragmentation chain transfer polymerization, atom transfer radical polymerization, and nitroxide-mediated polymerization.

While the technical simplicity of amphiphile self-assembly is very attractive, there are inherent quality control concerns that may limit translation into certain types of practical applications, especially in pharmaceutical settings. Most notably, batch-to-batch production and particle size control is often hard to reproduce. Multicomponent nanocomposites can be especially difficult to characterize at the atomic level. The dynamic nature of self-assembled systems makes it challenging to confirm size and structure throughout the entire time period of an application. If a self-assembled nanoparticle is not thermodynamically stable, it is important to evaluate its kinetic stability. For example, micelles typically dissociate into monomer components upon dilution below the critical micelle concentration, whereas vesicles are kinetically more stable and will remain intact for extended periods.

1.3 Common Applications

The following section describes in general terms the four major classes of applications (Scheme 1.10). Each application is illustrated with example receptor systems. The thermodynamic and kinetic properties that control receptor performance are also discussed.

1.3.1 Separations

Various biomolecule analysis and purification methods use a molecular recognition process to separate the target analyte from a mixture, often by translocating it into a separate phase of matter. Performance parameters include receptor/biomolecule affinity, biomolecule selectivity, rates of complex association/dissociation, separation efficiency, batch size, feasibility of continuous flow processes, and capacity for regeneration.

Affinity chromatography is a liquid chromatography process that employs an affinity agent as the stationary phase to selectively retain a target biomolecule.[39] In chromatography, a selectivity factor, α, is defined by eqn (1.1), where k' represents the retention factors for two analytes X and Y.

$$\alpha = k'_X / k'_Y \tag{1.1}$$

A fundamental goal in affinity chromatography is to maximize the difference in binding affinity for X over Y, effectively increasing the value of α. Affinity columns are often used in combination with sensitive detection methods like mass spectrometry to create analytical detection assays. Shown in Scheme 1.11 is the common "step-elution" mode of operation which requires a selective recognition system with large α value. The separation mixture (usually a complex biological mixture) is passed through the column and the targeted biomolecule is selectively retained for subsequent elution under weakened binding conditions. A typical configuration employs a column filled with packing material that exhibits low non-specific binding such as agarose or cellulose. Newer methods incorporate silica particles or

Scheme 1.10 Major applications of receptors for biomolecules.

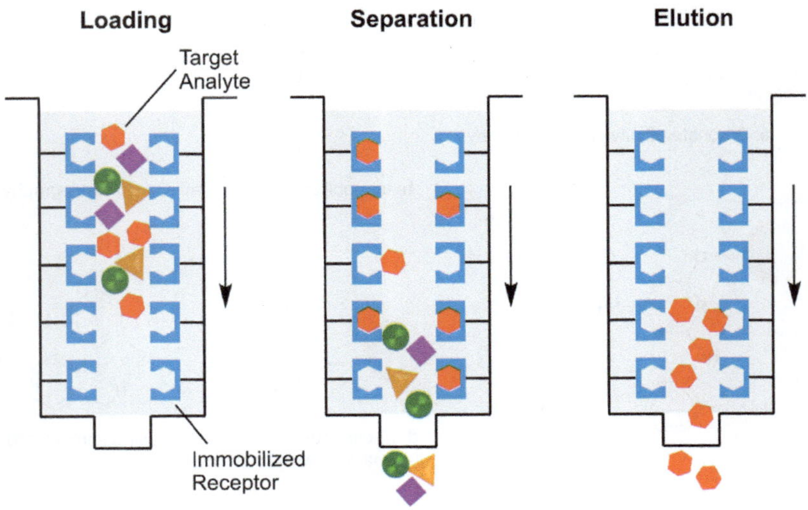

Scheme 1.11 Affinity chromatography involves three major steps: loading, separation, and elution.

monolithic materials that are capable of withstanding high flow rates and pressures. The affinity agents range from biopolymers, such as proteins and nucleic acids, to small organic dyes and Lewis acids. Antibodies are often used to isolate and identify drug molecules or metabolites, and sugar binding lectins are effective for the separation of glycoconjugates.

The robustness of synthetic receptors makes them attractive candidates as stationary phases for chromatography, but only a few systems are presently utilized for biomolecule separations.[39] Boronate affinity chromatography takes advantage of the reversible covalent association of diol-containing carbohydrates with boronic acids (Scheme 1.12a). Boronate columns are used extensively for the clinical quantification of glycated hemoglobin as an assessment of long-term diabetes management. They also have been used to separate glycoproteins from complex mixtures. Immobilized metal ion affinity chromatography exploits the reversible coordination of metal cations like Ni^{2+}, Zn^{2+}, Cu^{2+}, or Fe^{3+} by biomolecule targets such as amino acids, peptides, proteins, or nucleic acids.[40] The metal cations are immobilized on the stationary phase of the column by a covalently bonded chelating group (Scheme 1.12b), and the method is used extensively to purify proteins that contain multiple histidine residues (His-tagged proteins). The configuration of the stationary phase can be modified to produce new platforms for separations. A recent example employed magnetic beads coated with immobilized Zn^{2+} coordination complexes to remove bacterial cells from blood samples.[41]

The composition of the Lewis acid stationary phase can be varied. Metal-oxide affinity chromatography typically employs beads of titanium dioxide as the stationary phase to retain acidic biomolecules, especially phosphorylated

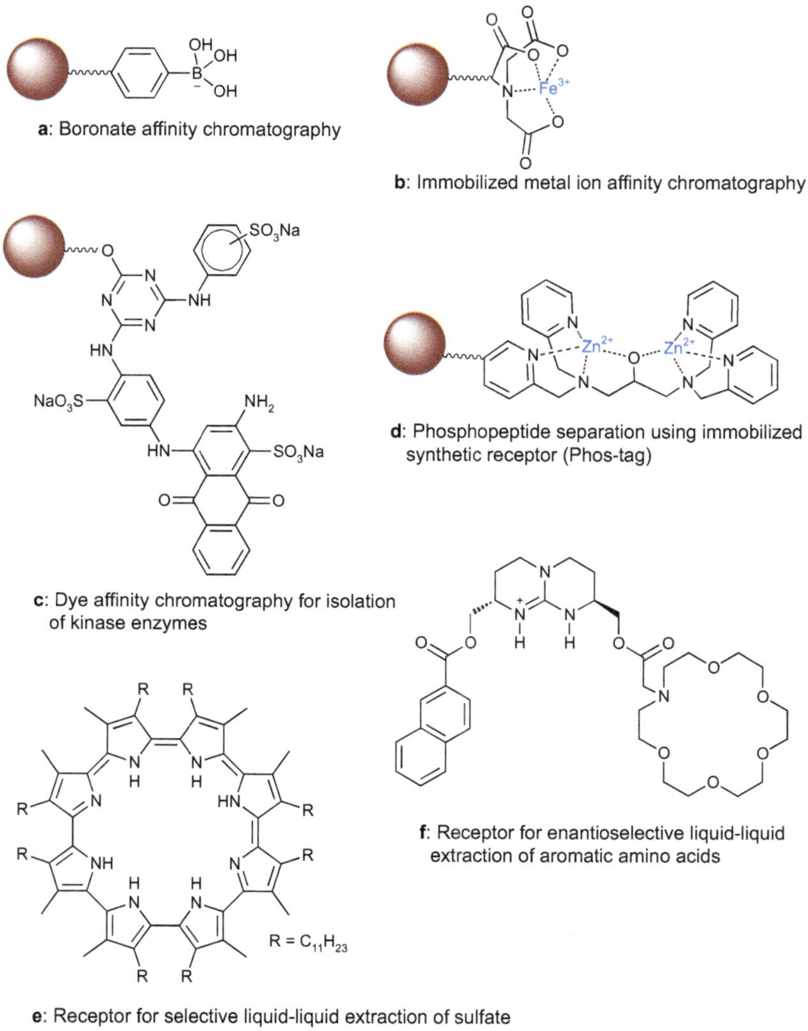

Scheme 1.12 Synthetic receptors for separations.

peptides. Dye affinity chromatography was developed in the late 1960s after the serendipitous discovery that stationary phases coated with immobilized triazine dyes selectively retain kinase enzymes (Scheme 1.12c). Follow-up work screened a large number of organic dyes from the textile industry and found examples that retained blood proteins and various enzymes. In recent years molecularly imprinted polymers[42] and aptamers have been examined as stationary phases for affinity chromatography of biomolecules and it is likely that future studies will show other types of synthetic receptors to be effective.[43]

There is a need in phosphoproteomics research for high-throughput methods that enrich digested samples containing phosphorylated proteins and peptides for subsequent analysis by mass spectrometry. In addition to the affinity chromatography methods described above, there are newer enrichment strategies using phosphoprotein-binding domains (PBDs). Commercial kits are available that use proteins with PBDs that specifically recognize the amino acid sequences surrounding phosphorylated Ser, Thr, or Tyr residues. The high cost and poor durability of these proteins has motivated efforts to develop synthetic receptors as PBDs. One example is the commercially available Phos-tag system (Scheme 1.12d), which has been used in a range of phosphopeptide separation and identification procedures, such as affinity chromatography, electroblotting, and gel electrophoresis.[40]

Liquid/liquid extraction is a classic separation method that typically uses a lipophilic synthetic receptor to extract a water soluble analyte into a liquid organic phase or an aqueous phase that is rich in PEG.[44,45] The ability of a biomolecule to partition between two phases is determined by the partition coefficient, K_p, which is defined in eqn (1.2), where C_T and C_B are the concentration of the solute in the top and bottom phase, respectively.

$$K_p = C_T/C_B \tag{1.2}$$

For receptor-mediated extraction, the extraction constant K_{ex} is the product of the association constant, K_a, and the partition coefficient of the receptor/analyte complex, K_p eqn (1.3).

$$K_{ex} = K_a \times K_p \tag{1.3}$$

Selectivity is the ratio of extraction constants, and receptor mediated extraction can be used to separate analytes with similar shapes and charges. For example, the cyclo[8]pyrrole receptor in Scheme 1.12e can selectively remove sulfate from an aqueous mixture containing other anions, such as nitrate, that are predicted by the Hofmeister series to be easier to extract (Chapter 4). Enantioselective extraction is another important goal and requires chiral receptors that can selectively bind to one enantiomer. The chiral receptor shown in Scheme 1.12f is able to enantioselectively extract aromatic amino acids using a three-point molecular recognition process.

Once extraction is achieved, a subsequent challenge is to reclaim the analyte and to do so in a continuous process. One solution is to convert the extraction process into a liquid membrane transport system. As shown in Scheme 1.13, the receptor resides in the organic membrane, which can be a bulk organic liquid or an organic liquid retained within a thin, porous polymer sheet. There are many examples of synthetic receptors that facilitate the transport of small hydrophilic biomolecules such as amino acids, nucleotides, sugars, and carboxylates, through liquid or supported liquid membranes.[46] With liquid membrane transport there is typically a bell shaped relationship between transport rate and affinity. High-affinity transporters can exhibit lower transport rates because release of the analyte from the membrane into the receiving phase becomes rate-limiting. Under these conditions the approximation that transport selectivity is proportional to extraction selectivity is no longer valid. Finally, co-transport and

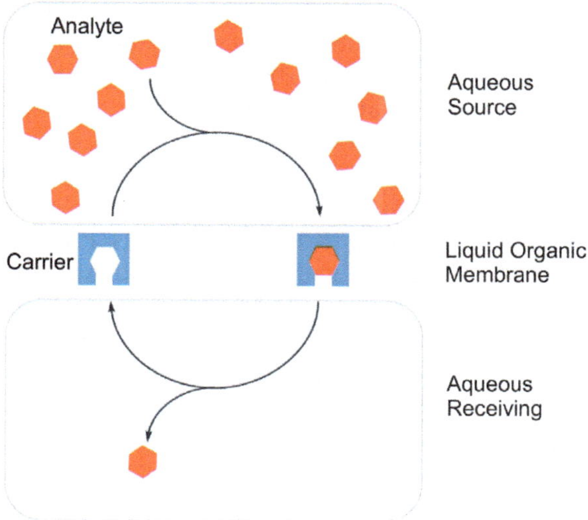

Scheme 1.13 Liquid membrane transport.

counter-transport schemes can be designed to couple the movement of a second analyte across the membrane and push transport of the original analyte uphill against a concentration gradient.

1.3.2 Imaging and Sensing

The related fields of biomolecule imaging, sensing, and diagnostics are growing rapidly and having a broad impact in many sectors of modern society. The remarkable growth is fueled by a synergistic cycle of innovation by academics developing new paradigms and commercial vendors producing turn-key instrumentation for easy operation. The ideal technology is label free and amenable to non-invasive evaluation of samples with no preliminary preparation. However, the opaqueness of many biomedical samples requires the development of exogenous contrast agents or indicators (collectively called probes) that can (a) be delivered to a desired internal site, and (b) interact with an external detection device to produce a signal that can be recorded and interpreted. The development of molecular or nanoparticle probes for imaging or sensing applications is a complex multiparameter task that often involves close collaboration between different types of scientists and engineers. The various technologies have hugely different spatiotemporal requirements with length scales ranging from nanometer to centimeter and time scales ranging from a femtosecond to days.

The focus here is on probes with ability to recognize biomolecules and report a signal for imaging, sensing, and diagnostics. The most common technologies are microscopy and diagnostics methods and they predominantly use fluorescent reporter groups. The synthetic chemistry to make fluorescent probes is well established and they can be stored for long periods and utilized quite safely. Furthermore, the technology to produce and detect fluorescent signals is mature with many advantages, such as high sensitivity, wide spectral range of wavelengths, and miniaturized instrumentation. Fluorescence is an intellectually rich optical phenomenon, and it is possible to engineer a wide array of probe configurations that can report analyte quantities and also provide information about changes in analyte structure and dynamics. Each specific application will have its own performance requirements, such as signal detection sensitivity, image contrast and resolution, and perhaps a need for multiplex detection of multiple wavelengths. Furthermore, probe performance properties should be considered, such as synthetic accessibility, probe brightness, chemical stability, photochemical stability, photosensitization ability, and toxicity. The imaging performance requirements become more stringent as samples become larger and more complicated. Perhaps the most demanding situation is in the field of molecular imaging, where the sample is a living subject and the optical signal has to penetrate through skin and tissue. There are various signal interference problems including scattering of the light (decreases with longer wavelength), absorption of the light by endogenous biomolecules (also decreases with longer wavelength), and the subsequent autofluorescence. These problems are minimized when the optical imaging wavelength is 650–900 nm, the so-called near-infrared window. But even with an optimized near-infrared wavelength the useful signal penetration depths are a few millimeters and thus the imaging methods are only suitable for shallow sites. A particularly exciting emerging application is fluorescence-guided surgery, where fluorescent imaging probes are used to identify nerves and tumor margins and help guide surgeons during resection. But for deeper tissue penetration, other molecular imaging modalities are more appropriate, such as ultrasound, nuclear imaging, magnetic resonance imaging (MRI), or X-ray. Each modality has its strengths and weaknesses regarding sensitivity and resolution and there is a need for injectable contrast agents to improve image quality and provide molecular-level information. The contrast agents may be targeted or non-targeted and a current trend is to develop probes with multiple labels that enable imaging by two or more modalities (*e.g.*, optical and MRI).

Shown in Scheme 1.14 are four of the most common ways that a synthetic receptor is used as a molecular or nanoparticle probe for imaging, sensing, or diagnostics. The first system is a targeted contrast agent (Scheme 1.14a). It is simply a synthetic receptor with an attached reporter group that continually emits a detectable signal. Targeted contrast agents are used increasingly for microscopic imaging of cells and tissues sections, and also for non-invasive molecular imaging of living subjects. Scheme 1.15a and b shows two probes that have been developed for the imaging of cancer and bacterial infection, respectively.[47,48] Targeted fluorescent probes are valuable in microarray diagnosis technology, where they are used to label and identify different species within a spatially arranged microarray. Even weakly

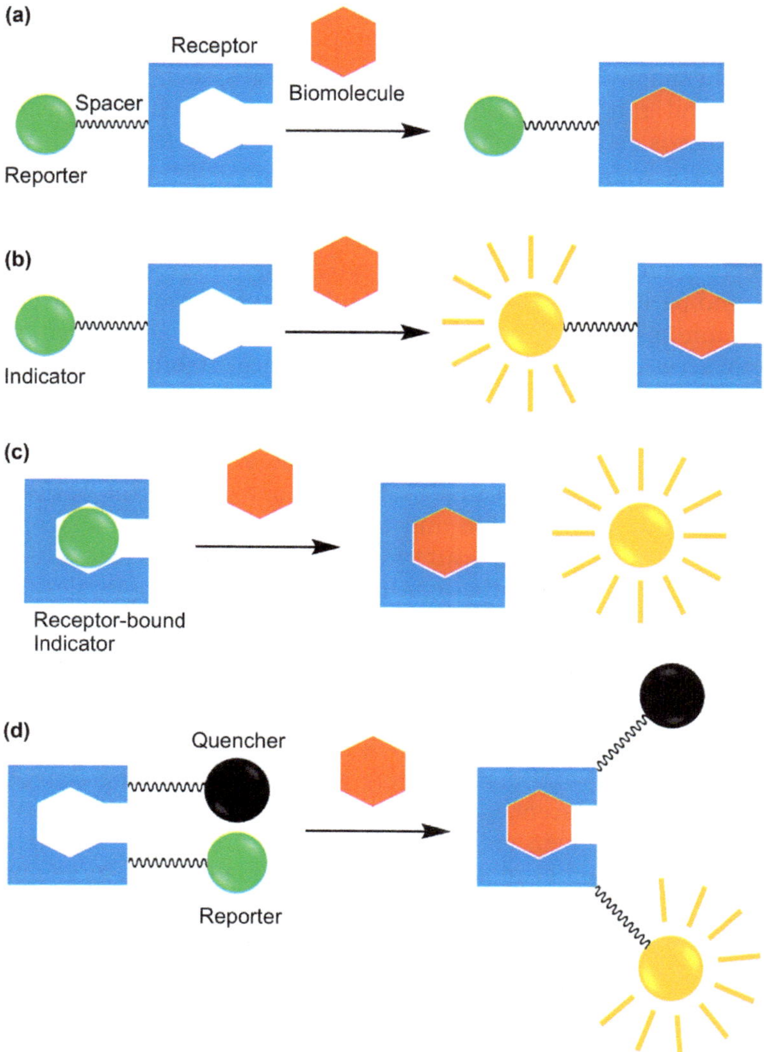

Scheme 1.14 Supramolecular systems for imaging and sensing.

selective receptors can produce unique thermodynamic or kinetic signal patterns that can be identified by appropriate pattern recognition algorithms.[49]

The second molecular design (Scheme 1.14b) is a selective indicator (sometimes called a chemosensor). In this case the signal from an attached chromophore is modulated by a molecular recognition event. The example in Scheme 1.15c is a well-known fluorescent sensor for glucose.[50] Compared to a targeted contrast agent, a "switch-on" indicator exhibits improved sensitivity due to the lower background signal caused by the silent off-target probe. Technically it is better to produce indicator systems that change

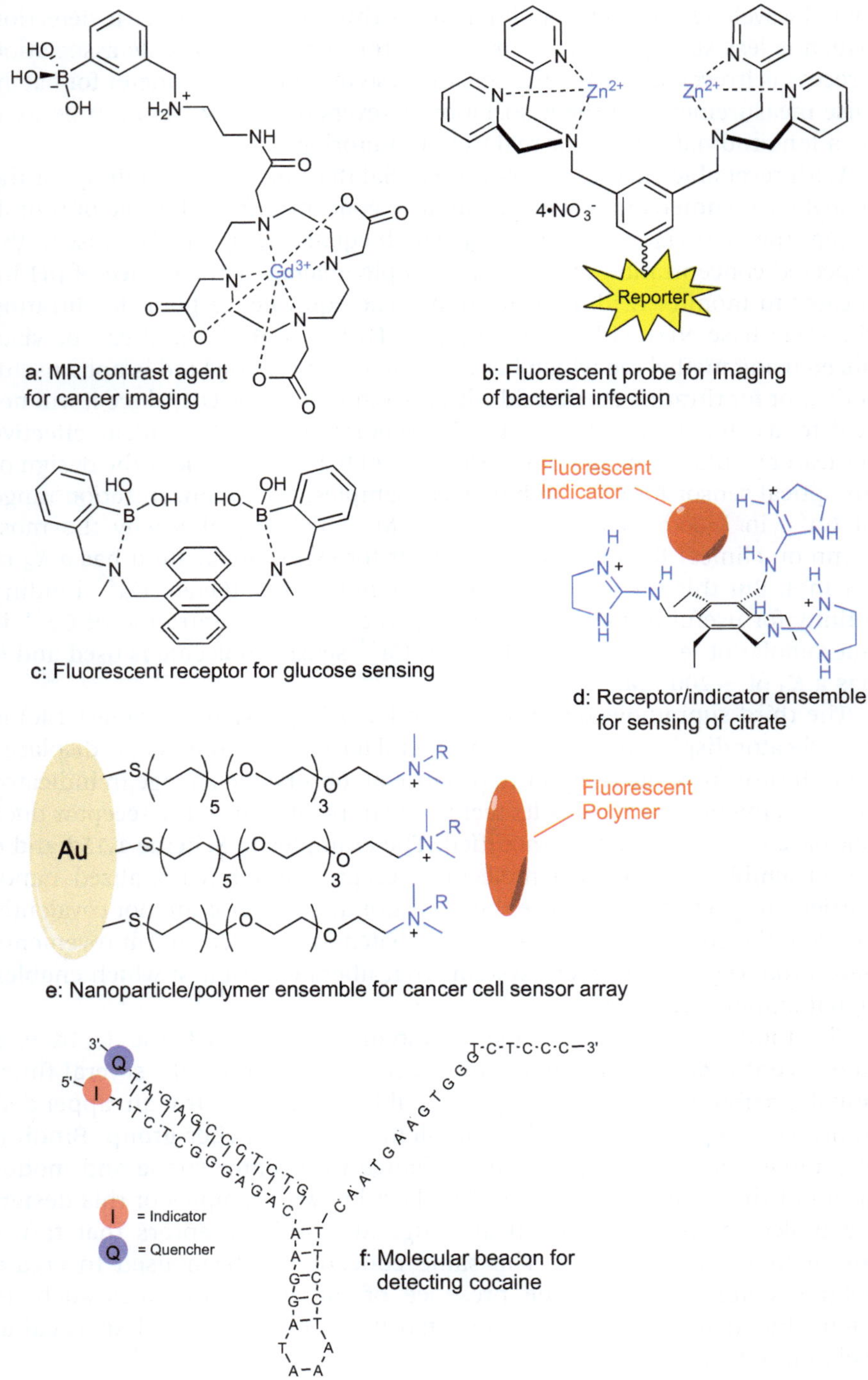

Scheme 1.15 Synthetic receptors used for imaging and sensing.

signal wavelength rather than intensity as this enables ratiometric detection which is less susceptible to systematic error. If the biomolecule association process is irreversible, then the sensing system acts as a dosimeter for batch-wise measurements. If the association is reversible, the system can be used as a longitudinal sensor for real-time monitoring.

With reversible chemosensors it is crucial that the receptor affinity for the analyte is commensurate with the analyte concentration. The rule of thumb is that the dissociation constant K_d (which equals $1/K_a$) must be close to the expected concentration of analyte. A simple analogy is the choice of pH indicator to monitor acid/base titrations. The equivalence point for titrating the weak base NH_3 with the strong acid HCl is ~pH 5.1 and can be visualized using methyl red which has a pK_a of 5.0. But methyl red is not a useful indicator for titrating acetic acid with the strong base NaOH, since it will not register a color change at the equivalence point of ~8.8. Instead, an effective indicator would be thymol blue with a pK_a of 9.2. Now consider the design of an optical sensor for Ca^{2+} in biological samples.[51] The concentration range of Ca^{2+} in blood plasma is 1.0–3.2 mM, which explains why the most common clinical fluorescent Ca^{2+} sensor for extracellular fluid has a K_d of 1.1 mM. But this fluorescent sensor does not have sufficient Ca^{2+} binding affinity for useful intracellular imaging where the concentration of Ca^{2+} is micromolar or less. Instead, a different Ca^{2+} sensor molecule is used and it has a K_d of ~200 nM.

The third supramolecular imaging and sensing design (Scheme 1.14c) is an indicator displacement assay. In general terms, the biomolecule displaces an indicator from the receptor to produce an observable change in indicator signal. A major advantage is its technical simplicity, since the receptor does not have to be synthetically modified. The examples in Scheme 1.15d and e are ensembles based on a molecular receptor and functionalized nanoparticle, respectively.[49,52] Since the indicator and receptor are not covalently bonded, the ensemble is best suited for batch assays. Recent improvements have produced displacement systems that liberate a catalyst which enables signal amplification.[53,54]

The fourth biomolecule sensing system shown in Scheme 1.14d is a fluorescent indicator comprised of a receptor structure with several functional attributes. It is typically a flexible molecule with an appended fluorescent reporter group and complementary quencher group. Binding of a target biomolecule produces a change in receptor shape and modulation of the fluorescence signal. The best known examples of this design are molecular beacons, which are oligonucleotide receptors that target complementary sequences. The same concept has been used to create aptamers able to detect the presence of target biomolecules such as adenosine triphosphate (ATP) or pharmacological agents like cocaine (Scheme 1.15f).[55]

Finally, it is worth noting that analytical chemosensing can be achieved using other forms of signal transduction, such as an electrochemical response or a mass response recorded by a quartz crystal microbalance.[56]

1.3.3 Catalysis

The research topic of enzyme mimicry has a rich history within supramolecular chemistry, as researchers have spent several decades preparing and studying synthetic models of enzyme active sites.[57] In recent years, the academic topic of synthetic enzymes has begun to merge with transition-metal catalysis, and, more recently, with organocatalysis. The current state is a connected network of emerging technologies encompassing homo- and heterogeneous catalysis, asymmetric synthesis, and green chemistry.[58] Important catalyst performance parameters include the amount of reaction acceleration, catalyst turnover number, and degree of asymmetric induction.[59] Ideally, the enantioselectivity for an asymmetric reaction pathway should be >50, which is the ratio needed to obtain 99% ee.

The goal of "biomimetic chemistry" is to produce new chemistry using mechanisms that resemble processes exhibited by living systems. In the context of synthetic enzymes, the aim is to accelerate chemical reactions of selected substrates using concepts from enzymology, such as lock and key association to form a catalyst/substrate complex, transition state stabilization, covalent stabilization of intermediates, and general acid or base catalysis (Scheme 1.16).[60] The classic Michaelis–Menten model of enzyme catalysis includes two system-specific parameters, V_{max}, the maximum reaction velocity, and K_m, the Michaelis–Menten constant, which is equal to the substrate concentration when the rate is 1/2 V_{max}. For fast binding catalyst/substrate complexes, K_m is equal to $1/K_a$. Thus, K_m for synthetic enzymes can be controlled by rational molecular design.

Classical efforts to produce synthetic enzymes *de novo* start with structural information about the crucial catalytic residues within enzyme active sites. The advent of advanced computer modeling methods has greatly enhanced

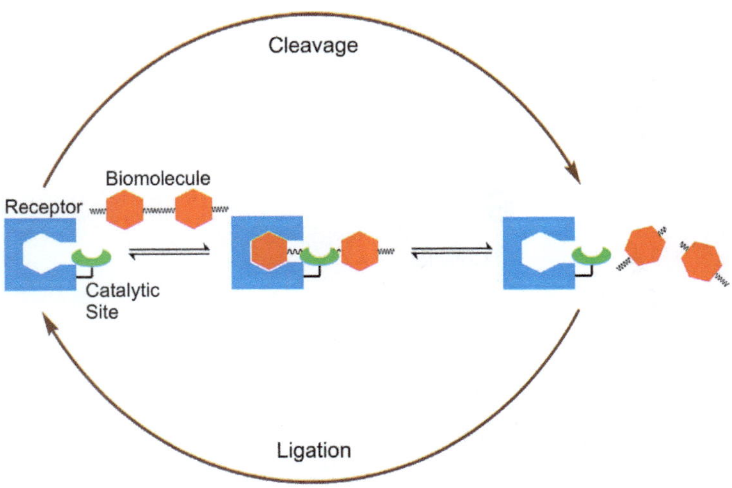

Scheme 1.16 Supramolecular catalysis of ligation or cleavage reactions.

the capability to rationally design an enzyme mimic, but it is worth emphasizing that it is much more challenging to model a catalyst/substrate transition-state than a ground-state receptor/guest complex. Catalyst fabrication has typically involved multistep synthetic procedures that append catalytic residues to organic container molecules such as cyclodextrins, crown ethers, calixarenes, or porous dendrimers. The illustrative example in Scheme 1.17 shows a high-generation dendrimer with a pyridoxamine cofactor at the core.[61] The system is able to accelerate transamination reactions 1000 times faster than free pyridoxamine. In recent years the receptor scaffold options have expanded with the development of rapid self-assembly methods that use metal–ligand interactions to produce porous hollow cages that can encapsulate and constrain reactive substrates.

An alternative way to develop an effective catalyst based on a synthetic receptor is to use combinatorial synthesis and screening methods. In short, the goal is to create a large library of candidate structures and then screen them to find a member with appropriate catalytic performance. This approach works well for molecules that can be synthesized rapidly in combinatorial fashion, such as synthetic peptides[62] or oligonucleic acid aptamers.[63] An example of this approach to catalyst discovery is the work of Reymond and coworkers, who examined libraries of peptide dendrimers (Scheme 1.18a) with structures containing a judicious mixture of hydrophobic side chains to encourage guest binding and polar groups for reaction catalysis. Screening studies revealed library members with strong esterase activity, including enantioselective catalysts.[64]

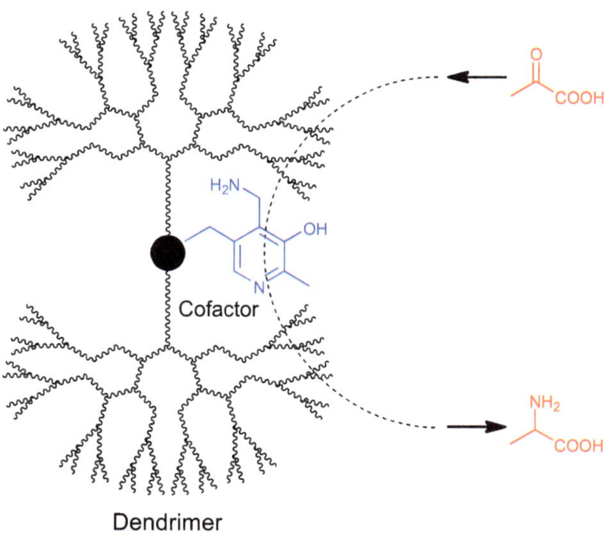

Scheme 1.17 Synthetic pyridoxamine dendrimer as transaminase mimic.

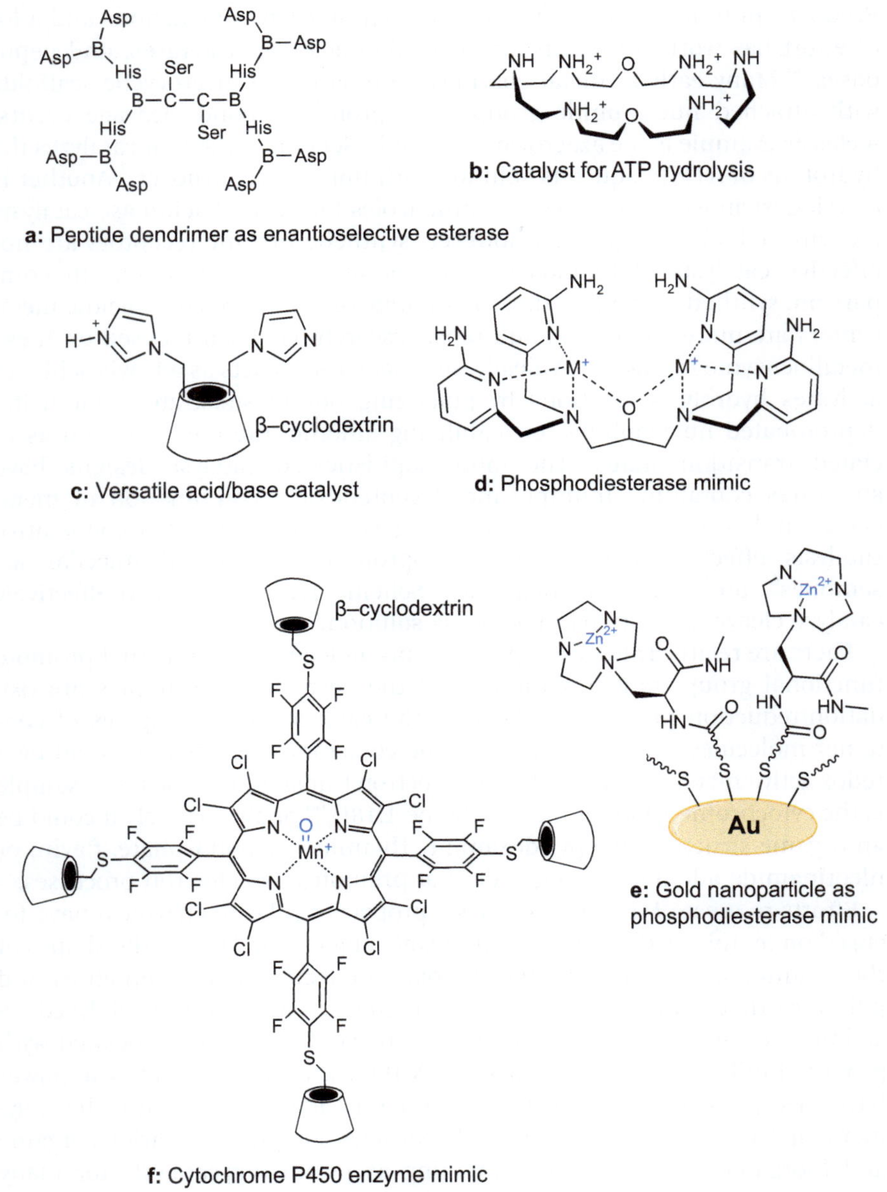

Scheme 1.18 Synthetic receptors for catalysis.

The most common biomolecule reactions that are promoted by synthetic catalysts are cleavage processes (Scheme 1.16). In contrast, examples of biomolecule ligation reactions are fairly rare, especially in water.[65] A common drawback with ligation catalysts is limited catalyst turnover due to product inhibition. This occurs often because the ligated product molecule has higher affinity for the catalyst than the two smaller reactant molecules.

Product inhibition is typically not a problem in cleavage reactions and a lot of effort has worked towards artificial ribonucleases, esterases, and peptidases.[66] Many early artificial enzymes were based on macrocyclic scaffolds with attached nucleophilic groups that promoted bond cleavage events. A classic example is the azacrown receptor in Scheme 1.18b that catalyzes the hydrolysis of ATP in aqueous solution with impressive turnover.[2] Another is a cyclodextrin with two appended imidazoles for general acid/base catalysis (Scheme 1.18c).[66] In general, however, synthetic organic receptors are not effective catalysts of biomolecule hydrolysis reaction in water. In comparison, synthetic receptors that incorporate one or more Lewis acidic metal centers are more successful, and most research has been focused on these metalloenzyme mimics.[67] Typically, the metal cation acts as a Lewis acid and activates hydrolysis reactions by polarizing bound substrates, stabilizing deprotonated nucleophiles, or stabilizing anionic intermediates and associated transition states. The more sophisticated catalyst designs have structures containing multiple metal centers or a combination of metal center and secondary hydrogen bonding sites that produce a cooperative catalysis effect (Scheme 1.18d).[68] Appropriately designed micellar assemblies[69] and coated nanoparticles (Scheme 1.18e)[70] can also effectively catalyze cleavage reactions in aqueous solution.

There are relatively few examples of supramolecular catalysts that promote functional group transformations, and the best known examples are oxidation/reduction processes.[66] Most of the catalysts are conjugates of container molecules with appended enzyme cofactors. The cofactor could be a redox active metal that promotes one electron transfer processes (an example is the cytochrome P450 mimic in Scheme 1.18f),[66] or alternatively it could be an organic structure like pyridoxamine, thiamine pyrophosphate, flavin, or nicotinamide adenine dinucleotide that promotes two electron processes.

Efforts to create MIPs with catalysis properties (plastic enzymes) have focused on templates that are transition-state analogues. That is, the shapes of the binding sites created by the template molecules have geometries and polarities that match reaction transition states.[71] Most of the catalyzed reactions are cleavage processes that tend to avoid problems associated with product inhibition. Efforts to produce MIPs with increased catalytic power have incorporated transition metals as Lewis acid catalysts into the polymeric matrix.[71] In addition, nanoscale particles composed of various organic and inorganic composites are emerging as promising catalysts for many types of reactions involving biomolecules (nanozymes).[72]

Future research into enzyme mimics is expected to include efforts to produce catalysts whose activity is regulated by effector molecules or external stimuli.[73] Another likely direction is next-generation designs of molecular self-replication due to autocatalysis; that is, a synthetic receptor that catalyzes a reaction that produces another copy of itself, leading to exponential growth. One of the motivations for this field of research is to find possible explanations for the emergence of life from the primordial soup of molecules that existed on early earth.

1.3.4 Pharmaceutical Activity

The majority of small-molecule pharmaceuticals (drugs) work by binding tightly and selectively to a crevice within a medically relevant protein. The protein crevice could be an enzyme active site, receptor binding site, allosteric site, or binding interface with a partner biopolymer. The drug discovery pathway to find the perfect compound that binds tightly and selectively to a protein crevice is long and arduous. The traditional medicinal chemistry approach employs an iterative cycle of synthesis and screening protocols to generate empirical trends that guide follow-up studies. The search can be augmented by a wide range of predictive tools that help narrow the choice of candidate molecules for synthesis and testing. Binding site docking algorithms utilize computational models to predict association strengths based on non-covalent interactions. There is no doubt that this part of the drug discovery process is enhanced by a deep understanding of the fundamental concepts that control supramolecular assembly.[74]

Increasingly, the field of supramolecular chemistry is able to produce synthetic receptors that recognize biomolecules effectively under physiological conditions. Inspired by this success, researchers are beginning to devise novel supramolecular strategies to elicit pharmacological outcomes. Especially attractive are synthetic receptors with the capacity to modulate the four protein/biomolecule association systems illustrated in Scheme 1.19. In principle, this can be achieved by targeting the protein or the biomolecule partner. The affinity has to be strong enough to saturate the target biomolecule at dosing concentrations, which are typically micromolar or less. Selective binding of the target molecule is, of course, only one of the many necessary properties that have to be optimized in order to create a useful pharmaceutical agent. Eventually the pharmaceutical performance of a synthetic receptor has to be judged by the full array of classic ADMET parameters. In this regard it is inherently easier to optimize the pharmaceutical properties of a small molecule than a relative large nanoparticle, and the task becomes even more challenging if the nanoparticle is a non-covalent assembly of multiple components.

Examples of small-molecule pharmaceuticals that block protein/protein association are already known,[75] which supports the feasibility of future efforts to achieve this type of pharmacological effect in a rational way using synthetic receptors. Molecules that can cover a large area of exposed protein surface are expected to be most effective and the design criteria to achieve this are discussed further in Chapter 9. Prototype protein receptor systems include extended organic scaffolds such as porphyrins with appended arms, nanoparticle molecularly imprinted polymers,[76] and core–shell nanoparticles coated with a multicomponent monolayer. In terms of pharmaceutical performance, the most clinically advanced synthetic recognition systems for protein surfaces are aptamers. As stated above, pegaptanib was the first aptamer-based therapeutic agent approved for treating age-related macular degeneration. Pegaptanib works by binding tightly to vascular

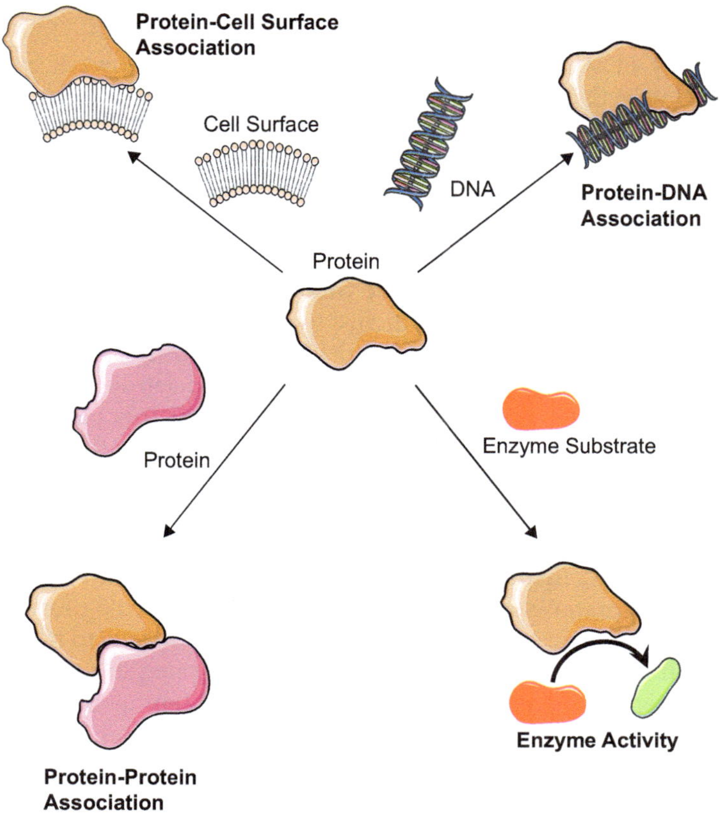

Scheme 1.19 Pharmaceutically relevant protein/biomolecule association systems.

endothelial growth factor (VEGF) within the intravitreous region of the eye and inhibiting VEGF binding to cell surface receptors, thus preventing the formation of new blood vessels in the eye.[77]

In certain cases, the pharmaceutical target is a cell surface that contains many copies of a protein and the pharmaceutical effect is enhanced if binding is increased. This situation invites a targeting strategy based on multivalency, which is expected to increase affinity and also improve cell selectivity. Most published examples of multivalent targeting use a nanoparticle that is coated with multiple copies of a targeting ligand such as a vitamin, peptide, or carbohydrate. The individual ligand may only have moderate affinity for the membrane-bound target, but the multivalent nanoparticle binds strongly to multiple copies of the target on the cell surface. Some of the structural factors that control affinity and selectivity of multivalent binding systems are discussed in Chapter 2. A key variable is receptor abundance on the cell surface. In some cases, binding alone produces a desired pharmaceutical outcome, like blocking viral infection using nanoparticles coated with carbohydrates. In other cases, the multivalent targeting triggers a cell entry process such as endocytosis and thus it can be used for selective drug delivery. A good example is the targeting of folate-coated liposomes to cancer cells that overexpress the folate receptor.[78]

The same multivalency concepts are applicable if the targeting ligand on the nanoparticle is a synthetic receptor.[79] This strategy was employed in a recent study of micelle nanocarrier delivery of *cis*-platin drug to cancer.[80] The drug delivery was enhanced when the surface of the nanocarrier was decorated with multiple boronic acid groups (Scheme 1.20a) that targeted over-expressed sialic acid residues on the surface of cancer cells. The boronic acid coated nanocarrier was able to deliver comparatively higher doses of the anticancer drug to solid tumors and reduce the growth rate of melanoma in animal models.

Nucleic acids are attractive pharmaceutical targets for synthetic receptors since they are intimately involved in cell replication and transcription processes and ultimately cell proliferation. The surface of DNA and RNA presents a rich assortment of binding targets including the anionic phosphate backbone, the large hydrophobic surfaces of the nucleobases, and the hydrogen bonding motifs on the exposed edges of the major and minor grooves of the double-stranded helix. The many different types of synthetic receptors that achieve effective nucleic acid recognition are described in Chapter 7. Arguably the most impressive programmable system with significant biological activity is the nucleic acid binders produced by Dervan and coworkers. The group has developed a series of small molecule pyrrole-imidazole polyamides that bind to double-stranded nucleic acids with extremely high affinity and exquisite sequence selectivity. The recognition is based on hydrogen bonding patterns created by the alternating pyrrole and imidazole groups within the receptor structure (Scheme 1.20b). Sequence specificity is determined by pairing rules that maximize hydrogen bonding with the donor-acceptor sites that are exposed on the edges of the stacked nucleobases. For example, an imidazole-pyrrole pair distinguishes G · C from C · G and also the A · T/T · A base pairs, while a pyrrole–pyrrole binding pair is selective for A · T over G · C.[81] These receptors have been used to block numerous RNA/DNA–protein interactions that control cell function, including RNA polymerase II,[82] nuclear factor-κB DNA binding,[83] and have been shown to control gene expression in a xenograft animal tumor model.[84] Studies of *in vivo* efficacy have encountered obstacles concerning delivery, toxicity, and slow clearance.[85] However, the modularity of the polyamide design allows manipulation of the ADMET properties, and thus this class of synthetic receptors has promise for continued pharmaceutical development.

Another way that synthetic receptors can produce pharmaceutical activity is to alter the cellular distribution of a specific biomolecule. In principle, it is an attractive way to treat cells that have become pathological due to an abnormal accumulation of lipophilic biomolecules. An example is 2-hydroxypropyl-β-cyclodextrin (2HPBCD) (Scheme 1.20c), a synthetic receptor for cholesterol with promising pharmaceutical activity. It was recently reported that 2HPBCD has efficacy against a mouse model of Niemann–Pick type C (NPC) disease, a rare but fatal neurodegenerative disorder caused by a defect

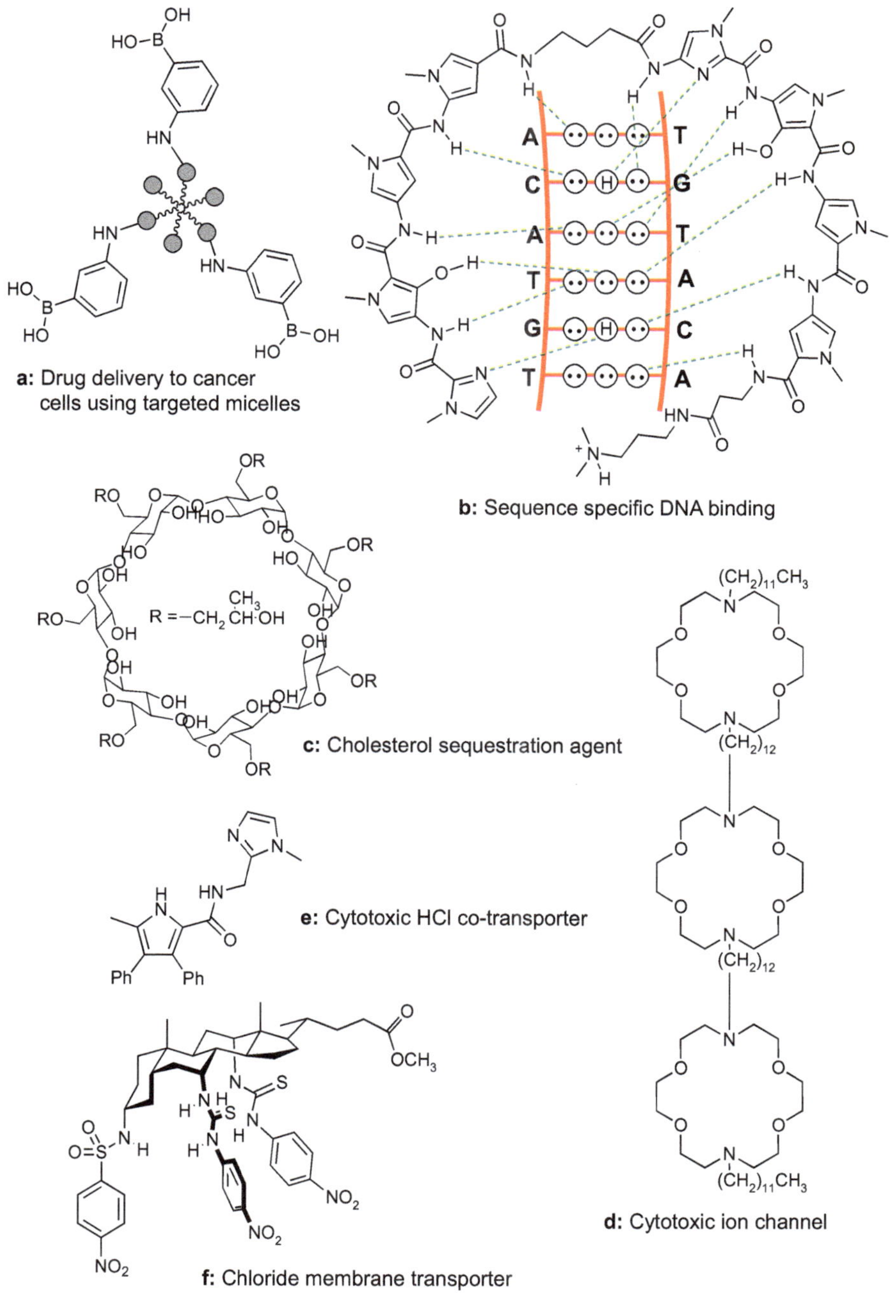

Scheme 1.20 Synthetic receptors with pharmaceutical activity.

in the NPC1 protein and characterized by widespread intracellular accumulation of cholesterol.[86] Currently, there is no effective treatment for NPC disease, which primarily affects children. Recent experiments using a mouse model of the disease found that systemic treatment with 2HPBCD significantly prolonged lifespan by allowing trapped cholesterol within the late endosome/lysosome to be released into the cell cytosol where it was normally metabolized.[87] Ongoing studies are attempting to determine exactly how 2HPBCD induces its therapeutic effect, but the reversal of abhorrent cholesterol trafficking is strongly implicated.

It is well known that healthy living cells maintain a tightly regulated transmembrane distribution of ion concentrations. This balance is maintained by the collective action of a series of endogenous active and passive transport systems, and disruption of this ion distribution leads to cell malfunction. There are logical reasons to develop pharmacological agents that can alter transmembrane ion gradients; they can either become a method to kill pathological cells, or a way to restore ion gradients in genetically abnormal cells. Ionophores are chemicals that associate with ions and many facilitate ion transport across cell membranes. Historically the term ionophore referred to cation binding molecules, but these days many synthetic anion binding ionophores are also known.[88] Quite a few natural products are known to kill cells by permitting unregulated ion flow across the plasma membrane.[89] These compounds are produced by microbes and include the ionophore antibiotics, a wide range of natural products that promote ion transport by mechanisms that resemble the classic channel and carrier processes shown in Scheme 1.21. In some cases, the ionophore antibiotics form well-defined 1 : 1 lipophilic complexes with metal cations (*e.g.*, monensin A and valinomycin) and they act as membrane transport carriers. In other cases (*e.g.*, gramicidin A) they exhibit classic channel behavior, whereas yet others self-assemble within membranes to create pores (*e.g.*, gramicidin S and alamethicin). Sometimes the pore assembly process involves molecular components within the target membrane. An example is

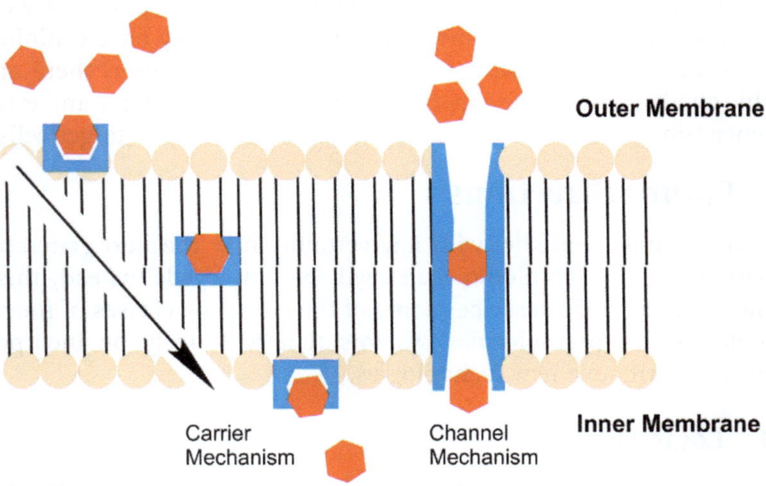

Scheme 1.21 Channel and carrier mechanisms for membrane transport.

amphotericin B, an antifungal drug that binds with ergosterol, a component of fungal cell membranes, and forms a transmembrane channel that promotes ion leakage (Chapter 10).

Over the years, supramolecular chemists have prepared a large number of synthetic ionophores and determined if they transport ions across membranes by carrier, channel, or pore mechanisms (Chapter 3). Many of these compounds have been tested for cytotoxicity. In general, simple crown ethers are not highly toxic molecules. However, there are examples of channel-forming systems having significant toxicity.[90] For example, Gokel's hydrophiles (crown ether oligomers with channel transport behavior; Scheme 1.20d) are toxic to mammalian and bacterial cells.[91] Amino-substituted cyclodextrins have also been shown to exhibit strong antimicrobial activity and an ability to promote K^+ efflux from both Gram-positive and Gram-negative bacteria.[92] It appears that many synthetic cation ionophores are likely to be cytotoxic and the pharmaceutical challenge is to target them to sites of disease such as cancer or microbial infection.

Anion-binding ionophores have also been shown to exhibit a range of interesting pharmaceutical properties. For example, the naturally occurring prodigiosins exhibit antimicrobial, immunosuppressive, and anticancer activities.[89] These compounds are known to promote HCl co-transport, but how this relates to their pharmacological properties is not exactly clear. Nonetheless, synthetic receptors with HCl co-transport ability have been shown to exhibit interesting anticancer properties (Scheme 1.20e).[89]

An alternative pharmaceutical goal using ionophores is not to kill cells, but rather to restore ion gradients in genetically abnormal cells. Certain genetic conditions, including Bartter syndrome, Best disease, and cystic fibrosis are caused by defects in endogenous cell membrane transport channels. It has been suggested that exogenous addition of synthetic chloride transporters may restore transmembrane chloride gradients to normal levels. A large number of synthetic anion transporters have been tested in model membranes and cell culture, and many have exhibited potent chloride transport ability (Scheme 1.20f).[93,94] However, there are formidable challenges to solve concerning tissue specific delivery and extended residence time in the appropriate apical membrane of the target cells.

1.4 Future Directions

It is not possible to predict with any certainty how the next generation of synthetic receptors for biomolecules will be employed. Instead, this concluding section briefly describes some of the emerging classes of functional molecules and supramolecular systems that will likely be incorporated within transformative new technologies.

1.4.1 Logic Devices

The most complicated known machine in the universe is the human brain and there is an active community of scientists and engineers trying to create molecule-based computers.[95] The nanoscale dimensions of biomolecules make them attractive building blocks for bottom-up fabrication of high-density information devices for storage and calculation. There are already a number of biomolecule sensing systems that use logic gates to report analyte levels in biomedical samples.[96] The sensors only produce a signal when there is a precise number of multiple analytes, thus providing sophisticated information about the state of a living biological sample. There is also a growing number of synthetic receptor systems that can selectively modulate biomolecule association events.[97] Merging these topics will undoubtedly lead to biomolecular logic devices that mimic the complex signaling pathways of cells.

1.4.2 Biomolecule Responsive Materials

An extension of logic devices is the field of responsive materials for controlled release and drug delivery.[98] Current methods for drug release usually rely on passive diffusion of the drug molecules from an inert vehicle into the physiological environment. There are many prototype examples of release systems that can be triggered by external stimuli, such as light, heat, or change in pH, and increasingly there are new classes of advanced materials with properties that respond to the presence of biological molecules.[99] State of the art examples include glucose-responsive systems that trigger the release of pharmaceutically relevant proteins[100] or cells,[101] and synthetically modified protein tubes that are activated by intracellular ATP to disassemble and release encapsulated guests.[102]

1.4.3 Drug Delivery

One of the unpleasant realities of drug development is the fact that many chemical entities that bind strongly to a validated pharmaceutical target *in vitro* do not become useful drugs because they cannot reach the target *in vivo*. A dogma within drug discovery is Lipinski's rule of five,[103] which is a set of simple structural criteria that help predict if a drug candidate is likely to be orally active. A corollary of these rules is that high molecular weight and/or highly charged molecules are unlikely to cross cell membranes, which means that they will be ineffective against intracellular targets. This includes biomolecular structures such as peptides, antibodies, enzymes, nucleotides, nucleic acids, aptamers, and many carbohydrates. Furthermore, many of these structures are susceptible to enzyme degradation in the bloodstream. Thus, there is a need for drug delivery vehicles that will protect these fragile agents and deliver them into the cytoplasm of target cells. To date, there has been considerable progress in the area of transfection and gene delivery *in vitro* using supramolecular assemblies such as cationic polymers and amphiphiles.[51] The next major challenge is *in vivo* organ-specific delivery. A reason for great optimism is the rapid ongoing development of biocompatible, self-assembled nanocomposites with multiple attributes that allow cell targeting, *in vivo* imaging, and drug delivery.[104] Similarly, there are elegant new designs of delivery molecules, such as "molecular umbrellas", which are facially amphiphilic receptors that can encapsulate biomolecules in aqueous solution and then invert polarity as they pass through the lipophilic core of a cell membrane.[105]

1.4.4 Biomolecule-fueled Molecular Machines

The molecular machines that operate inside cells, like the ATP-powered motor proteins that drive the contraction of muscle fibers in animals or transport biological cargo along microfilaments, have captured the imaginations of many supramolecular chemists. The recent literature contains a diverse array of brilliantly designed molecules with machine-like motions such as rotors, switches, and shuttles.[106] But the functional performances of these primitive, early-generation molecules are very modest when compared to their much larger and more complicated biological counterparts. The challenge of fabricating biomimetic supramolecular systems that can convert biomolecule binding energy (input) into coherent mechanical movement (output) seems daunting.[107] At present the most advanced projects utilize base-pairing of nucleic acids as the binding event that triggers macroscale movement,[108,109] or they prepare nanomechanical devices using the rules of DNA origami.[110]

The long-term goal is to produce nanoscale robots for various applications in artificial life, synthetic biology, and nanomedicine.

Acknowledgement

We are grateful for funding support from the National Science Foundation (Arlington, VA, USA) (CHE1401783).

References

1. H.-J. Schneider, *Applications of Supramolecular Chemistry*, CRC Press, 2012.
2. J.-M. Lehn, *Science*, 1985, **227**, 849.
3. C. J. Pedersen, *Angew. Chem., Int. Ed. Engl.*, 1988, **27**, 1021.
4. H. J. Schneider, *Angew. Chem., Int. Ed.*, 2009, **48**, 3924.
5. D. Chandler, *Nature*, 2005, **437**, 640.
6. P. T. Corbett, J. Leclaire, L. Vial, K. R. West, J.-L. Wietor, J. K. Sanders and S. Otto, *Chem. Rev.*, 2006, **106**, 3652.
7. A. Barnard and D. K. Smith, *Angew. Chem., Int. Ed.*, 2012, **51**, 6572.
8. S. Song, Y. Qin, Y. He, Q. Huang, C. Fan and H. Y. Chen, *Chem. Soc. Rev.*, 2010, **39**, 4234.
9. F. A. A. Paz, J. Klinowski, S. M. Vilela, J. P. Tomé, J. A. Cavaleiro and J. Rocha, *Chem. Soc. Rev.*, 2012, **41**, 1088.

10. T. Dieckmann, E. Suzuki, G. K. Nakamura and J. Feigon, *RNA*, 1996, **2**, 628.
11. B. Strehlitz, C. Reinemann, S. Linkorn and R. Stoltenburg, *Bioanal. Rev.*, 2012, **4**, 1.
12. A. Geiger, P. Burgstaller, H. vonderEltz, A. Roeder and M. Famulok, *Nucleic Acids Res.*, 1996, **24**, 1029.
13. K. A. Yang, M. Barbu, M. Halim, P. Pallavi, B. Kim, D. M. Kolpashchikov, S. Pecic, S. Taylor, T. S. Worgall and M. N. Stojanovic, *Nat. Chem.*, 2014, **6**, 1003.
14. A. D. Keefe, S. Pai and A. Ellington, *Nat. Rev. Drug Discovery*, 2010, **9**, 537.
15. Q. L. Liu, C. Jin, Y. Y. Wang, X. H. Fang, X. B. Zhang, Z. Chen and W. H. Tan, *NPG Asia Mater.*, 2014, **6**, e95.
16. S. Kruspe and U. Hahn, *Angew. Chem., Int. Ed.*, 2014, **53**, 10541.
17. M. Zhao, T. Wu, X. Xiao, Y. Liu and X. Su, *Anal. Bioanal. Chem.*, 2013, **405**, 5679.
18. C. Feng, S. Dai and L. Wang, *Biosens. Bioelectron.*, 2014, **59**, 64.
19. K. Schlosser and Y. Li, *Chem. Biol.*, 2009, **16**, 311.
20. B. P. Gray and K. C. Brown, *Chem. Rev.*, 2014, **114**, 1020.
21. M. T Dohm, R. Kapoor and A. E Barron, *Curr. Pharm. Des.*, 2011, **17**, 2732.
22. L. K. Pilsl and O. Reiser, *Amino Acids*, 2011, **41**, 709.
23. R. Schirhagl, *Anal. Chem.*, 2014, **86**, 250.
24. S. S. Wang, J. Ye, Z. J. Bie and Z. Liu, *Chem. Sci.*, 2014, **5**, 1135.
25. H. Q. Zhang, *Polymer*, 2014, **55**, 699.
26. C. S. Mahon and D. A. Fulton, *Nat. Chem.*, 2014, **6**, 665.
27. Y. Hoshino, H. Koide, T. Urakami, H. Kanazawa, T. Kodama, N. Oku and K. J. Shea, *J. Am. Chem. Soc.*, 2010, **132**, 6644.
28. H. F. Gao, *Macromol. Rapid Commun.*, 2012, **33**, 722.
29. S. H. Medina and M. E. H. El-Sayed, *Chem. Rev.*, 2009, **109**, 3141.
30. J. Khandare, M. Calderon, N. M. Dagia and R. Haag, *Chem. Soc. Rev.*, 2012, **41**, 2824.
31. T. Satoh, *Soft Matter*, 2009, **5**, 1972.
32. J. Hu and S. Liu, *Acc. Chem. Res.*, 2014, **47**, 2084.
33. S. C. Zimmerman, I. Zharov, M. S. Wendland, N. A. Rakow and K. S. Suslick, *J. Am. Chem. Soc.*, 2003, **125**, 13504.
34. K. R. Raghupathi, J. Guo, O. Munkhbat, P. Rangadurai and S. Thayumanavan, *Acc. Chem. Res.*, 2014, **47**, 2200.
35. A. M. Caminade, A. Ouali, M. Keller and J. P. Majoral, *Chem. Soc. Rev.*, 2012, **41**, 4113.
36. R. Ghosh Chaudhuri and S. Paria, *Chem. Rev.*, 2012, **112**, 2373.
37. A. E. Nel, L. Mädler, D. Velegol, T. Xia, E. M. Hoek, P. Somasundaran, F. Klaessig, V. Castranova and M. Thompson, *Nat. Mater.*, 2009, **8**, 543.
38. M. E. Davis, Z. G. Chen and D. M. Shin, *Nat. Rev. Drug Discovery*, 2008, **7**, 771.

39. D. S. Hage, J. A. Anguizola, C. Bi, R. Li, R. Matsuda, E. Papastavros, E. Pfaunmiller, J. Vargas and X. Zheng, *J. Pharm. Biomed. Anal.*, 2012, **69**, 93.
40. L. Beltran and P. R. Cutillas, *Amino Acids*, 2012, **43**, 1009.
41. J. J. Lee, K. J. Jeong, M. Hashimoto, A. H. Kwon, A. Rwei, S. A. Shankarappa, J. H. Tsui and D. S. Kohane, *Nano Lett.*, 2014, **14**, 1.
42. K. Yoshimatsu, T. Yamazaki, Y. Hoshino, P. E. Rose, L. F. Epstein, L. P. Miranda, P. Tagari, J. M. Beierle, Y. Yonamine and K. J. Shea, *J. Am. Chem. Soc.*, 2014, **136**, 1194.
43. K. Yoshimatsu, B. K. Lesel, Y. Yonamine, J. M. Beierle, Y. Hoshino and K. J. Shea, *Angew. Chem., Int. Ed.*, 2012, **51**, 2405.
44. T. Oshima, A. Suetsugu and Y. Baba, *Anal. Chim. Acta*, 2010, **674**, 211.
45. P. G. Mazzola, A. M. Lopes, F. A. Hasmann, A. F. Jozala, T. C. V. Penna, P. O. Magalhaes, C. O. Rangel-Yagui and A. Pessoa Jr, *J. Chem. Technol. Biotechnol.*, 2008, **83**, 143.
46. L. D. Nghiem, P. Mornane, I. D. Potter, J. M. Perera, R. W. Cattrall and S. D. Kolev, *J. Membr. Sci.*, 2006, **281**, 7.
47. W. M. Leevy, S. T. Gammon, H. Jiang, J. R. Johnson, D. J. Maxwell, E. N. Jackson, M. Marquez, D. Piwnica-Worms and B. D. Smith, *J. Am. Chem. Soc.*, 2006, **128**, 16476.
48. S. Geninatti Crich, D. Alberti, I. Szabo, S. Aime and K. Djanashvili, *Angew. Chem., Int. Ed.*, 2013, **52**, 1161.
49. A. Bajaj, O. R. Miranda, I. B. Kim, R. L. Phillips, D. J. Jerry, U. H. Bunz and V. M. Rotello, *Proc. Natl. Acad. Sci. U. S. A.*, 2009, **106**, 10912.
50. T. D. James, K. Sandanayake and S. Shinkai, *Angew. Chem., Int. Ed.*, 1994, **33**, 2207.
51. H.-J. Schneider, *Supramolecular Systems in Biomedical Fields*, Royal Society of Chemistry, 2013.
52. A. Metzger, V. M. Lynch and E. V. Anslyn, *Angew. Chem., Int. Ed.*, 1997, **36**, 862.
53. S. L. Wiskur, H. Ait-Haddou, J. J. Lavigne and E. V. Anslyn, *Acc. Chem. Res.*, 2001, **34**, 963.
54. G. Ghale and W. M. Nau, *Acc. Chem. Res.*, 2014, **47**, 2150.
55. M. N. Stojanovic and D. W. Landry, *J. Am. Chem. Soc.*, 2002, **124**, 9678.
56. B. Wang and E. V. Anslyn, *Chemosensors: Principles, Strategies, and Applications*, John Wiley & Sons, 2011.
57. L. Marchetti and M. Levine, *ACS Catal.*, 2011, **1**, 1090.
58. J. Meeuwissen and J. N. Reek, *Nat. Chem.*, 2010, **2**, 615.
59. M. Raynal, P. Ballester, A. Vidal-Ferran and P. W. van Leeuwen, *Chem. Soc. Rev.*, 2014, **43**, 1660.
60. M. Raynal, P. Ballester, A. Vidal-Ferran and P. W. van Leeuwen, *Chem. Soc. Rev.*, 2014, **43**, 1734.
61. L. Liu and R. Breslow, *J. Am. Chem. Soc.*, 2003, **125**, 12110.
62. E. A. C. Davie, S. M. Mennen, Y. Xu and S. J. Miller, *Chem. Rev.*, 2007, **107**, 5759.
63. K. Schlosser and Y. Li, *Chem. Biol.*, 2009, **16**, 311.

64. T. Darbre and J. L. Reymond, *Acc. Chem. Res.*, 2006, **39**, 925.
65. K. Severin, D. H. Lee, A. J. Kennan and M. R. Ghadiri, *Nature*, 1997, **389**, 706.
66. R. Breslow, *Artificial Enzymes*, John Wiley & Sons, 2006.
67. T. Heinisch and T. R. Ward, *Curr. Opin. Chem. Biol.*, 2010, **14**, 184.
68. M. Zhao, H.-B. Wang, L.-N. Ji and Z.-W. Mao, *Chem. Soc. Rev.*, 2013, **42**, 8360.
69. J. Zhang, X.-G. Meng, X.-C. Zeng and X.-Q. Yu, *Coord. Chem. Rev.*, 2009, **253**, 2166.
70. F. Mancin, L. J. Prins and P. Scrimin, *Curr. Opin. Colloidal Interface Sci.*, 2013, **18**, 61.
71. M. Resmini, *Anal. Bioanal. Chem.*, 2012, **402**, 3021.
72. H. Wei and E. Wang, *Chem. Soc. Rev.*, 2013, **42**, 6060.
73. M. J. Wiester, P. A. Ulmann and C. A. Mirkin, *Angew. Chem., Int. Ed.*, 2011, **50**, 114.
74. C. Bissantz, B. Kuhn and M. Stahl, *J. Med. Chem.*, 2010, **53**, 5061.
75. T. L. Nero, C. J. Morton, J. K. Holien, J. Wielens and M. W. Parker, *Nat. Rev. Cancer*, 2014, **14**, 248.
76. S.-H. Lee, Y. Hoshino, A. Randall, Z. Zeng, P. Baldi, R.-a. Doong and K. J. Shea, *J. Am. Chem. Soc.*, 2012, **134**, 15765.
77. E. W. Ng, D. T. Shima, P. Calias, E. T. Cunningham Jr., D. R. Guyer and A. P. Adamis, *Nat. Rev. Drug Discovery*, 2006, **5**, 123.
78. P. P. Deshpande, S. Biswas and V. P. Torchilin, *Nanomedicine*, 2013, **8**, 1509.
79. G. F. Whyte, R. Vilar and R. Woscholski, *J. Chem. Biol.*, 2013, **6**, 161.
80. S. Deshayes, H. Cabral, T. Ishii, Y. Miura, S. Kobayashi, T. Yamashita, A. Matsumoto, Y. Miyahara, N. Nishiyama and K. Kataoka, *J. Am. Chem. Soc.*, 2013, **135**, 15501.
81. S. White, J. W. Szewczyk, J. M. Turner, E. E. Baird and P. B. Dervan, *Nature*, 1998, **391**, 468.
82. F. Yang, N. G. Nickols, B. C. Li, G. K. Marinov, J. W. Said and P. B. Dervan, *Proc. Natl. Acad. Sci. U. S. A.*, 2013, **110**, 1863.
83. J. A. Raskatov, J. L. Meier, J. W. Puckett, F. Yang, P. Ramakrishnan and P. B. Dervan, *Proc. Natl. Acad. Sci. U. S. A.*, 2012, **109**, 1023.
84. J. A. Raskatov, N. G. Nickols, A. E. Hargrove, G. K. Marinov, B. Wold and P. B. Dervan, *Proc. Natl. Acad. Sci. U. S. A.*, 2012, **109**, 16041.
85. T. W. Synold, B. Xi, J. Wu, Y. Yen, B. C. Li, F. Yang, J. W. Phillips, N. G. Nickols and P. B. Dervan, *Cancer Chemother. Pharmacol.*, 2012, **70**, 617.
86. B. Liu, *Clin. Lipidol.*, 2012, **7**, 289.
87. C. D. Davidson, N. F. Ali, M. C. Micsenyi, G. Stephney, S. Renault, K. Dobrenis, D. S. Ory, M. T. Vanier and S. U. Walkley, *PLoS One*, 2009, **4**, e6951.
88. H. Valkenier, C. J. Haynes, J. Herniman, P. A. Gale and A. P. Davis, *Chem. Sci.*, 2014, **5**, 1128.
89. I. Alfonso and R. Quesada, *Chem. Sci.*, 2013, **4**, 3009.

90. P. L. Boudreault, M. Arseneault, F. Otis and N. Voyer, *Chem. Commun.*, 2008, 2118.
91. G. W. Gokel and S. Negin, *Adv. Drug Delivery Rev.*, 2012, **64**, 784.
92. H. Yamamura, K. Suzuki, K. Uchibori, A. Miyagawa, M. Kawai, C. Ohmizo and T. Katsu, *Chem. Commun.*, 2011, **48**, 892.
93. A. P. Davis, D. N. Sheppard and B. D. Smith, *Chem. Soc. Rev.*, 2007, **36**, 348.
94. P. A. Gale, *Acc. Chem. Res.*, 2011, **44**, 216.
95. A. P. De Silva, *Molecular Logic-based Computation*, Royal Society of Chemistry, 2012.
96. K. S. Hettie, J. L. Klockow and T. E. Glass, *J. Am. Chem. Soc.*, 2014, **136**, 4877.
97. D. A. Uhlenheuer, K. Petkau and L. Brunsveld, *Chem. Soc. Rev.*, 2010, **39**, 2817.
98. M. A. C. Stuart, W. T. S. Huck, J. Genzer, M. Muller, C. Ober, M. Stamm, G. B. Sukhorukov, I. Szleifer, V. V. Tsukruk, M. Urban, F. Winnik, S. Zauscher, I. Luzinov and S. Minko, *Nat. Mater.*, 2010, **9**, 101.
99. T. Miyata, M. Jige, T. Nakaminami and T. Uragami, *Proc. Natl. Acad. Sci. U. S. A.*, 2006, **103**, 1190.
100. T. Miyata, in *Biomedical Applications of Hydrogels Handbook*, ed. R. M. Ottenbrite, K. Park and T. Okano, Springer New York, 2010.
101. H. Liu, Y. Li, K. Sun, J. Fan, P. Zhang, J. Meng, S. Wang and L. Jiang, *J. Am. Chem. Soc.*, 2013, **135**, 7603.
102. S. Biswas, K. Kinbara, T. Niwa, H. Taguchi, N. Ishii, S. Watanabe, K. Miyata, K. Kataoka and T. Aida, *Nat. Chem.*, 2013, **5**, 613.
103. C. A. Lipinski, *Drug Discov. Today: Technol.*, 2004, **1**, 337.
104. C. Sanchez, K. Shea and S. Kitagawa, *Chem. Soc. Rev.*, 2011, **40**, 453.
105. V. Janout, B. Jing, I. V. Staina and S. L. Regen, *J. Am. Chem. Soc.*, 2003, **125**, 4436.
106. E. Kay, D. A. Leigh and F. Zerbetto, *Angew. Chem., Int. Ed.*, 2007, **46**, 72.
107. D.-H. Qu and H. Tian, *Chem. Sci.*, 2013, **4**, 3031.
108. B. Yurke, A. J. Turberfield, A. P. Mills, F. C. Simmel and J. L. Neumann, *Nature*, 2000, **406**, 605.
109. F. J. Martinez-Veracoechea, B. M. Mognetti, S. Angioletti-Uberti, P. Varilly, D. Frenkel and J. Dobnikar, *Soft Matter*, 2014, **10**, 3463.
110. A. Kuzuya and Y. Ohya, *Acc. Chem. Res.*, 2014, **47**, 1742.

CHAPTER 2

Design and Synthesis of Synthetic Receptors for Biomolecule Recognition

KATHARINE L. DIEHL, JAMES L. BACHMAN, BRETTE M. CHAPIN, RAMAKRISHNA EDUPUGANTI, P. ROGELIO ESCAMILLA, ALEXANDRA M. GADE, ERIK T. HERNANDEZ, HYUN HWA JO, AMBER M. JOHNSON, IGOR V. KOLESNICHENKO, JAEBUM LIM, CHUNG-YON LIN, MARGARET K. MEADOWS, HELEN M. SEIFERT, DIANA ZAMORA-OLIVARES AND ERIC V. ANSLYN*

Department of Chemistry and Biochemistry, 1 University Station A1590, Austin, TX 78712, USA
*Email: anslyn@austin.utexas.edu

2.1 The Foundations of Host–Guest Chemistry: Preorganization and Complementarity

The lock-and-key principle, first proposed by Emil Fischer, has served as a cornerstone to describe molecular recognition.[1] Fischer suggested that enzymes recognize substrates in a fashion similar to a key fitting a lock, specificity being determined by the appropriate fit of the substrate. These observations are the intellectual foundation for two fundamental concepts in supramolecular chemistry: *complementarity* and *preorganization*. Strong

Monographs in Supramolecular Chemistry No. 14
Synthetic Receptors for Biomolecules: Design Principles and Applications
Edited by Bradley D. Smith

Published by the Royal Society of Chemistry, www.rsc.org

binding between host and guest occurs when there is a good match of guest shape and host binding site (complementarity), and when this match is achieved with little conformational rearrangement (preorganization), resulting in attractive pairwise interactions between host and guest. In the case of biomolecule recognition, these interactions include correct pairing between hydrogen bond donors/acceptors, electrostatic forces, π-effects, induced interactions, and the hydrophobic effect (Figure 2.1).

In addition to matching of molecular shapes, complementarity supposes proper spacing of the binding sites. If steric strain is involved, binding affinity is weakened. Donald Cram considered both pairing of binding sites and spacing in his formulation of the principle of complementarity: "to complex, hosts must have binding sites which can simultaneously contact and attract the binding sites of the guests without generating internal strains or strong nonbonded repulsions".[2]

Binding is more likely to occur when the host and guest structures are preorganized for complexation. According to Cram, "the more highly hosts and guests are organized for binding and low solvation prior to their complexation, the more stable will be their complexes".[2] In other words, there is an energetic penalty if binding requires the structures to undergo significant conformational changes or if solvent must be removed from the binding site.

Macrocyclic polyether host molecules are classic examples of these predictive principles. Scheme 2.1 compares the affinities of pentaethyleneglycol dimethylether (EG5), 18-crown-6, and [2.2.2]cryptand for K^+.[3] EG5 exhibits the weakest binding: a greater entropic cost is required to adopt a conformation that wraps around the cation. 18-Crown-6 is the next weakest binding host because the crown ether's most stable conformation is a rectangle shape with two methylene groups directed inside the cavity.

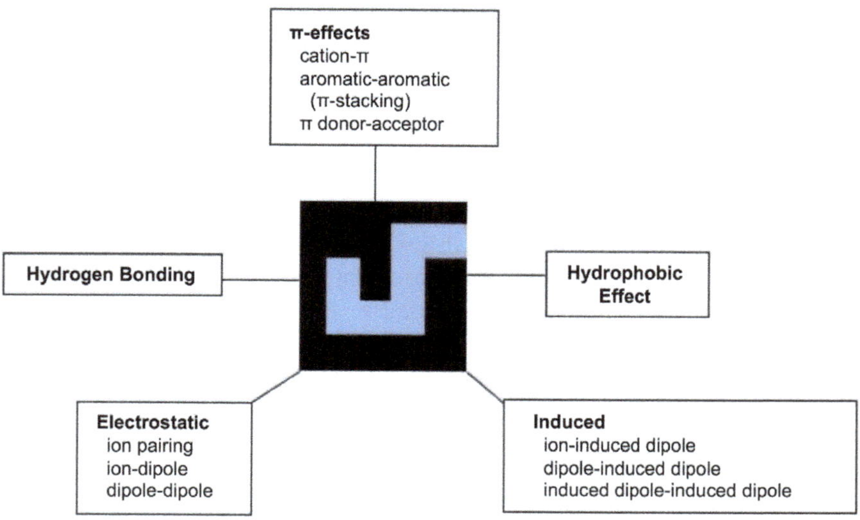

Figure 2.1 Intermolecular interactions involved in complexation.

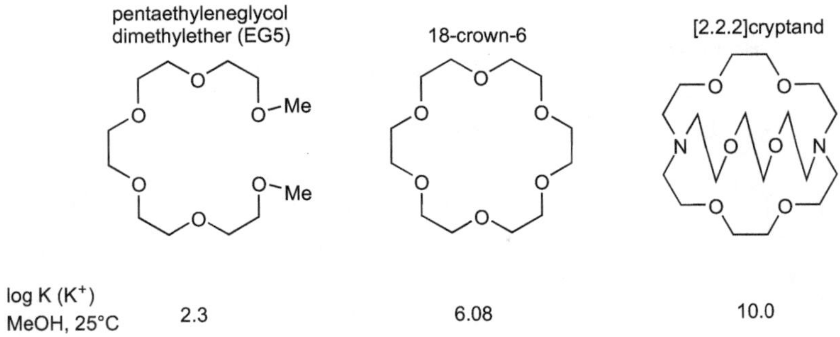

Scheme 2.1 Binding affinities of cyclic and acyclic polyethers for K^+.

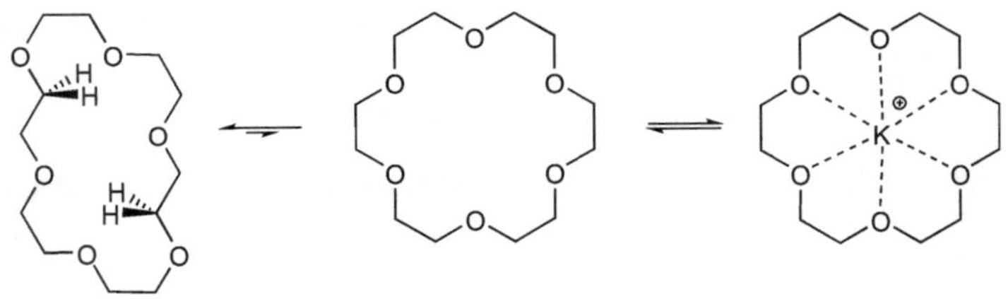

Scheme 2.2 Conformational change required for K^+ binding in 18-crown-6.

A modest conformational change is required to orient the oxygen atoms towards the interior (Scheme 2.2).[2] Potassium cation binding with [2.2.2]cryptand increases by four orders of magnitude when compared to 18-crown-6. The additional polyethylene glycol bridge restricts the conformational freedom and provides two additional oxygens for K^+ coordination.

The series of hydrogen bonding hosts in Scheme 2.3 provide another demonstration of the effectiveness of preorganization and complementarity. In this case, the guest is the barbiturate drug, barbital.[4] The acyclic hosts **2.2** and **2.3** have linked 2,6-diamidopyridine motifs that complement the two hydrogen bonding edges of the barbital guest. They both bind barbital better than host **2.1** which only has one 2,6-diamidopyridine motif. But the champion by far is the highly preorganized macrocyclic receptor **2.4** developed by Andrew Hamilton.

In addition to macrocycles, preorganized hosts can be developed by clever design of U-shaped molecules that create a rigid cleft with convergent binding sites. One of the first synthetic molecular building blocks for this purpose was Kemp's triacid, developed by Julius Rebek. Two of these building blocks are utilized in the U-shaped host **2.5** (Scheme 2.4), which has convergent imide motifs that form hydrogen bonds with an adenosine guest.[5] The preorganized cleft shape of **2.5** is maintained by the conformational rigidity of the Kemp's triacid.[6]

Preorganization *via* conformational constraints is exemplified by "pinwheel" building blocks, hexasubstituted benzene derivatives that produce cleft-shaped receptors. Pinwheel receptors, such as **2.6**, have been developed

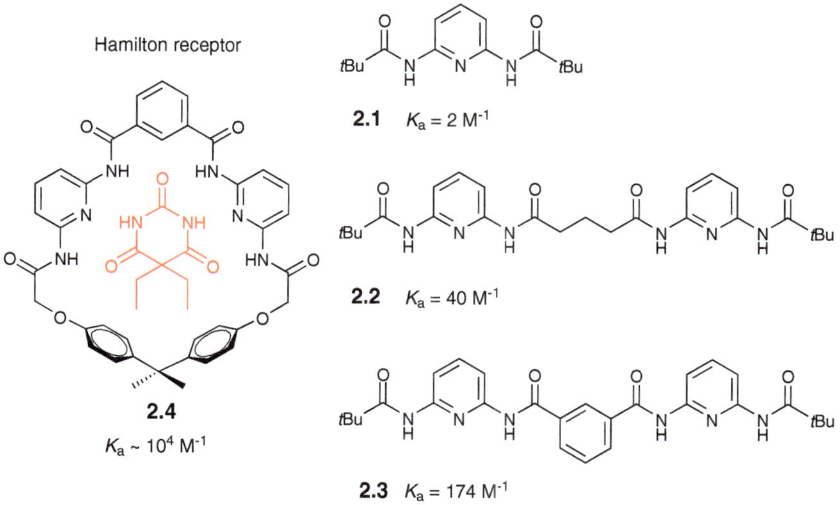

Scheme 2.3 Barbital binding by macrocyclic Hamilton receptor and deconstructed acyclic hosts.

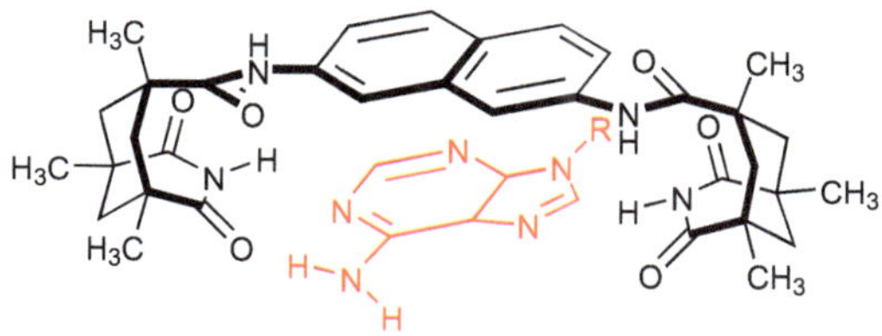

2.5

Scheme 2.4 Adenosine binding by cleft-shaped receptor.

by the Anslyn group to target several biologically active molecules.[7] As shown in Scheme 2.5, the facial preference of the substituents at the 1,3,5 positions alternate from the substituents at the 2,4,6 positions. Receptor **2.6** with six guanidinium groups was designed to bind with anionic guests such as inositol-1,4,5-triphosphate (IP_3). Strong binding of IP_3 was measured by a competition assay with 5-carboxyfluorescein, giving an association constant (K_a) value of 4.7×10^5 M^{-1} in water.

An alternative design principle, which contrasts with preorganization, is the concept of "induced-fit" binding. Koshland developed this term as a refinement of Fischer's lock-and-key model, and used it to rationalize the considerable conformational changes enzymes undergo upon binding.[8] Supramolecular chemists have also embraced induced-fit as a means to achieve complexation and to design new receptors. For example, Steed and coworkers demonstrated that pyridinium derivative **2.7** with a pinwheel host structure adopts an inward conformation when it associates with Br^- (Scheme 2.6).[9]

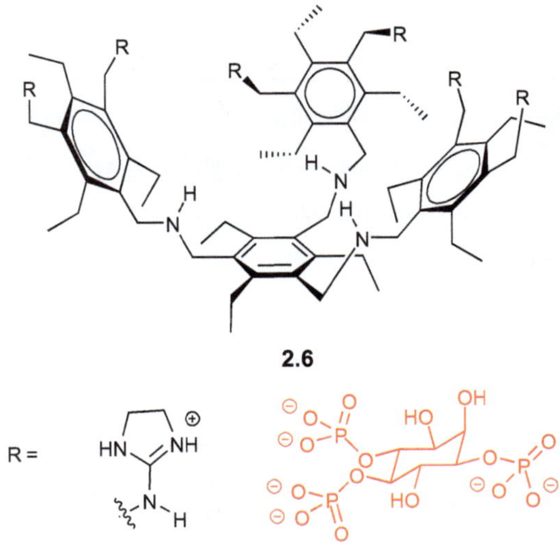

Scheme 2.5 Pinwheel host **2.6** with six guanidinium groups binds strongly to inositol-1,4,5-triphosphate.

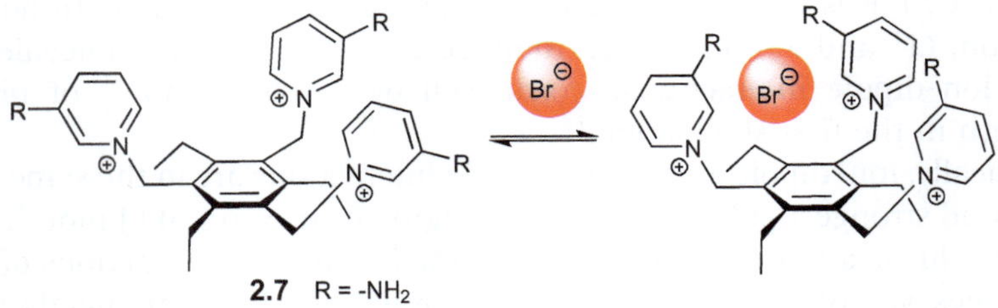

Scheme 2.6 Inward binding conformation of pinwheel host is induced by the presence of Br^- guest.

As a group of fundamental concepts, the lock-and-key model, complementarity, preorganization, convergence, and induced-fit are very useful guiding principles for effective receptor design.

2.2 Strengths of Molecular Recognition Interactions

Complementary binding infers pairwise interactions of weak forces between the host and guest. Knowledge of the strength of non-covalent interactions such as ion–ion interactions, ion–dipole interactions, dipole–dipole interactions, hydrogen bonding, cation–π interactions, and van der Waals forces can assist in designing hosts and estimating overall affinities. Non-covalent interactions have a wide range of values, from strong ion–ion bonds of several hundred kJ mol^{-1} to the very weak van der Waals forces of less than a few kJ mol^{-1} (Figure 2.2).[3,10] The strength of non-covalent interactions depends on several factors, such as solvent and temperature, and very rarely equal a covalent bond. For example, bond dissociation energies of H–CH_3

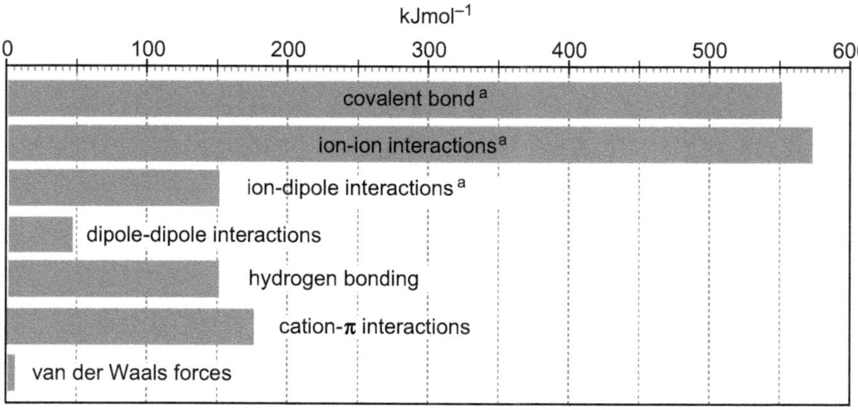

Figure 2.2 Strengths of covalent bonds and non-covalent interactions. [a]Single covalent bond or monovalent interaction.

and $H_3C–CH_3$ are 439 kJ mol^{-1} and 377 kJ mol^{-1}, respectively.[11] The strongest ionic single bond is Li–F with a gas state bond energy of 578 kJ mol^{-1}, but despite its large heat of formation and high melting point of 845 °C, LiF is easily dissolved in polar solvents like water. In aqueous solution, Li^+ and F^- ions are surrounded by layers of water molecules that form ion–dipole interactions, worth well over 100 kJ mol^{-1} of binding strength in the first shell alone.[12]

Typically, ion–dipole and dipole–dipole interactions are in the same group of second strongest polar interactions ranging from 5 to 150 kJ mol^{-1}.[10] The ionic radius is an important factor, and the ion–dipole interactions of polar molecules with organic cations are generally weaker than with metal cations. For example, gas-phase free energies for binding of water or *N*,*N*-dimethylacetamide to tetramethylammonium cation are four to ten times smaller than the corresponding binding to Li^+.[13]

Hydrogen bond interactions are separated into three classes according to energy and heteroatom–heteroatom distance; strong hydrogen bonds (60–120 kJ mol^{-1}; 2.2–2.5 Å), moderate hydrogen bonds (15–60 kJ mol^{-1}; 2.5–3.2 Å), and weak hydrogen bonds (<15 kJ mol^{-1}; up to 4 Å). The strongest known hydrogen bond is the F–H···F$^-$ hydrogen bond (163 kJ mol^{-1}), and it is so strong that it is considered a quasi-covalent bond.[14] The strength of a hydrogen bond is highly dependent on the D–H···A angle and typically reaches a maximum when the angle is 180°. This directional dependence means that hydrogen bonds are very effective interactions for controlling host–guest binding selectivity.

The cation–π interaction is a quadrupole–ion interaction (Figure 2.3) and includes association of an alkali metal or organic cation with a localized or delocalized π system. Arene complexes of transition metals such as Ag^+, Fe^+, In^+, *etc.*, are not included in this group because the bonding involves d orbitals on the metal cation.[10] The strengths of cation–π interactions range from 39 kJ mol^{-1} for the gas phase $Me_4N^+\cdots C_6H_6$ interaction to

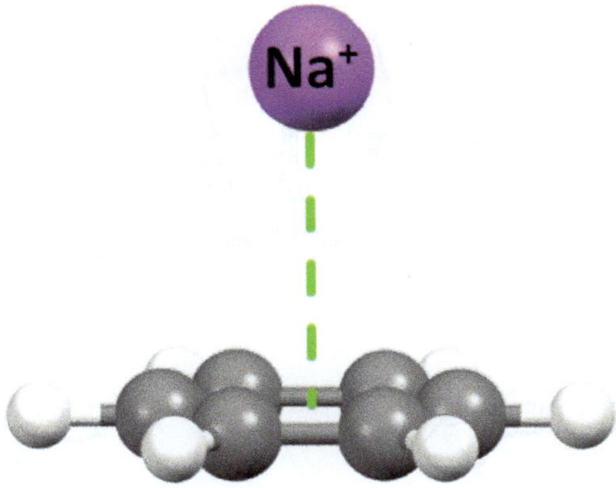

Figure 2.3 Cation···π interaction.

160 kJ mol^{-1} for the $Li^+\cdots C_6H_6$ interaction.[12,13,15] The $K^+\cdots C_6H_6$ interaction of 80 kJ mol^{-1} is even stronger than the $K^+\cdots OH_2$ ion–dipole interaction (75 kJ mol^{-1}).[16]

The weakest non-covalent interaction in Figure 2.2 involves dispersion or van der Waals forces (<5 kJ mol^{-1}) that arise from induced dipole moments caused by the proximity of a polarized molecule. For example, each CH_2 unit in a small alkane molecule can contribute an average of 4 kJ mol^{-1} to dispersion forces.[10]

One way to quantify non-covalent interactions is to design conformationally restricted molecules and evaluate the effect of intramolecular interactions on the conformational equilibrium.[17] The molecules act as balance frames with weighing platforms, and thus they are called *molecular torsion balances*. Figure 2.4 shows three molecular torsion balances that can exchange between two distinct conformations (folded or unfolded) *via* rotation of a sigma bond that connects a rigid platform (black) with an unsymmetrical arm (green). The free energy difference between unfolded and folded conformers can be calculated by measuring the equilibrium constant K and using eqn (2.1). The energy difference is assumed to correspond to the change in non-covalent interactions within the molecule and the surrounding solvent.

$$\Delta G^\circ = -RT \ln K \tag{2.1}$$

The concept of molecular torsion balances can be traced to early work by Ōki and his coworkers in the 1970s (Figure 2.4A).[18] They examined restricted bond rotation within 1,9-disubstituted triptycene derivatives caused by steric hindrance of substituents R_1 and R_2 and studied many types of non-covalent interactions including CH···O, CH···arene, and oxygen/halogen···arene interactions. Wilcox *et al.* coined the phrase "molecular torsion balance" in 1994 and used the conformational system shown in Figure 2.4B to investigate interactions with aromatic surfaces.[19] Recently, the Shimizu group introduced the new torsional balance shown in Figure 2.4C.[20]

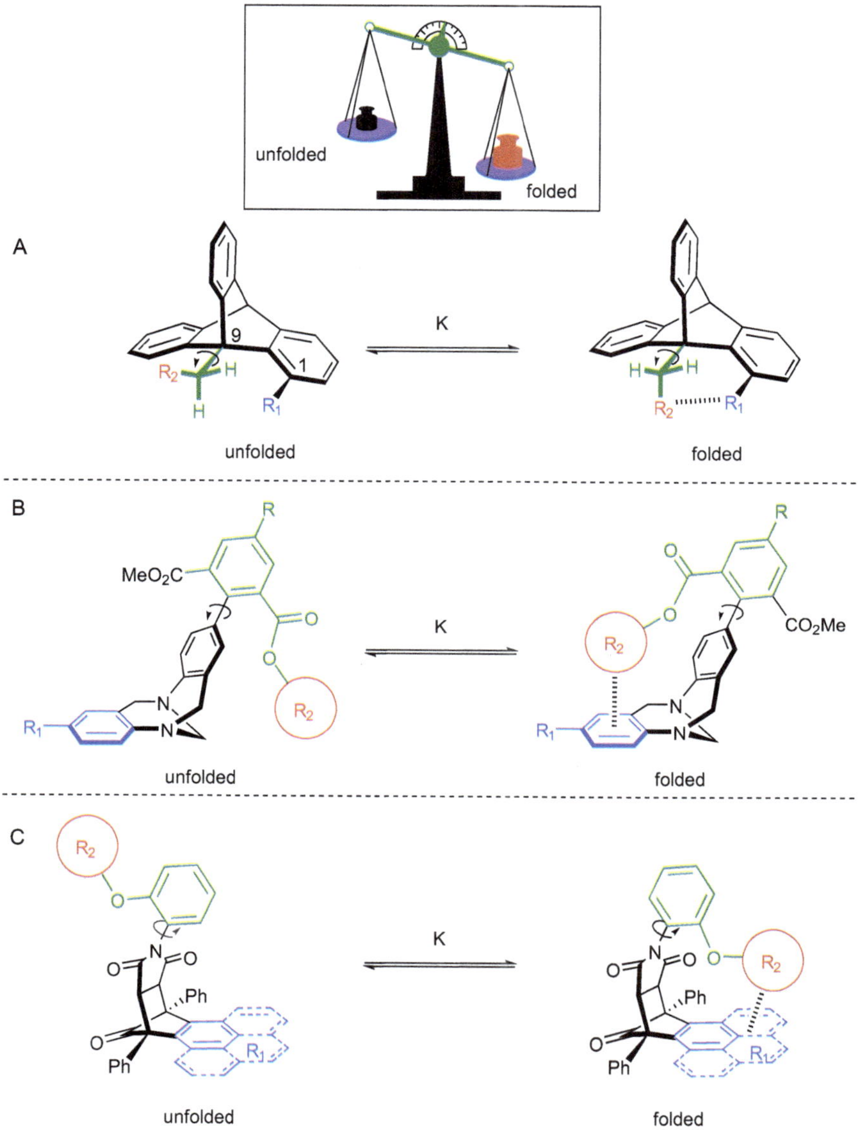

Figure 2.4 Three types of molecular torsion balances designed by the groups of: (A) Ōki, (B) Wilcox, and (C) Shimizu.

2.3 Solvent Dependence of Binding Interactions

The strengths of non-covalent binding interactions are highly dependent on the solvent. Any presentation of a binding constant should always include the identity of the solvent with a general understanding of its polarity and related properties.

It is very difficult to rationalize specific solvation effects, but some general concepts are broadly applicable. One is the solvophobic effect as a driving force that brings together hosts and guests with expulsion of solvent between the entities (Figure 2.5). Solvophobic theory states that the free energy change of complex formation (ΔG) is a linear function of solvent surface tension (γ).[21] In polar protic solvents, the solvent molecules released upon host–guest binding have strong enthalpic interactions with the bulk media. Furthermore, release of ordered solvent molecules commonly leads to an increase in entropy. The hydrophobic effect is especially important for biomolecule recognition in aqueous solution, and it is the dominant force for the folding of proteins into well-defined, three-dimensional structures.

The insight gained by correlating host–guest complex stability with general solvent properties is illustrated by considering two case studies. The data set in Table 2.1 [22] corresponds to the binding of nonpolar pyrene guest by cyclophane host **2.8** (Scheme 2.7). The association constants correlate with the solvent polarity parameter $E_T(30)$ and increase with the polarity of the solvent. The second data set (Table 2.2)[23] shows that the binding of 18-crown-6 with K^+ depends on the surface tension of the solvent.[24,25] Both data sets provide quantitative empirical measures of the solvophobic effect.

As a general rule, electrostatic-driven association of host and guest improves in nonpolar and particularly aprotic solvents. According to Coulomb's Law (eqn (2.2)) the extent of ion pairing inversely depends on

Figure 2.5 Host–guest association in solution.

Table 2.1 Log K and ΔG° values for complexation of pyrene by cyclophane host **2.8** in various solvents that differ in polarity as expressed by $E_T(30)$ values, $T = 303$ K.

Solvent	log K	ΔG° kcal mol^{-1}	$E_T(30)$ kcal mol^{-1}
Water	6.8	−9.4	63.0
Methanol	4.6	−6.4	55.5
Ethanol	4.4	−6.1	51.9
Acetone	3.1	−4.3	42.2
Dimethyl sulfoxide	2.8	−3.9	45.0
Dimethylformamide	2.2	−2.9	43.8
Dichloromethane	2.1	−2.9	41.4
Tetrahydrofuran	1.9	−2.7	37.4
Chloroform	1.6	−2.3	39.1
Benzene	1.1	−1.5	34.5
Carbon disulfide	1.0	−1.3	32.6

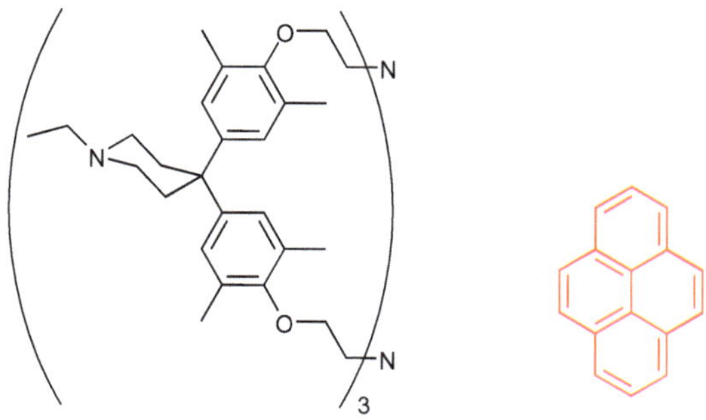

2.8 pyrene

Scheme 2.7 Structures of cyclophane host and pyrene guest.

Table 2.2 Log K values for complexation of K^+ by 18-crown-6 in various solvents that differ in surface tension.

Solvent	log K	Surface tension N m^{-1}
Water	2.0	0.072
Dimethyl sulfoxide	3.7	0.043
Dimethylformamide	4.3	0.035
Acetonitrile	5.7	0.028
Acetone	6.0	0.023
Methanol	6.2	0.022

dielectric constant (ε) of the solvent. Thus, a smaller ε increases the bond energy between ion pairs.

$$E = \frac{q_1 q_2}{4\pi\varepsilon_0 \varepsilon r^2} \tag{2.2}$$

However, it is important to realize that the dielectric constant is a bulk solvent property and not a measure of explicit coordination or hydrogen bonding ability. For example, the stability of a hydrogen bonded host–guest complex is greatly affected by the hydrogen bonding capabilities of the individual solvent molecules.

2.4 Reversible Covalent Bonding

Typically, organic synthesis uses irreversible covalent bonds to generate reaction products that are under kinetic control. In contrast, reversible covalent bonds give products based on thermodynamic control. Chart 2.1 [26,27] lists reversible covalent processes that have been utilized by supramolecular chemists in two distinct ways. Often, they are used to create reversible host–guest recognition processes where the directionality and strength of the covalent bond provides excellent functional group selectivity. Furthermore,

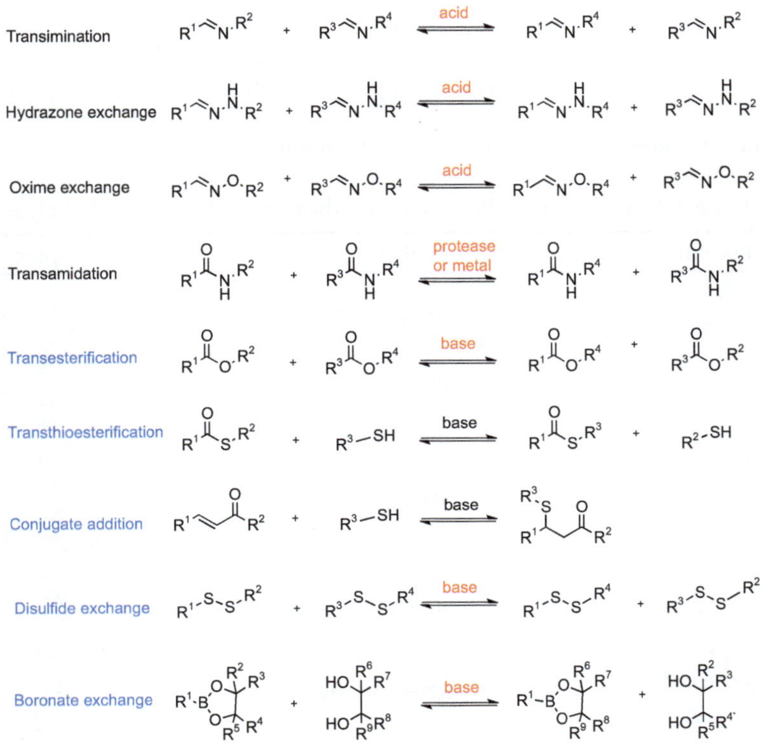

Chart 2.1 Reversible covalent reactions.

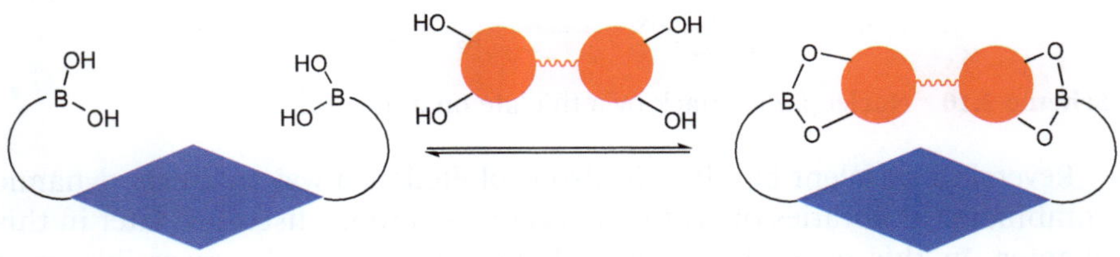

Scheme 2.8 Reversible covalent bonding of diol-containing guest using boronic acids. Adapted with permission from ref. 28, © 2010 American Chemical Society.

the structural change can be exploited as a signaling mechanism for design of optical indicators. The best known group of synthetic receptors that use reversible covalent binding are the boronic acids which react with diols and related functional groups (Scheme 2.8),[28] and also form dative bonds with various nucleophiles and Lewis bases (Scheme 2.9).[26–31] Another group of synthetic receptors employs an electron-deficient carbonyl group to achieve reversible covalent bonding. Trifluoroacetyl ketones have been used as indicators for various nucleophilic analytes including amines, anilines, alcohols, and thiols, as have aldehydes and the other carbonyl derivatives shown in Table 2.3.[32,33]

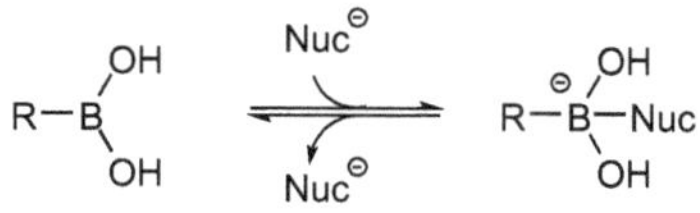

Scheme 2.9 Reversible nucleophilic attack on a boronic acid.

Table 2.3 Common carbonyl-based indicators and associated analytes.

Receptor	Analyte	Complex
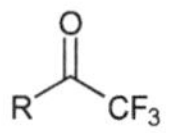	RNH_2, R_2NH	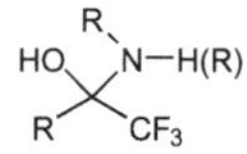
	ROH	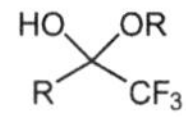
	RSH	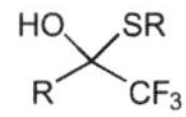
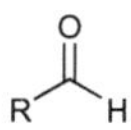	SO_3H^-	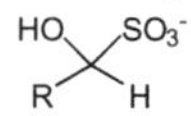
	$RHNNH_2$	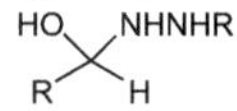
	HCN	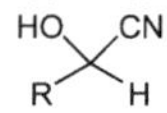
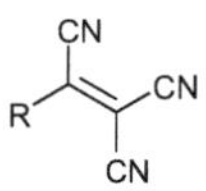	RNH_2, R_2NH	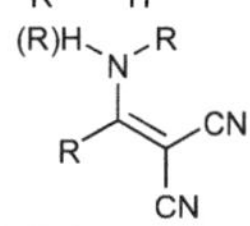

Scheme 2.10 Nucleophilic attack by a thiolate on a disulfide.

Reversible covalent bonding is also exploited as a way to create dynamic combinatorial libraries of synthetic receptors, a topic discussed later in this chapter. In this case, the covalent bonding needs to be reversible on a timescale that allows the combinatorial reaction to reach thermodynamic equilibrium.[34] Often the reaction is conducted in the presence of a catalyst or under certain experimental conditions that must be compatible with the reactivity of the host and guest(s).[26,33] For example, the exchange of disulfide bonds is rapid under basic conditions (Scheme 2.10),[35] but is sufficiently slow in acid to allow isolation of a thermodynamically assembled receptor.

2.5 From Plastic Models to Modern Computational Methods

Before the advent of computational modeling, the design of synthetic receptors largely depended on chemical intuition augmented by examination of physical models that mimicked the relative sizes of the bonds and atoms.

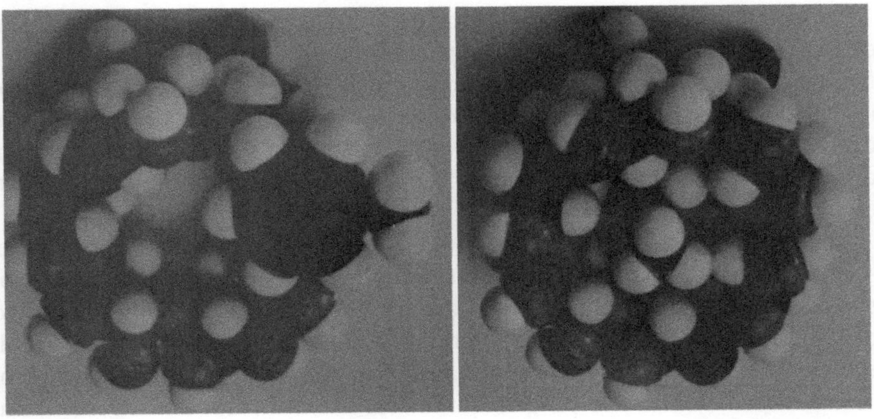

Figure 2.6 CPK models of α-cyclodextrin showing toluene outside (*left*) and inside (*right*) the cavity.
Reproduced with permission from ref. 38, © 1971 American Chemical Society.

For example, plastic CPK models (named after the pioneer users, Corey, Pauling, and Koltun) were a common tool for gauging the size and shape of a receptor binding site, or to rationalize observed binding efficiencies.[36,37] Figure 2.6 shows a photograph of CPK models for the binding of toluene by α-cyclodextrin.[38]

In the early days of computational modeling, the binding of cationic guests to organic hosts was modeled as a purely electrostatic interaction. This allowed the application of comparatively simple molecular mechanics calculations. However, the application of these simple models was rather difficult due to uncertainties in assigning local microenvironments and solvation effects.[37] Subsequent improvements by the groups of Jorgensen[39] and Kollman[40] led to computational methods that could generate quantitatively correct relative free energies of binding.[40] However, most calculations were performed on relatively rigid hosts and small or rigid guests. For more flexible or larger molecules it is necessary to combine the calculations with an extensive conformational search, in order to guarantee that the lowest energy conformation is found. Two such algorithms are MINTA[41] and the second-generation mining minima algorithm (M2),[42] both developed in the late 1990s. MINTA runs an extensive conformational search using a Monte Carlo method and calculates the conformational free energies of all conformers without the need for expensive free-energy simulations.[41] The mining algorithm calculates the binding free energy from standard chemical potentials, using the CHARMM force field and an implicit solvation model.[42] The M2 method calculates both enthalpy and entropy terms, and has been used to study entropy–enthalpy compensation.[42] Both MINTA and M2 allow for the ranking of possible receptors based on how well they are likely to bind the target molecule.

Another challenge is the automatic generation of receptor scaffolds.[43] In drug design, computer driven design methods are widely used, and range from virtual screening of known compounds to *de novo* design without prior knowledge of previously synthesized active ligands. While the problem of designing synthetic receptors for biomolecules is essentially the inverse of drug design, comparatively few programs exist for the computational design of synthetic receptors.

The computer program CAVEAT was designed by Bartlett and first described in 1994.[43,44] CAVEAT was originally developed for the design of peptides and peptidomimetics as enzyme inhibitors. It has also been applied to the design of molecular receptors for ions and small organic molecules.[43,45,46] CAVEAT is a vector-based program for the identification of molecular scaffolds. The program relies on the user input of a vector relationship between two or more bonds. One mode of operation is the search for a framework that exhibits the correct geometry to link functional groups or receptor fragments that have been placed in appropriate positions for binding the guest. The user defines the attachment vectors where the bond to the framework is to be formed. CAVEAT then searches the specified database for structures that contain bonds to substituents (including hydrogen) that match the defined vectors within given tolerances. The structures can then be grouped based on similarity to simplify the output. After clustering, each framework is represented by only one structure.

CAVEAT was used for the design of a glucopyranose receptor, as outlined in Scheme 2.11.[47] A conformational search was performed on glucopyranose derivative **2.9**, followed by geometry optimization using HF//6-31G*. A pair of vectors corresponding to the methyl–aryl bonds was then defined. CAVEAT

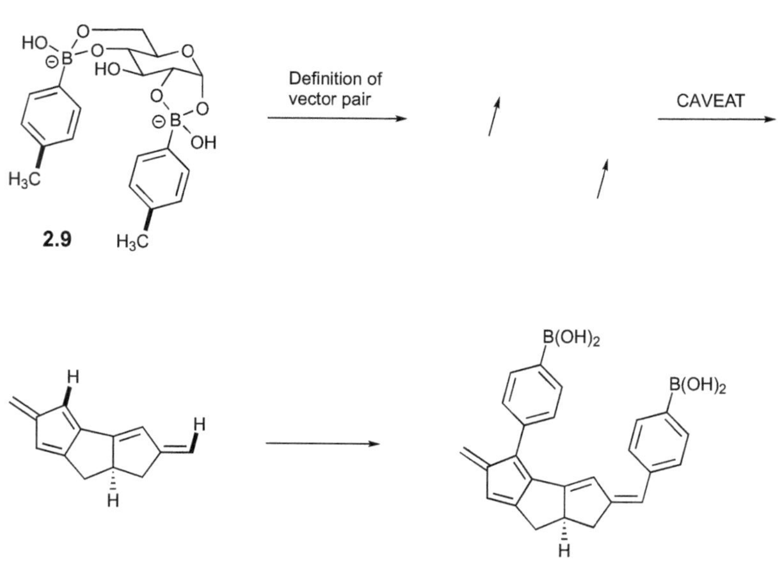

Scheme 2.11 Design of a glucopyranose receptor using CAVEAT.

then searched the TRIAD database, a computer-generated database of tricyclic hydrocarbons, for structures that contained bonds matching the vector pair. Many results were eliminated manually due to high complexity or because they contained highly strained ring systems. Minor changes were made to the skeleton to allow for easy synthesis, better solubility, and to allow for later derivatization. Finally, compound **2.10** (Scheme 2.12)[47] was synthesized and complex formation with glucose was confirmed by nuclear magnetic resonance spectroscopy (NMR) and mass spectroscopy. Fluorescence experiments gave an apparent dissociation constant of 2.5×10^{-5} M. The receptor exhibited a 400-fold greater affinity for glucose over any of the other common sugars tested.

Wang and coworkers used a different vector-based approach for the design of a boronic acid-based receptor for dopamine.[48] Starting from a crystal structure of dopamine, they assigned possible binding vectors to a dopamine–boronic acid complex. A virtual library of compounds containing a boronic acid, a carboxylic acid and a hydrophobic linker was generated and the vectors were compared to the ideal binding vectors. The proposed binding mode is shown in Scheme 2.13.[48] The synthesized compounds showed binding constants of 520–940 M^{-1} and good selectivity over epinephrine.

HostDesigner, developed by Hay and coworkers, is another program for the interactive *de novo* design of synthetic receptors.[49] HostDesigner contains two algorithms, LINKER and OVERLAY. LINKER starts with a known

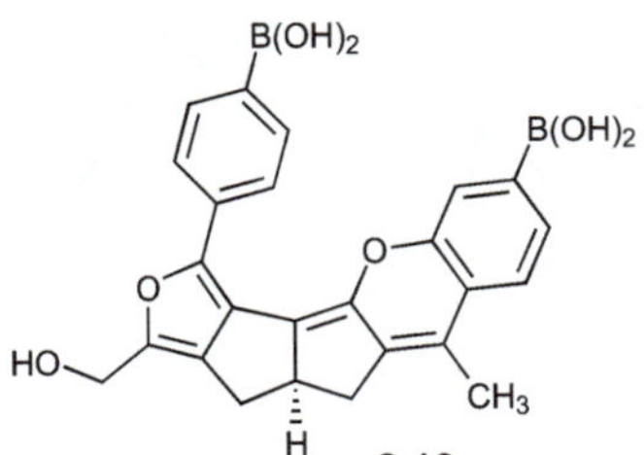

2.10

Scheme 2.12 Structure designed using CAVEAT.

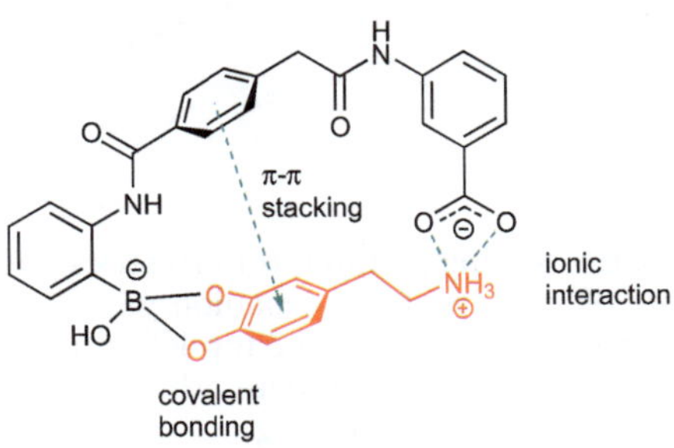

Scheme 2.13 Dopamine binding by Wang's receptor.

complex, where ligands are placed in optimal positions relative to the guest. This starting geometry can either come from crystal structures or computer simulations. LINKER then links the ligands at defined positions using a library of linker fragments. For each host, all possible conformers are generated. In each conformer, guest molecules are placed in ideal positions relative to the individual binding sites in the original fragments. The conformers are then scored by calculating the root mean square deviation (RMSD) between the atoms of these guest molecules. The principle is illustrated for receptor binding of an oxoanion guest in Figure 2.7.[50] The OVERLAY algorithm functions in a manner similar to CAVEAT. Again, a starting complex and two bonding vectors for the attachment of a linker are defined. These vectors are then matched to a library of possible linking fragments. Scoring is achieved by calculating the RMSD between the vectors of the original complex and the linking fragment. HostDesigner was originally made for the design of metal cation receptors and has also been applied to the design of hosts for anions.[50,51]

In 2014, Hay and coworkers used HostDesigner to design a set of novel, hydrogen bonding hosts for a neutral guest molecule, namely oxygen

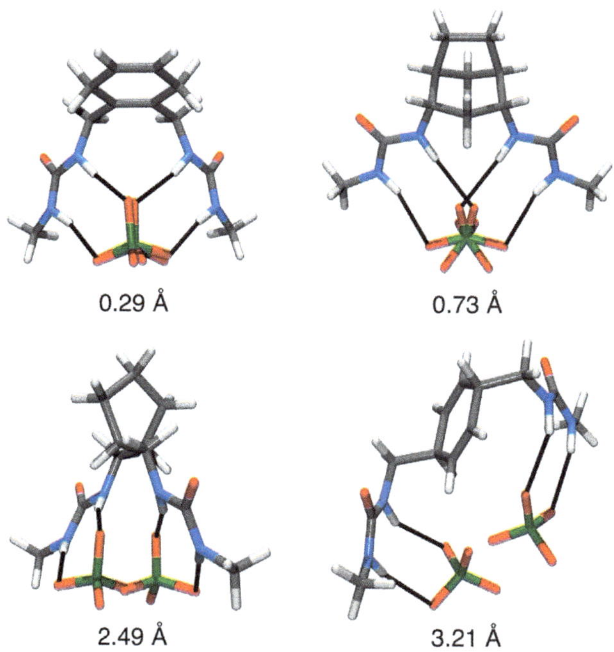

Figure 2.7 Example of the scoring method from HostDesigner's LINKER algorithm. The degree of overlap of the oxoanion guests of the two original fragments determines the score. The top left structure with a RMSD = 0.29 is the highest scoring candidate, the bottom right structure with RMSD = 3.21 is the lowest scoring.
Reproduced with permission from ref. 50, © 2006 American Chemical Society.

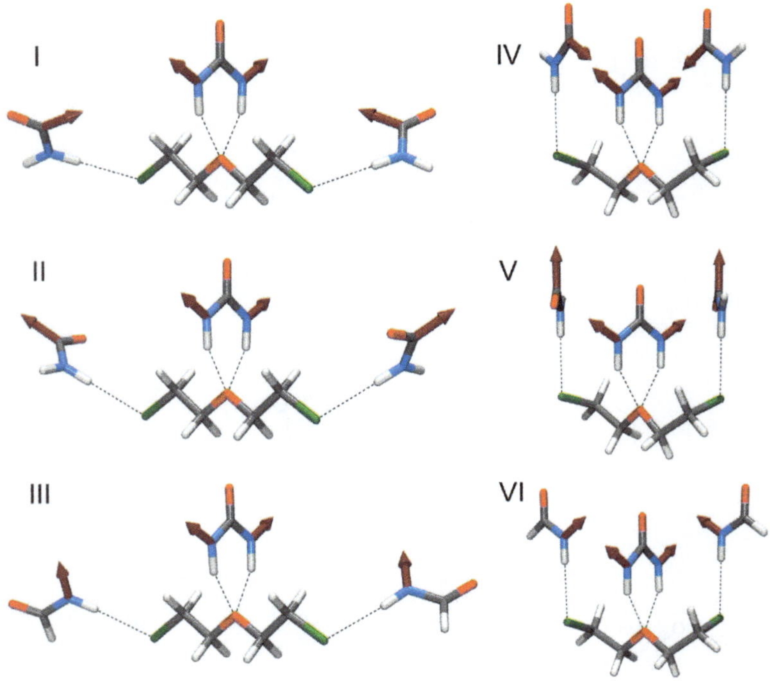

Figure 2.8 Input fragments for generating a host for oxygen mustard. Arrows indicate where new bonds will be formed.
Reprinted from ref. 52 copyright 2014, with permission from Elsevier.

mustard.[52] An urea was chosen as the central host motif to hydrogen bond with the oxygen atom of the guest. Likewise, host amide groups were chosen as hydrogen bond donors for interaction with the terminal chlorine atoms on the guest. Six input fragments (Figure 2.8)[52] were generated and connected using structures from a hydrocarbon library. The generated hosts were scored by calculating the RMSD as described earlier. The top 600 structures were subjected to molecular mechanics analysis, and the top 30 structures were retained. A final consideration of synthetic accessibility led to a candidate list of five host structures (Scheme 2.14).[52]

ConCept was developed by Gilson and coworkers and uses an evolutionary algorithm for computer aided host design.[53] Unlike CAVEAT and HostDesigner, ConCept does not rely on user-defined interactions. Instead, the guest is probed using appropriate molecule fragments and a map of hydrogen bonding ability and possible nonpolar interactions is generated. According to this map, appropriate fragments from the library are linked and the generated structures are scored using the CHARMM force field. The first generation of receptors is chosen using a metropolis Monte Carlo criterion and further elaborated by mutation or addition. Typically, 10 cycles of modification, scoring, and selection are applied. The M2 algorithm is used for the final evaluation of the candidate host structures. To demonstrate the

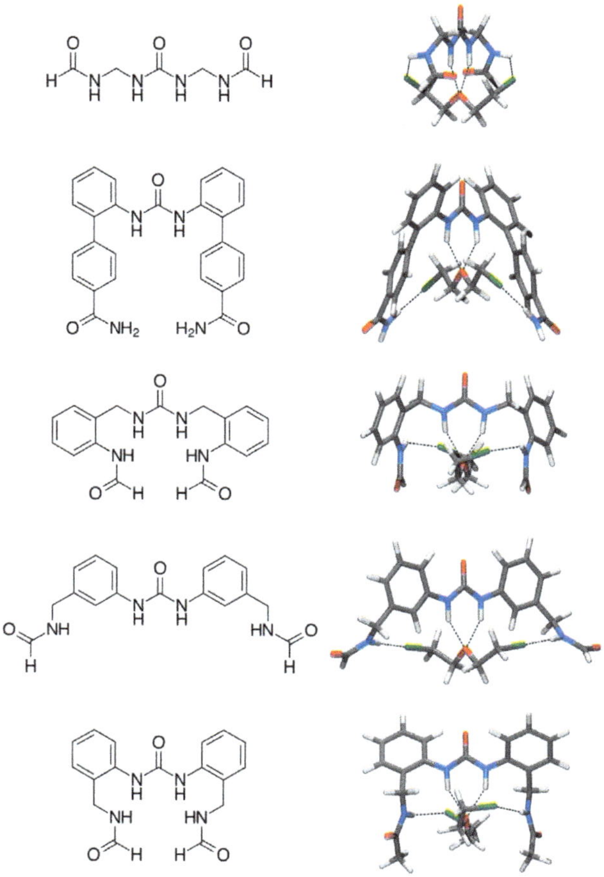

Scheme 2.14 Top five synthetically accessible host structures for oxygen mustard generated using HostDesigner.
Reprinted from ref. 52 copyright 2014, with permission from Elsevier.

potential of the program, Gilson and coworkers generated a set of potential adenine receptors, although they have not yet been tested experimentally.[53]

2.6 Common Synthetic Receptor Scaffolds

The idea of designing a synthetic organic receptor from scratch using computational algorithms is very alluring. But in reality the vast majority of synthetic receptors produced over the last three decades are synthetic derivatives of a small number of organic scaffolds. The six examples in Scheme 2.15[53] have the following common attributes (a) they are readily available from relatively cheap sources; (b) chemical methods to modify the structures are well-established; and (c) they have highly preorganized structures that make them good host molecules. The following sections provide a short overview of each of these receptor scaffolds.

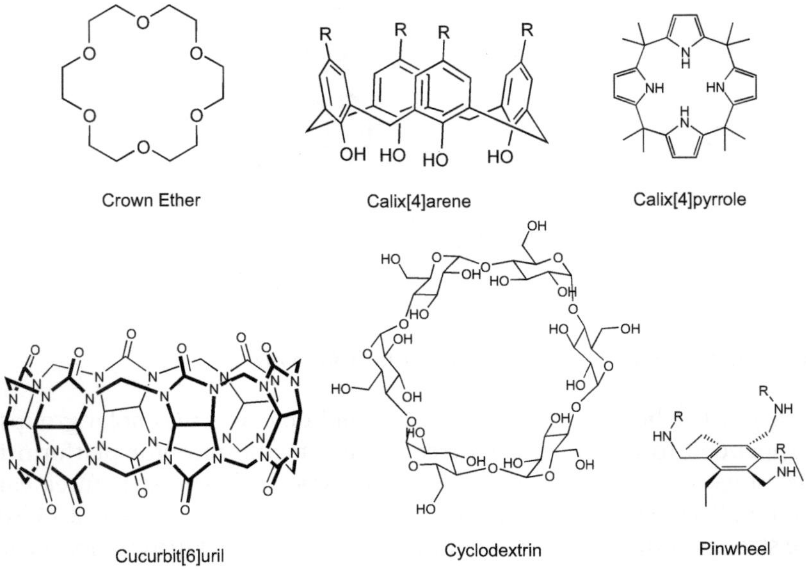

Scheme 2.15 Common scaffolds for synthetic organic receptors.

2.6.1 Crown Ethers

In 1967, Charles Pedersen recognized the ability of dibenzo-18-crown-6 to complex with K^+, a landmark event in supramolecular chemistry.[54] Since his discovery, a myriad different crown ethers have been synthesized to complex with a variety of different cations.[55] Subsequent work by Lehn and Sauvage led to cryptands, which are three-dimensional derivatives of crown ethers with much better metal cation binding affinities and selectivities.[56] The binding properties of crown ethers and cryptands are discussed further in Chapter 3. In the context of biomolecule recognition, it is worth noting that neutral crown ethers have strong affinities for metal cations in organic solvents, but not in aqueous solution. For instance, 18-crown-6 binding of K^+ has an association constant of 10^6 M^{-1} in methanol, but only 10^2 M^{-1} in water.[57] Guest affinities in water can be increased by structural modifications that provide hydrophobic stacking and electrostatic interactions. For example, Chen *et al.* prepared tetrasulfonated 1,5-dinaphtho-32-crown-8, **2.11** (Scheme 2.16), and showed that it binds the cationic guest nicotinamide adenine dinucleotide (NAD^+) in water with a K_a value of 2.2×10^3 M^{-1}.[58]

2.6.2 Calix[*n*]arenes

Calixarenes are cyclic structures formed by the condensation of a *p*-substituted phenol, resorcinol or pyrogallol with an aldehyde. The size of the ring can vary and they can adopt different conformations. For example, calix[4]arene (**2.12**) is shown in a cone conformation with inward pointing

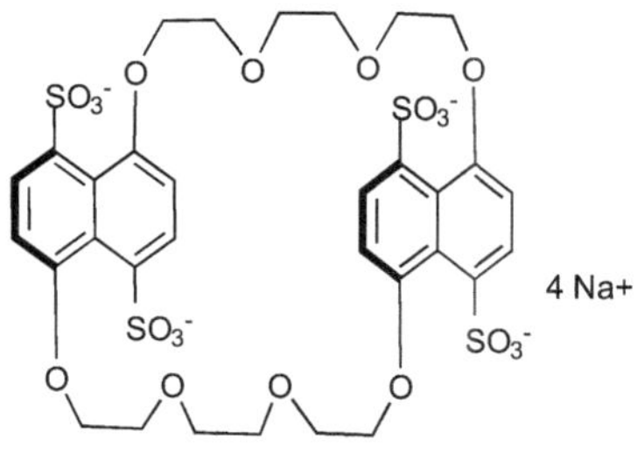

2.11

Scheme 2.16 Water soluble crown ether host for NAD^+.

hydroxyls. The bowl-shaped cavity is hydrophobic, whereas polar groups can be appended to the rims, thus creating a versatile receptor platform for a wide range of cation, anion, and neutral guests.[59] The structures can be functionalized with polyether units to form calixcrowns, such as **2.13**, which have strong affinity for metal or ammonium cations.[60] Further modifications can produce effective receptors of biomolecules. For example, Lee *et al.* developed a protein microarray based on the calixcrown **2.14**. The calixcrown was immobilized to a glass surface and shown to bind proteins *via* the lysine ammonium residues (Figure 2.9).[61]

2.6.3 Calix[*n*]pyrroles

Calixpyrroles are related to calixarenes, but they use functionalized pyrroles as building blocks (Scheme 2.17).[62–64] They typically adopt a conical conformation, and their ability to bind anions was first reported by Sessler in 1996.[65] The recognition capabilities can be expanded with structural modification. For example, covalent attachment of a cation binding unit can produce a receptor with the ability to simultaneously bind a cation and an anion.[65] Receptor **2.15** is a calixcrown linked to a calixpyrrole and is capable of selectively extracting cesium nitrate as an ion pair from aqueous solution.[62] Other analytical applications use immobilized calix[4]pyrroles like **2.16** and **2.17** for high-performance liquid chromatography (HPLC) separation of oligonucleotides and peptides.[63] In terms of selective drug delivery, a calix[4]pyrrole derivative (**2.18**) was recently shown to transfer a Pt(II) center to the nucleobase nitrogen of an associated nucleotide.[64]

2.6.4 Cucurbit[*n*]urils

Cucurbit[*n*]urils (CB*n* or sometimes Q*n*) are formed from the condensation of glycoluril and formaldehyde under acidic conditions. In the early 1980s, Mock *et al.* discovered a wide variety of uses for cucurbiturils as container molecules.[66] They can associate strongly with large organic or metal cations

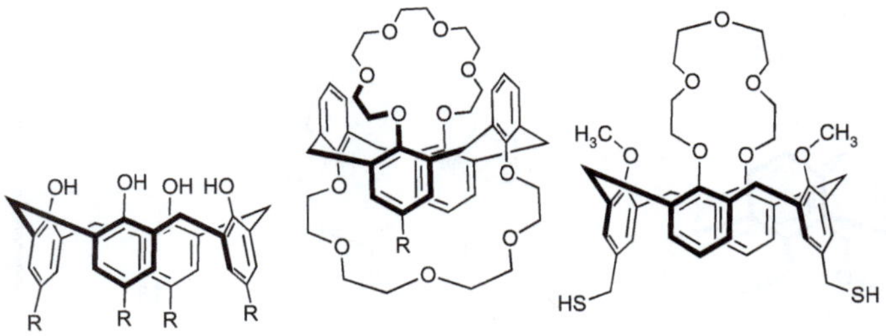

2.12 **2.13** **2.14**

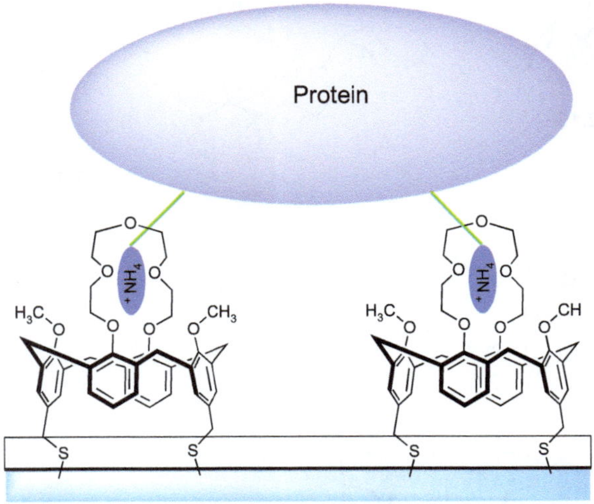

Figure 2.9 Protein immobilization on glass slides with calix[4]crown-5 derivatives. Adapted with permission from ref. 61, © 2003 Wiley.

in water and association constants up to 10^{15} M^{-1} have been reported. The primary interactions are hydrogen bonding or ion–dipole interactions with the carbonyl oxygens at the rims of the macrocycle. The hydrophobic interior suggests that cucurbiturils are promising delivery vehicles for various hydrophobic drugs.[67] Among the CB*n* homologues, CB8 has the ability to encapsulate two guests, yielding stable ternary complexes. The complex of methyl viologen (MV) **2.19** inside CB8 is able to accommodate a second electron-rich guest such as 2,6-dihydroxynaphthalene (**2.20**) to form an encapsulated charge transfer complex (Scheme 2.18).[68] Biological guests, such as the side chains of amino acids, peptides and tagged-proteins, can also be recognized by the CB8:MV complex.[69] Urbach *et al.* reported binding constants of 4.3×10^{4} M^{-1} between the CB8:MV complex and Trp, which was selective over Phe and Tyr.[70]

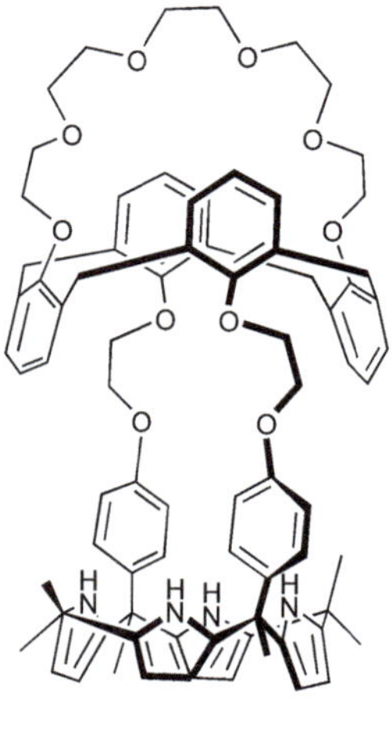

2.15

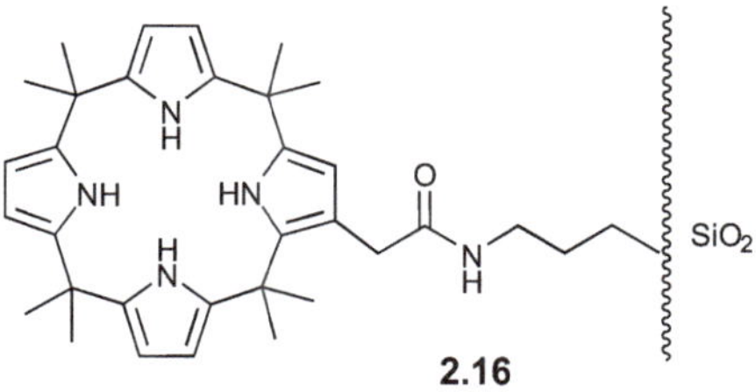

2.16

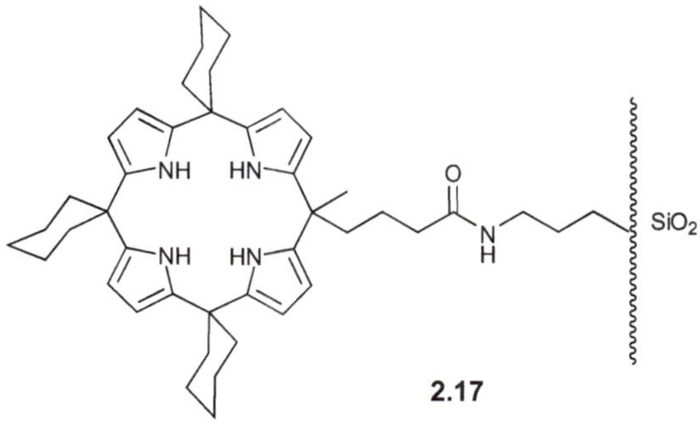

2.17

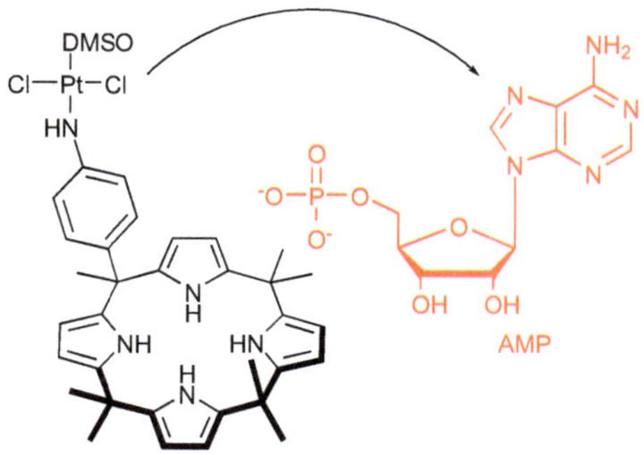

2.18

Scheme 2.17 Structures showing receptor **2.15** (a calixcrown linked to a calixpyrrole), receptors **2.16** and **2.17** immobilized on silica gel, and calixpyrrole **2.18** delivering Pt(II) drug to adenosine monophosphate (AMP).

2.6.5 Cyclodextrins

Cyclodextrins are water-soluble container molecules formed from α-1,4 linked glucose monomers in various ring sizes (Scheme 2.19).[3] They are produced enzymatically and used often in the food and pharmaceutical industries. They have also been studied extensively for a wide variety of supramolecular applications and many examples are described throughout this book.[71] The three most common structures are α-, β-, and γ-cyclodextrin (CD) with six, seven, and eight glucose units, respectively. The diameter of the hydrophobic cavity determines the size of the encapsulated guest, and the largest, γ-CD, can accommodate two guest molecules. The upper and

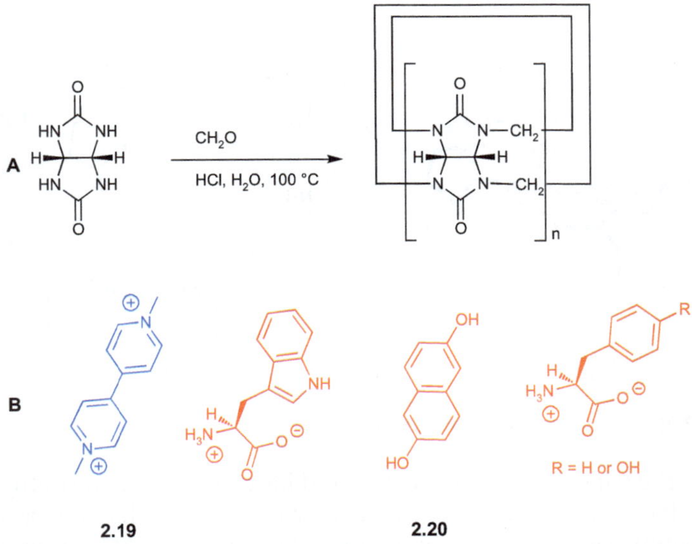

Scheme 2.18 (A) Schematic representation of the synthesis of CB8 under acidic conditions. (B) CB8 forms ternary complexes with electron-deficient methyl viologen (**2.19**) and a second electron-rich partner (**2.20**). Adapted with permission from ref. 68, © 2001 Wiley.

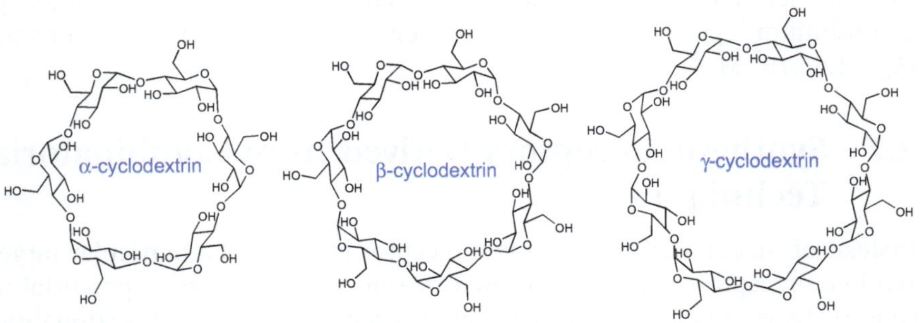

Scheme 2.19 The three most common cyclodextrins.

lower rims can be covalently modified to attach catalytic functional groups or optical indicators. Inclusion of attached fluorescent dyes within the cavity leads to a modified signal and becomes the basis of sensing assays. For example, γ-CD with an appended pyrene dye can form a dimeric self-inclusion complex with a pyrene excimer emission that is disrupted upon binding of steroids or terpenes inside the cavity.[72]

2.6.6 Pinwheel Scaffolds

A number of acyclic molecules with fixed conformations have been developed as highly preorganized cleft-shaped receptors. One of the best

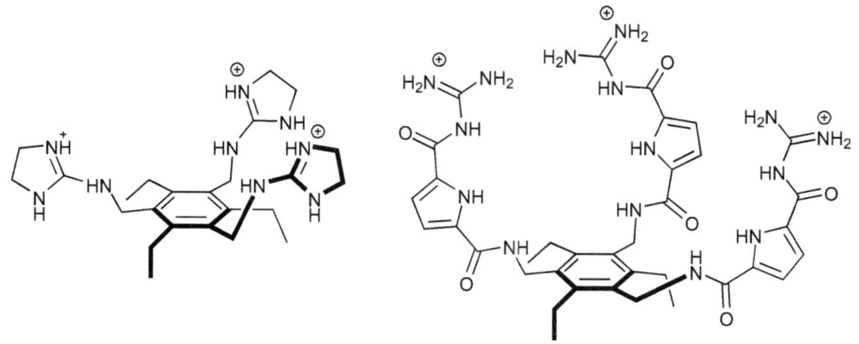

2.21 2.22

Scheme 2.20 Structures of pinwheel hosts.

known is the tripodal "pinwheel" scaffold based on the 2,4,6-triethylbenzene platform (Scheme 2.20).[73,74] The Anslyn group showed that a host with three guanidiniums can selectively bind citrate ($K_a = 6.9 \times 10^3$ M^{-1}) over carboxylic acids, phosphates, sugars, and salts in water. The strong affinity relies on ion pairing and is greatly decreased by the presence of buffer salts ($K_a < 10^2$ M^{-1}).[73] As a proof of principle, receptor **2.21** was used successfully in an indicator displacement assay to detect citrate in sport drinks. Later work by Schmuck and coworkers produced receptor **2.22** with three acylguanidinium pyrroles and a higher citrate association constant ($K_a = 1.6 \times 10^5$ M^{-1}).[74]

2.7 Synthetic Receptors Derived from Combinatorial Techniques

Instead of meticulously designing a receptor tailored to a particular target, synthetic receptors can be discovered by screening a library of potential receptors. Large molecular libraries that encompass a variety of functionalities ("chemical space") can be rapidly generated using combinatorial synthetic methods.[75] A chemist can increase the chances of finding a successful molecule by increasing the library number and structural diversity, and as such, combinatorial chemistry is often employed in high-throughput drug discovery.

Depending on the manner in which the components are mixed, combinatorial libraries can be generated to give either individual receptors of known composition or mixtures of receptors. Automatic parallel synthesizers enable the simultaneous synthesis of multiple compounds of discrete and known structure from the same components (Figure 2.10A).[76] Using parallel synthesis, Schmuck *et al.* created a library of 24 guidininiocarbonyl-pyrrole receptors for stereoselective recognition of amino acids and peptides (Scheme 2.21).[77] The side chains of the receptors were varied by coupling to a variety of amines. Parallel synthesis produces receptors of known

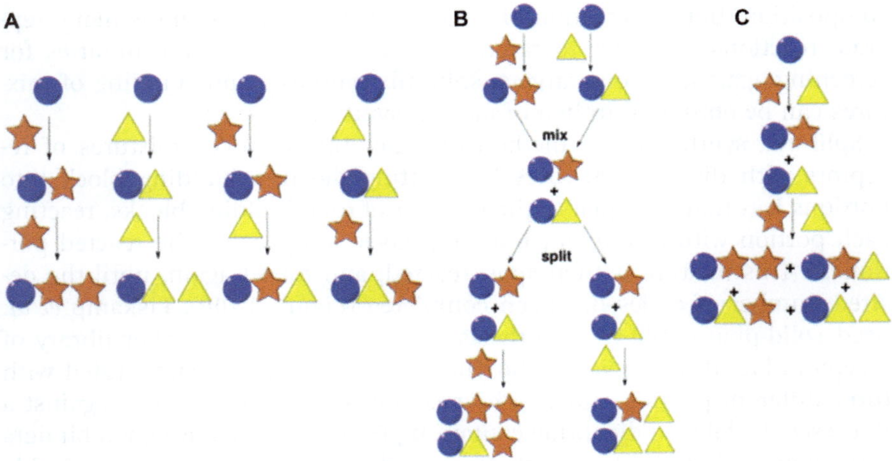

Figure 2.10 (A) Parallel synthesis method of combinatorial library generation. (B) Split-mix method of combinatorial library generation, also called split-and-pool. (C) Combinatorial library generated through coupling of mixtures.

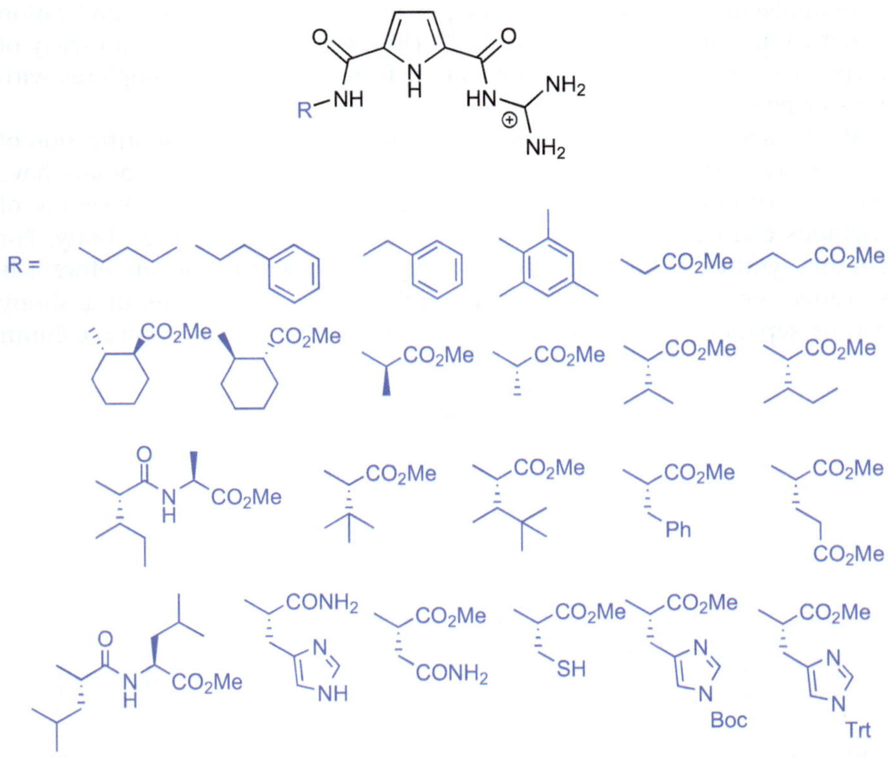

Scheme 2.21 Guanidiniocarbonyl-pyrrole receptors generated using parallel synthesis. All of the R groups are directly attached to the amide nitrogen.

composition, but the creation of large diverse libraries requires many separate reactions. In order to rapidly create diverse and large libraries for screening against various targets, split-mix synthesis and coupling of mixtures can be employed in lieu of parallel synthesis.

Split-mix synthesis is a method of generating complex mixtures of receptors with diverse structures by splitting the first building block into portions less than or equal to the number of total building blocks, reacting each portion with a different building block, and mixing the reacted portions.[78] This mixture is then split, reacted, and mixed again, until the desired number of cycles has been completed (Figure 2.10B). Liskamp *et al.* used solid-phase split-mix synthesis to create a 46 656 member library of receptors based on a triazacarboxylic acid (TAC) scaffold substituted with three different peptide arms.[79] The receptor library was screened against a fluorescently labeled dipeptide containing D-Ala-D-Ala. Twelve good binders were identified, and their structures were determined to have a remarkable amount of sequence consensus (Scheme 2.22).[79]

Another approach to generating complex mixtures of receptors utilizes mixtures of building blocks during the reaction step (Figure 2.10C). Each reactant is added at a concentration that allows for equimolar incorporation into the receptor structure. Wessjohann and Rivera created a library of supramolecular receptors by a one-pot multicomponent macrocyclization using tri-functional building blocks.[80] They were able to create a variety of cryptands, cages, and cryptophanes that formed inclusion complexes with ions or neutral molecules.

When dealing with large libraries and complex mixtures, identification of the best receptors can be difficult. High-throughput automated assays have emerged to evaluate the information from library screening. A variety of methods can be used to ascertain the active components in a library. For parallel synthesis, the identity of a library component is known before it is screened. For complex mixtures, the high affinity components in a library can be separated from weak binders by differential retention on a column

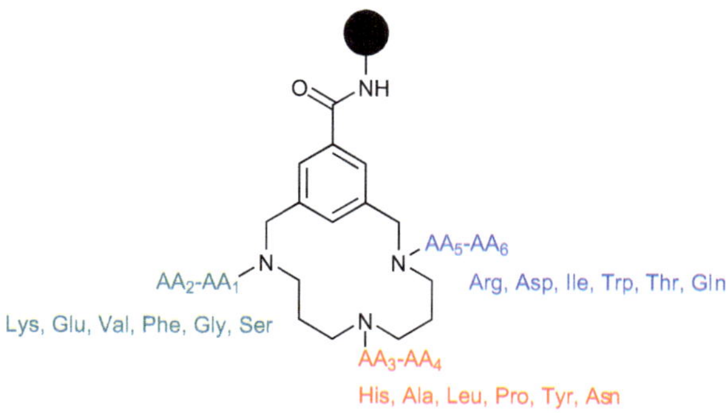

Scheme 2.22 Peptide substituted-TAC generated by split-mix synthesis. The library consists of amino acids at the specified positions, resulting in $6^6 = 46\,456$ members.

containing the immobilized molecular target.[81] Alternatively, strategies that create sublibraries can be used, such as positional screening,[81] iteration,[82] and tagging.[83] The generation of sublibraries allows for their chemical replication, meaning the libraries can be used for selection protocols that compete the best binding candidates against each other.[84]

In biology, molecular recognition is often achieved by large biopolymers that can fold and create high-affinity binding pockets. Hence, it makes sense to investigate unnatural biopolymer sequences as synthetic receptors for biological targets. Additionally, large libraries of oligonucleotides, peptides and proteins, and carbohydrates can be readily synthesized and screened for biomolecule recognition. Indeed, the use of building blocks to generate diverse structures parallels the biological processes of mutation and natural selection.[84] Diverse biopolymer libraries can be generated in a variety of ways that are chemical,[76] biological,[85] or biosynthetic.[86] It is especially advantageous to harness biosynthetic machinery for both the generation of molecular diversity and the selection of molecules that have high affinity for the biological target. Libraries of random biopolymer sequences can be winnowed by exposure to the biological target and suitable selection pressure, such as an affinity assay. Artificial evolution of a complex library mimics the diversity created through evolution but is accelerated by a faster rate of diversity generation.

Two major methods of artificial evolution are phage display and SELEX (systematic evolution of ligands by exponential enrichment). Phage display is an *in vivo* method of generating diverse libraries where a viral DNA programs the expression of a peptide with a particular sequence on the bacteriophage coat protein.[85] A heterogeneous mixture of phages, each with a unique DNA insert and therefore expressing a particular peptide, can be used as a combinatorial library. Affinity purification of the library against the biological target leads to capture of the best binding phages. The library is copied, passing along any mutations in the encoding DNA, and amplified when the bacteriophage infects a bacterial host and replicates. This creates a new library for further affinity purification. Peptide sequences are identified from the DNA sequences of the selected phages. Such libraries have been used to generate human antibody fragments *in vitro*,[87] and to generate peptide ligands for drug discovery.[88] Weiss *et al.* developed a biosensor for prostate-specific membrane antigen (PSMA) by screening a library of phage-displayed peptides in order to isolate an unnatural polypeptide that selectively binds PSMA.[89] The chosen peptide was displayed on a virus and used for the fabrication of nanowires as part of a PSMA-detecting device.

SELEX is an *in vitro* method where large libraries of oligonucleotides are synthesized chemically and exposed to a biological target (Figure 2.11).[84] A library is screened against a biological target for affinity, and the binding sequences retained are replicated using polymerase chain reaction (PCR). This creates a new library for further affinity purification, until multiple cycles of selection and amplification lead to purification of a small number of sequences specific to the biological target. Single-stranded (ss) nucleic

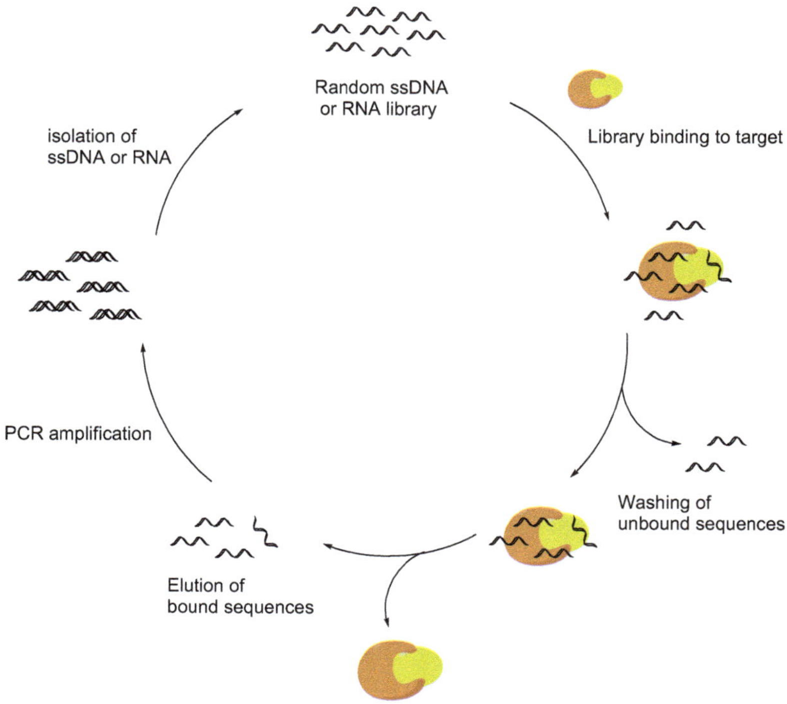

Figure 2.11 SELEX protocol for artificial evolution of a complex biopolymer library.

acids that show high affinity and specificity for a target are known as aptamers. Using SELEX, aptamers have been generated for a variety of applications, such as affinity purification, biosensors, and therapeutics.[90] Szostak and Lorsch developed RNA aptamers specific for vitamin B_{12} with affinities of 10^8 M^{-1}.[91] In 2004 the United States Food and Drug Administration approved an aptamer therapeutic for macular degeneration within the eye, known as pegaptanib (MACUGEN®).[92]

2.8 Dynamic Combinatorial Libraries for Receptor Discovery

Dynamic combinatorial libraries (DCLs)[93,94] are self-programming. Instead of synthesizing each potential receptor in combinatorial fashion, DCLs utilize reversible building blocks for potential receptors. These building blocks connect *via* reversible interactions, whether covalent or non-covalent, under thermodynamic control. All the possible combinations of building blocks comprise the virtual dynamic library,[95] with the actual concentration of each species determined by the system equilibrium. The addition of a template perturbs the system equilibrium, and the reversible interactions between the building blocks reach a new thermodynamic minimum, thereby producing receptors for the template. Often the structures of these receptors are unpredictable and they are sometimes hard to isolate.

In principle, a plethora of diverse building blocks with different functional groups and geometries could be combined to produce the most thermodynamically preferred receptors. But many factors limit the choice of building blocks. Incompatible functional groups may react irreversibly, either among library members themselves or with the template. Precipitation of certain building blocks or their ensuing equilibrium assemblies would act as an independent thermodynamic sink and likely produce an undesired equilibrium bias. Lastly, choosing building blocks of similar reactivity maintains the "fairness" of the library.

For a combinatorial library to be dynamic, it should exist under thermodynamic control, and therefore interactions among building blocks must be reversible. Chart 2.1 in Section 2.4 lists typical reversible covalent bonds, and enzymes can render other covalent bonds reversible. Proteases catalyze amide formation and hydrolysis,[96] and aldolases catalyze aldol condensations and their reversal.[97] However, enzymes prefer certain substrates over others, potentially introducing a non-thermodynamic bias to the system.[98] With the building blocks interacting reversibly within a system, all the possible combinations of building blocks form a virtual dynamic library.[95] Figure 2.12 [93] shows energy wells (energy minima) for several components of a virtual library, with the greater concentrations of components in the deeper, more stable wells. Any changes to the system, such as temperature, pressure, pH, solvent composition, or solvent volume, will shift the equilibrium constants for each virtual library component, resulting in changes to the relative concentrations. Likewise, addition of a template compound also perturbs the system equilibrium. Under

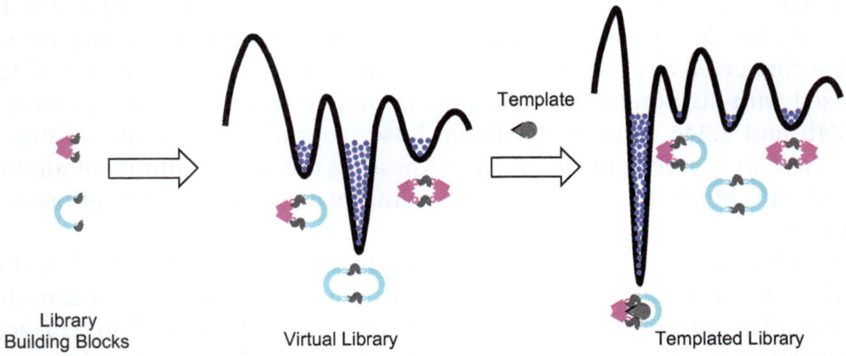

Figure 2.12 Cartoon depiction of building blocks assembling in a dynamic system, with their respective concentrations dependent upon the depth of the thermodynamic wells. Persistent template binding produces a new low-energy library member, and the equilibrium shifts towards the amplified member.
Adapted with permission from ref. 93, © 2010 Wiley-VCH Verlag GmbH & Co. KGaA.

thermodynamic control, only binding that reduces the free energy of the system will persist, resulting in the library producing effective receptors (Figure 2.12). A virtual component of the pre-template library may be amplified and become a major template-binding component. Static combinatorial libraries cannot provide such receptor amplification.

Comparison of the post-template library to the pre-template virtual library yields the new and effective receptors. Common forms of analysis include HPLC, mass spectrometry (MS), NMR, and X-ray crystallography. The conditions for analysis may change the properties of the library, unintentionally shifting the equilibrium. Therefore, the equilibrated system must first be rendered irreversible in some manner, such as changing pH, precipitating library components, removing catalysts by filtration, or oxidizing or reducing the system.

A DCL to find inhibitors of hen egg-white lysozyme exploited the large size of the enzyme to separate high-affinity templated molecules from low-affinity non-templated library members. Passing the template-equilibrated library through size exclusion chromatography (SEC) resulted in retention of the unbound library components while the enzyme-bound molecules passed through. Denaturing the protein in acetonitrile freed the bound molecules and the organic soluble substances were analyzed by liquid chromatography (LC)/MS to determine their building-block composition. Without the lysozyme template, no library component was detected by LC/MS, indicating that SEC retained all non-templated virtual library members.[99]

A dynamic combinatorial library comprised of a reactive macrocycle and multiple linker molecules produced a dimeric receptor exhibiting admirable affinity for sulfate anion in aqueous solution. The basis of the library was cyclohexapeptide **2.23** (Scheme 2.23), which binds a single sulfate dianion (Na_2SO_4) as a sandwich complex with an affinity of 10^5 M^{-2} in 20% methanol/deuterated water.[100] Thus, there was expectation that two covalently bonded cyclohexapeptides should bind sulfate better as long as they were connected by an appropriate linker. Addition of sulfate to a DCL of **2.24** mixed with six potential dithiol linkers produced an amplified amount of **2.24b** and **2.24c**. Studies on dimer **2.24c** determined a sulfate affinity of 10^6 M^{-1} in 2 : 1 acetonitrile/water, surpassing the sulfate affinity of another covalently bonded cyclohexapeptide dimer with a "designed" adipic acid linkage.

A subsequent DCL was created by mixing **2.25** (Scheme 2.24),[100] a dual linkage variant of **2.24**, with four dithiols, and led to amplified **2.25c$_2$** and **2.25d$_2$**. The latter exhibited an affinity for sulfate of 10^8 M^{-1} in 2 : 1 acetonitrile/water. In addition to defining the optimal linker length, orientation, and flexibility, the DCL templation process facilitated the receptor synthesis.

2.9 Cooperativity

Cooperativity is a phenomenon that arises from multiple interactions acting as a collective whole. Specifically, cooperativity is defined as occurring when

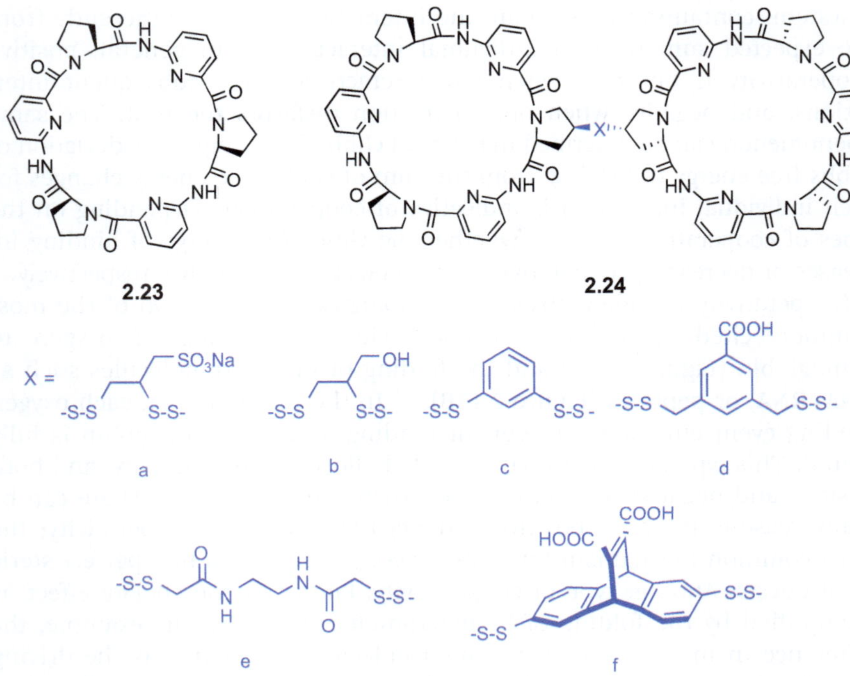

Scheme 2.23 Cyclic monomeric host **2.23** and dimeric host **2.24**, where X is one of the disulfide linkages a to f.

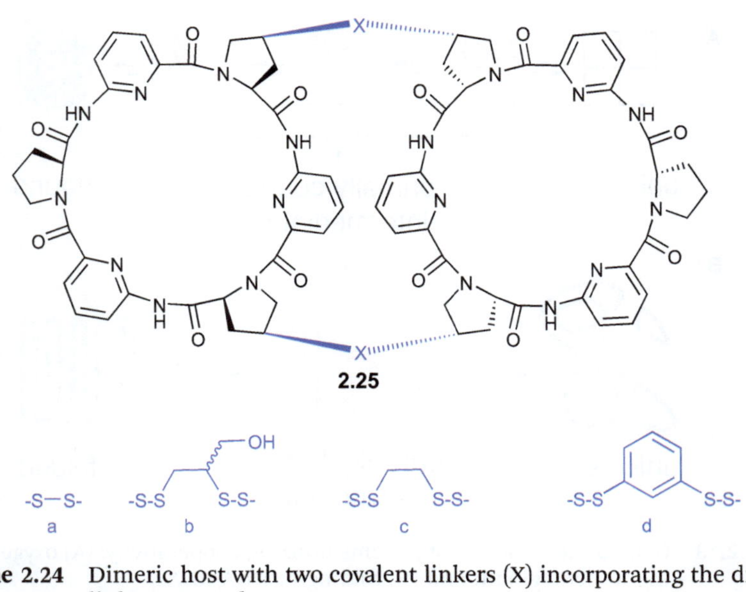

Scheme 2.24 Dimeric host with two covalent linkers (X) incorporating the disulfide linkages a to d.

a system, containing more than one interaction, behaves differently from the expected sum of each individual interaction.[101] In general, positive cooperativity is observed when one interaction favors subsequent interactions, and negative when one interaction disfavors the next. The same phenomenon can be described in terms of Gibbs free energy: any deviation of Gibbs free energy of binding from the sum of Gibbs free energy changes for each individual interaction is indicative of cooperativity. Depending on the types of cooperativity, and on whether the Gibbs free energy of binding increases or decreases, cooperativity can be positive or negative, respectively.

Cooperativity is widely observed in biological systems. Two of the most common cited examples of cooperativity are binding of oxygen to hemoglobin (Figure 2.13A) and the folding of biomacromolecules such as DNA, RNA, or peptides (Figure 2.13B).[101] In the first example, each oxygen binding event enhances subsequent binding until the hemoglobin is fully bound. This type of cooperativity is called allosteric cooperativity, and both positive and negative allosteric cooperativity are observed.[102] There can be many reasons for both positive and negative allosteric cooperativity; the most common mechanism for negative cooperativity is a non-perfect steric fit of guests. Another type of cooperativity, known as the chelate effect, is exemplified by the folding of biomacromolecules.[103] In this example, the difference in intramolecular and intermolecular interaction is the driving force for the assembly. Regardless of the type of cooperativity, a hallmark of positive cooperativity is "all-or-nothing behavior" with a sigmoidal binding isotherm and a sharp transition (Figure 2.14).[101] This means that the host is either primarily in the unbound or bound state, and little of the partially

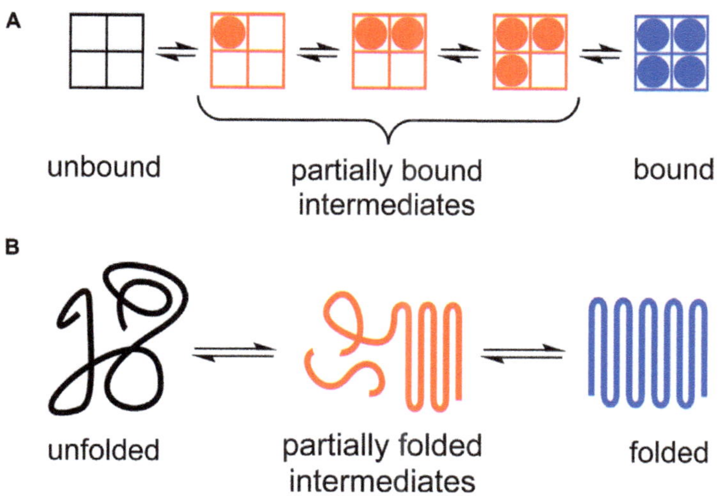

Figure 2.13 General illustrations of systems utilizing cooperativity: (A) oxygen binding to hemoglobin and (B) folding of DNA/RNA.
Reproduced with permission from ref. 101, © 2009 Wiley.

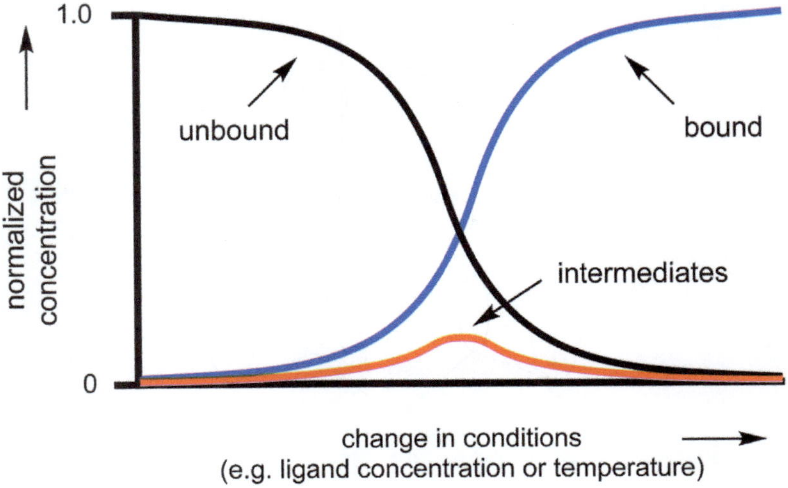

Figure 2.14 Speciation profile of a system exhibiting positive cooperativity. Note the sharp transition and the low build-up of intermediates. Reproduced with permission from ref. 101, © 2009 Wiley.

bound intermediates are present. In other words, host binding is extremely sensitive to the concentration of guest. This heightened binding sensitivity is widely observed in biology, where the intermediate species/structures can cause unwanted effects.

Additional terms that are pertinent to this discussion are *multivalency* and *avidity*. The concept of multivalency is defined as the joining of multiple binding sites on one entity to multiple binding sites on another.[104] While affinity refers to the strength of an individual monovalent interaction, avidity is used to describe the overall strength produced by the multiple binding interactions.[105] Therefore, avidity is the term used in the context of multivalency to describe the overall increase in binding resulting from an increase in the number of binding sites. Often the goal in biomolecule recognition is a synthetic receptor that employs multivalency, which results in high avidity, and positive cooperativity, which results in high sensitivity.

Many published reports have explored the origin of cooperativity. Cyclodextrin receptors are especially attractive as models for studying cooperative binding interactions. Zhang and Breslow investigated ditopic ligand binding interactions with monovalent and divalent cyclodextrin receptors (Scheme 2.25).[103] Negative cooperativity was observed for the binding of monovalent receptors, whilst the divalent receptor showed positive cooperativity. It is worth noting here that in the case of negative cooperative monovalent receptor binding, the cooperativity is allosteric in nature. In contrast, the tethering of two receptors leading to positive cooperativity embodies the idea of chelate cooperativity.

While the positive cooperativity observed with tethering of receptors is often attributed to favorable enthalpic strengthening, the entropic

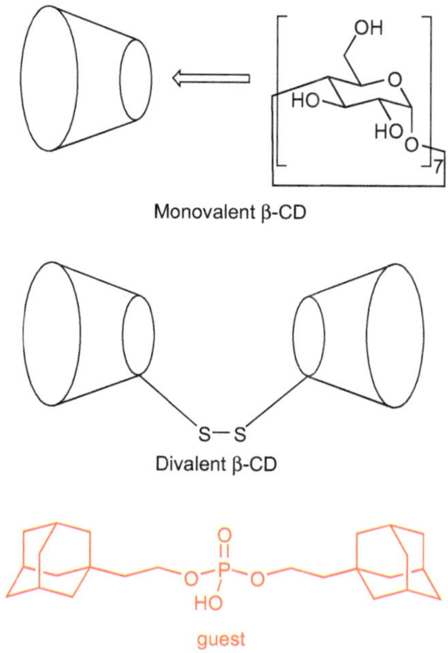

Scheme 2.25 Monovalent and divalent β-cyclodextrin (CD) receptors and guest.

effects must be considered when analyzing the thermodynamics of cooperativity.[103,106,107] Often, entropy related factors like the loss of residual motions and binding modes are frequently cited as the cause of negative cooperativity.[108] However, as noted by Jencks,[109] the translational entropic factors for a tethered host could just as well contribute to positive cooperativity if tethering causes a complete binding interaction (Figure 2.15).[107] In reality, the thermodynamics of cooperativity is often complicated by both situations.

2.10 Entropy–Enthalpy Compensation

Entropy–enthalpy compensation describes the case in which receptor modification results in a change in the enthalpy of binding that is matched by an opposite change in the entropy of binding. Hence, there is essentially no difference in the overall binding free energy of the system (eqn (2.3)):

$$\begin{aligned} \Delta G &= \Delta H - T\Delta S \\ \Delta\Delta H &\equiv \Delta H_2 - \Delta H_1, T\Delta\Delta S \equiv T\Delta S_2 - T\Delta S_1 \\ \Delta\Delta H &\approx T\Delta\Delta S \Rightarrow \Delta\Delta G \approx 0 \end{aligned} \tag{2.3}$$

In other words, modifying a receptor to strengthen its intermolecular interactions with the guest results in stronger bonds (*i.e.* enthalpically

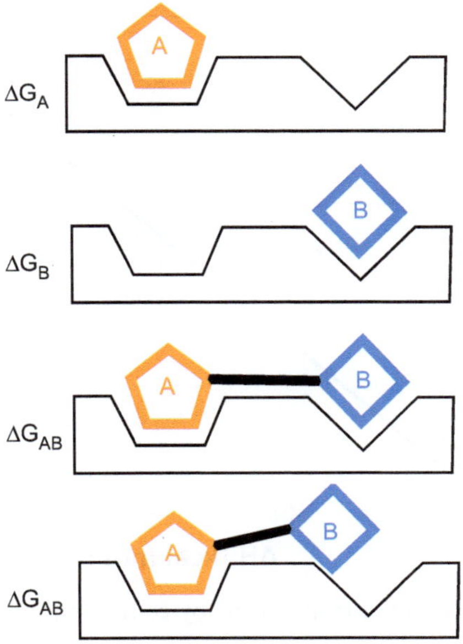

Figure 2.15 Model of binding interaction caused by tethering, with the bottom two images showing complete and incomplete binding interaction, respectively.
Reproduced with permission from ref. 107, © 2007 National Academy of Sciences, U.S.A.

favored); however, the modification produces a ligand–receptor complex with fewer degrees of freedom (*i.e.* entropically disfavored). The net result is a negligible gain in association constant. While this physical description of compensation is simple and intuitive, it does not explain every case.[110] Solvent reorganization has been proposed as another source of entropy–enthalpy compensation.[111]

Compensation is demonstrated graphically by plotting $T\Delta S$ against ΔH, yielding a linear regression with a slope near unity (Figure 2.16). It can also be revealed by plotting ΔS *vs.* ΔH, yielding a linear regression with a slope that is referred to as the compensation temperature. The compensation temperature is often ascribed physical meaning. For example, in a temperature-dependence study of enzymes isolated from fish that normally live at quite different temperatures, the compensation temperature was found to be within the range of living temperatures for each particular species.[112]

In order to detect entropy–enthalpy compensation, the entropic and enthalpic contributions to binding must be measured under different conditions (*e.g.* temperatures, guest structures, or receptor structures) and plotted against one another, as in Figure 2.16. These components can be obtained by taking multiple measurements of K_a at different temperatures and generating a van't Hoff plot (ln K_a *vs.* temperature). However, this

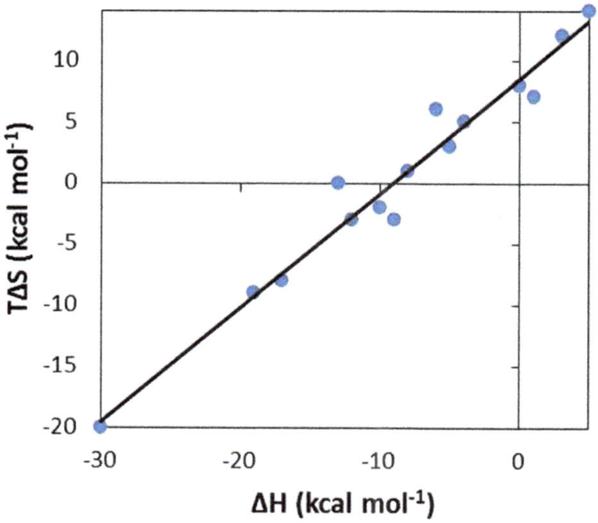

Figure 2.16 Example of an entropy–enthalpy compensation plot.

method has largely been supplanted by the use of isothermal calorimetry (ITC), which is capable of measuring K_a and ΔH in a single experiment. From these two parameters, ΔG and $T\Delta S$ can be calculated.

There are numerous examples in the literature of entropy–enthalpy compensation, and the concept has been reviewed extensively.[113] The current consensus is that this phenomenon is pervasive in ligand binding. However, there are critics who cite error in ITC measurements, failure to correctly propagate error in calculations, and other problems with conventional methodologies as possible sources of this apparent effect and argue that compensation is not as ubiquitous as generally thought.[114]

Entropy–enthalpy compensation makes the rational design of high-binding receptors a difficult task. For example, compensation has been observed in the binding of *p*-sulfonatocalix[*n*]arenes to a variety of guests in water. The data plotted in Figure 2.17[115] includes *p*-sulfonatocalix[*n*]arenes ($n = 4$, 6, and 8) as the receptors and a number of organic and metal cation guests. Despite the structural diversity of the guests, a linear relationship between ΔH and $T\Delta S$ was obtained with a slope of 1.10, which is indicative of compensation. Many of the organic guests interacted with the receptor through electrostatic interactions, hydrogen bonding, and van der Waals interactions, which all contribute favorably to the enthalpy of binding (lower left-hand corner of the plot); however, there was a greater entropic penalty due to enhanced complex rigidity.

It is sometimes possible to circumvent entropy–enthalpy compensation. The entropy of binding has two main components: a favorable component from desolvation, which results in the release of solvent molecules,

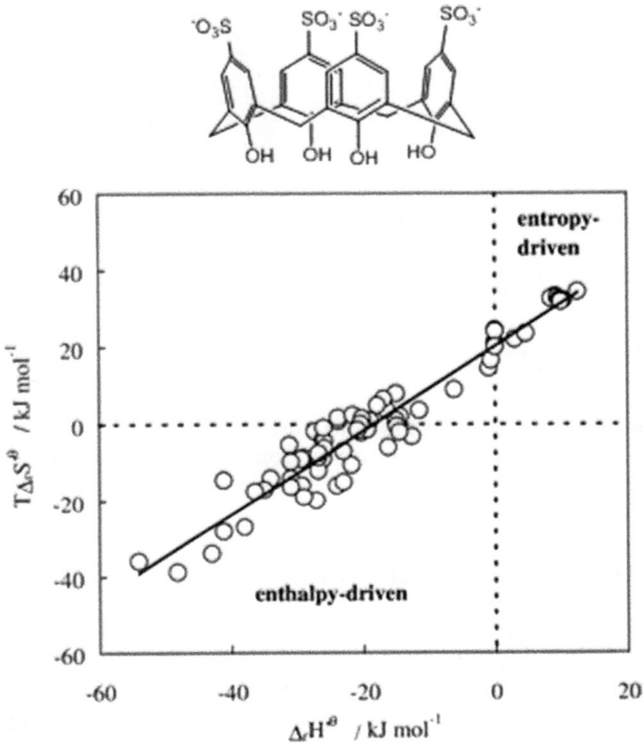

Figure 2.17 Compensation behavior exhibited by a variety of guests to *p*-sulfonatocalix[*n*]arenes in water.
Reproduced from ref. 115 with permission from the Royal Society of Chemistry.

and an unfavorable component due to a decrease in conformational degrees of freedom. This unfavorable component can be minimized by designing receptor/guest partners that do not change conformation upon binding. For example, a synthetic binding system was developed by Rekharsky and coworkers that overcomes entropy–enthalpy compensation.[116] The system involves inclusion of ferrocene guests in the hydrophobic pocket of cucurbit[7]uril (Scheme 2.26). The extremely strong association (10^9–10^{15} M^{-1}) is enthalpically favored with only a minor entropic penalty due to the rigidity and extensive desolvation of both the receptor and the guest.

Approaching receptor design from a computational standpoint by trying to calculate the enthalpic contribution to binding is quite challenging.[117] Even when enthalpies can be calculated accurately, they do not always correlate well with binding free energies.[118] However, it is now possible to compute binding free energies directly, which can be validated experimentally using ITC.[113]

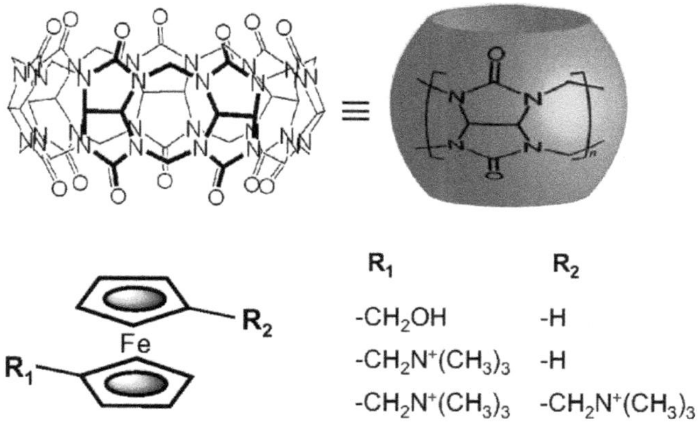

Scheme 2.26 Structure of cucurbit[7]uril host and ferrocence guests. Reproduced with permission from ref. 116, © 2007 National Academy of Sciences, U.S.A.

2.11 Cross-reactive Arrays

The conventional approach to creating synthetic receptors, a "lock-and-key" design as highlighted throughout this chapter, requires the synthesis of a specific receptor for each analyte of interest. Despite many advantages, there are shortcomings with this approach, particularly in the context of sensing. For each analyte, a different and selective receptor needs to be synthesized.

An alternative approach to chemical and biosensing employs an array of synthetic indicators that perform like the human nose or tongue.[119] In this situation, each indicator is not required to have high selectivity toward any given analyte. Rather, arrays of many different indicators respond by giving a pattern which is different for each analyte. Thus, the array response pattern acts as a unique fingerprint for the analyte.[120] In the past few decades, arrays and microarrays have become increasing diverse tools for biological studies. Optical microarrays were widely applied to identify and detect ions, vapors, organic small neutral molecules, and biomacromolecules.[121]

Differential sensing in conjunction with pattern-based recognition is used in two different contexts. First, it describes a system where the output signal from each sensor is directly compared to the original array before analyte application, and the difference is used to evaluate and identify analytes using pattern recognition protocols (Figure 2.18).[120] The alternative setup is a system where the all indicator elements bind differently with each analyte to yield a distinctive response pattern, as shown in Figure 2.19.[120,121] The collection of indicators in the array enables fingerprinting of the analyte.

The Rotello group used an indicator array to identify and differentiate cell types as normal, cancerous, or metastatic.[122] They used a family

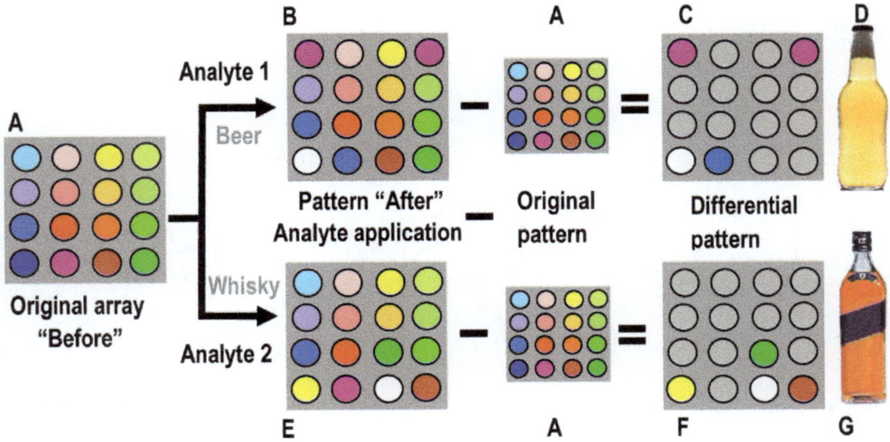

Figure 2.18 Differential sensing using a hypothetical array-based sensor. To a 16-element array A, analyte 1 (later identified as a beer) and analyte 2 (whisky) are applied. The response patterns B, E are used directly, or the pattern A is subtracted to obtain differential patterns C, F. Reproduced from ref. 120 with permission from the Royal Society of Chemistry.

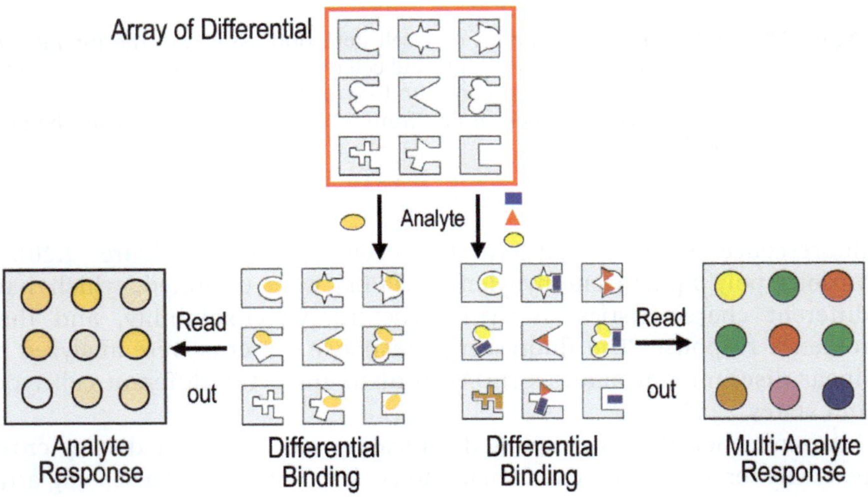

Figure 2.19 Array-based sensor utilizing analyte binding by differential receptors. Regardless of whether the analyte is a single- or multi-component analyte, receptors bind a number of analytes, but each receptor binds the analytes differently, thus providing a differential response. Reproduced from ref. 120 with permission from the Royal Society of Chemistry.

of conjugated fluorescent polymers as receptors in the cell-sensing array. The polyvalent interaction of the polymer with the cell surfaces detects the difference between various cell types resulting in a characteristic

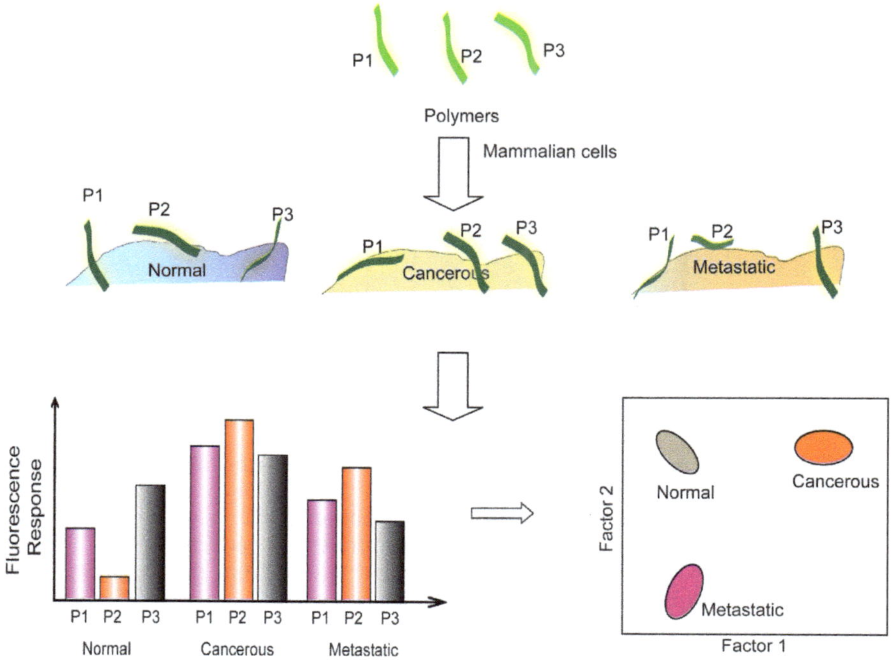

Figure 2.20 Schematic presentation of a cell detection assay and the interactions between polymers (ribbons) and cell types (surface). Fluorescence of the polymer is quenched upon binding to the cell surface.
Reproduced with permission from ref. 122, © 2010 American Chemical Society.

fluorescence response pattern for each cell type (Figure 2.20).[122] Various poly(*p*-phenylenethynylenes) (PPEs) were exploited, which have different characteristics in their structure, water-solubility, and fluorescence response. The fluorescence response patterns were analyzed by linear discriminant analysis (LDA) to distinguish the different cell types and states.

Anzenbacher also has exploited indicating receptors that display cross-reactivity for specific kinds of small molecules. They created a sensing array for carboxylates based on indicators using the octamethylcalix[4]pyrrole scaffold **2.26** (Scheme 2.27).[123] The array could discriminate between carboxylate-bearing drugs and biological molecules with a 100% classification success rate.

The Anzenbacher and Issacs groups have developed a sensing array using functionalized cucurbiturils **2.27** and **2.28** to differentiate a number of organic bases and drugs (Scheme 2.28).[124] Using LDA, the array was capable of distinguishing tobacco alkaloids, nitrosamines, and various biological amines.

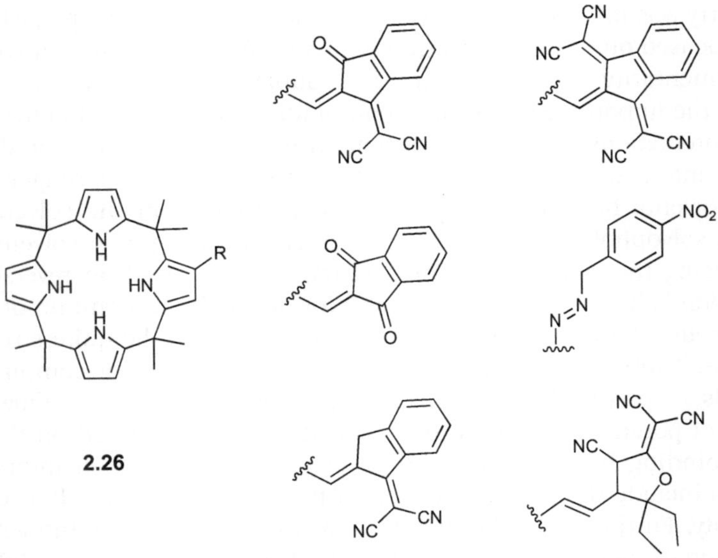

2.26

Scheme 2.27 Calixpyrrole scaffold used to create carboxylate indicators.

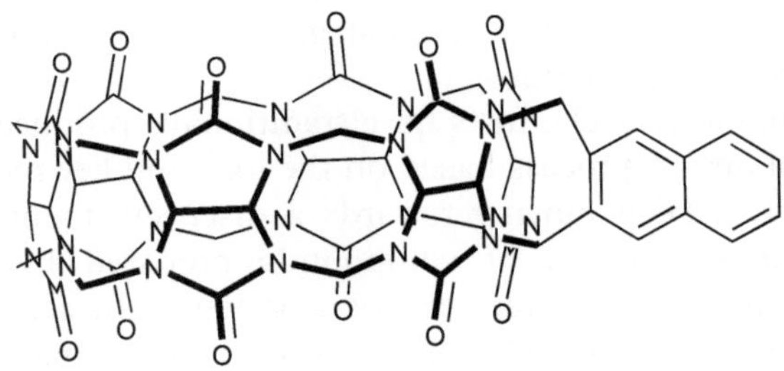

2.27

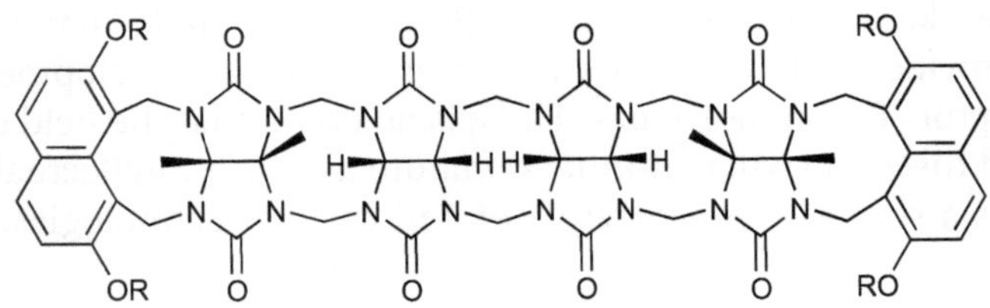

R = $(CH_2)_3SO_3Na$

2.28

Scheme 2.28 Cucurbituril hosts used to target nitrosamines.

2.12 Conclusion

This chapter has highlighted several important design concepts for the creation of synthetic receptors. Early on, the field of supramolecular chemistry was primarily influenced by the lock-and-key principle, with most effort focused on designing receptors for specific guests based on structural complementarity and preorganization. Supramolecular chemists learned quickly the importance of solvation on binding interactions. In effect, host–guest interactions are in competition with both host–solvent and guest–solvent interactions. Electrostatic-based interactions are stronger in less polar solvents because there is less competition from the solvent. Conversely, solvophobic effects are accentuated in polar protic solvents, with water being the extreme. In recent years, designs based on reversible covalent bonds have become quite useful for molecular recognition of certain biomolecules because their enthalpic interactions can be quite strong and selective. Another ongoing advance is the development of computational methods to predict binding free energies. Some techniques allow a collection of potential receptors to be scored and ranked based on their predicted binding free energies, and other methods have been improved in order to include the energetic effects of preorganization as well as complementarity. Furthermore, the design of synthetic receptors is influenced by the notion of cooperativity, in which multiple interactions behave differently than the sum of their individual interactions. A long-term goal is to produce synthetic binding systems that exhibit positive cooperativity, where the collective system produces stronger host–guest binding than the sum of the expected individual interactions.

While supramolecular chemists have traditionally produced synthetic receptors using a design process based on the lock-and-key model, it is likely that the field will keep moving towards a discovery mentality, in which chemists evaluate libraries of compounds prepared by combinatorial methods. The library techniques can be augmented by screening and evolutionary processes, as with SELEX. The use of DCLs enables high-affinity receptors and/or guests to "design" themselves. This approach may be a way to circumvent the tyranny of enthalpy–entropy compensation.

The remarkable diversity of naturally occurring molecular recognition systems provides vivid inspiration for the continued development of synthetic receptors for biomolecules. There is no doubt that the field of receptor design and discovery will continue to mature as a rich intellectual endeavor and also as a source of practically useful molecular technologies.

References

1. *The Lock and Key Principle: The State of the Art – 100 Years On*, ed. J. P. Behr, Perspectives in Supramolecular Chemistry, Wiley, Chichester, West Sussex, 1994, vol. 1.
2. D. J. Cram, *Angew. Chem., Int. Ed. Engl.*, 1986, **25**, 1039.
3. J. W. Steed and J. L. Atwood, *Supramolecular Chemistry*, Wiley, Chichester, West Sussex, 2nd edn, 2009.
4. J. M. McGrath and M. D. Pluth, *J. Org. Chem.*, 2013, **79**, 711.

5. J. Rebek Jr, L. Marshall, R. Wolak, K. Parris, M. Killoran, B. Askew, D. Nemeth and N. Islam, *J. Am. Chem. Soc.*, 1985, **107**, 7476.
6. B. Askew, P. Ballester, C. Buhr, K. S. Jeong, S. Jones, K. Parris, K. Williams and J. Rebek Jr, *J. Am. Chem. Soc.*, 1989, **111**, 1082.
7. K. Niikura, A. Metzger and E. V. Anslyn, *J. Am. Chem. Soc.*, 1998, **120**, 8533.
8. D. E. Koshland, *Angew. Chem. Int. Ed. Engl.*, 1995, **33**, 2375.
9. K. J. Wallace, W. J. Belcher, D. R. Turner, K. F. Syed and J. W. Steed, *J. Am. Chem. Soc.*, 2003, **125**, 9699.
10. G. A. Jeffrey, *An Introduction to Hydrogen Bonding (Topics in Physical Chemistry)*, Oxford University Press, Oxford, 1997.
11. S. J. Blanksby and G. B. Ellison, *Acc. Chem. Res.*, 2003, **36**, 255.
12. R. W. Taft, F. Anvia, J.-F. Gal, S. Walsh, M. Capon, M. C. Holmes, K. Hosn, G. Oloumi, R. Vasanwala and S. Yazdani, *Pure Appl. Chem.*, 1990, **62**, 17.
13. M. Meot-Ner and C. A. Deakyne, *J. Am. Chem. Soc.*, 1985, **107**, 469.
14. F. Hibbert and J. Emsley, *Adv. Phys. Org. Chem.*, 1990, **26**, 255.
15. C. A. Deakyne and M. Meot-Ner, *J. Am. Chem. Soc.*, 1985, **107**, 474.
16. J. Sunner, K. Nishizawa and P. Kebarle, *J. Phys. Chem.*, 1981, **85**, f14.
17. I. K. Mati and S. L. Cockroft, *Chem. Soc. Rev.*, 2010, **39**, 4195.
18. M. Ōki, *Acc. Chem. Res.*, 1990, **23**, 351.
19. B. Bhayana and C. S. Wilcox, *Angew. Chem., Int. Ed.*, 2007, **46**, 6833.
20. P. Li, C. Zhao, M. D. Smith and K. D. Shimizu, *J. Org. Chem.*, 2013, **78**, 5303.
21. O. Sinanoglu, *Molecular Associations in Biology*, ed. B. Pullman, Academic Press, New York, 1968, ch. 27, Solvent Effects on Molecular Associations, p. 427.
22. C. Reichardt and T. Welton, *Solvents and Solvent Effects in Organic Chemistry*, Wiley-VCH, Weinheim, Germany, 4th edn, 2011.
23. D. B. Smithrud and F. Diedericb, *J. Am. Chem. Soc.*, 1990, **112**, 339.
24. M. Rekharsky and Y. Inoue, *Supramolecular Chemistry From Molecules to Nanomaterials*, ed. P. A. Gale and J. W. Steed Wiley, 2012, vol. 1, Concepts, Solvation Effects in Supramolecular Recognition.
25. R. M. Izatt, K. Pawlak, J. S. Bradshaw and R. L. Bruening, *Chem. Rev.*, 1995, **95**, 2529.
26. G. Gasparini, M. Dal Molin, A. Lovato and L. J. Prins, *Supramolecular Chemistry: From Molecules to Nanomaterials*, ed. P. A. Gale and J. W. Steed, Wiley, Sussex, 2012, vol. 4, Supramolecular Catalysis, Reactivity, and Chemical Biology, Dynamic Covalent Chemistry, p. 1497.
27. J. P. Lorand and J. O. Edwards, *J. Org. Chem.*, 1959, **24**, 769.
28. A. E. Hargrove, R. N. Reyes, I. Riddington, E. V. Anslyn and J. L. Sessler, *Org. Lett.*, 2010, **12**, 4804.
29. S. D. Bull, M. G. Davidson, J. M. H. van den Elsen, J. S. Fossey, A. T. A. Jenkins, Y. Jiang, Y. Kubo, F. Marken, K. Sakurai, J. Zhao and T. D. James, *Acc. Chem. Res.*, 2013, **46**, 312.

30. G. Springsteen and B. Wang, *Tetrahedron*, 2002, **58**, 5291.
31. R. Nishiyabu, Y. Kubo, T. D. James and J. S. Fossey, *Chem. Commun.*, 2011, **47**, 1106.
32. G. J. Mohr, *Chem. – Eur. J.*, 2004, **10**, 1082.
33. Y. Jin, Y. C. Yu, R. J. Denman and W. Zhang, *Chem. Soc. Rev.*, 2013, **42**, 6634.
34. Y. Jin, Q. Wang, P. Taynton and W. Zhang, *Acc. Chem. Res.*, 2014, **47**, 1575.
35. S. Otto, R. L. E. Furlan and J. K. M. Sanders, *J. Am. Chem. Soc.*, 2000, **122**, 12063.
36. D. J. Cram, R. C. Helgeson, L. R. Sousa, J. M. Timko, M. Newcomb, P. Moreau, F. de Jong, G. W. Gokel, D. H. Hoffman, L. A. Domeier, S. C. Peacock, K. Madan and L. Kaplan, *Pure Appl. Chem.*, 1975, **43**, 327.
37. I. O. Sutherland, *Chem. Soc. Rev.*, 1986, **15**, 63.
38. R. Breslow, *Adv. Chem. Ser.*, 1971, **100**, 21.
39. W. L. Jorgensen, J. K. Buckner, S. Boudon and J. Tirado-Rives, *J. Chem. Phys.*, 1988, **89**, 3742.
40. P. Kollman, *Chem. Rev.*, 1993, **93**, 2395.
41. I. Kolossváry, *J. Am. Chem. Soc.*, 1997, **119**, 10233.
42. C.-E. Chang and M. K. Gilson, *J. Am. Chem. Soc.*, 2004, **126**, 13156.
43. D. Drueckhammer, *Chemosensors; Principles, Strategies, and Applications*, ed. B. Wang and E. V. Anslyn, Wiley, Hoboken NJ, 2011, ch. 5, Scaffold Design Using Computational Chemistry, p. 69.
44. G. Lauri and P. Bartlett, *J. Comput.-Aided Mol. Des.*, 1994, **8**, 51.
45. H. Huang and D. G. Drueckhammer, *Chem. Commun.*, 2006, **28**, 2995.
46. C. Lin and D. G. Drueckhammer, *New J. Chem.*, 2006, **30**, 1725.
47. W. Yang, H. He and D. G. Drueckhammer, *Angew. Chem., Int. Ed.*, 2001, **40**, 1714.
48. S. Jin, M. Li, C. Zhu, V. Tran and B. Wang, *ChemBioChem*, 2008, **9**, 1431.
49. B. P. Hay and T. K. Firman, *Inorg. Chem.*, 2002, **41**, 5502.
50. V. S. Bryantsev and B. P. Hay, *J. Am. Chem. Soc.*, 2006, **128**, 2035.
51. B. P. Hay, *Chem. Soc. Rev.*, 2010, **39**, 3700.
52. B. P. Hay, C. Jia and J. Nadas, *Comput. Theor. Chem.*, 2014, **1028**, 72.
53. W. Chen and M. K. Gilson, *J. Chem. Inf. Model.*, 2007, **47**, 425.
54. C. J. Pedersen, *J. Am. Chem. Soc.*, 1967, **89**, 7017.
55. G. W. Gokel, W. M. Leevy and M. E. Weber, *Chem. Rev.*, 2004, **104**, 2723.
56. B. Dietrich, J. M. Lehn and J. P. Sauvage, *Tetrahedron Lett.*, 1969, **10**, 2885.
57. J. D. Lamb, R. M. Izatt, C. S. Swain and J. J. Christensen, *J. Am. Chem. Soc.*, 1980, **102**, 475.
58. L. Chen, H.-Y. Zhang and Y. Liu, *J. Org. Chem.*, 2012, 77, 9766.
59. M. G. Gorbunova, P. V. Bonnesen, N. L. Engle, E. Bazelaire, L. H. Delmau and B. A. Moyer, *Tetrahedron Lett.*, 2003, **44**, 5397.
60. S. B. Nimse and T. Kim, *Chem. Soc. Rev.*, 2013, **42**, 366.
61. K. Lee, Y. W. Cho, T. Matsui, I.-C. Kang, T.-S. Kim and M. H. Han, *Proteomics*, 2003, **3**, 2289.

62. S. K. Kim, G. I. Vargas-Zúñiga, B. P. Hay, N. J. Young, L. H. Delmau, C. Masselin, C.-H. Lee, J. S. Kim, V. M. Lynch, B. A. Moyer and J. L. Sessler, *J. Am. Chem. Soc.*, 2011, **134**, 1782.
63. J. L. Sessler, P. A. Gale and J. W. Genge, *Chem. - Eur. J.*, 1998, **4**, 1095.
64. G. Cafeo, G. Carbotti, A. Cuzzola, M. Fabbi, S. Ferrini, F. H. Kohnke, G. Papanikolaou, M. R. Plutino, C. Rosano and A. J. P. White, *J. Am. Chem. Soc.*, 2013, **135**, 2544.
65. P. A. Gale, J. L. Sessler, V. Král and V. Lynch, *J. Am. Chem. Soc.*, 1996, **118**, 5140.
66. L. Isaacs, *Chem. Commun.*, 2009, **6**, 619.
67. K. Kim, N. Selvapalam, Y. H. Ko, K. M. Park, D. Kim and J. Kim, *Chem. Soc. Rev.*, 2007, **36**, 267.
68. H.-J. Kim, J. Heo, W. S. Jeon, E. Lee, J. Kim, S. Sakamoto, K. Yamaguchi and K. Kim, *Angew. Chem., Int. Ed.*, 2001, **113**, 1574.
69. A. R. Urbach and V. Ramalingam, *Isr. J. Chem.*, 2011, **51**, 664.
70. J. J. Reczek, A. A. Kennedy, B. T. Halbert and A. R. Urbach, *J. Am. Chem. Soc.*, 2009, **131**, 2408.
71. J. Szejtli, *Chem. Rev.*, 1998, **98**, 1743.
72. H. Ikeda, *Artificial Receptors for Chemical Sensors*, ed. V. M. Mirsky and A. K. Yatsimirsky, Wiley-VCH, Weinheim, 2011, ch. 4, Fluorescent Cyclodextrins as Chemosensors for Molecule Detection in Water, p. 113.
73. A. Metzger and E. V. Anslyn, *Angew. Chem., Int. Ed.*, 1998, **37**, 649.
74. C. Schmuck, *Coord. Chem. Rev.*, 2006, **250**, 3053.
75. M. A. Gallop, R. W. Barret, W. J. Dower, S. P. A. Fodor and E. M. Gordon, *J. Med. Chem.*, 1994, **37**, 1385.
76. R. A. Houghten, *Proc. Natl. Acad. Sci. U. S. A.*, 1985, **82**, 5131.
77. U. E. Hackler, M. Dechantsreiter, M. Heil and C. Schmuck, *J. Lab. Autom.*, 2001, **6**, 51.
78. R. A. Houghten, *Annu. Rev. Pharmacol. Toxicol.*, 2000, **40**, 273.
79. C. Chamorro, J.-W. Hofman and R. M. J. Liskamp, *Tetrahedron*, 2004, **60**, 8691.
80. D. G. Rivera and L. A. Wessjohann, *J. Am. Chem. Soc.*, 2006, **128**, 7122.
81. R. A. Houghten, C. Pinilla, J. R. Appel, S. E. Blondelle, C. T. Dooley, J. Eichler, A. Nefzi and J. M. Ostresh, *J. Med. Chem.*, 1999, **42**, 3743.
82. J. Blake and L. Litzi-Davis, *Bioconjugate Chem.*, 1992, **3**, 510.
83. J. C. Chabala, *Curr. Opin. Biotechnol.*, 1995, **6**, 632.
84. S. E. Osborne and A. D. Ellington, *Chem. Rev.*, 1997, **97**, 349.
85. G. P. Smith and V. A. Petrenko, *Chem. Rev.*, 1997, **97**, 391.
86. K. W. Uphoff, S. D. Bell and A. D. Ellington, *Curr. Opin. Struct. Biol.*, 1996, **6**, 281.
87. G. Winter, A. D. Griffiths, R. E. Hawkins and H. R. Hoogenboom, *Annu. Rev. Immunol.*, 1994, **12**, 433.
88. B. K. Kay, A. V. Kurakin and R. Hyde-DeRuyscher, *Drug Discovery Today*, 1998, **3**, 370.
89. J. A. Arter, J. E. Diaz, K. C. Donavan, T. Yuan, R. M. Penner and G. A. Weiss, *Anal. Chem.*, 2012, **84**, 2776.

90. D. H. Bunka and P. G. Stockley, *Nat. Rev. Microbiol.*, 2006, **4**, 588.
91. J. R. Lorsch and J. W. Szostak, *Biochemistry*, 1994, **33**, 973.
92. E. W. M. Ng, D. T. Shima, P. Calais, E. T. Cunningham, D. R. Guyer and A. P. Adamis, *Nat. Rev. Drug Discovery*, 2006, **5**, 123.
93. J. N. H. Reek and S. Otto, *Dynamic Combinatorial Chemistry*, Wiley-VCH, Weinheim, 2010.
94. A. Hermann, *Org. Biomol. Chem.*, 2009, 7, 3195.
95. J. M. Lehn, *Chem. - Eur. J.*, 1999, **5**, 2455.
96. P. G. Swann, R. A. Casanova, A. Desai, M. M. Frauenhoff, M. Urbancic, U. Slomczynska, A. J. Hopfinger, G. C. Le Breton and D. L. Venton, *Pept. Sci.*, 1996, **40**, 617.
97. R. J. Lins, S. L. Flitsch, N. J. Turner, E. Irving and S. A. Brown, *Tetrahedron*, 2004, **60**, 771.
98. P. T. Corbett, J. Leclaire, L. Vial, K. R. West, J. L. Wietor, J. K. M. Sanders and S. Otto, *Chem. Rev.*, 2006, **106**, 3652.
99. Z. Fang, W. He, X. Li, Z. Li, B. Chen, P. Ouyang and K. Guo, *Bioorg. Med. Chem. Lett.*, 2013, **23**, 5174.
100. Z. Rodriguez-Docampo, E. Eugenieva-Ilieva, C. Reyheller, A. Belenguer, S. Kubik and S. Otto, *Chem. Commun.*, 2011, **47**, 9798.
101. C. A. Hunter and H. L. Anderson, *Angew. Chem., Int. Ed.*, 2009, **48**, 7488.
102. A. P. Bisson, C. A. Hunter, J. C. Morales and K. Young, *Chem. - Eur. J.*, 1998, **4**, 845.
103. B. Zhang and R. Breslow, *J. Am. Chem. Soc.*, 1993, **115**, 9353.
104. J. D. Badjić, A. Nelson, S. J. Cantrill, W. B. Turnbill and J. F. Stoddart, *Acc. Chem. Res.*, 2005, **38**, 723.
105. I. M. Roitt, J. Brostoff and D. K. Male, *Immunology*, Mosby, St. Louis, 6th edn, 2001, 72.
106. M. Mammen, S.-K. Choi and G. M. Whitesides, *Angew. Chem., Int. Ed.*, 1998, **37**, 2754.
107. A. Hughes and E. V. Anslyn, *Proc. Natl. Acad. Sci. U. S. A.*, 2007, **104**, 6538.
108. D. H. Williams and M. S. Westwell, *Chem. Soc. Rev.*, 1998, **27**, 57.
109. W. P. Jencks, *Proc. Natl. Acad. Sci. U. S. A.*, 1981, **78**, 4046.
110. V. Lafont, A. A. Armstrong, H. Ohtaka, Y. Kiso, L. M. Amzel and E. Freire, *Chem. Biol. Drug Des.*, 2007, **69**, 413.
111. E. Grunwald and C. Steel, *J. Am. Chem. Soc.*, 1995, **117**, 5687.
112. H. Gutfreund, *Kinetics for the Life Sciences*, Cambridge University Press, Cambridge UK, 1995, 246.
113. J. D. Chodera and D. L. Mobley, *Annu. Rev. Biophys.*, 2013, **42**, 121.
114. D. M. Ford, *J. Am. Chem. Soc.*, 2005, **127**, 16167.
115. N. Douteau-Guevel, F. Perret, A. W. Coleman, J. Morel and N. Morel-Desrosiers, *J. Chem. Soc., Perkin Trans. 2*, 2002, **2**, 524.
116. M. Rekharsky, T. Mori, C. Yang, Y. Ko, N. Selvapalam, H. Kim, D. Sobransingh, A. Kaifer, S. Liu, W. Chen, S. Moghaddam, M. Gilson, K. Kim and Y. Inoue, *Proc. Nat. Acad. Sci. U. S. A.*, 2007, **104**, 20737.
117. G. Schneider, *Nat. Rev. Drug Discov.*, 2010, **9**, 273.

118. H. Böhm and G. Klebe, *Angew. Chem., Int. Ed.*, 1996, **32**, 2588.
119. K. J. Albert, N. S. Lewis, C. L. Schauer, G. A. Sotzing, S. E. Stitzel, T. P. Vaid and D. R. Walt, *Chem. Rev.*, 2000, **100**, 2595.
120. P. Anzenbacher, P. Lubal, P. Bucek, M. A. Palacios and M. E. Kozelkova, *Chem. Soc. Rev.*, 2010, **39**, 3954.
121. E. V. Anslyn, *J. Org. Chem.*, 2007, **72**, 687.
122. A. Bajaj, O. R. Miranda, R. Phillips, I.-B. Kim, D. J. Jerry, U. H. F. Bunz and V. M. Rotello, *J. Am. Chem. Soc.*, 2010, **132**, 1018.
123. Y. Liu, T. Minami, R. Nishiyabu, Z. Wang and P. Anzenbacher, *J. Am. Chem. Soc.*, 2013, **135**, 7705.
124. T. Minami, N. A. Esipenko, B. Zhang, M. E. Kozelkova, L. Isaacs, R. Nishiyabu, Y. Kubo and P. Anzenbacher, *J. Am. Chem. Soc.*, 2012, **134**, 20021.

CHAPTER 3

Synthetic Receptors for Alkali Metal Cations

GEORGE W. GOKEL*[a,b,c] AND JOSEPH W. MEISEL[a,b]

[a] Center for Nanoscience, University of Missouri–St. Louis, 1 University Blvd. Saint Louis, MO 63121, USA; [b] Department of Chemistry and Biochemistry, University of Missouri–St. Louis, 1 University Blvd. Saint Louis, MO 63121, USA; [c] Department of Biology, University of Missouri–St. Louis, 1 University Blvd. Saint Louis, MO 63121, USA
*Email: gokelg@umsl.edu

3.1 Complexation of Alkali Metal Cations

This chapter concerns the quest to develop complexing agents, indicators, receptors, and transporters for alkali metal cations. In one sense, alkali metal cations are essentially featureless spheres that differ in size and, as a result, in charge density. Notwithstanding the lack of complexity, the sodium cation comprises nearly 50% of the molar composition of seawater and is a component found in cells. The potassium cation is far less abundant, but these two cations are critical to life and closely regulated in biological systems. Sodium cation is abundant in extracellular fluids, in which K^+ is present at a relatively low concentration. Within most cells, however, the concentrations are reversed and K^+ is far more abundant. While there is no "average" mammalian cell, typical concentrations for Na^+_{in} (5–10 mM) and Na^+_{out} (~140 mM) are mirrored by K^+_{in} (~140 mM) and K^+_{out} (5–10 mM). Simply being able to conveniently monitor cellular ion concentrations is an important justification for much of the research described here.

Monographs in Supramolecular Chemistry No. 14
Synthetic Receptors for Biomolecules: Design Principles and Applications
Edited by Bradley D. Smith

Published by the Royal Society of Chemistry, www.rsc.org

Another motivation is to conduct model studies that provide insight into the mechanism of biological cation receptors and membrane ion channels. The past decade has seen remarkable advances in structural understanding of membrane transport proteins, and the associated supramolecular mechanisms that are used to distinguish between sodium, potassium, and calcium ions. This rapidly growing body of knowledge is based on early work that elucidated the structure and function of biological ionophores – low-molecular-weight natural products that associate with metal cations and carry them across membranes. Chart 3.1 shows the X-ray crystal structure of monensin A, a polyether ionophore from *Streptomyces cinnamonensis* that is used as an antibiotic in animal feeds.[1] The sodium cation is coordinated by inward-directed ether oxygens and the nonpolar CH residues are exposed on the surface, which makes the complex lipophilic. Solid-state structures like these have served as inspiration for synthetic mimics and they also provide key insight into the operation of biological ion channel transport systems. Highlighted in Chart 3.2 are the important metal cation coordination interactions that control binding selectivity within a biological transmembrane Na^+ pump and K^+ channel.[2] The function of structural features like the "channel selectivity filter" are most easily rationalized using the rules of metal cation coordination that are explained in the following sections.

3.1.1 Coordination Complexes, Receptors, and Dynamics

It has now been more than 100 years since Alfred Werner won the 1913 Nobel Prize for his discovery that neutral ligands could "coordinate" with metal ions.[3] In a complex such as $Co(NH_3)_6{}^{3+}$, the neutral ammonia molecules are strongly held and form a very stable, mixed metal–organic cation (Figure 3.1). Such a compound is commonly referred to as an inorganic complex. The elements of the complex are linked in a stable array that

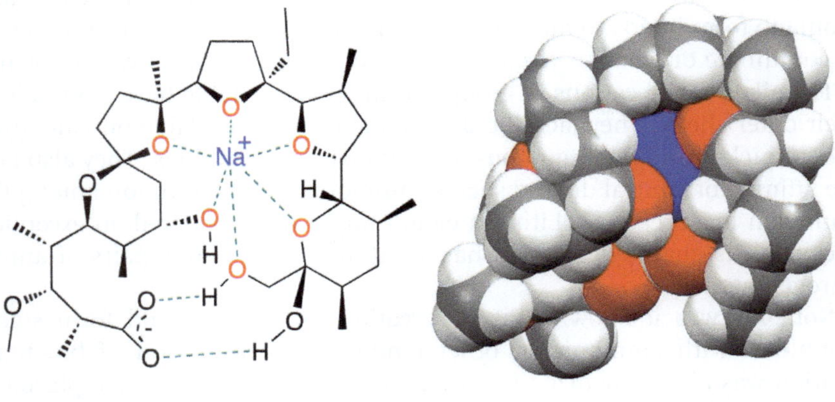

Chart 3.1 Solid-state structure of the polyether ionophore, monensin A, binding Na^+.

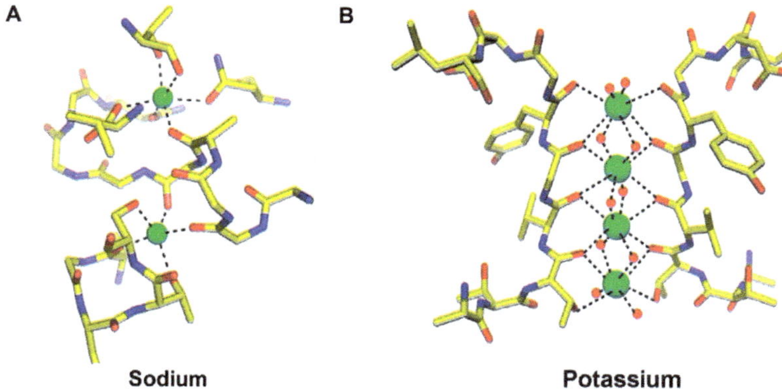

Chart 3.2 Partial structures of two biological ion channels showing: (A) Two Na^+ binding sites in the LeuT Na^+-dependent pump (Protein Data Bank (PDB) code 2A65). (B) Four K^+ binding sites in the KcsA K^+ channel (PDB code 1K4C).
(From *Science*, 2005, **310**, 1461. Reprinted with permission from AAAS.)[2]

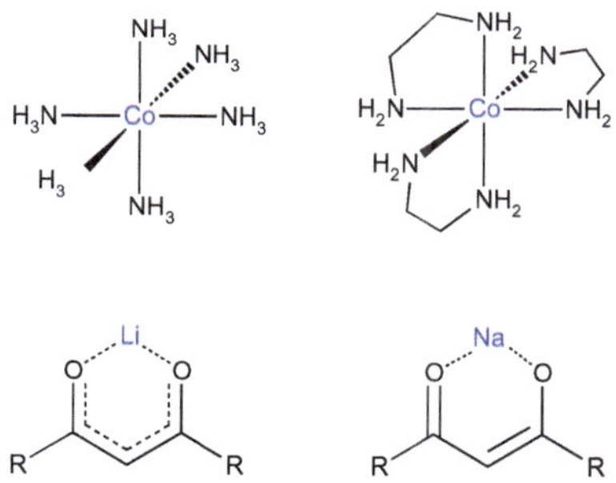

Figure 3.1 Coordination compounds and bidentate complexes.

comprises a single cation. Thus, the ability of neutral donors such as ammonia to form bonds to metal cations was well established while the means to coordinate or complex alkali metal cations was an unresolved problem.

The alkali metal cations of group I in the periodic table have s-orbitals in their outer shells. They lack the directional orbitals available on zinc, iron, cobalt, nickel, and the numerous other transition metal ions. They also lack the affinity for neutral donors such as ammonia that is common among the transition metal cations. Lithium cation might be considered an exception because it is charge dense and has an affinity for the lone pairs of amine nitrogen atoms.

Not only was it known that such cations as cobalt could form stable complexes with ammonia or other amines, but the notion of bidentate ligation was also recognized (Figure 3.1). When two donors were placed at a distance of two or three atoms from each other, the formation of a five- or six-membered ring facilitated complex formation. Compounds such as 1,2-diaminoethane ($H_2NCH_2CH_2NH_2$) readily formed complexes with transition metal cations. Two ethylenediamine nitrogen atoms could replace two separate molecules of ammonia (or other amine) and complex formation was favored by proximity of the second nitrogen once the first nitrogen had formed a bond.

A similar type of coordination was known for alkali metals with the enolate anions of 1,3-dicarbonyl compounds. The compounds are as much salts as coordination complexes and they are unstable with respect to reactivity with electrophiles. Electron-deficient boron and aluminum were known to coordinate or complex neutral ethers, but assemblies as stable as these do not form with alkali metal cations such as sodium or potassium.

In the early 1960s, a chemist named Charles Pedersen working at the DuPont Company in Delaware, USA wanted to prepare a new complexing or binding agent for divalent cations such as vanadyl ion.[4] He reasoned that the most effective complexing agent would take advantage of two negative charges in a relatively rigid scaffold. The design would use catechol (1,2-dihydroxybenzene) as a key element. One of the two hydroxyl groups would be used to connect to a linker that would attach to a second catechol. The two remaining hydroxyl groups are ionizable and would provide the charges to neutralize a divalent cation. The overall assembly would be a bidentate ligand. The connector or linker element chosen was bis(2-chloroethyl) ether. The ether oxygen in the linker would provide additional, albeit weaker, coordination to the bound cation.

In order to synthesize the desired complexing agent, one of the two catechol hydroxyl groups would require protection. The standard method of protection used dihydropyran to form the tetrahydropyranyl ether. This reaction certainly provides the monoprotected species, but some unreacted starting material remains present. When this mixture was carried forward, the desired product was obtained and eventually patented. Some neutral material was isolated from this reaction which appeared to interact with metal cations even though the compound was uncharged. The unusual compound was, of course, the now well-known dibenzo-18-crown-6 molecule (Figure 3.2).[5] This single example sired thousands of variants, became the origin from which the field of supramolecular chemistry eventually emerged, and earned Pedersen a Nobel Prize.[6]

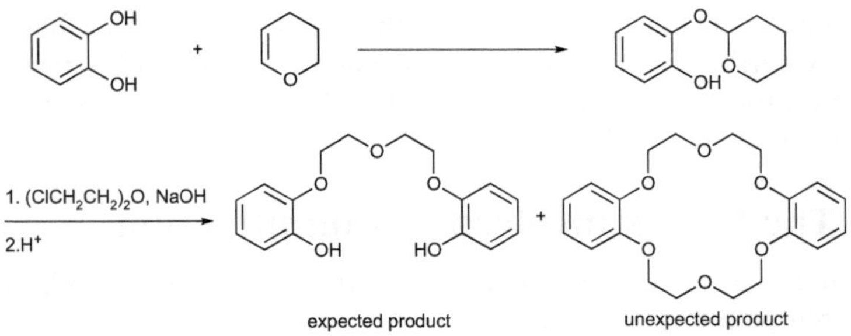

Figure 3.2 The chemistry leading to the first crown ethers.

Figure 3.3 Solid-state structure of dibenzo-18-crown-6 binding K^+ (Cambridge Structural Database: BEBFAP).

Dibenzo-18-crown-6 readily binds and forms stable complexes with Na^+, K^+, and other monovalent cations.[7] Since the binder is neutral, the cation complex must be associated with an anion. The Cambridge Structural Database (CSD) currently lists more than 300 complexes of this single crown ether. The structure shown in Figure 3.3 is dibenzo-18-crown-6 complexing a potassium cation. Many hundreds of structures are now known and they differ in many ways, including conformations, metal-donor distances, and even whether all available donors are involved in the complexation process. What remains similar about most successful alkali metal cation binders is that they have multiple Lewis basic donor sites and the scaffold that presents the donors to the cationic guest possess at least pre-organization, if not rigidity.

Following the publication of Pedersen's papers, a stunning array of new structural types and new compounds were prepared and reported in the literature.[8,9] These included the cryptands and cavitands devised respectively by Lehn[10,11] and Cram,[12,13] who shared the Nobel Prize[14,15] with Pedersen. Despite the vast number and even innovative design of many new structures, in some cases little information except the preparation was reported. Initially, this was not only justifiable but valuable because the limits of macrocycle chemistry remained to be defined. It was unclear which rings could be prepared, what heteroatoms or combinations of heteroatoms could be tolerated in the various rings, and what organic structures could be fused to the rings.

As structural variety evolved, greater attention was paid to issues such as complexation and decomplexation rates, the strengths of binding with various ions and different heteroatoms, the roles of cation size and rigidity, the effect of counterion, and many other features.[16] The answers to these questions are clearer today, owing to the extensive work that has been reported. Nevertheless, it is useful to step back and to consider how we think about receptor design and the extent to which our assumptions are rooted in experimental data.

3.2 The Forces Involved in Complexation

Every first-year chemistry student learns that free energy is related to enthalpy, temperature, and entropy by eqn (3.1).

$$\Delta G = \Delta H - T\Delta S \tag{3.1}$$

It is worth considering the importance of the individual components of this equation. The free energy, ΔG, is the quantity that we most often measure or determine. It ultimately defines and reflects the overall chemical system. When we design a receptor, we think about what donor groups may be appropriate. We also consider how they should be placed within a scaffold and how rigid that scaffold should be. One knows, for example that an aliphatic ether oxygen is a stronger Lewis base donor than is a diaryl ether oxygen. The equilibrium binding constants, expressed as $\log_{10} K_S$ for the reaction of the salt with the macrocycle to give a complex, for K^+Cl^- with 18-crown-6 and with dibenzo-18-crown-6 in 100% CH_3OH solution are 6.08 and 4.35, respectively. These values differ by $\sim$100 fold. This is expected, since the aromatic ring is in resonance with the ether oxygen and oxygen's electron density and therefore Lewis basicity (or donicity) is diminished. A second consequence of this resonance is that the oxygen atom rehybridizes from sp^3 to sp^2. This means that it is a p-orbital that is available for interaction with a metal cation. If dibenzo-18-crown-6 is planar, the two p-orbitals adjacent to each benzene ring are perpendicular to the plane of the macrocycle. We see the consequence of hybridization effects in the solid-state structure shown in Figure 3.3. Even the presence of two aromatic rings does not prevent dibenzo-18-crown-6 from adjusting to the further changes in hybridization that occur upon complexation.

This discussion has focused exclusively on geometric and electronic factors relevant to one of the earliest and simplest metal cation binding receptors. Of course, these issues all fall within the concept of enthalpy, ΔH. What about the solvent? We noted that the binding or complexation constants reported above were determined in anhydrous methanol. The solvent in which the complexation reaction is conducted raises two separate questions. What role is played by solvent as a competitor for the receptor? If the solvent is a better donor than the receptor, binding will be difficult to detect or nonexistent. Additionally, we need to consider that the macrocycle is solvated, the salt is solvated, and the macrocycle–salt complex is solvated. In the case of a complex such as shown in Figure 3.3, the counter anion may be associated with the complexed cation or it may be individually solvated and comprise a solvent-separated ion pair. These solvent issues are also essentially enthalpic as they concern the interaction of the solvent with the solutes.

Another concern is the extent to which solvent molecules reorganize upon complexation. Initially, solvent surrounds the salt or its individual components (*i.e.* cation and anion) as well as the macrocycle or other complexing agent. When complexation occurs, a combination of solvent reorganization about each of the components and the new species must occur, as well as a reorganization within the bulk solvent. The importance of the latter will vary with concentration. When we determine the equilibrium constant, K, where $\Delta G = -RT \ln K$, it tells us about ΔG, but it does not sort the contributions of enthalpy and entropy. Of course, calorimetric measurements can reveal the enthalpic contribution. The reality, however, is that analysts who measure energies and the individuals who design and synthesize receptors rarely reside in the same laboratory.

3.2.1 Enthalpy–Entropy Compensation

An issue that is difficult to incorporate into design criteria, but that has been extensively studied is called "enthalpy–entropy compensation."[17] This work can hardly be done justice in any short discussion, but essentially it means that as the organization of the complex increases, so does solvent organization. An increase in binding strength means a favorable change in ΔH and a more organized cation–host complex. The solvent also becomes more organized, but greater solvent order diminishes ΔG because the term $T\Delta S$ is negative. In short, there is usually a limit to how much ΔG can be improved by enhancing ΔH by factors such as altering donor placement, ring size, *etc.*, because the concurrent increase in solvent order (ΔS) will limit the overall gain possible. A problem faced by chemists is that we can often predict changes that will enhance enthalpy, but conceptualizing entropy of solvent reorganization is a daunting problem.

3.2.2 Drawing Structural Conclusions from Binding Constants

The problem of dealing with ΔH and ΔS when assessing ΔG is apparent in a study conducted in the author's laboratory. The goal of the study was to demonstrate the effect of side arm π-donors on a ring-bound cation.[18] The 4,13-diaza-18-crown-6[19] framework was used as the scaffold and side arms of similar size but increasing π-electron density were appended.[20] The four most closely related compounds had *n*-propyl, allyl, propargyl, and cyanomethyl side arms. The compounds under study are illustrated in Figure 3.4. The Na^+ binding constants (log K_S) were measured in anhydrous MeOH and found to increase as follows: $CH_2CH_2CH_3$, 2.86; $CH_2CH{=}CH_2$, 3.04; and $CH_2C{\equiv}CH$, 3.61. This systematic increase in K_S, and therefore ΔG, seemed to clearly confirm that a secondary π-donor interaction was contributing to the complexation strength.

Binding data were obtained at temperatures ranging from 15 °C to 41 °C. This permitted a van't Hoff analysis that could separate the binding strength (as ΔG) into the enthalpic (ΔH) and entropic (ΔS) components. The values

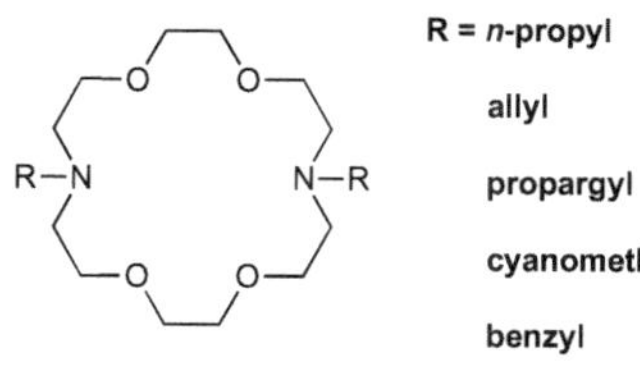

Figure 3.4 Two-armed diaza-18-crown-6 derivatives having three atom side arms.

obtained for ΔH were *n*-propyl, -2.8 kcal mol^{-1}; allyl, -3.6 kcal mol^{-1}; and propargyl, -5.0 kcal mol^{-1}. The value of ΔH for the cyanomethyl side-armed compound ($R = CH_2C{\equiv}N$) was -4.9 kcal mol^{-1}, although its binding constant was only 2.69 compared to the propargyl isostere. The difference was that $T\Delta S$ for propargyl was ~0 and -1.2 cal K^{-1} mol^{-1} for cyanomethyl. In fact, all four of these compounds were weaker binders than 18-crown-6 (log $K_S = 4.34$, $\Delta H = -7.4$ kcal mol^{-1}). In this case, a comparison of binding constants in terms of enthalpy would have been misleading and incorrect.

A solid-state structure was obtained for the propargyl-side-armed compound and the side arms were extended from the macrocycle. There was no indication of any additional complexation as a result of π–cation interactions. The solid-state structure of dibenzyldiaza-18-crown-6 likewise showed no secondary, cation–π interaction. Cation–π interactions were later confirmed for various macrocycles containing π-donor side arm elements, as discussed in Section 3.10.3.

3.3 A Brief Overview of Complexation Constants

Binding constants and thermodynamic parameters have been recorded in many laboratories. Izatt and coworkers have collated them in several reviews.[12,13,21] Data are reported for hundreds of macrocycles with a vast array of cations. Binding and thermodynamic parameters have been determined by a variety of methods and in numerous solvents or solvent mixtures. Some of the data are presented in Table 3.1. The data presented here are intended to give the reader only the most general concept of the values and variations. Data have been chosen only for systems that were studied in methanol solution. When several values have been reported, the average is presented here.

Several facts are apparent concerning the simple 3*n*-crown-*n* compounds. They generally bind K^+ ion more strongly than either Na^+ or Li^+ despite any apparent size correspondence. In contrast, binding by 18-crown-6 diminishes for the larger cation Rb^+ compared to K^+. The binding strength reduction caused by the fusion of a benzene group to 18-crown-6 that was alluded to in Section 3.1 is borne out by the data in Table 3.1.

Binding of alkali metal cations by cryptands is achieved with both greater strength and selectivity, by dint of size fit within the three-dimensional encapsulating agent. The data available for binding Li^+ by 3*n*-crown-*n* compounds suggest complexation constants in the range 0–100 (log $K_S = 2$). Cryptands 2.1.1 and 2.2.1 bind Li^+ with log K_S values of ~5–8.[12,13] The smaller cryptand shows stronger Li^+ binding than might be expected, presumably because the five donor atoms can adjust to the cation's preferred coordination geometry and partly because Li^+ is a stronger Lewis acid than the other alkali metal cations. Thus, its affinity for nitrogen is greater than that of potassium, for example.

However, it is interesting to note that the cryptand three-dimensional framework also limits access to the cavity. Thus K^+ is bound more strongly

Table 3.1 Homogeneous complexation constants and thermodynamic parameters[a] determined in methanol.[b]

	Cation	Log K_S	ΔH kJ mol^{-1}	ΔS J K^{-1}mol^{-1}
12-Crown-4	Li^+	~0	−3	
12-Crown-4	Na^+	1.4	−13	−14
12-Crown-4	K^+	1.7	−2	−41
15-Crown-5	Li^+	1.2	−11	−14
15-Crown-5	Na^+	3.3	−22	
15-Crown-5	K^+	3.4	−33	−54
18-Crown-6	Li^+	~0	−12	
18-Crown-6	Na^+	4.4	−33	−32
18-Crown-6	K^+	6.1	−56	−72
18-Crown-6	Rb^+	5.3	−51	−68
Benzo-18-crown-6	K^+	5.1		
Dibenzo-18-crown-6	K^+	5.0		
21-Crown-7	Na^+	2.1	−43	−112
21-Crown-7	K^+	4.3	−36	−40
2.1.1[cryptand]	Li^+	6–8		
2.1.1[cryptand]	Na^+	6.4		
2.1.1[cryptand]	K^+	2.3		
2.2.1[cryptand]	Li^+	5.2		
2.2.1[cryptand]	Na^+	9.5		
2.2.1[cryptand]	K^+	8.5		
2.2.2[cryptand]	Na^+	7.9		
2.2.2[cryptand]	K^+	10.6	−71	−36
3.2.2[cryptand]	K^+	4.8		
3.3.2[cryptand]	K^+	>7		

[a]Complexation data were taken from various sources and, in some cases, the numbers reported here are the average of 5–10 reported values.
[b]Data reported in anhydrous 100% methanol where possible. Some data were obtained with an electrolyte added to the methanol solution.

than Na^+ by [2.2.2]cryptand, but less strongly than Na^+ by [2.2.1]cryptand. The larger [3.2.2] and [3.3.2]cryptands show diminished K^+ binding compared to perfectly sized [2.2.2]cryptand.

3.4 The Dynamics of the Complexation Process

The equilibrium constant for complexation is related to the rates at which complexation and decomplexation occur. We may write eqn (3.2) as follows:

$$K = \frac{k_{\text{complex}}}{k_{\text{decomplex}}} = \frac{k_{\text{bind}}}{k_{\text{release}}} \tag{3.2}$$

Dynamic behavior is typically less of an issue with inorganic complexes such as $Co(NH_3)_6{}^{3+}$ (see Section 3.1.1). In contrast, complexes of crown ethers are generally stable when crystallized, but remain in equilibrium with the metal ions when in solution. Moreover, the dynamics of complexation as well as the equilibrium constants are appreciably affected by the medium in which the association occurs.

Consider what might be called the simplest class of crown ethers: $(CH_2CH_2O)_n$. The compounds commonly placed in this class are those in which *n* ranges from 4 to about 20, but the most common compounds by far in this family are 12-crown-4, 15-crown-5, 18-crown-6, and 21-crown-7. Each of these compounds complexes alkali metal cations such as Li^+, Na^+, K^+, and Cs^+. The strengths of binding vary to some extent according to the size complementarity between ring and cation, to some extent according to the associated counter-anion, and always depending upon the solvent in which the complexation reaction is conducted.

It is worth noting at the outset that organization of the donor groups within a ring structure is important. This is referred to as the "macrocyclic effect". If the binding of K^+ by 18-crown-6 is compared to that of $CH_3O(CH_2CH_2O)_5CH_3$, binding by the former is enormously stronger despite the fact that the numbers of carbons and oxygens are the same and the structures are so similar. This is due to the macrocyclic effect or to "pre-organization" as posited by Cram.[22] Simply put, the entropic cost of organizing the donors into a suitable arrangement is paid by synthesis, so it is not part of the binding equation.[12,13]

A popular concept concerning simple macrocycles is the so-called "hole–size relationship". The notion is that binding is best when there is a near or exact size correspondence between the donor group array and the metal cation. Thus, the prediction would be that Li^+ is best bound by 12-crown-4, that Na^+ is best bound by 15-crown-5, that K^+ is best bound by 18-crown-6, and so on. While this notion is esthetically pleasing, the situation is more complicated. Unfortunately, the hole–size concept is so appealing that it is quoted in text books even today. Of course, a good correspondence between donors and a cation to be complexed is generally a good strategy if the ion to be bound will be held by a relatively rigid scaffold.

For the 3*n*-crown-*n* family, 18-crown-6 is the superior binder in all cases. The graph in Figure 3.5 shows this clearly.[23] From 12-crown-4 to 24-crown-8, peak binding for Na^+, K^+, NH_4^+, and Ca^{2+} all show maxima with 18-crown-6. Binding of Na^+, K^+, and NH_4^+ by 12-crown-4 are all weak and about the same as for Li^+, at least in anhydrous methanol solution (not shown).

It seems very reasonable that a size relationship would be obtained for these ions and macrocycles. It does not, for a variety of reasons. At the small end, Li^+ favors tetrahedral binding and the 12-crown-4 macrocycle cannot adjust to that geometry. Sodium cation is bound by 15-crown-5 with a substantial association constant ($10^{4.35}$). Even so, Na^+ is normally solvated by six donors in bulk solution and the five donors present in 15-crown-5 do not meet this requirement. 18-Crown-6 offers six donors, albeit not in an ideal octahedral array.

Within the family of 3*n*-crown-*n* structures, the binding of K^+ by 18-crown-6 is substantially better than for any of the other combinations of ring size and cation. An important advantage of the ethylenedioxy unit in the macrocycle is its conformational flexibility and lack of unfavorable interactions. The ethylene units can adopt a gauche conformation with little steric stress. This would not be the case for example, for propylenedioxy

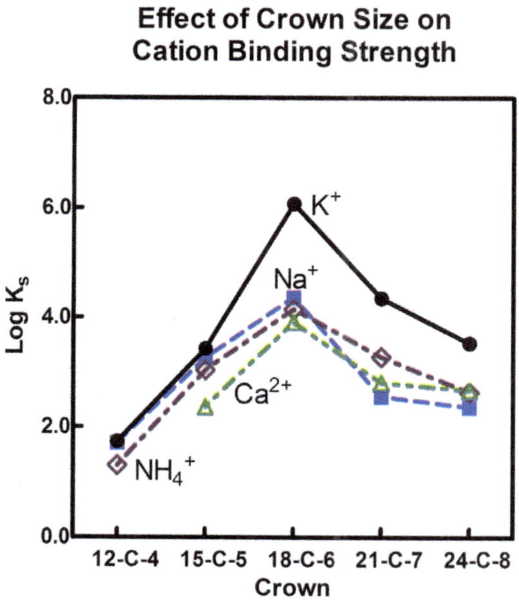

Figure 3.5 Binding constants determined in 100% methanol solution for 3*n*-crown-*n* compounds where $n = 4–8$.

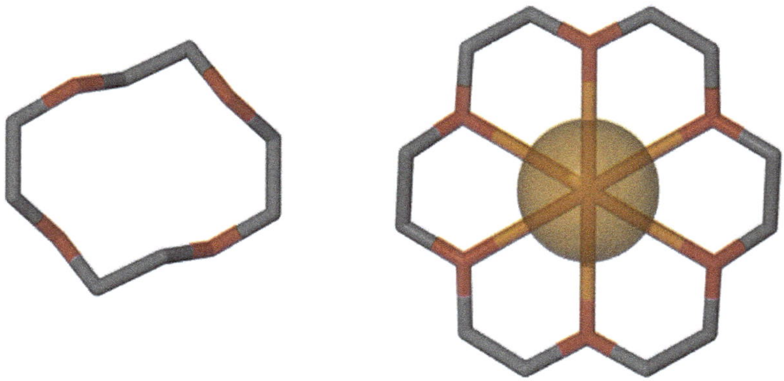

Figure 3.6 Solid-state structures of uncomplexed 12-crown-4 (Cambridge Structural Database (CSD): TOXCDP) (left) and K^+ ion complexed by 18-crown-6 (CSD: KTHOXD) (right).

(–$OCH_2CH_2CH_2O$–). The presence of three, rather than two, methylene units introduces conformational issues not present with ethylenedioxy (–OCH_2CH_2O–). Moreover, when 18-crown-6 binds K^+, the size fit is good, the ethylene units are not conformationally compromised, and the six donor groups are focused on the cation with three oxygens above the plane and three below. This minimizes unfavorable dipolar interactions to some extent. The complementary fit and oxygen alternation are both clear in the crystal structure shown in Figure 3.6.

When a macrocycle is smaller than the target cation, a sandwich structure may form rather than a simple 1 : 1 complex. When Na^+ is bound by either 12-crown-4[24] or aza-12-crown-4,[25] the cation is sandwiched between two rings (Figure 3.7). In the latter complex, the NH bonds are oriented at an angle of 90°, presumably to minimize the dipoles.

3.4.1 Solvent Effect on Complexation

Before considering the amazing variation in crown structures, some additional discussion of solvent effects is in order. If there can be a summarizing statement concerning the effect of solvent on complexation, it is that cation-receptor binding will be strongest in the least polar solvent. The data presented in Figure 3.8 show the decrease in binding for 18-crown-6 as the

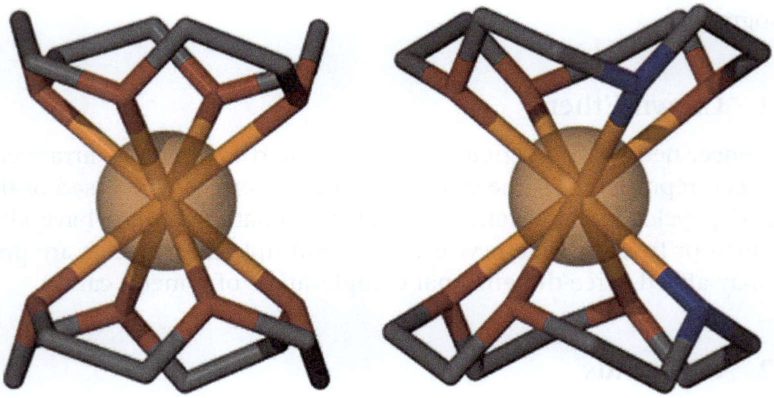

Figure 3.7 Solid-state structures of $(12C4)_2 \cdot Na^+$ (Cambridge Structural Database (CSD): BEYHES) (left) and $(Aza\text{-}12C4)_2 \cdot Na^+$ (CSD: FEHDOL) (right).

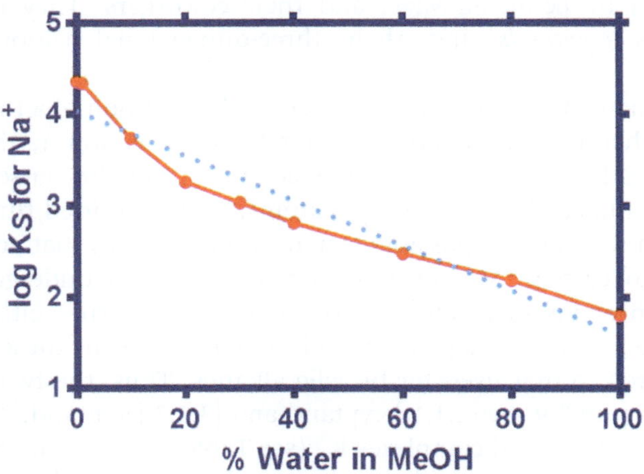

Figure 3.8 Solvent dependence of 18-crown-6 · Na^+ binding in methanol and water.

percentage of water in methanol is increased. In anhydrous methanol, K_S for Na^+ is $10^{4.35}$. In water it is only 10^2. This change is not surprising in light of the difference in polarity between the two solvents. The dielectric constants for methanol and water are 40 and 80, respectively. The dotted line in the graph shows that the dependence is nearly linear, although there is no fundamental thermodynamic basis for this. In fact the exponential variation is masked to some extent by the fact that the data are plotted as decadic logarithms.

3.5 Crown Ethers and Related Receptor Types

Literally thousands of crown ethers have been prepared and reported in the literature. New and often complex crowns are reported to this day. Surveys of structural variations have been reported in review articles and lengthy monographs.

3.5.1 Crown Ethers

In essence, nearly all imaginable ring sizes and heteroatom arrangements have been reported. In some cases, two or more crowns are fused or linked. Aliphatic, cycloalkyl, aromatic, and heteroaromatic residues have all been appended or fused to macrocycles. When multiple macrocycles are present, they may afford three-dimensional complexation of a metal cation.

3.5.2 Cryptands

The most effective and ingenious way to achieve three-dimensional binding of a cation is to include a third donor-group-containing strand in the structure. In fact, this was achieved almost simultaneously with the findings of Pedersen by Lehn, Sauvage, and their coworkers. They named these compounds *cryptands* and their three-dimensional cation complexes *cryptates*.

As with many highly original discoveries, the addition of a third strand to a crown ether may seem obvious in retrospect. However, the design of cryptands had a goal of three-dimensionality from the outset. The three strands are linked through nitrogen atoms. This permitted three links to a single atom without the possible stereochemical issues that accompany attachment to carbon. Moreover, nitrogen's ability to undergo low-energy structural inversion means that its lone pair may be turned either inward or outward. Lehn and Sauvage suggested nomenclature for these compounds that is related to that used for bicyclic alkanes. Thus, the two compounds shown in Figure 3.9 are [2.1.1.]cryptand and [3.2.2.]cryptand.[26]

When [2.2.2]cryptand complexes K^+ (see Table 3.1), it completely envelops it and provides six oxygen and two nitrogen donors. The solid-state structure of the complex is shown in Figure 3.10. The K^+ favors 8-coordination and the

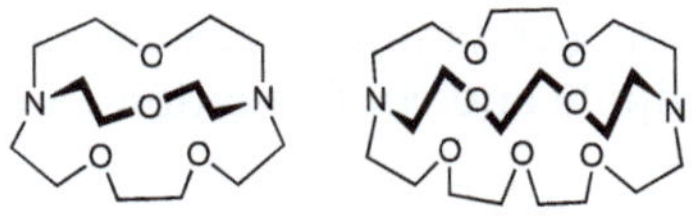

[2.1.1] **[3.2.2]**

Figure 3.9 Structures of [2.1.1]cryptand and [3.2.2]cryptand.

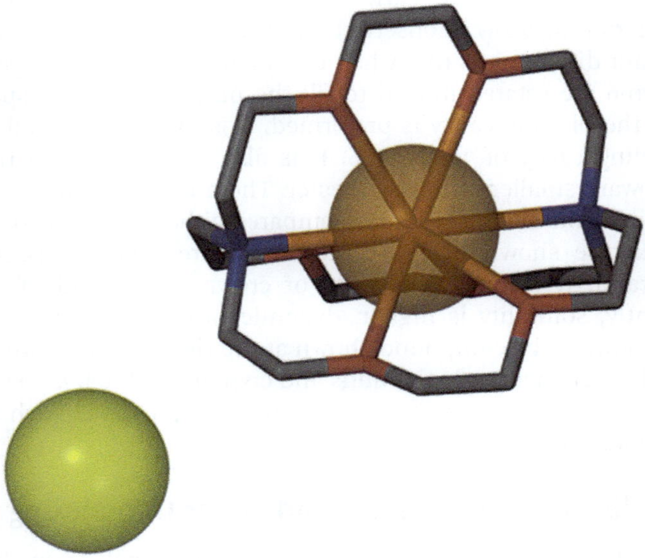

Figure 3.10 Solid-state structure of [2.2.2]cryptand complexing KI.

array of donors is such that the equilibrium constant for binding KI in methanol is $>10^8$ M^{-1}. Of course, there is a consequence to the high binding constant. In order for the cation to enter or exit the cryptand, there must be significant conformational adjustment so that the cation can pass among the chains. In crown ethers, both "sides" of the macrocycle are open so cation entry and egress are fast. When a cation is bound by a crown ether, the associated anion may play a role as a donor group in which it occupies an apical vacancy. In cryptate complexes, the anion is isolated from the cation, as seen in the solid-state structure of Figure 3.10.

The rates of complexation and decomplexation are not major considerations in the formation of strong cation complexes. However, dynamic behavior is of critical importance in the biological context. Although the binding rates can be quite rapid, they have been determined for several crowns and cryptands. The differences in rates are significant. For 18-crown-6 and Na^+ in water, the values are $k_{bind} = 2.2 \times 10^8$ M^{-1} s^{-1} and $k_{release} = 3.4 \times 10^7$ M^{-1} s^{-1}. In contrast, the binding and release rates for [2.2.2]cryptand are 10^5 M^{-1} s^{-1} and 27 M^{-1} s^{-1}, respectively.[27] As might be expected, when the cation is bound within the enveloping cryptand, its release and replacement by solvent is resisted by both ion-dipole and conformational energies.

3.5.3 Spherands

Perhaps the strongest complexation of cations has been achieved by Cram and coworkers, who used the pre-organization principle to define rigid cavities of precise size and donor group positions.[28] Examples are shown in Figure 3.11. The structure at the left is an early spherand that has six oxygen donors held in a D_{3d} alternating array as observed for 18-crown-6 when bound to a cation. An important difference is that when unbound, two of the methylene groups of 18-crown-6 are rotated inward to fill the macrocycle's void space. In the spherand, the binding cavity is preformed. It is, of course, smaller than the cation-binding cavity of 18-crown-6. It is also more rigid and the binding strength toward smaller cations is greater. The presence of multiple aromatic rings also presents solubility issues compared either to crowns or cryptands.

The structure shown in the center of Figure 3.11 has been called a hemispherand. It contains elements of crown ethers and of spherands. Consequently, solubility is higher, dynamics are higher, and ion binding strength is lower. The compound shown at the right of the figure is a hybrid that combines elements of spherands and cryptands. When $n = m = 2$, a total of nine donor groups are present that can envelop cations such as K^+ that are octa-coordinate in polar solvents.

3.5.4 Calixarenes, Cavitands, and Other Complexing Agents

A number of macrocyclic systems are known that result from the condensation of an electron-rich arene with a carbonyl compound. Examples include the condensation of pyrrole with acetone to give compounds that are now called calixpyrroles. Furan condenses readily with acetone to give a macrocycle having four aromatic oxygen donor groups within it. Likewise, phenol, resorcinol, and pyrogallol all condense with aldehydes to give calixarenes, resorcinarenes, and pyrogallolarenes, respectively. Figure 3.12 illustrates these classes of compounds.

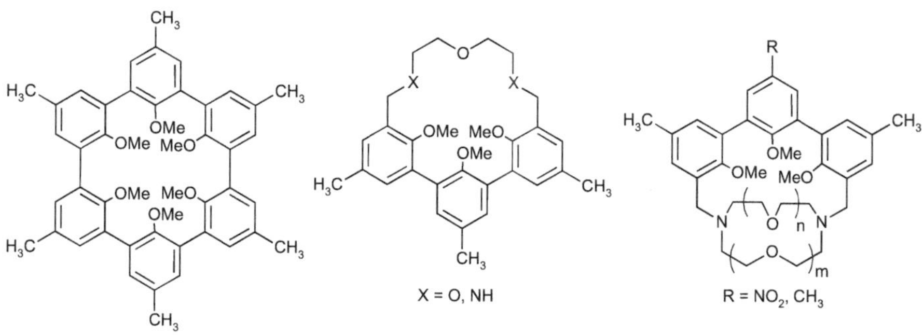

Figure 3.11 Left: spherand; center: hemispherand; right: crown-hemispherand.

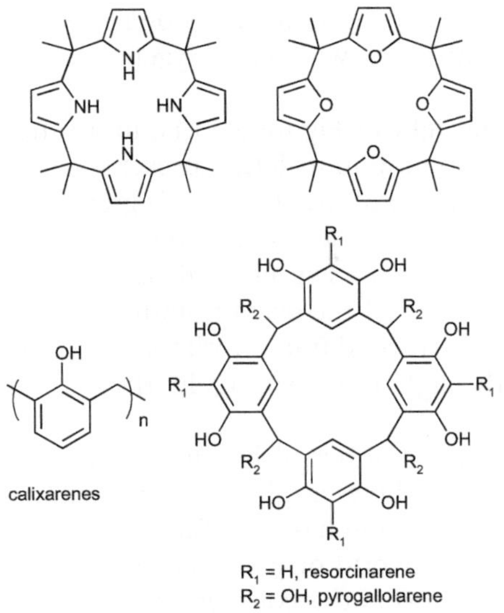

Figure 3.12 Macrocyclic compounds formed by acid-catalyzed, multiple condensations.

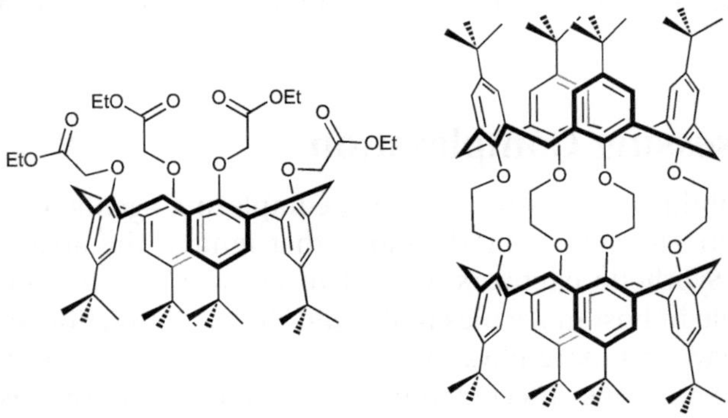

Figure 3.13 Calixarene receptor molecules.

The pyrrole–acetone condensation was known in the late 19th century,[29] and the furan–acetone condensation was discovered in the 1950s. Likewise, the calixarenes date from the 1950s[30] and their chemistry has been extensively reviewed by Gutsche in three monographs[31–33] and by Mandolini and Ungaro in a fourth.[34] A further monograph describes both calixarenes and resorcinarenes.[35]

It was not until the 1970s when the so-called "petrolite procedure"[36] was developed that the area exploded. A dependable method to make tetra-*t*-butylated calix[4]arene led to numerous structural elaborations. Many of these were directed to cation complexation. Two cation-binding calixarene variants are shown in Figure 3.13. The compound on the left provides Na^+ with octacoordination that involves both ether and ester carbonyl oxygen atoms. The success of this binder depends on the calixarene being in the so-called cone conformation, in which all oxygens are on the same side of the molecule.

The binder on the right of Figure 3.13 is, in a sense, a four-strand cryptand. As with the compound to its left, it presents a cation with eight oxygen donors. The overall rigidity of the structure is greater than the tetraester and the aralkyl ether oxygens are less polar than the ether/ester combination. This compound favors K^+ in a symmetrical eight fold complexation in which the cation is in the center of the overall binding site. The compound at the right also presents a fairly rigid framework for entry and egress of the K^+ ion, which may pass through the opening in the four arenes to access the binding site.

As with spherands, a number of hybrid calixarenes have been prepared and studied. Figure 3.14 shows a calix crown hybrid in which a calixarene in the 1,3-alternate conformation has been fused to a dibenzocrown.[37] A K^+ ion is enveloped within the hybrid or pseudo-cryptand. Perchlorate ion (ClO_4^-, not shown in Figure 3.14) is the counter-anion. It is somewhat disordered and does not interact directly with the calix-bound cation. Two hydrophobic chains are present, which enhance the compound's solubility in nonpolar solvents. Numerous other examples of calixarene hybrids have been reported. One of these is noted below in the section describing cation–π interactions.

3.6 Assessing Complexation

It was the ability of receptors, binders, complexing agents, or—as chemists use the term—ligands to bind cations that ignited the area of host–guest chemistry. Synthetic chemists wanted to know what structural variations were possible in host molecules that could then still function as hosts. Many chemists, having a more physical or analytical inclination, wanted to know how these variations affected binding strengths, selectivities, and rates.

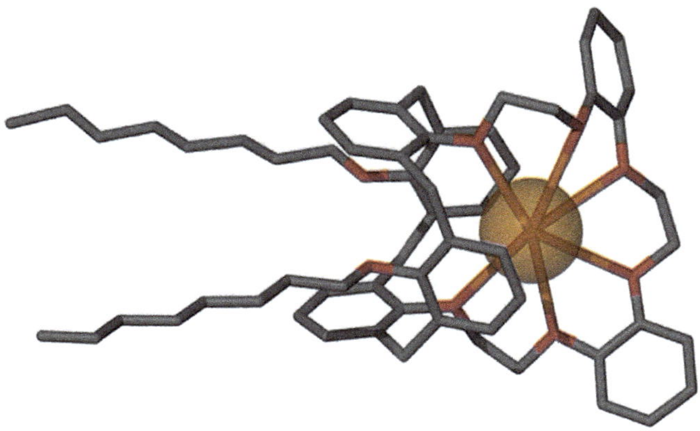

Figure 3.14 K^+ complexation by a calix-crown.

A range of methodologies has been applied to these problems and are noted very briefly below.

3.6.1 Calorimetry

Although there are obvious caveats related to assessing binding strength and assuming it corresponds directly to enthalpy, in many cases there is no other information available. In fact, numerous methods have been used to assess the effect of structural changes on binding. By far the best of these is to use calorimetric measurements to obtain the binding constant and the contributions of enthalpy and entropy to the free energy.[38] Izatt, Christensen, Bradshaw, and their coworkers have made an extensive study of macrocycle binding by using calorimetry, and much of it is published in review articles.[39–41]

Calorimetric studies are of undeniable value and were undertaken early in macrocycle history. A practical limitation is that such studies require instrumentation and expertise that is often inaccessible to the synthetic chemists preparing novel structures for specific purposes. A number of methods have been developed to assay binding strengths and selectivities. Such methodology may not be directly comparable to other techniques or to calorimetry, but it provides design guidance. Moreover, the relative binding strengths and selectivities are informative if experimental conditions are consistent over the group of compounds studied.

3.6.2 Ion-selective Electrode Studies

Ion-selective electrodes are commercially available and convenient to use. Complexation constants can be determined by using them.[42] Sodium-selective glass electrodes are particularly convenient and generally safe to use in a variety of solvents. Monovalent cation electrodes can be used in water or, with appropriate conditioning, in methanol. In nonpolar solvents, the danger is the extraction of the ionophore from the electrode tip, rendering it useless. It is worth noting that data obtained from titration calorimetry and by ion-selective electrode methods are generally similar, although certainly not always identical. Calorimetry gives comprehensive information about binding. The ion-selective electrode method can do as well, but it relies on van't Hoff plots, which are inherently less accurate.

3.6.3 Salt Extraction and Colorimetry

One of the early methods to assay binding was simple extraction.[43] In this method, an aqueous solution of sodium picrate is shaken with a solution of the binder in an immiscible organic solvent. The partitioning of the cation is then assessed colorimetrically. It is assumed that the concentration of the yellow picrate anion is equal to the concentration of cation and that the more picrate that is extracted corresponds to stronger cation binding.

Alternatively, an extraction system of D_2O:$CDCl_3$ could use *t*-butylammonium cation and the partition observed by nuclear magnetic resonance.[44]

As noted, the experimental conditions must be kept consistent throughout the study.[45] If the organic solvent is $CHCl_3$ in one case and CH_2Cl_2 in another, a comparison may be informative. If the solvent is changed from chlorocarbon to ether, however, the uncertainty increases. It should also be noted that although extraction has been widely used, there is no "analytical standard" for the method. One variable is the salt itself. For example, sodium picrate is both hard to dry and hard to dissolve completely in water. Addition of picric acid to a sodium hydroxide solution is more convenient, but usually requires excess base. If one uses a small or large excess of base, the ionic strength of the aqueous phase will differ and this may affect the comparability of extraction results. This notwithstanding, much information has been gleaned from these relatively simple measurements.

3.6.4 Comparison of Binding Assays

A reasonable question is whether there is a correlation between extraction constant and homogeneous binding constant data. In general, there is. Of course, the values cannot be directly compared numerically. Homogeneous binding constants are usually expressed as log K_S and data obtained by extraction are reported as percentages. An obvious variable is the issue of the solvent for homogeneous study compared to the two-phase system required for extraction.

In an early study in our laboratory, we sought to compare the values obtained by different methods in an attempt to answer the question of comparative values.[46] We prepared a family of azacrown derivatives variously substituted at nitrogen by CH_2CO–R in which R was an ester or an amide. The crowns were 12-, 15-, or 18-membered. We measured the extraction constants, the homogeneous complexation constants, and the transport rates in a $CHCl_3/H_2O$ bulk membrane system. The data are shown in Figure 3.15, in which scaled data are plotted as a function of the transport rate. A general correlation is apparent, although r^2 for the lines are only 0.80 and 0.86.

3.6.5 Bulk Membrane Transport Rates

The work described above involved measurements of cation transport through a chloroform bulk membrane. In such experiments, two aqueous phases are separated by an immiscible phase, such as chloroform or dichloromethane. The fact that the density of CH_2Cl_2 is greater than that of water is advantageous. A simple U-shaped glass tube can be filled above the bend with chlorocarbon and water added to the arms above it.[47,48] The host molecule is dissolved in the chlorocarbon and salt is charged into one arm of the tube. As transport proceeds, the salt concentration diminishes in the first arm and increases in the second. By taking aliquots at various time intervals and determining the ion concentration in the second side arm, a rate can be determined.

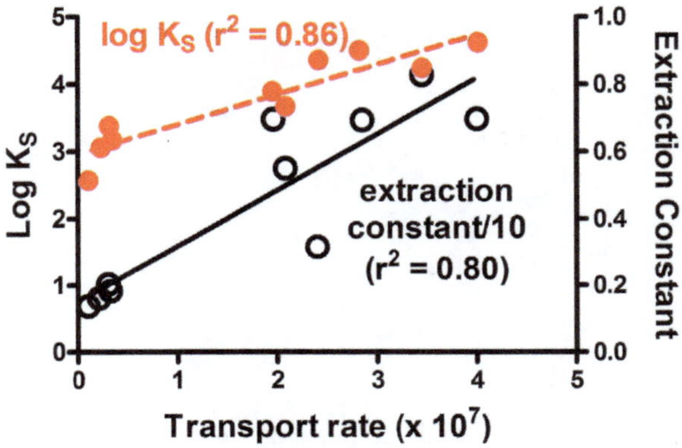

Figure 3.15 Comparison of homogeneous binding and extractions constants with transport rate.

An alternative apparatus that has proved to be very convenient uses a beaker and a glass tube. As with a U-tube, a dense, water-immiscible solvent is used as the membrane and charged into the beaker. The glass tube is inserted and aqueous solutions can be added to the space within the tube and in the area external to the tube. This apparatus is simple to assemble and is more readily stirred than the U-tube, although the principles are the same.

The driving force in these transport experiments is the concentration gradient. Thus, the rate can be assayed only until equal ion concentrations are present in both arms, *i.e.*, when the system reaches equilibrium. The change in concentration may be assessed by various techniques. Transport of Na^+ or K^+ can be detected in the receiving phase by use of a Na^+- or monovalent cation-selective electrode. The counter-anion is typically Cl^- in such experiments. Alternatively, the alkali metal cation may be paired with yellow picrate anion. The appearance of picrate in the receiving arm can be quantitated colorimetrically. As with extraction constant data, results can be compared directly only when all experimental variables are identical.

3.6.6 Liposomal Membrane Studies

Bulk organic membranes have been used extensively to evaluate the transport capabilities of various complexing agents. In order to evaluate a transporter in a more biological context, liposomal or vesicular systems are preferable. Liposomes are typically prepared from phospholipids, which can be purchased commercially.[49] Various techniques are available for their formation, but essentially, the phospholipids are sonicated in aqueous suspension to form liposomes, which are typically spheres having an outer bilayer boundary. A typical phospholipid that forms liposomes and a schematic of the liposome formed are shown in Figure 3.16.

Liposomes can be formed in aqueous media having the desired salt concentrations. They can be filtered to a uniform size and separated from

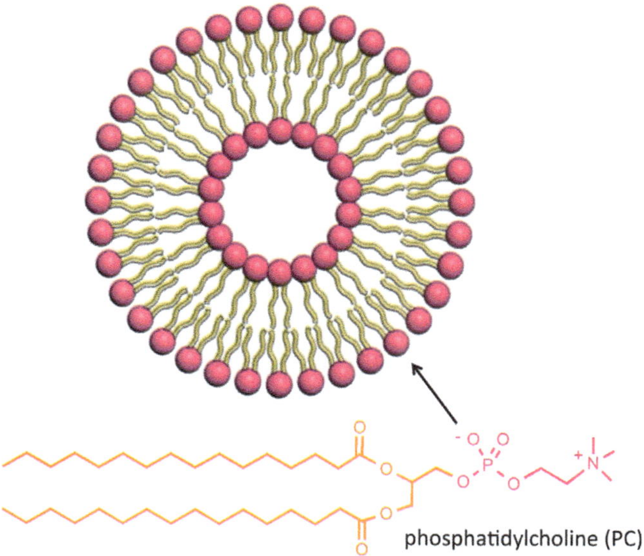

Figure 3.16 Schematic representation of a liposome (top) and a typical phospholipid (bottom).

the medium in which they were formed. The transport of ions out of a liposome into the external aqueous media can be detected in a variety of ways. Perhaps the most convenient is the use of ion-selective electrodes (see above). Transport of ions into empty vesicles is also possible, but it requires an entrapped fluorescent dye to act as a reporter.

3.7 Controlling Transport and Molecular Switching

Complexing agents that are used as transporters are defined by their inherent complexation and decomplexation rates. For example, a transporter containing an amine nitrogen donor that complexes K^+ could be induced to release it in the receiving phase if the latter is acidic. Protonation of N: to form $^+$N–H removes the electron pair donor and places a positive charge proximate to the bound cation. In order to make this sort of transport economical, a mechanism would be required to deprotonate the carrier so that it could again bind the cation in the source phase.

3.7.1 Redox Switching

An alternative approach is to introduce a switching mechanism. This was explored by using a variety of redox-active residues attached to various macrocycles. The general principle was that reduction of an electron-deficient arene such as nitrobenzene[50] or anthraquinone[51] would enhance the electron density of an apical binding site. This would enhance the strength with which the ring-bound cation was held, but it could be

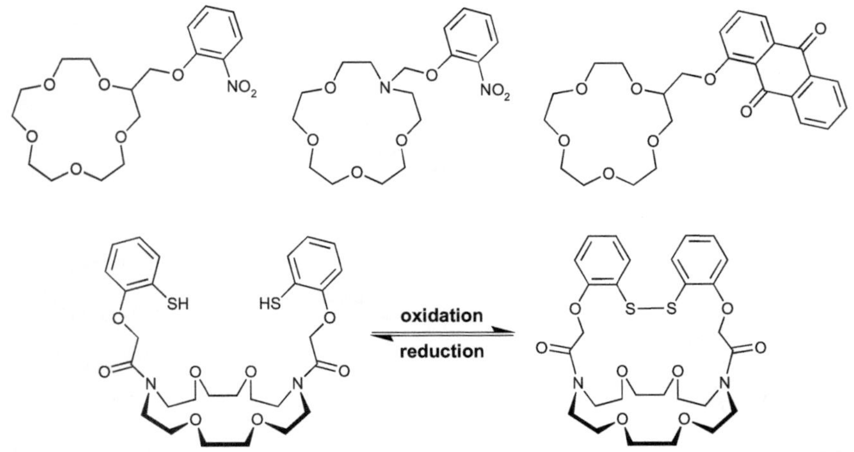

Figure 3.17 Redox-switched molecular receptors.

reversed by electrochemical oxidation. The principle was demonstrated with *ortho*-nitrobenzene as a side arm. Electrolytic reduction afforded a radical anion that was a stronger binder than the neutral compound or than the reduced *para*-nitrophenyl derivative. Figure 3.17 shows four compounds that were used in a similar fashion.

A disadvantage of the nitrobenzene redox couple is that it is water-sensitive. Thus, the anthraquinone side arm was incorporated into a similar receptor. A transport device that included redox capability was prepared and transport was successfully enhanced in water, albeit with oxygen excluded. Several macrocycles were prepared that used anthraquinone as the redox-active species. Transfer of an electron to the quinone residue enhanced the donicity of the ketone, which served as an apical donor for the ring-bound cation. Maintenance of a reduction potential on one side of the transport apparatus and an oxidation potential on the other permitted cation transport to occur at a rate higher than observed for controls.

Other redox-switchable systems have been developed. An example is an intriguing crown-to-cryptand equilibrium involving thiol to sulfide ring closure under oxidative conditions.[52,53] This interesting example also illustrates two significant design issues. First, the side arms are attached as amides, which limits their conformational mobility. Second, when the cryptand is formed from the bis(crown), the third strand donors comprise a disulfide linkage. This is both somewhat frail and likely to favor cations softer than the alkali metal cations.

3.7.2 Photo-switched Transport

Azobenzene, C_6H_5–N=N–C_6H_5, has been an extremely popular module in receptor chemistry because the molecule in the ground state has a *trans* (*E*) N=N bond that can be photoswitched to the *cis* (*Z*) state. When the *trans–cis*

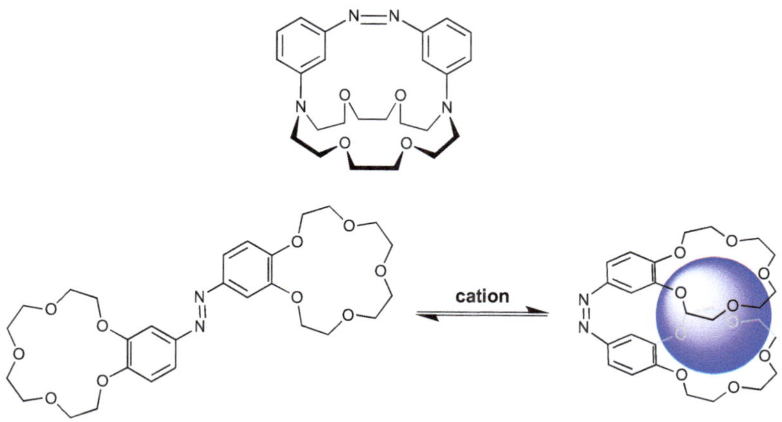

Figure 3.18 Examples of host molecules that can be photoswitched.

switch occurs, the relative positions of the two arenes are dramatically changed.

Shinkai and coworkers demonstrated photoswitched transport using this principle in the 1980s.[54,55] The bis(benzocrown) shown in Figure 3.18 can readily be prepared from 4-nitrobenzo-15-crown-5 by treatment with zinc. This leads to the azobenzene crown dimer. Although a 15-crown-5 is not the ideal binder for a K^+ cation, a stable 15-crown-5 sandwich complex is known. Upon photoswitching, *trans* to *cis* isomerization of the azobenzene places the macrocycles in position to bind K^+ in this sandwich arrangement. A thermal reaction returns *cis*-azobenzene to its ground-state *trans* arrangement in which the macrocycles act independently of each other.

3.8 Ammonium Cations and Structural Effects

Our understanding of geometric and conformational requirements in macrocyclic compounds has been informed by the effect of structural variations on ammonium ion binding.[56] This is discussed only briefly here as the present chapter is focused on alkali metal cations. The key difference between ammonium (NH_4^+) ions and group 1 metal cations is that the latter are spherical and the former are tetrahedral. Ammonium ions require donor groups placed appropriately to form directional hydrogen bonds, whereas the geometrical requirements for donors to alkali metal cations are more forgiving.

An important early study was undertaken by Cram and coworkers, which demonstrated the effect of apparently small changes in 18-membered ring macrocycles. The binding of ammonium ion, illustrated in Figure 3.19, requires the formation of three hydrogen bonds. The fourth N^+–H is perpendicular to the approximate plane of the macrocycle. As noted above, 18-crown-6 is symmetrical, it lacks significant butane-type *gauche*

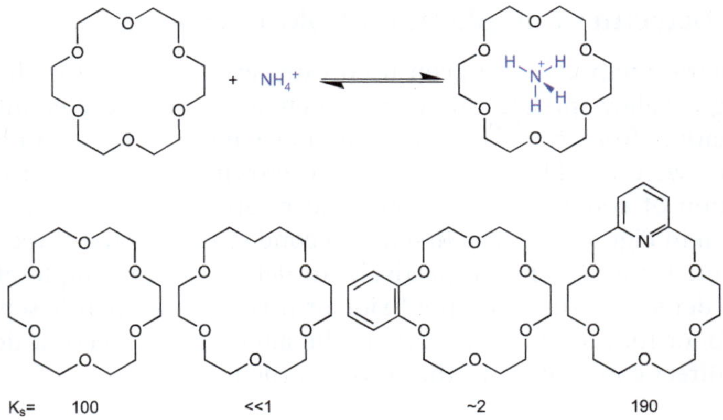

Figure 3.19 Relative NH_4^+ binding strengths for 18-membered ring macrocycles.

interactions owing to its $-OCH_2CH_2O-$ subunits, and it can complex cations in a conformation that alternates the oxygen dipoles.

The lower row of Figure 3.19 shows four structural variations on the 18-crown-6 structure. Ammonium binding has been arbitrarily set to 100 for 18-crown-6. As illustrated in the upper row, three hydrogen bonds can form with 18-crown-6 and there is good size compatibility. Moreover, the ammonium ion may bind with oxygens 1, 3, and 5 or with 2, 4, and 6. These are equivalent arrangements and the two possibilities are favorable for binding. Replacement of an oxygen atom by a methylene group eliminates one of these possibilities and also results in unfavorable conformational issues. Ammonium ion binding drops dramatically—by more than a thousand fold.

Fusion of a catechol residue to the macrocycle also diminishes NH_4^+ binding. The two alternate binding arrays (1, 3, 5 *vs.* 2, 4, 6) are still possible, but two of the oxygen atoms are conjugated to benzene, they are less basic, and the p orbital on the aryl oxygens is poorly positioned to H bond with NH_4^+. Introduction of a pyridine group does not compromise the macrocyclic oxygen atoms and the greater basicity of pyridyl nitrogen probably forms a stronger hydrogen bond to NH_4^+. This leads to a modest, but significant, two fold increase in complex stability relative to 18-crown-6.

3.9 Photoresponsive Receptors

Extensive work has been conducted to develop receptor molecules that could serve as indicators of cation complexation, much as phenolphthalein gives a colorimetric indication of pH change. The advantage of a visible change is that it can be monitored by very inexpensive ultraviolet-visible spectrometers accessible in nearly all laboratories. The advantage of fluorescence is its great sensitivity. As might be expected, colorimetric indicators were generally developed earlier than were fluorescent receptors, both of which are exemplified below.

3.9.1 Detecting Complexation Colorimetrically

Colorimetric methods have been used extensively in receptor chemistry.[57] The bright yellow picrate anion has been used to detect counter-cation complexation from the dawn of supramolecular chemistry. Colorimetric indicators were quickly incorporated into crown ethers and permitted the observation of complexation at low concentrations. Three early structures are shown in Figure 3.20. The first and second structures use three- and two-dimensional interactions, respectively, to detect cation complexation. The electron density of the quinonoid side arm is altered when it serves as an apical donor for a ring-bound cation. The alteration in electron density occurs by direct conjugation in the second structure.

3.9.2 Detecting Complexation Fluorometrically

Three examples of fluoroionophores are shown in Figure 3.21. In the upper and middle structures, an aza-15-crown-5 residue serves to capture the cation. Once bound, it interacts with the dye either by conjugation[58] or with a donor element, as in the coumarin lactone oxygen[59] shown in the middle of the figure. The diphenylpyrazolone crown (top) has the advantage of fluorescing in the near-infrared range. This avoids the shorter wavelength background autofluorescence in biological systems caused by such residues as the indole group in tryptophan. Indeed, a recent report describes the use of an azacrown-linked dye as a K^+ probe in vital cells.[60]

The ionophore shown at the bottom of Figure 3.21 is a hybrid crown ether–calixarene which functions differently from the molecules described above. In this case, complexation by the K^+ cation alters the receptor's conformation to bring the two pyrene residues into proximity. This leads to excimer formation and a strong signal—near 470 nm. This particular compound shows fluorescence quenching in the presence of Pb^{2+}, which leads to separation of the two pyrenes. When K^+ is added, Pb^{2+} is displaced, the system is no longer heavy-atom quenched, and the result is an "on–off switchable" sensor.

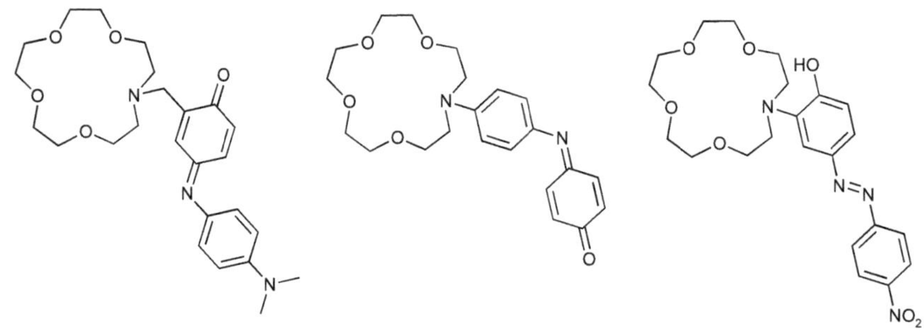

Figure 3.20 Crown ether-derived colorimetric sensors: "chromoionophores".

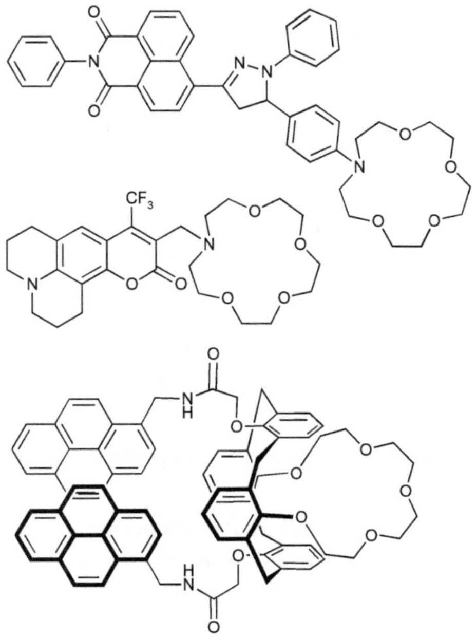

Figure 3.21 Fluoroionophores based on crown ethers and calixarenes.

3.10 Cation Transport

Living cells are surrounded by barriers that may be bilayer membranes, cell walls, double membranes, or other ensembles that protect their integrity. Cells have evolved in such a way that their interior salt concentrations are well regulated and different from the medium in which they live. Biological transport of alkali metal cations is described briefly at the start of this chapter and approaches to ion regulation by synthetic amphiphiles are addressed in this section.

3.10.1 Design of a Dynamic Cation Carrier

A vast literature exists that describes the use of macrocycles in various forms as carriers that transport cations across membrane barriers of one sort or another. As noted above, transport can give an indication of binding and selectivity. It can also be switched to make it more effective. Although thiol to disulfide redox coupling can affect the three-dimensional stability of a peptide or protein, nature generally relies on inherent properties such as conformation and conformational change to regulate rapid processes.

The design of a dynamic ion carrier requires careful consideration of the conflicting duality of selective binding caused by three-dimensional encapsulation *vs.* diminished dynamics. A cryptand would be ideal if the transported cation could be released. Strategies to switch carrier properties include reduction of an oxidized carrier, oxidation of a reduced carrier, or

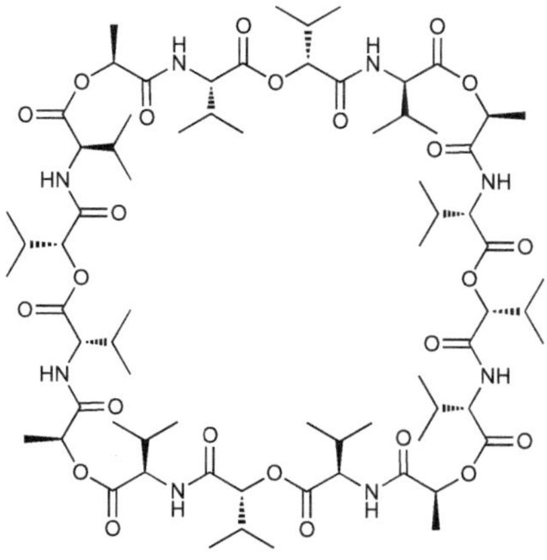

Figure 3.22 Chemical structure of the cyclic peptide K^+ carrier valinomycin.

acidification of a basic carrier, all at the receiving phases. To be sure, evolution has produced many approaches in the long history of life. One biological solution to this dilemma is the remarkable K^+ transporter valinomycin (Figure 3.22). It is a bacterial peptide that uses a combination of amide and ester groups to coordinate potassium cation, which fits perfectly into the folded 36-membered ring. The complex has been described as having a "tennis ball seam" arrangement about the cation.[61]

The inherent contradiction between binding strength and dynamics led Okahara and coworkers[62] and the author,[63] independent of each other, to add donor-containing side arms to macrocycles. The notion was that a three-dimensional, cryptand-like complex would have the advantage of binding strength and selectivity while the lack of a covalent attachment would permit dynamic binding rates. It was, of course, anticipated that neither binding nor selectivity would be better than that observed for cryptands. Likewise, slower binding than observed for 18-crown-6 was a reasonable expectation. This notwithstanding, a family of complexing agents that offered reasonable binding strengths, selectivity, and dynamics was worthy of study and development. We called these compounds "lariat ethers"[64] because of our concept that they could "rope and tie" a cation.

3.10.2 Lariat Ethers

Lariat ethers are crown ethers that have dynamic ion binding and release behavior. They sport one or more side arms that can function like the "third strand" in cryptands. The flexibility of the structure retains overall dynamics while the side arm donors provide additional binding strength and selectivity.

Figure 3.23 shows two lariat ethers[65] that differ in the way that the side arms are attached to the macrocycle. The compound labeled "carbon-pivot" is attached at a carbon atom, with the result that a racemic pair of structures

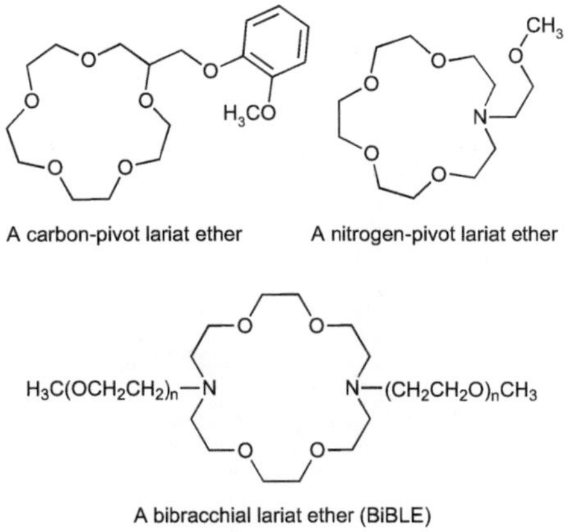

Figure 3.23 Top: single-armed carbon-pivot and nitrogen-pivot lariat ethers; bottom: a two-armed or bibracchial nitrogen-pivot lariat ether.

is formed. Such attachment leads to an inherent sidedness of the molecule which is not an issue in the nitrogen-pivot structures. Nitrogen undergoes rapid stereochemical inversion so a cation can enter the macroring from either side of the structure and the side arm can adjust accordingly. The two side arms of bibracchial lariat ethers can interact with a ring-bound cation, either from the same or from opposite sides, depending on the size of the ring, the size of the cation, and the lengths of the side arms.

Complexation of alkali metal cations by lariat ethers was strongly reminiscent of complexation by cryptands. The diaza-18-crown-6 bibracchial lariat ether having twin $CH_2CH_2OCH_3$ side arms, appended to nitrogen-bound both Na^+ and K^+, but the geometry of binding differed.[66] When the bound cation was Na^+, both side arms were on the same side of the macrocycle. When K^+ was ring-bound, the side arms were above and below the ring plane. The latter structure is illustrated in Figure 3.24. It is of note that the cation interacts exclusively with the oxygen and nitrogen donors and the iodide counter-anion is excluded from the primary solvation sphere.

The binding constants for the lariat ethers show cooperation between the ring and the side arm. It is especially interesting to compare aza-15-crown-5, which has a $(CH_2CH_2O)_3CH_3$ side arm with aza-18-crown-6, which has a $(CH_2CH_2O)_2CH_3$ side arm. Both compounds have one nitrogen and seven oxygen atoms. Their complexation constants (log K_S in methanol) with Na^+ ion are identical: 4.3. A comparison of Na^+ and K^+ binding for azalariat ethers having 12-, 15-, or 18-membered rings and various side arm lengths is presented in Table 3.2.

3.10.3 Cation–π Interactions with Alkali Metal Cations

One study that may be considered a triumph of crown ether/lariat ether chemistry is the search for clear evidence of cation–π interactions. The study

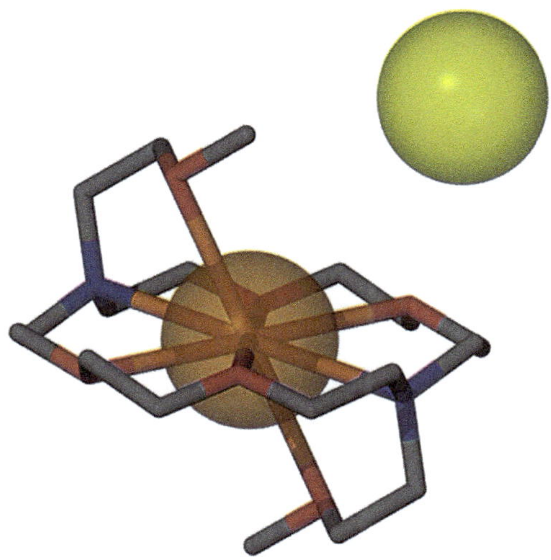

Figure 3.24 Solid-state structure of 4,13-diaza-18-crown-6 having two methoxyethyl side arms attached to nitrogen and binding KI.

Table 3.2 Sodium and potassium cation binding[a] by lariat ethers expressed as $\log_{10} K_S$.

	Aza-12-crown-4		Aza-15-crown-5		Aza-18-crown-6	
Side arm	Na^+	K^+	Na^+	K^+	Na^+	K^+
$CH_2CH_2OCH_3$	3.25	2.73	3.88	3.95	4.58	5.67
$(CH_2CH_2O)_2CH_3$	3.60		4.54	4.68	4.33	6.07
$(CH_2CH_2O)_3CH_3$	3.64	3.85	4.32	4.91	4.28	5.81
$(CH_2CH_2O)_4CH_3$	3.76		4.15	5.25	4.27	5.86

[a]Data determined in anhydrous methanol solution with chloride salts.

described above in Section 3.2.2 used 4,13-diaza-18-crown-6 macrocycles that had allyl, propargyl, or benzyl groups attached to the macroring nitrogen atoms. These compounds failed to provide evidence for any cation–π interaction. However, the simple expedient of changing the spacing between nitrogen and the π element led to much more rewarding results. The compounds prepared and studied are shown in Figure 3.25.

Several of these compounds are closely related to molecules that were prepared and studied years before. They differ from them by a single methylene group. In the earlier compounds, the arene or π-element was separated by a single methylene group from the macrocycle. In the successful cases, everything else was identical, but the spacing was ethylene rather than methylene. Of course, the latter corresponds to the spacing encountered between nitrogen and the arene in such amino acids as phenylalanine, tyrosine, and tryptophan.

Two classes of two-armed or bibracchial lariat ethers were studied in one group, the π-element was an arene. In the second group, the π-element was a double or triple bond. The complexation process was studied both in

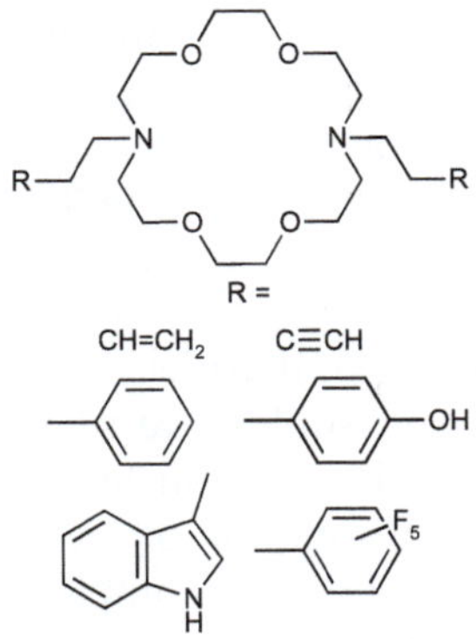

Figure 3.25 Bibracchial lariat ethers containing π-donor side arms.

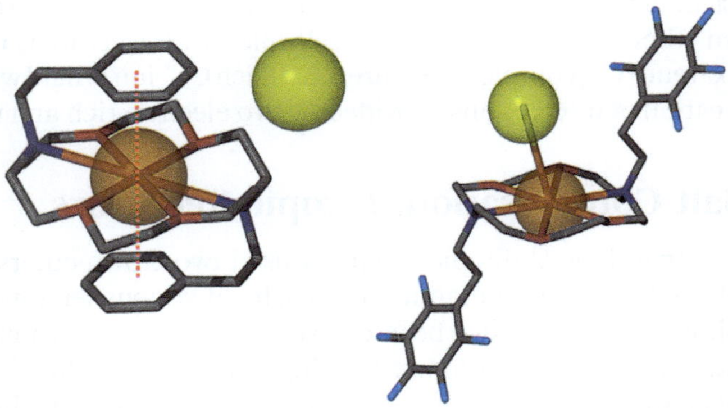

Figure 3.26 Solid-state structures of phenyl (Cambridge Structural Database (CSD): OCABEZ) (left) and pentafluorobenzyl (CSD: OCACIE) (right) side-armed bibracchial lariat ethers binding KI.

solution[67,68] and in the solid state. These studies have been reviewed,[69] but examples are included here to demonstrate the principles. The initial studies focused on side arms that contained arenes present in the 20 common amino acids. These include phenyl, 4-hydroxyphenyl, and indole. The solid-state structure of the benzene compound is shown in Figure 3.26. The structure shown in the left panel of Figure 3.26 is the K^+I^- complex of *N,N′*-bis(2-phenylethyl)-4,13-diaza-18-crown-6. The potassium cation is situated in the center of the macrocycle and is in contact with all six heteroatoms. The side arms are positioned so that one phenyl ring is above the cation and the other below. A line drawn from the center of each arene to the other arene passes through the center of the cation.

Studies of KI complexation by the related 4-hydroxyphenyl- and indole-terminated side arms revealed structures that were essentially similar. An interesting issue that arose concerned the indole compound. Computational studies predicted that the benzene ring rather than the pyrrole would be the most potent π-donor in indole. This is likely to be true, but the cation was bound by the pyrrole π-system rather than the six-membered ring. We inferred that the π-donicity might be weaker in the five-membered ring, but that proximity compensated for it.

The conformations and geometries of the K^+ complexes of the three lariats having phenyl, 4-hydroxyphenyl, and indole side arms were remarkably similar. It seemed possible that the cation–π interaction was not the force defining geometry, but that the similarities could result from crystal packing forces. In order to test this possibility, we prepared the pentafluorophenyl derivative. Fluorine is a little larger than hydrogen, but we felt it was not sufficiently so to alter the conformation or crystal packing if either of those was the key issue. In fact, a solid-state complex was obtained. As expected, the Lewis acidic pentafluorophenyl residue was turned away from the ring-bound cation and iodide served as the apical donor group. The structure is shown in the right panel of Figure 3.26.

Cation–π interactions have been observed in the solid-state structures of other crown ether complexes. An esthetically pleasing structure is the C_2 symmetric calixarene · Cs_2 complex in Figure 3.27. Each Cs^+ ion is held with a crown ether with cation–π interactions provided by two electron rich aryl rings.

3.11 Salt Complexation: Ditopic Receptors

Early crown ether-derived ditopic receptors used two crown ethers separated by covalent linkages to bind diammonium salts of various lengths.[70,71] More recently, ditopic receptors have been devised that have a crown ether cation binding site and a tris-arene anion binding site (Figure 3.28).[72,73] The isophthalic acid diamides were first reported as anion binders by Crabtree *et al.*[74,75] Such systems have been reviewed[76] and will not be discussed here, except to note that further discussion may be found in Chapter 4 of this book.

3.12 Pore-forming Amphiphiles

Numerous complexing and transport agents have been devised and studied in various applications. Many simple complexing agents can probably

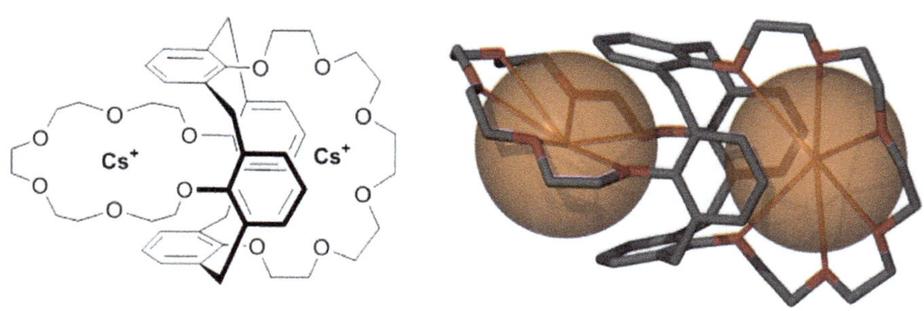

Figure 3.27 Solid-state structure of a calixarene · Cs_2 complex (Cambridge Structural Database: RADBUT).

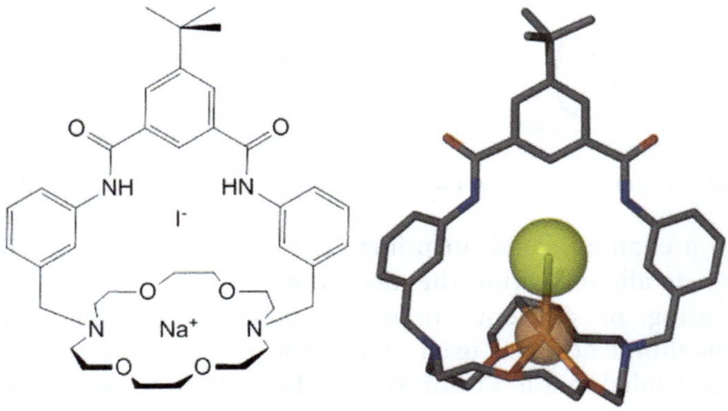

Figure 3.28 A ditopic receptor binding both Na^+ and I^- (Cambridge Structural Database: IBUKUM).

facilitate the transport of ions across bulk organic, liposomal, or other types of membranes. A considerable challenge was to design a compound that would transport ions across bilayers in a fashion that mimicked the function of natural channels. This was a particularly vexing problem for two reasons. First, despite more than a century of study,[77] there was no solid-state structure of a channel protein that could guide the chemist hoping to mimic the function. A structure did not emerge until 1998.[78] Second, in the absence of structural information, only estimates could be made about the mechanisms of conduction and gating. The latter is a property of protein channels that present "open–close" behavior.

3.12.1 A Cyclodextrin-based Channel

As early as 1982, Tabushi and colleagues devised a compound that inserted into bilayer membranes and showed conductance properties.[79] They used β-cyclodextrin as their scaffold. It has an internal opening of 7.8 Å, which should be sufficient to pass common alkali metal ions, even within an aqueous first solvation shell. Because β-cyclodextrin is constructed of seven glucose molecules, 14 secondary and 7 primary hydroxyl groups are present. Because they are less sterically hindered, the primary hydroxyl groups can be substituted more readily than can be the secondary ones. Moreover, cyclodextrin is a relatively rigid scaffold, so the substituted and unsubstituted groups will be on opposite sides of the torus. The compound that was prepared for study is shown in Figure 3.29.

Tabushi and coworkers estimated that that the cyclodextrin was ~6 Å thick and the chains extended from it by ~18 Å. If two of these molecules inserted into opposite leaflets of a bilayer, the pendant chains would span the estimated 30–35 Å thickness of the insulator regime, and 14 hydroxyl groups on each side would interact with water and/or lipid head groups to stabilize the assembly. Lateral motion within a bilayer leaflet is typically

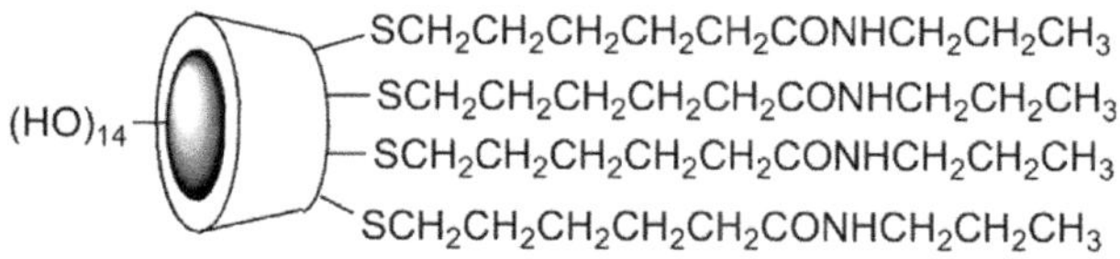

Figure 3.29 An ion-conducting channel based on the cyclodextrin scaffold.

facile and a channel could, in principle, form when two molecules in opposite leaflets aligned within the membrane.

The pairing of divalent copper or cobalt ions with a dihydroxybenzenedisulfonic acid was used to monitor transport colorimetrically. The authors concluded that ion transport was faster than for ion carriers, which is consistent with channel function. Of course, Cu^{2+} and Co^{2+} are not abundant biological cations, but the early demonstration of this principle was a landmark in the development of synthetic ion channels. Indeed, the cyclodextrin scaffold was developed much later by Gin and coworkers, who reported facilitated anion transport.[80]

3.12.2 Synthetic Amphiphilic Ionophores: Synthetic Ion Channels

In considering how to design a cation-conducting channel, one must think about dynamics more than binding strength. Thinking about selectivity must be, to use a biological term, everted. If an element within the channel is selective for a particular ion, it is likely to bind it rather than transport it. Another consideration is that membranes are remarkably heterogeneous in composition and variable in structure and thickness.[81] The design of an artificial channel required a number of estimates about structural requirements and even guesses about mechanism. The goal, of course, was to prepare a compound that could conduct the abundant alkali metal ions through a bilayer.

During the late 1980s, considerable effort was focused on the mechanism or mechanisms by which proteins conducted ions, how they opened and closed, the origins of their selectivity profiles, and how rectification was achieved. Much information had been accumulated, but there was no clear mechanism. Despite this uncertainly, our laboratory, along with those of Lehn[82] and Fyles,[83] began an effort to emulate this remarkable biological function. All three laboratories chose an 18-crown-6 derivative as the central scaffold. The compound designed by Lehn and by Fyles was a tetracarboxy crown that was assembled in part from two tartaric acid residues. In our laboratory, we used 4,13-diaza-18-crown-6 units on the principle that linkages to nitrogen would be relatively flexible.

The Lehn and Fyles designs are illustrated in Figure 3.30. In both cases, it was anticipated that the transient ion would pass through the macrocycle. Both compounds are structured as tubes, which seems reasonable for a molecule that should function like a tunnel. Both molecules have polar

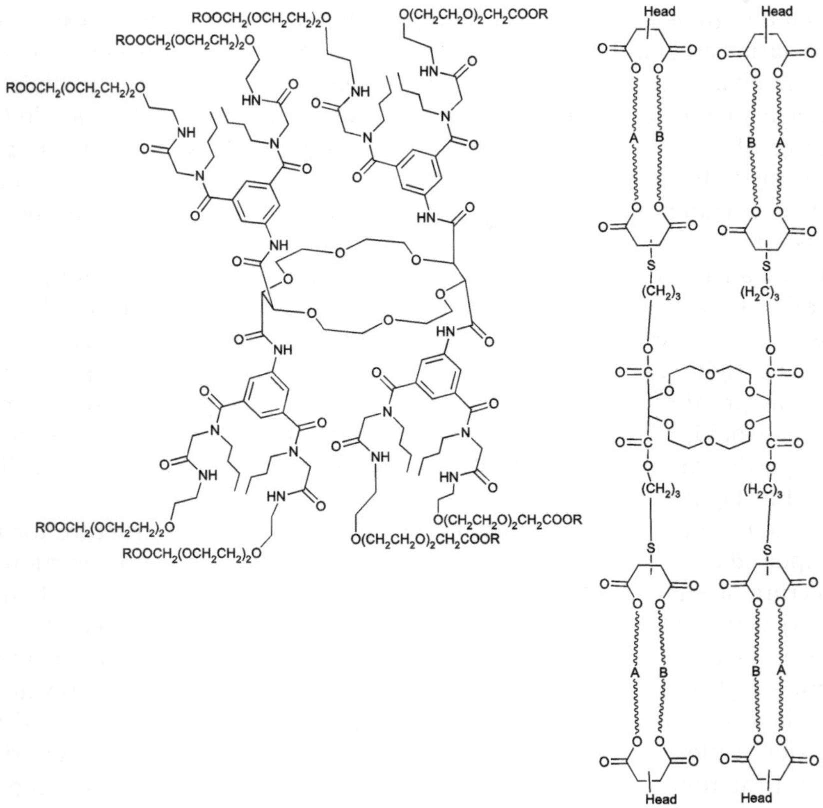

Figure 3.30 Channel designs reported by Lehn (left) and by Fyles (right) and their coworkers.

residues at their extremities. In the Lehn design, these are ester-terminated diethylene glycols. The efficacy of this channel was unclear in the initial publication and later work used a cyclodextrin as the central scaffold in compounds called molecular "bouquets."[84]

The Fyles approach used monosaccharide head groups ("head" in Figure 3.30) and the spacers identified as "A" and "B" in Figure 3.30 were alkyl groups. Vesicles were prepared that included carboxyfluorescein. Changes in proton concentration and its effect on fluorescence were monitored to assay the effects of the synthetic amphiphiles. These studies used the naturally occurring bacterial channel gramicidin and amphotericin as controls. It was found that when the spacer chains were both hydrocarbons, proton transport was less effective than when one chain was octylene and the other was triethylene glycol. Intuitively, one might expect this to be the case as a cation will require less energy to make shorter "jumps" between donors. Conversely, every position that stabilizes the transient cation should slow its movement. When the solid-state structure of the KcsA channel was reported,[77] the channel interior was likened to a hydrophobic tube.

Our design of a synthetic ion channel followed in part from our experience with lariat ethers. We found that covalent connections through nitrogen linkers ensured molecular flexibility. Increased rigidity is an enthalpic advantage in the design of a complexing agent. Because $\Delta G = -RT \ln K$, a change of ~1.4 kcal mol^{-1} in ΔG results in a 10 fold increase in binding constant K. In the design of a channel, sufficient binding is needed to recognize a cation, but the interaction must be weak enough to permit transport.

Using an approximate thickness of 30–35 Å for the insulator regime of the membrane, we designed a tris(macrocycle) that would use crown ethers as head groups. We chose to incorporate a third macrocycle in the center, but not as a rigid scaffold. The concept was that a cation would find it energetically prohibitive to "jump" the full distance. The presence of the macrocycle would, we surmised, lower the energy of transit. This proved to be only partly true as discussed below. The initial target compound is illustrated in Figure 3.31.

As Lehn and Fyles had done, we envisioned a tubular structure for the compounds we eventually called *hydraphiles*. The schematic portion of structure of Figure 3.31 shows the envisioned conformation of the channel (right) within the bilayer. Extensive biophysical studies revealed that the central macrocycle was parallel rather than perpendicular to the membrane's acyl chains. This seemed sensible since it lowered the polarity at the midplane of the bilayer. In addition, it permitted some solvation of the transient cation without requiring it to pass directly through the macroring. A confirmation of this conclusion involved replacement of the central macrocycle with a diaza-12-crown-4 residue, which is too small for Na^+ to pass through. Sodium transport was reduced but not blocked.

Figure 3.32 shows a number of compounds that have shown at least some characteristics of channel function. The structures are identified in the figure as **A–H** and represent a remarkable range of designs. Compound **A** is a

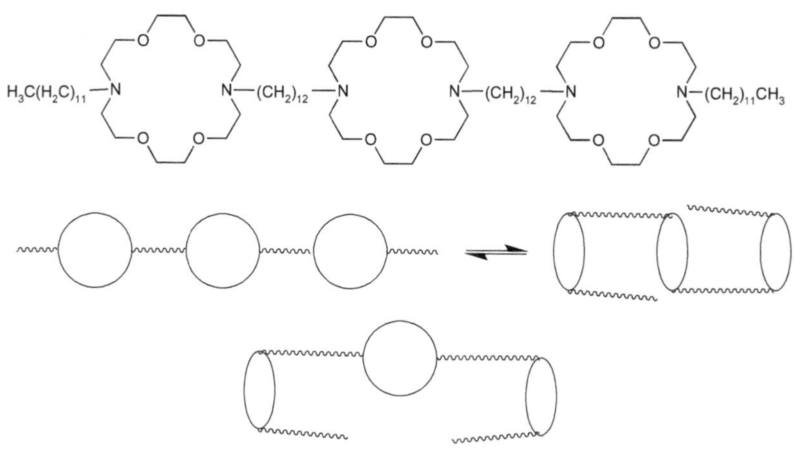

Figure 3.31 The hydraphile channel concept.

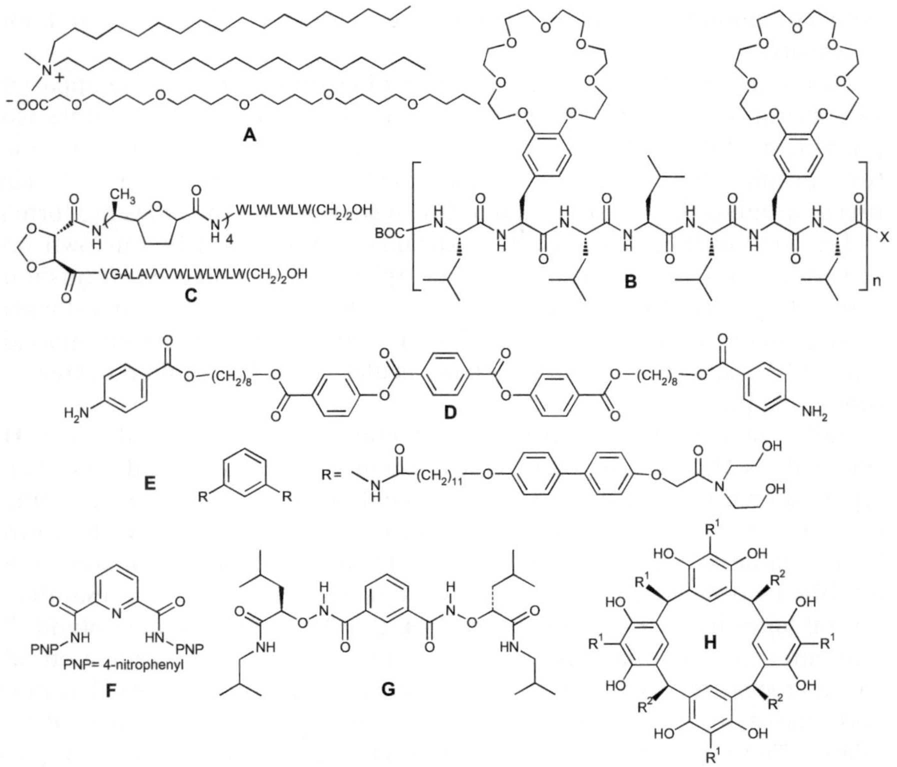

Figure 3.32 An array of synthetic amphiphiles that show channel-like function.

self-assembled system that was designed by Kobuke and coworkers. Although the ion conductance was relatively low, it was reproducibly observed and clear open–close behavior was apparent.

Compounds **B** and **C**, designed in the laboratories of Voyer[85,86] and Koert,[87] respectively, took inspiration from natural systems. The repeating tryptophan (W)-leucine (L) sequences in **C** are similar to those present in the natural channel-forming peptide gramicidin. The scaffold chosen for **C** was inspired by work that resulted in synthetic porins or peptide bundles reported by DeGrado[88] and by Montal.[89] Voyer's design relied on the α-helicity of the peptide. The macrocycles were placed at distances $(i+4)$ that would ensure their formation of a tubular structure adjacent to the peptide. Because the rings are relatively large, ion selectivity was not high.

Structures **D** and **E** reflect a recognition in Fyles's[90] and the author's[91,92] laboratories, respectively, that structures simpler than our original designs could exhibit channel-like behavior. It is interesting that extensive efforts were made to devise complex structures before simpler structures were attempted. In retrospect, it seems apparent that the earliest channels must have been membrane-active amphiphiles rather than the exquisitely complex proteins than serve the channel function today. Despite their simplicity, both compounds **D** and **E** showed open–closed conductances and ion selectivity.

Two of the most remarkable examples of channel function are apparent in structures **F** and **G**. Li and coworkers reported that the simple isophthalic acid derivative, **G**, showed many characteristics of channel function, despite the fact that it cannot possibly span a membrane.[93,94] The presumption is that these compounds can stack within the bilayer to form a viable conductance pathway. This surmise is reinforced by our own observation that tris-arene **F**, an anion complexing agent, likewise forms ion-conducting pores that exhibit open–close behavior.[95] We posited a stacked arrangement of these essentially flat molecules that are spaced approximately 4 Å apart. As such, only 8–10 molecules would be required to create a functional pore.

Quite a number of molecules are represented by the structure shown as **H**. When R^1 is **H**, the compound is a resorcinarene. Kobuke and coworkers reported channel function for a compound **H** in which R^2 was a normal octadecyl chain.[96] These authors presented evidence that their **H**-derivative formed dimeric channels—one molecule in each leaflet of the bilayer—that exhibited open–close behavior. In work in our laboratory, we examined several long-chain (R^2) resorcinarenes (**H**, $R^1 = H$)[97] and several cavitands.[98] The latter are essentially resorcinarenes in which hydroxyl groups on adjacent arenes are linked by methylene groups. We interpreted the behavior of both classes of structures to form aggregates in the upper leaflet of the bilayer. The pore opening was therefore far too large for the ions to pass through the center of the macrocycle, as was clearly the case for the Kobuke resorcinarene. We inferred that the resorcinarenes formed aggregated pores because of evidence obtained with pyrogallolarenes (**H**, $R^1 = OH$, $R^2 =$ alkyl). For compound **H** ($R^1 = OH$, $R^2 =$ *n*-undecyl), pores formed, corresponding in size (as determined by conductance) to five macrocycles side-by-side.[99] When cholesterol, which "rigidifies" membranes, was added, pores formed that corresponded in size to tetramers and the stability increased. When the alkyl chains of **H** are branched (**H**, $R^1 = OH$, $R^2 =$ 3-pentyl),[100] unique nanotubes form[101] that also show conductance properties.[102] In the latter case, it appears that two hexamers of tetramers stack to form a conducting opening in the bilayer.

Three very recent reports from the laboratories of Barbiou[103] and Fyles[104] deserve note. The first of these is a collaborative effort that has resulted in a gramicidin mimic. The second is a family of 17 synthetic ionophores based on a cyclodextrin scaffold.

It seems clear from the diversity of channel-forming structures that more than one mechanism must operate by which pores form in membranes and how ions are transported through it. Fyles has recently addressed this mechanistic problem using oligoesters (**D** derivatives in Figure 3.32) as the model system.[105] This is, at the time of writing, the state of the art in terms of our understanding of mechanism.

3.12.3 Hydraphile Biological Activity

Extensive studies have shown that hydraphiles possess significant biological activity due to their ability to disturb ion balance (homeostasis). They are toxic to bacteria and to mammalian cells and at low concentrations they can dramatically alter the primary/lateral root architectural balance in a plant model.[106] Leevy and coworkers[107–109] proposed that the mammalian cell toxicity could be utilized for direct injection chemotherapy.[110] Small molecules, such as ethanol and acetic acid have been used on the principle that their toxicity to a tumor would be greatest if completely localized. The ready diffusion of small molecules engenders collateral tissue damage. In contrast, hydraphiles diffuse slowly and therefore are of potential value in this application.[111] The localized toxicity was demonstrated in a mouse model[112] and the tumor cell death visualized by fluorescence molecular imaging.[113]

3.13 Future Challenges

In the author's opinion, bilayer membranes constitute the most remarkable self-assembled supramolecular system known. It is incredible to think that less than 40 years ago, membranes were viewed by many as being simple and homogeneous. It was only in 1972 that the "fluid mosaic" model for membrane structure was proposed.[114] More recently, it was recognized that some segments of membranes had structures that held together in "lipid rafts."[115–117] Further study has led to an understanding that there are many microdomains[118–120] within a bilayer and that these have affinities that correspond to the proteins that target them.[121] The chemist has much to offer in terms of understanding membrane structure and function. From a practical perspective, synthetic cation receptors that can insert into membranes and alter their function are likely to have valuable pharmaceutical properties. As brain research accelerates, there will be increased interest in switchable ion channel systems that can be activated remotely using laser light or an electrical impulse. Similarly, there is still an unmeet need in cell biology for improved fluorescent molecular probes that can selectively report Na^+ and K^+ levels inside living cells. The possibilities for developing synthetic cation receptors that have valuable analytical and biological function represents a vast landscape of fertile soil that will no doubt be plowed for decades to come.

References

1. D. Łowicki and A. Huczyński, *BioMed Res. Int.*, 2013, 742149.
2. E. Gouaux and R. MacKinnon, *Science*, 2005, **310**, 1461.
3. Q.-Q. Wang, R. A. Begum, V. W. Day and K. Bowman-James, *Polyhedron*, 2013, **52**, 515.
4. C. J. Pedersen, US Pat., 3 361 778, 1968.

5. C. J. Pedersen, *J. Am. Chem. Soc.*, 1967, **89**, 7017.
6. C. J. Pedersen, in *Nobel Lectures, Chemistry 1981–1990*, ed. T. Frängsmyr and B. G. Malmström, World Scientific Publishing Co., Singapore, 1992, pp. 419–437.
7. *Cation Binding by Macrocycles*, ed. Y. Inoue and G. W. Gokel, Marcel Dekker, Inc., New York, 1990.
8. G. W. Gokel and S. H. Korzeniowski, *Macrocyclic Polyether Syntheses*, Springer-Verlag, Berlin, 1982.
9. J. S. Bradshaw, K. E. Krakowiak and R. M. Izatt, *Aza-Crown Compounds*, Wiley, New York, 1993.
10. B. Dietrich, J. M. Lehn and J. P. Sauvage, *Tetrahedron Lett.*, 1969, **34**, 2885.
11. B. Dietrich, P. Viout and J.-M. Lehn, *Macrocyclic Chemistry: Aspects of Organic and Inorganic Supramol. Chem.*, VCH, Weinheim, 1993.
12. D. J. Cram and J. M. Cram, *Science*, 1983, **219**, 1177.
13. D. J. Cram and J. M. Cram, *Container Molecules and Their Guests*, RCS, Cambridge, 1994.
14. J. M. Lehn, in *Nobel Lectures, Chemistry 1981–1990*, ed. T. Frängsmyr and B. G. Malmström, World Scientific Publishing Co., Singapore, 1992, pp. 495–511.
15. D. J. Cram, *Science*, 1988, **240**, 760.
16. R. M. Izatt, J. S. Bradshaw, S. A. Nielsen, J. D. Lamb, J. J. Christensen and D. Sen, *Chem. Rev.*, 1985, **85**, 271.
17. L. Liu and Q. X. Guo, *Chem. Rev.*, 2001, **101**, 673.
18. K. A. Arnold, A. M. Viscariello, M. Kim, R. D. Gandour, F. R. Fronczek and G. W. Gokel, *Tetrahedron Lett.*, 1988, 3025.
19. V. J. Gatto, S. R. Miller and G. W. Gokel, *Org. Synth.*, 1989, **68**, 227.
20. D. A. Gustowski, V. J. Gatto, J. Mallén, L. Echegoyen and G. W. Gokel, *J. Org. Chem.*, 1987, **52**, 5172.
21. R. M. Izatt, K. Pawlak, J. S. Bradshaw and R. L. Bruening, *Chem. Rev.*, 1991, **91**, 1721.
22. D. J. Cram, T. Kaneda, R. C. Helgeson, S. B. Brown, C. B. Knobler, E. Maverick and K. N. Trueblood, *J. Am. Chem. Soc.*, 1985, **107**, 3645.
23. G. W. Gokel, D. M. Goli, C. Minganti and L. Echegoyen, *J. Am. Chem. Soc.*, 1983, **105**, 6786.
24. E. Mason and H. A. Eick, *Acta Crystallogr., Sect. B: Struct. Crystallogr. Cryst. Chem.*, 1982, **38**, 1821.
25. B. D. White, K. A. Arnold, R. L. Garrell, F. R. Fronczek, R. D. Gandour and G. W. Gokel, *J. Org. Chem.*, 1987, **52**, 1128.
26. B. Dietrich, *Comprehensive Supramolecular Chemistry*, Pergamon Press, 1996, ch. 4, vol. 1, pp. 153–212.
27. G. W. Liesegang, A. Vasquez, N. Purdie and E. M. Eyring, *J. Am. Chem. Soc.*, 1977, **99**, 3240.
28. E. Maverick and D. J. Cram, *Comprehensive Supramolecular Chemistry*, Pergamon Press, 1996, ch. 5, vol. 1, pp. 213–244.
29. A. Baeyer, *Ber. Dtsch. Bot. Ges.*, 1886, **19**, 2184.
30. R. G. Ackman, W. H. Brown and G. F. Wright, *J. Org. Chem.*, 1955, **20**, 1147.

31. C. D. Gutsche, *Calixarenes*, RSC, Cambridge, 1989, vol. 1, p. 223.
32. C. D. Gutsche, *Calixarenes Revisited: (Monographs in Supramol. Chem.)*, RSC, Cambridge, 1998, p. 233.
33. C. D. Gutsche, *Calixarenes: An Introduction (Monographs in Supramol. Chem.)*, RSC, Cambridge, 2nd edn, 2008, p. 248.
34. *Calixarenes in Action*, ed. L. Mandolini and R. Ungaro, World Scientific Publishing Company, Hackensack, NJ, 2000.
35. W. Sliwa and C. Kozlowski, *Calixarenes and Resorcinarenes: Synthesis, Properties, and Applications*, Wiley-VCH Verlag GmbH & Co., Weinheim, 2009.
36. C. D. Gutsche, *Org. Synth.*, 1990, **68**, 234.
37. V. Lamare, J.-F. Dozol, F. Ugozzoli, A. Casnati and R. Ungaro, *Eur. J. Org. Chem.*, 1998, 1559.
38. *Principles of Thermal Analysis and Calorimetry*, ed. P. Haines, RSC, Cambridge, UK, 2002.
39. R. M. Izatt, J. S. Bradshaw, K. Pawlak, R. L. Bruening and B. J. Tarbet, *Chem. Rev.*, 1992, **92**, 1261.
40. R. M. Izatt, K. Pawlak, J. S. Bradshaw and R. L. Bruening, *Chem. Rev.*, 1995, **95**, 2529.
41. X. X. Zhang, J. S. Bradshaw and R. M. Izatt, *Chem. Rev.*, 1997, **97**, 3313.
42. C. J. Pedersen and H. K. Frensdorff, *Angew. Chem., Int. Ed.*, 1972, **11**, 16.
43. H. K. Frensdorff, *J. Am. Chem. Soc.*, 1971, **93**, 4684.
44. E. P. Kyba, R. C. Helgeson, K. Madan and G. W. Gokel, *J. Am. Chem. Soc.*, 1977, **99**, 2564.
45. Y. Takeda, *Top. Curr. Chem.*, 1984, **121**, 1.
46. J. C. Hernandez, J. E. Trafton and G. W. Gokel, *Tetrahedron Lett.*, 1991, **32**, 6269.
47. B. C. Pressman, E. J. Harris, W. S. Jagger and J. H. Johnson, *Proc. Natl. Acad. Sci. U. S. A.*, 1967, **58**, 1949.
48. B. C. Pressman, *Fed. Proc*, 1968, **27**, 1283.
49. D. D. Lasic, *Liposomes from Physics to Applications*, Elsevier, Amsterdam, 1993.
50. D. A. Gustowski, L. Echegoyen, D. M. Goli, A. Kaifer, R. A. Schultz and G. W. Gokel, *J. Am. Chem. Soc.*, 1984, **106**, 1633.
51. L. M. Echeverria, M. Delgado, V. J. Gatto, G. W. Gokel and L. Echegoyen, *J. Am. Chem. Soc.*, 1986, **108**, 6825.
52. S. Shinkai, K. Inuzuka and O. Manabe, *Chem. Lett.*, 1983, 747.
53. S. Shinkai, K. Inuzuka, K. Hara, T. Stone and O. Manabe, *Bull. Chem. Soc. Jpn.*, 1984, **57**, 2150.
54. S. Shinkai, K. Shigematsu, Y. Kusano and O. Manabe, *J. Chem. Soc., Perkin Trans. 1*, 1981, 3279.
55. S. Shinkai and O. Manabe, in *Host Guest Complex Chemistry III*, ed. F. Vögtle and E. Weber, 1984, p. 68–104.
56. D. J. Cram and K. N. Trueblood, *Top. Curr. Chem.*, 1981, **98**, 43.
57. M. Takagi and K. Ueno, *Top. Curr. Chem.*, 1984, **121**, 39.
58. K. Rurack, U. Resch-Genger, J. L. Bricks and M. Spieles, *Chem. Commun.*, 2000, 2103.

59. J. Bourson, J. Pouget and B. Valeur, *J. Phys. Chem.*, 1993, **97**, 4552.
60. S. Ast, T. Schwarze, H. Muller, A. Sukhanov, S. Michaelis, J. Wegener, O. S. Wolfbeis, T. Korzdorfer, A. Durkop and H. J. Holdt, *Chem. – Eur. J.*, 2013, **19**, 14911.
61. M. R. Truter and M. A. Bush, *J. Chem. Soc., Chem. Commun.*, 1970, 1439.
62. K. Ping-Lin, M. Miki and M. Okahara, *Chem. Commun.*, 1978, 504.
63. G. W. Gokel, D. M. Dishong and C. J. Diamond, *J. Chem. Soc., Chem. Commun.*, 1980, 1053.
64. G. W. Gokel and L. Echegoyen, in *Advances in Bio-organic Frontiers*, ed. H. Dugas, Springer Verlag, Inc., 1990, Issue 1, pp. 116–141.
65. G. W. Gokel and O. F. Schall, *Comprehensive Supramolecular Chemistry*, ed. G. W. Gokel, Pergamon, Oxford, 1996, vol. 1, pp. 97–152.
66. R. D. Gandour, F. R. Fronczek, V. J. Gatto, C. Minganti, R. A. Schultz, B. D. White, K. A. Arnold, D. Mazzocchi, S. R. Miller and G. W. Gokel, *J. Am. Chem. Soc.*, 1986, **108**, 4078.
67. S. L. De Wall, L. J. Barbour and G. W. Gokel, *J. Am. Chem. Soc.*, 1999, **121**, 8405.
68. S. L. De Wall, E. S. Meadows, L. J. Barbour and G. W. Gokel, *J. Am. Chem. Soc.*, 1999, **121**, 5613.
69. G. W. Gokel, L. J. Barbour, R. Ferdani and J. X. Hu, *Acc. Chem. Res.*, 2002, **35**, 878.
70. I. O. Sutherland, *Pure Appl. Chem.*, 1989, **61**, 1547.
71. I. O. Sutherland, in *Advances in Supramolecular Chemistry*, ed. G. W. Gokel, JAI Press, Greenwich, CT, 1990, vol. 1, pp. 66–108.
72. A. V. Koulov, J. M. Mahoney and B. D. Smith, *Org. Biomol. Chem.*, 2003, **1**, 27.
73. J. M. Mahoney, G. U. Nawaratna, A. M. Beatty, P. J. Duggan and B. D. Smith, *Inorg. Chem.*, 2004, **43**, 5902.
74. K. Kavallieratos, S. R. de Gala, D. J. Austin and R. H. Crabtree, *J. Am. Chem. Soc.*, 1997, **119**, 2325.
75. K. Kavallieratos, C. M. Bertao and R. H. Crabtree, *J. Org. Chem.*, 1999, **64**, 1675.
76. P. A. Gale, *Acc. Chem. Res.*, 2011, **44**, 216.
77. B. Hille, *Ionic Channels of Excitable Membranes*, Sinauer Associates, Sunderland, MA, 3rd edn, 2001.
78. D. A. Doyle, J. M. Cabral, R. A. Pfuetzner, A. Kuo, J .M. Gulbis, S. L. Cohen, B. T. Chait and R. MacKinnon, *Science*, 1998, **280**, 69.
79. I. Tabushi, Y. Kuroda and K. Yokota, *Tetrahedron Lett.*, 1982, **23**, 4601.
80. N. Madhavan, E. C. Robert and M. S. Gin, *Angew. Chem., Int. Ed.*, 2005, **44**, 7584.
81. *The Structure of Biological Membranes*, ed. P. L. Yeagle, CRC Press, Boca Raton, FL, 2012.
82. L. Lehn and J.-M. Jullien, *Tetrahedron Lett.*, 1988, **29**, 3803.
83. V. E. Carmichael, P. J. Dutton, T. M. Fyles, T. D. James, J. A. Swan and M. Zojaji, *J. Am. Chem. Soc.*, 1989, **111**, 767.

84. J. Canceill, L. Jullien, L. Lacombe and J.-M. Lehn, *Helv. Chim. Acta*, 1992, **75**, 791.
85. N. Voyer and M. Robataille, *J. Am. Chem. Soc.*, 1995, **117**, 6599.
86. P. L. Boudreault, M. Arseneault, F. Otis and N. Voyer, *Chem. Commun.*, 2008, 2118.
87. A. Schrey, A. Vescovi, A. Knoll, C. Rickert and U. Koert, *Angew. Chem., Int. Ed.*, 2000, **39**, 900.
88. K. S. Åkerfeldt, J. D. Lear, Z. R. Wasserman, L. A. Chung and W. F. DeGrado, *Acc. Chem. Res.*, 1993, **26**, 191.
89. M. Montal, M. S. Montal and J. M. Tomich, *Proc. Natl. Acad. Sci. U. S. A.*, 1990, **87**, 6929.
90. J. K. W. Chui, T. M. Fyles and H. Luong, *Beilstein J. Org. Chem.*, 2011, **7**, 1562.
91. W. Wang, R. Li and G. W. Gokel, *Chem. Commun.*, 2009, 911.
92. W. Wang, R. Li and G. W. Gokel, *Chem. – Eur. J.*, 2009, **15**, 10543.
93. X. Li, B. Shen, X.-Q. Yao and D. Yang, *J. Am. Chem. Soc.*, 2007, **129**, 7264.
94. X. Li, Y. D. Wu and D. Yang, *Acc. Chem. Res.*, 2008, **41**, 1428.
95. C. R. Yamnitz, S. Negin, I. A. Carasel, R. K. Winter and G. W. Gokel, *Chem. Commun.*, 2010, **46**, 2838.
96. Y. Tanaka, Y. Kobuke and M. Sokabe, *Angew. Chem., Int. Ed.*, 1995, **34**, 693.
97. P. Ogirala, S. Negin, C. Agena, C. Schäfer, T. Geisler, J. Mattay and G. W. Gokel, *New J. Chem.*, 2013, **37**, 105.
98. I. Elidrisi, S. Negin, P. V. Bhatt, T. Govender, H. G. Kruger, G. W. Gokel and G. E. M. Maguire, *Org. Biomol. Chem.*, 2011, **9**, 4498.
99. R. Li, O. V. Kulikov and G. W. Gokel, *Chem. Commun.*, 2009, 6092.
100. O. V. Kulikov, S. Negin, N. P. Rath and G. W. Gokel, *Supramol. Chem.*, 2014, **26**, 506.
101. O. V. Kulikov, M. M. Daschbach, C. R. Yamnit and N. P. Rath, *Chem. Commun.*, 2009, 7497.
102. S. Negin, M. M. Daschbach, O. V. Kulikov, N. P. Rath and G. W. Gokel, *J. Am. Chem. Soc.*, 2011, **133**, 3234.
103. M. Barboiu, Y. Le Duc, A. Gilles, P. A. Cazade, M. Michau, Y. M. Legrand, A. van der Lee, B. Coasne, P. Parvizi, J. Post and T. Fyles, *Nat. Commun.*, 2014, **5**, 4142.
104. J. K. W. Chui and T. M. Fyles, *Org. Biomol. Chem.*, 2014, **12**, 3622.
105. T. M. Fyles, *Acc. Chem. Res.*, 2013, **46**, 2847.
106. M. B. Patel, A. Stavri, N. S. Curvey and G. W. Gokel, *Chem. Commun.*, 2014, **50**, 11562.
107. W. M. Leevy, J. E. Huettner, R. Pajewski, P. H. Schlesinger and G. W. Gokel, *J. Am. Chem. Soc.*, 2004, **126**, 15747.
108. W. M. Leevy, S. T. Gammon, T. Levchenko, D. D. Daranciang, O. Murillo, V. Torchilin, D. Piwnica-Worms, J. E. Huettner and G. W. Gokel, *Org. Biomol. Chem.*, 2005, **3**, 3544.
109. W. M. Leevy, M. R. Gokel, G. B. Hughes-Strange, P. H. Schlesinger and G. W. Gokel, *New J. Chem.*, 2005, **29**, 205.

110. Y. Yamamoto, M. Yoshida, M. Sato, K. Sato, S. Kikuchi, H. Sugishita, J. Kuwabara, Y. Matsuno, Y. Kojima, M. Morimoto, A. Horiuchi and Y. Watanabe, *Int. J. Oncol.*, 2011, **38**, 33.
111. B. A. Smith, M. M. Daschbach, S. T. Gammon, S. Xiao, S. E. Chapman, C. Hudson, M. Suckow, D. Piwnica-Worms, G. W. Gokel and W. M. Leevy, *Chem. Commun.*, 2011, **47**, 7977.
112. B. A. Smith, S. T. Gammon, S. Xiao, W. Wang, S. Chapman, R. McDermott, M. A. Suckow, J. R. Johnson, D. Piwnica-Worms, G. W. Gokel, B. D. Smith and W. M. Leevy, *Mol. Pharmaceutics*, 2011, **8**, 583.
113. K. M. DiVittorio, W. M. Leevy, E. J. O'Neil, J. R. Johnson, S. Vakulenko, J. D. Morris, K. D. Rosek, N. Serazin, S. Hilkert, S. Hurley, M. Marquez and B. D. Smith, *ChemBioChem*, 2008, **9**, 286.
114. S. J. Singer and G. L. Nicolson, *Science*, 1972, **175**, 720.
115. L. J. Pike, *J. Lipid Res.*, 2003, **44**, 655.
116. S. Mishra and P. G. Joshi, *J. Neurochem.*, 2007, **103**(Suppl 1), 135.
117. P. S. Niemela, M. T. Hyvonen and I. Vattulainen, *Biochim. Biophys. Acta, Biomembr.*, 2009, **1788**, 122.
118. H. Zhang, Y. Liu, J. Xu, F. Zhang, H. Liang and X. Du, *Neuroscience*, 2013, **254**, 70.
119. M. Ibarguren, D. J. Lopez and P. V. Escriba, *Biochim. Biophys. Acta, Biomembr.*, 2014, **1838**, 1518.
120. A. K. M. M. Hasan, A. Hashimoto, Y. Maekawa, T. Matsumoto, S. Kushima, T. W. Ijiri, Y. Fukami and K. Sato, *Development*, 2014, **141**, 1705.
121. M. Luckey, *Membrane Structural Biology*, Cambridge University Press, Cambridge, UK, 2008.

CHAPTER 4

Synthetic Receptors for Small Organic and Inorganic Anions

STEFAN KUBIK

Department of Chemistry—Organic Chemistry, Kaiserslautern University of Technology, Erwin-Schrödinger-Straße, D-67663 Kaiserslautern, Germany
Email: kubik@chemie.uni-kl.de

4.1 Introduction

Small inorganic anions are typically not considered biomolecules, although biochemistry relies strongly on these species. The phosphate group is indispensable for the chemistry of life, for example, and exists primarily as hydrogenphosphate and dihydrogenphosphate anions or polyanionic anhydrides, often with an additional organic component. Other examples of common biologically relevant anions are chloride, iodide, and carbonate; also relevant are nitrate, cyanide, chromate, and arsenate, anions that can profoundly affect biosystems or biological processes.

These anions represent promising targets for studies that aim to control the biochemical pathways in which they are involved. In addition, it is crucial to monitor the presence of inorganic anions in the environment if they are toxic or otherwise problematic.[1] For example, the detection and selective extraction of nitrate and phosphate anions from wastewater is a way to prevent the eutrophication of rivers or lakes. These types of applications require anion receptors, compounds that associate with a specific anion and form a supramolecular complex. Such receptors act by sequestering an anion, reporting the presence of an anion by changing a physical property,[2–5] or transporting an anion across a lipid membrane.[6–13] For most

Monographs in Supramolecular Chemistry No. 14
Synthetic Receptors for Biomolecules: Design Principles and Applications
Edited by Bradley D. Smith

Published by the Royal Society of Chemistry, www.rsc.org

applications, synthetic anion receptors are more suitable than natural ones because of higher stability, more facile preparation in larger amounts, accessibility to a wider structural space, and the possibility of deliberately controlling the anion binding and/or detection mechanism. The corresponding field of research is nowadays termed anion coordination chemistry.

Guidelines for the design of anion receptors are straightforward to a first approximation. As with receptors for other types of substrates, anion receptors should be complementary in size and shape to the substrate in question and they should arrange suitable functional groups around the binding site. Favorable anion binding interactions include electrostatic attraction with positively charged receptor moieties, hydrogen bonding with donor groups in the receptor (NH, OH groups), or coordination bonds with Lewis acidic groups such as metal centers. Less frequently used, but sometimes rather efficient, are C–H···anion interactions,[14] halogen-bonding,[15,16] and anion–π interactions.[17,18]

Several additional factors need to be considered when designing anion receptors. First, the receptor design has to accommodate the wide diversity of anion structural shapes with geometries including spherical, linear, trigonal planar, tetrahedral, or even more complex shapes. Another aspect relates to the size of anions. Halides, for example, are larger than the corresponding isoelectronic alkali metal ions. As a consequence, they have a smaller charge-over-size ratio, causing their electrostatic interactions with a receptor to be weaker than those between a receptor and an isoelectronic cation. Furthermore, anions usually have higher free energies of hydration than cations with the same absolute charge and comparable size. If anion binding is to occur in water the combined strength of the receptor–anion interactions has to be sufficiently strong to overcome the energy required for release of the water molecules from the anion's hydration shell. Many anions are also involved in protonation equilibria in aqueous solution and receptor molecules have to be active in a pH window in which the anion is not fully protonated.

In solvents with lower dielectric constants anion–receptor interactions are usually stronger, and control over the anion protonation state is less problematic. In addition, anions are not well solvated by aprotic solvents such as acetone, acetonitrile, or dimethyl sulfoxide (DMSO). As a consequence, anion binding is stronger than in protic solvents such as methanol or ethanol.[19,20] Table 4.1 summarizes the properties of the anions treated in this chapter. The table contains the enthalpies of hydration of the respective anions. The absolute values of the Gibbs free energies of hydration are slightly lower, since the entropies of hydration are usually unfavorable.[21]

When only considering a single receptor interaction, smaller and less polarizable anions are bound stronger than larger, more polarizable ones. For example, hydrogen-bonding interactions to chloride anion are stronger than those to iodide. A receptor can overcome this selectivity bias if solvent effects, structural aspects of the receptor (*e.g.* cavity size) and/or a combination of different binding interactions favor the binding of a less strongly

Table 4.1 Ionic radii, enthalpies of solvation, and pK_a values of selected inorganic anions.[21]

Anion	r ppm	$-\Delta H^{\circ}_{hydr}$ kJ mol^{-1}	pK_a of the corresponding acid	Abundance at neutral pH[a] %
F^-	133	510	3.2	100
Cl^-	181	367	−7.0	100
Br^-	196	336	−9.0	100
I^-	220	291	−10.0	100
CN^-	191	493	9.2	2
NO_3^-	179	312	−1.3	100
HCO_3^-	156	384	6.4	92
CO_3^{2-}	178	1397	10.3	0
CH_3COO^-	162	425	4.7	100
$H_2PO_4^-$	200	522	2.2	39
HPO_4^{2-}	200	[b]	7.2	61
PO_4^{3-}	238	2879	12.3	0
$H_2AsO_4^-$	[b]	[b]	2.2	27
$HAsO_4^{2-}$	210	[b]	7.0	73
AsO_4^{3-}	248	[b]	11.5	0
HSO_4^{2-}	190	[b]	−3.0	0
SO_4^{2-}	230	1035	2.0	100
ClO_4^-	240	246	−10.0	100

[a]Relative amount of the anion in water at pH 7.4 based on its pK_a value. Other species present under these conditions are the protonated and/or deprotonated forms of the anion.
[b]Not available.

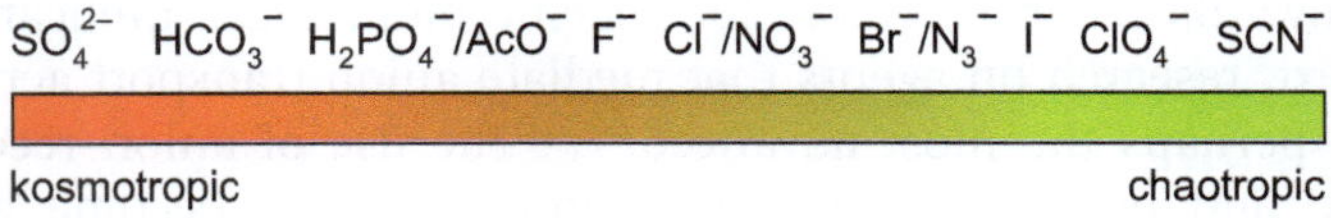

Figure 4.1 Typical Hofmeister series for a selection of common inorganic anions. Note that the order depends on the experiment used to determine the anion effect, on the counterion, and on concentration, so that anions can sometimes exchange position in this series.

coordinating anion. Interestingly, the intrinsic trend of anion selectivity generally follows the so-called Hofmeister series. This series was established in 1888 by the physiological chemist Franz Hofmeister who investigated the ability of salts to cause protein precipitation. He showed that strongly hydrated anions such as fluoride and sulfate decrease the solubility of a protein, while weakly solvated anions such as perchlorate and thiocyanate have the opposite effect. A related series can be established for cations, although the anion effect is more pronounced. Figure 4.1 depicts the Hofmeister series for a selection of inorganic anions.

The Hofmeister series has historically been associated with the ability of ions to modulate the hydrogen-bonding network of bulk water. Anions such as sulfate anions were considered to be structure makers (kosmotropes) and anions at the opposite end of the Hofmeister series structure breakers (chaotropes). Kosmotropes stabilize protein folding and therefore lead to the salting-out of a protein, whereas chaotropes destabilize protein structures, leading to the molten globule state, and consequently to a salting-in effect. Although the terms kosmotropes and chaotropes can still be found in the literature, it is generally accepted that ion properties in solution and their interactions with organic molecules are significantly more complex and the distinction of ions into structure makers and structure breakers is an over-simplification.[22,23]

In the context of anion recognition it is important to note that larger, weakly coordinating anions have an affinity for concave hydrophobic cavities. For example, iodide and perchlorate are known to bind inside α-cyclodextrin, and the perchlorate anion has been shown to associate with a water-soluble, resorcarene-based cavitand.[24] This may explain why chaotropic anions destabilize protein structures: the association of chaotropic anions with a hydrophobic surface in water diminishes the propensity of the surface to engage in hydrophobic interactions. The large exothermicity observed in some cases for anion binding to hydrophobic hosts suggests that the release of high-energy water from the receptor cavities could be the driving force for complex formation.[24]

Anion coordination chemistry has advanced considerably in the last 20–30 years and has provided many examples of selective and strong anion recognition. As a consequence, receptors for almost every relevant anion are known today, some of which possess remarkable affinity and selectivity, even in water and in the presence of competing anions. Increasingly, attention is turning towards anion receptors for specific biomedical applications.[25] In this context, research on agents that mediate anion transport across membranes is perhaps the most advanced.[6–13] The use of anion receptors for diagnostic and environmental applications has also become a realistic prospect.

Since this chapter aims to provide an overview of general principles and approaches in the field of anion coordination chemistry, it cannot be comprehensive. For detailed information about developments in the area the reader is referred to monographs[26–28] and numerous reviews, which are cited throughout this chapter. A valuable source of information is the series on recent developments in anion coordination chemistry, which is updated at least every 2 years (for the most recent review in this series, see Gale *et al.*[29]).

4.2 Strategies for Anion Coordination

4.2.1 Electrostatic Interactions Combined with Hydrogen Bonding

It is of historic interest that the first synthetic anion receptor described by Park and Simmons,[30] macrobicyclic receptor **4.1** (Chart 4.1), belongs to the family of receptors binding their substrate by a combination of electrostatic

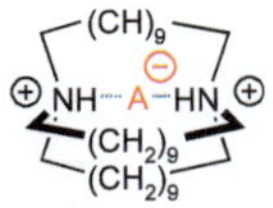

4.1

Chart 4.1

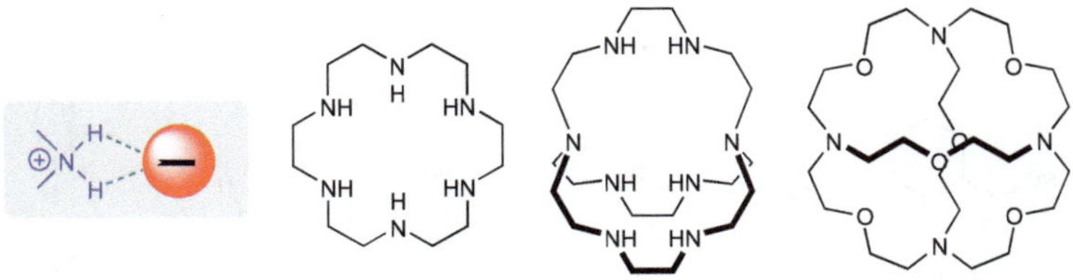

4.2 4.3 4.4

Chart 4.2

interactions and hydrogen bonding. Anion affinity of **4.1** is, however, very weak. For example, the chloride complex has a K_a of 4 M^{-1} in 50% aqueous deuteriotrifluoroacetic acid. Still, **4.1** has served well as a valuable prototype for many subsequently developed anion receptors.

The logical strategy to improve anion affinity involved systematic arrangement of more than two ammonium groups around a macrocyclic (**4.2**), macrobicyclic (**4.3**), or macrotricyclic cavity (**4.4**).[31–34] Examples of corresponding polyamines are shown in Chart 4.2, along with a schematic representation of the underlying binding interaction.

Binding studies using amine-based hosts are usually performed in aqueous solution by means of potentiometry to correlate the degree of protonation with complex stability. In general, anion affinity increases with the number of protonated amine groups, *i.e.* with the increase of positive charge on the receptor. Thus, electrostatic interactions greatly influence binding strength, although contributions from directed hydrogen bonds are also important, as evidenced by crystal structures of complexes between polyammonium receptors and anions.[31] A correlation exists between anion affinity and the size complementarity of macrocyclic polyammonium receptors and their anionic guests. Yet this correlation usually does not translate into substantial differences in binding constants because the flexibility of macrocyclic hosts allows them to adapt to guests of different sizes and shapes. Because of their higher structural rigidity and better defined cavities, macrobicyclic or macrotricyclic hosts such as **4.3** or **4.4** exhibit superior binding selectivities in comparison to macrocyclic analogs like **4.2**.

Guanidinium and amidinium groups also combine electrostatic interactions with hydrogen bonding for anion coordination.[35–37] In contrast to amines, both groups remain protonated over a wide pH range, including physiological pH, because their pK_a values typically range between 11 and 13. Compounds **4.5**, **4.6**, and **4.7** in Chart 4.3 are examples of receptors containing these binding motifs.

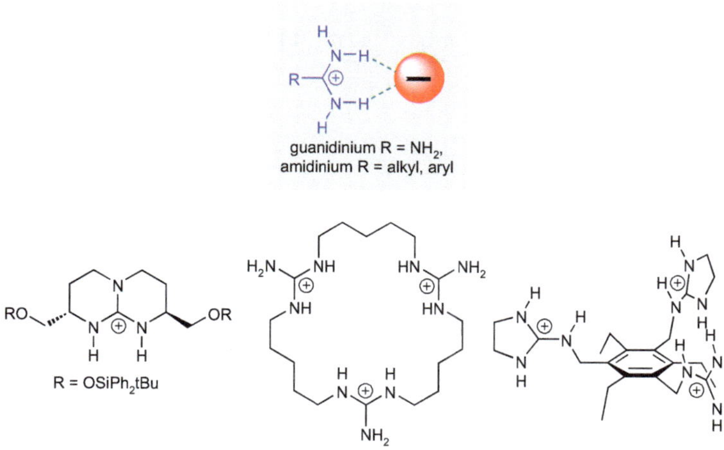

4.5 4.6 4.7

Chart 4.3

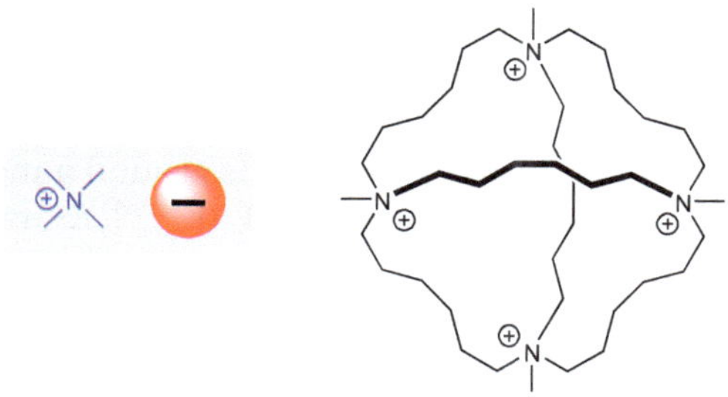

4.8

Chart 4.4

A disadvantage of monosubstituted guanidinium groups is that hydrogen-bonding interactions are not restricted to one side of the planar cation. Nature solves this problem of directionality by using secondary interactions to constrain the guanidinium group of an arginine residue within a protein binding site so that it only points two NH groups towards a guest anion. In synthetic receptors, the directionality problem is circumvented by using conformationally restricted bicyclic guanidinium residues, such as **4.5**.

Another disadvantage of guanidinium ions is strong solvation, which greatly weakens ion pairing with carboxylates or phosphates in aqueous solution ($K_a < 5\ M^{-1}$). Receptors therefore must combine two or more guanidinium moieties or other anion binding motifs to be active in aqueous solution.

4.2.2 Electrostatic Interactions

Receptors containing quaternary ammonium groups, such as macrotricyclic **4.8** (Chart 4.4), have the advantage that the total number of positive charges is independent of pH. However, the absence of hydrogen bond donors on the nitrogen atoms means that anion complexation has to rely solely on electrostatic interactions.

Receptor **4.8** has a sufficiently large cavity to include anions, as demonstrated by the crystal structure of the iodide complex.[38] However, anion affinity in water is significantly weaker when compared to polyammonium hosts with protonated amino groups,[39] indicating that charge-assisted hydrogen bonding enhances anion/receptor electrostatic interactions.

4.2.3 Hydrogen Bonding

Receptors containing amide groups mimic the coordination of anions to backbone or side-chain NH groups in proteins.[40–42] Shown in Chart 4.5 is a schematic representation of the underlying binding mode. Note that the strength of the NH hydrogen bond is amplified by an electrostatic interaction with the positive end of the amide dipole moment. Also shown are prototype receptors, with macrocyclic ones, such as **4.9**, again being generally less selective than cryptand-like analogs such as **4.10.**[43]

Such neutral receptors are generally only soluble in organic media and binding studies are therefore typically performed in DMSO, acetonitrile, acetone, or chloroform. The choice of the solvent has a strong impact on binding strength, with more stable complexes usually being observed in less polar solvents. In addition to solvent polarity, expressed quantitatively in terms of the dielectric constant ε, the solvent molecular properties also influence complex stability. DMSO, for example, does not solvate anions very well, but strongly interacts with the receptor hydrogen bond donors. Anion binding to these donors therefore requires (at least partial) displacement of DMSO molecules. This receptor desolvation penalty is much less pronounced in non-hydrogen bonding solvents, such as chloroform.

Another structural element that is widely used for anion recognition is the urea group.[44,45] More recently, squaramides have also gained importance.[46] The two adjacent NH residues in ureas pair with oxoanions *via* two parallel hydrogen bonds, but they also associate well with spherical halides. The ease of urea synthesis is a major technical advantage and provides easy access to

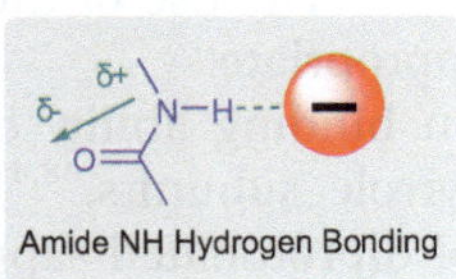

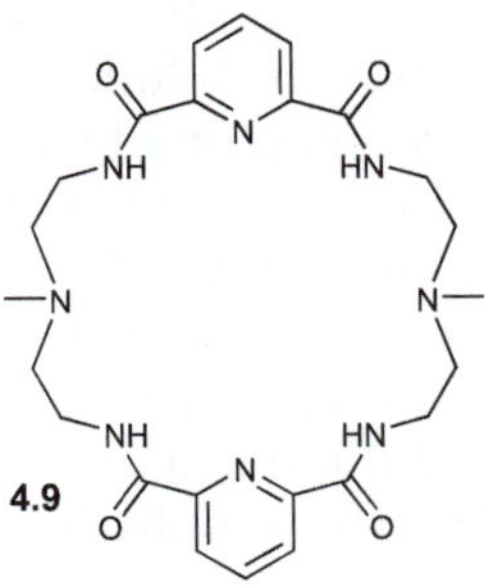

4.9

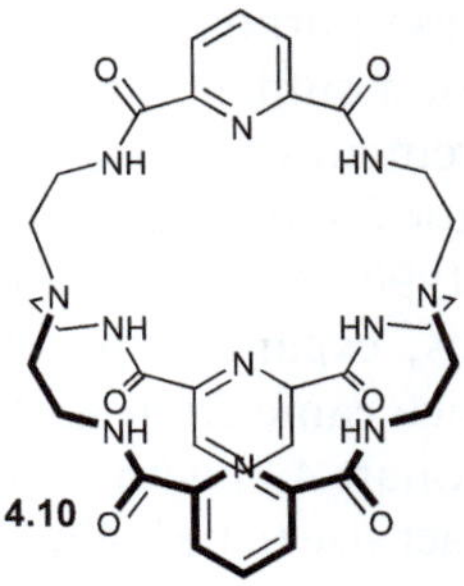

4.10

Chart 4.5

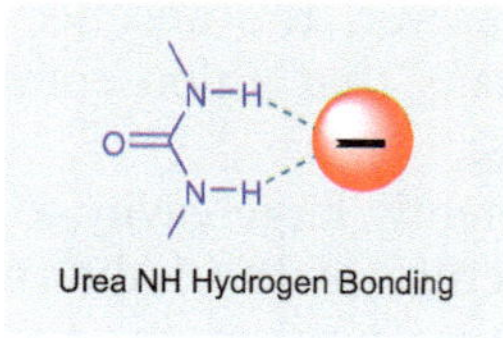

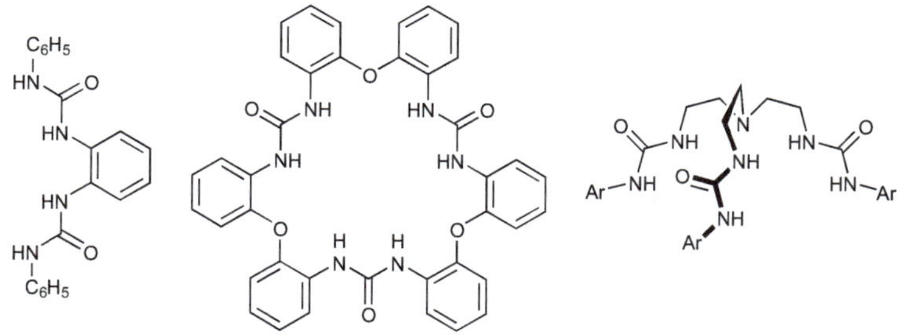

4.11 **4.12** **4.13a** Ar = pentafluorophenyl
4.13b Ar = 3-cyanophenyl

Chart 4.6

wide range of receptors, including acyclic (**4.11**), cyclic (**4.12**), and tripodal (**4.13**) structures containing more than one urea moiety (Chart 4.6).

The propensity of urea NH groups to interact with anions can be fine-tuned by the substituents on the nitrogen atoms. In general, electron-withdrawing groups, such as electron-poor aromatic rings increase NH acidity, and thus improve the ability of NH groups to act as hydrogen bond donors. However, a potential caveat is that increased acidification can lead to proton transfer to basic anions, especially in polar organic solvents such as DMSO or acetonitrile. Fluoride in particular causes deprotonation of acidic NH groups due to the formation of the stable $[HF_2]^-$ anion.[47] Stepwise processes are sometimes observed involving formation of an initial hydrogen-bonded complex followed by receptor deprotonation once fluoride is added in excess. Since deprotonation often results in profound changes in the receptor's ultraviolet/visual spectrum it can allow optical detection of anions in organic media. However, it has to be stressed that the ureas used for these purposes simply act as acid/base indicators that report the presence of a sufficiently strong base in solution, with no or only a secondary involvement of a molecular recognition phenomenon. The use of the term receptor for such urea derivatives is therefore inappropriate.

In 1990 the Sessler group discovered the anion binding ability of sapphyrins, expanded porphyrins containing five pyrrole subunits.[48,49] First evidence came from an X-ray structural analysis of diprotonated sapphyrin **4.14** (Chart 4.7) with a bound fluoride anion. While the fluoride anion has the exact size to be included into the sapphyrin cavity, giving rise to a coplanar arrangement of the anion and the heterocyclic receptor, larger anions such as chloride, carboxylates, or phosphate were shown to perch on top of the

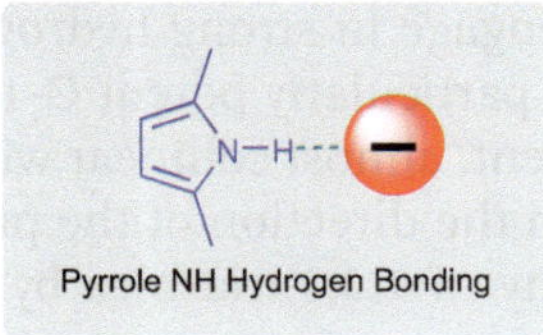

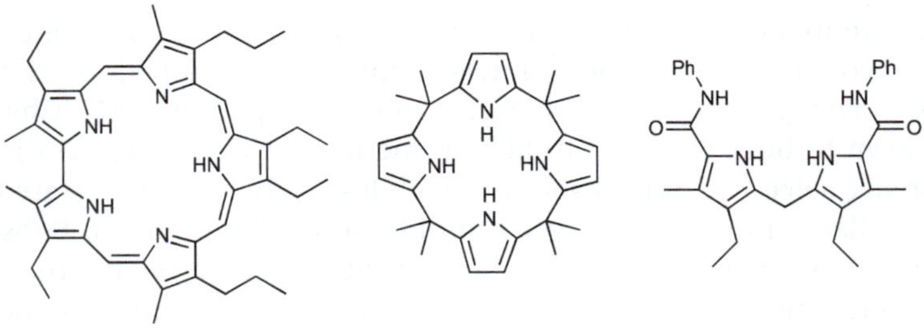

4.14 **4.15** **4.16**

Chart 4.7

sapphyrin ring. Anion coordination is also observed in solution, although the propensity of sapphyrins to self-aggregate complicates binding studies. Larger sapphyrin derivatives or heterosapphyrins containing other heterocyclic subunits possess significantly lower anion affinity than sapphyrin itself.[49]

A neutral macrocyclic tetrapyrrole that has found widespread use in anion coordination chemistry is the so-called calix[4]pyrrole **4.15** (Chart 4.7).[50] The name derives from the resemblance of this compound to calixarenes, not only in terms of structure but also in its conformational behavior. In solution, **4.15** preferentially adopts a conformation comparable to the 1,3-alternate conformation of calix[4]arenes. Upon anion binding, a conformational reorganization occurs, resulting in the cone conformation with all NH groups converging in one direction to allow their simultaneous interaction with an anion.

Advantages of calixpyrroles are the possibility of controlling anion affinity by stabilizing their conformations, by introducing additional binding sites, or by modulating the electronic properties of the heterocyclic subunits. Examples of corresponding receptors are calixpyrroles with a strap between two opposing aliphatic linking units[51] or ones with fluorinated pyrrole units.[52]

Interestingly, acyclic receptors such as dipyrrolylmethane **4.16** also possess substantial anion affinity.[49] Recently, related indole and indolocarbazole-based receptors have emerged as potent anion binders, whose binding benefits from the relatively high acidity of the NH group.[53] Longer chain oligoindolocarbazoles were shown to fold around anions producing helical structures with the guest inside the central cavity.[54] A significant practical advantage of these acyclic systems is their generally easier synthetic accessibility in comparison to their macrocyclic counterparts.

While all of the above anion receptors use acidic N–H or O–H groups for hydrogen bonding, it was recently shown that C–H groups, in particular ones in 1,2,3-triazoles, can engage in strong hydrogen-bonding interactions with an anion. Triazoles are particularly potent C–H hydrogen donors because of their large dipole moment, almost collinear with the C–H bond and with the positive end oriented in the direction of the proton.[14] A synthetic advantage is the facile accessibility of 1,2,3-triazoles by means of copper(I) catalyzed azide–alkyne cycloaddition.

The potential of 1,2,3-triazoles as anion binding sites is illustrated by the anion recognition properties of triazolophanes developed in the Flood group.[14] The prototype of these receptors, compound **4.17** (Chart 4.8), was shown to bind chloride in dichloromethane with a K_a of 1.1×10^7 M^{-1}. A clear size selectivity was observed; much less stable complexes are formed with smaller fluoride and larger bromide and iodide anions. Subsequent investigations involving systematic structural changes in the parent macrocycle uncovered a rich array of anion coordination properties.[14] Furthermore, the general concept was extended by demonstrating that the C–H group of cyanostilbenes also serves as a hydrogen bond donor for anion coordination when incorporated into a C_5 symmetric macrocycle.[55]

In addition to macrocyclic receptors, acyclic oligotriazoles have also been shown to bind anions *via* C–H···anion interactions.[54] An example is **4.18** (Chart 4.8), which was designed to adopt a helical arrangement with all the triazole C–H bonds directed towards a guest anion within the central cavity.

An unusual example of anion binding based on aliphatic C–H hydrogen bonds can be found in the cyclic glycoluril **4.19** (Chart 4.9), the so-called bambus[6]uril.[56] This macrocycle adopts a conformation with the individual 2,4-dimethylglycoluril moieties pointing upward or downward in an

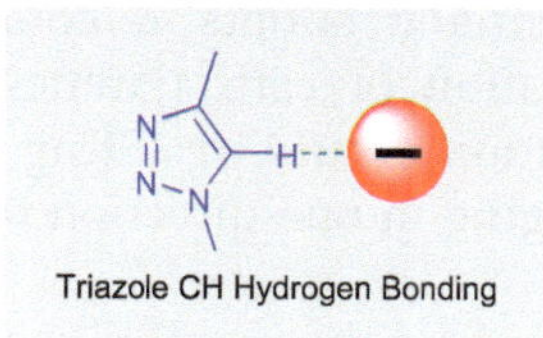

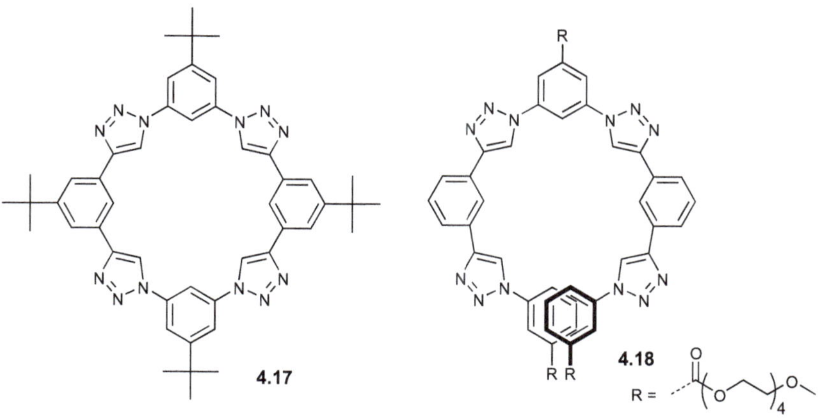

Chart 4.8

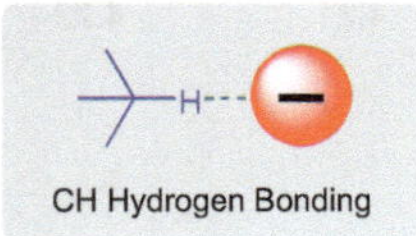

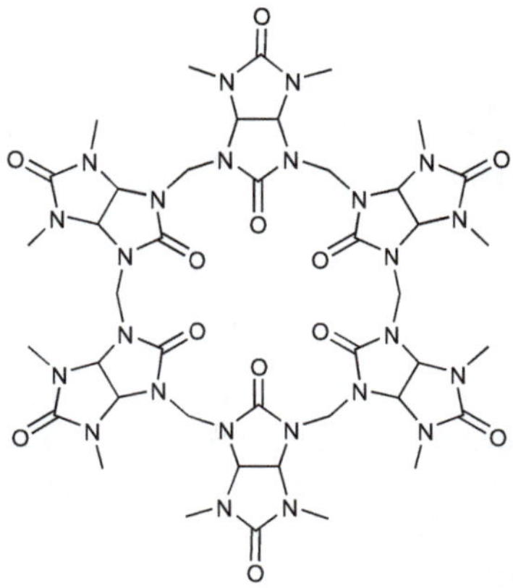

4.19

Chart 4.9

alternating fashion. As a result, the receptor cavity is lined by glycoluril C–H protons producing a substantial positive electrostatic potential at the cavity's inner surface, ideally suited to bind anions. The cavity size best matches the iodide anion, which is bound in acetonitrile/water, 1 : 1 (*v*/*v*) with an association constant of $8.9\times10^5\ M^{-1}$, quite high for a neutral receptor.[56]

4.2.4 Lewis Acid–Base Coordination

A general and versatile strategy to recognize anions relies on their Lewis basic character that allows specific and strong coordination to a suitable Lewis acid binding partner. Two types of Lewis acids are frequently used in the field of anion coordination chemistry: either boron in the form of organoboron compounds or transition metal ions. Advantages of Lewis acid–base coordination are the high degree of bond directionality combined with a bond strength that often allows anion binding to be efficient in water.

Anion coordination to the sp^2 boron center in trisubstituted organoboron compounds involves donation of an electron pair into the empty p orbital of the boron atom (dative bonding).[57] Because of the relatively weak Lewis acidity of boron, hard anions such as cyanide, hydroxide, or fluoride are particularly suited to bind to boron-based receptors, with most work concentrating on the fluoride anion.[58]

After initial reports demonstrating the feasibility of using organoboron compounds for fluoride recognition, *e.g.* the structural characterization of the potassium fluoride complex of crown ether **4.20** (Chart 4.10), extensive efforts were directed at the development of organoboron-derived fluoride probes.[57,58] For example, it was shown that fluoride anions quench the fluorescence of boronic acid **4.21** in buffered aqueous methanol solution at pH 5.5. As a consequence, fluoride can be detected and quantified in a concentration range between 50 mM and 3 mM. Alternatively, the ferrocene-derived boronic acid **4.22** allows electrochemical fluoride detection in water

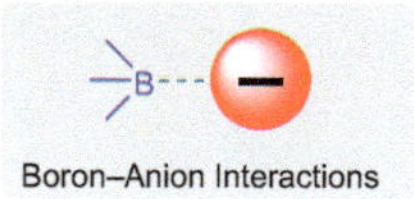

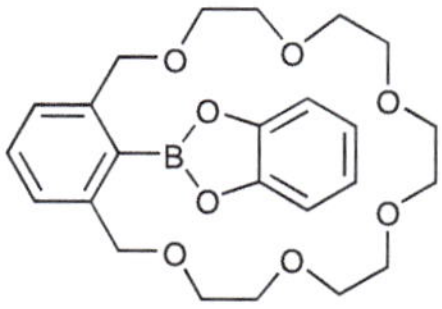

4.20

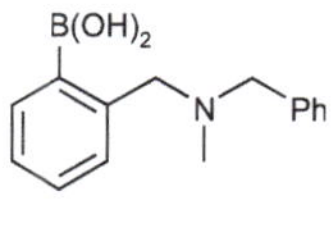

4.21

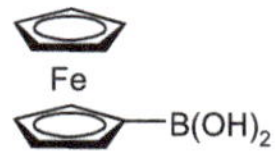

4.22

Chart 4.10

or water/methanol mixtures. Practically no interference by other anions was observed, with the exception of hydroxide anions that appear at high pH. More recently, antimony-based compounds were shown to be useful as fluoride receptors.[59]

An alternative way to bind anions to Lewis acid receptors involves the use of transition metal cations. It is worth mentioning that many anion receptors use metal cations as organizing elements, for example, to connect building blocks with appended anion binding sites, thus facilitating their cooperative action for anion recognition.[60,61] Here, the focus is only on receptors in which the metal centers interact directly with the anionic guest.[62–64]

Metal cations that are frequently used as Lewis acids for anion coordination are lanthanides(III), copper(II), or zinc(II). The unique luminescence properties of receptors containing lanthanide cations make them excellent anion detection probes.[65,66] Copper(II) is redox-active, which facilitates binding studies using potentiometric, absorption, or fluorescence methods, but it is not conducive for studies by nuclear magnetic resonance (NMR) spectroscopy, due to line broadening. In contrast, the binding properties of receptors with zinc centers can be readily investigated by NMR spectroscopy. Anion receptors containing transition metal cations as binding sites usually feature a molecular scaffold that can form a chelate-type complex with the metal center leaving at least one open coordination site for anion recognition. Examples of such hosts are shown in Chart 4.11.

An early example of a metal-containing anion receptor is the dicopper(II) cryptate **4.23**. In this compound, each tren subunit coordinates to a Cu^{2+} ion in a trigonal bipyramidal binding mode leaving one axial position available for anion binding. The anion incorporates between the two copper(II) centers and coordinates simultaneously with both metals. As a consequence, association favors anions whose size and shape allows them to bridge the metal ions (see Section 4.4.3). Mononuclear complexes have also been used for anion binding. In the trigonal complex **4.24** the metal serves as both a binding site and a structural organizational element to preorganize the positively charged ammonium groups so they can engage in secondary interactions with the anionic guest. Other type of metal complexes used for anion binding in aqueous solution are based on zinc(II)-cyclen complexes, *e.g.* **4.25**, or zinc(II)-dipicolylamine complexes,

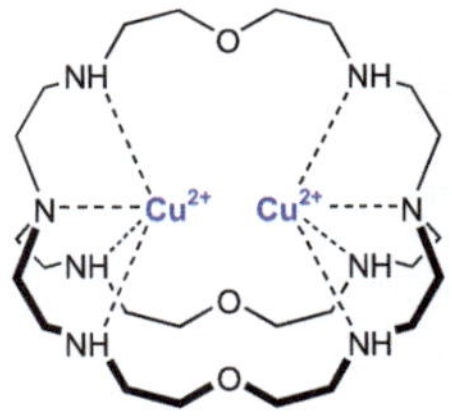

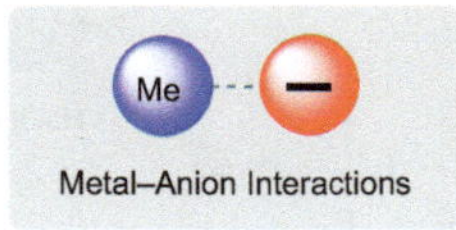

4.23

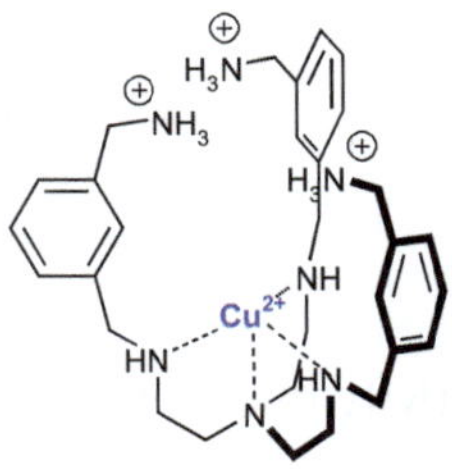

4.24

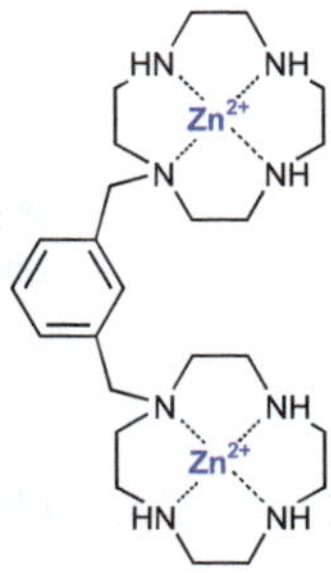

4.25

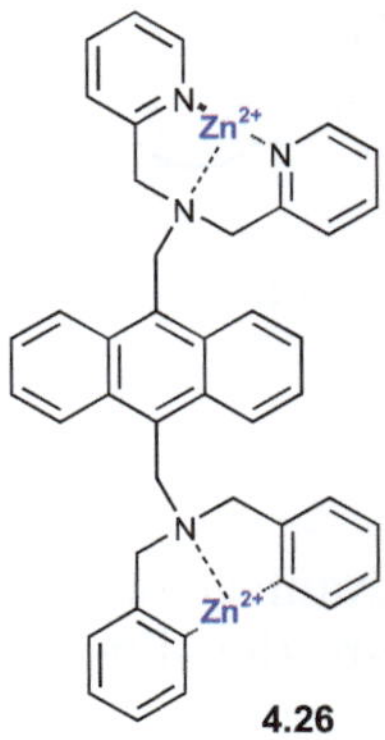

4.26

Chart 4.11

e.g. **4.26**. Both types of complexes exhibit very useful phosphate recognition properties.

4.2.5 Halogen Bonding

While chlorine, bromine, and iodine atoms in halocarbon molecules are usually considered to be electron-rich substituents, the distribution of electron density around these atoms is highly anisotropic, with an electropositive region located on the terminus of the carbon–halogen bond (σ-hole). Electrostatic interactions between an electron-donating species, such as an anion, and the electropositive region of a halogen substituent is termed halogen bonding.[67] While there is extensive evidence for the importance of halogen bonding in crystals, there are still very few examples utilizing this interaction for anion recognition in solution.[68] Selected receptors are shown in Chart 4.12.

The tripodal receptor **4.27** was found to selectively associate with halides (as their tetrabutylammonium salts) in acetone with binding constants for chloride up to 1.9×10^4 M^{-1} while only negligible binding of oxoanions was detected under the same conditions.[69] Receptors **4.28a,b** were shown to interact with iodide and bromide anions in CD_3OD/D_2O (9 : 1) with affinities ranging between 2.9×10^4 and 9.6×10^5 M^{-1}.[70] The bis(imidazolium) derivative **4.29** was used to obtain information about the thermodynamics of charge-assisted halogen bonding in acetonitrile.[71] Further research on halogen bonding for anion recognition is warranted since it has the

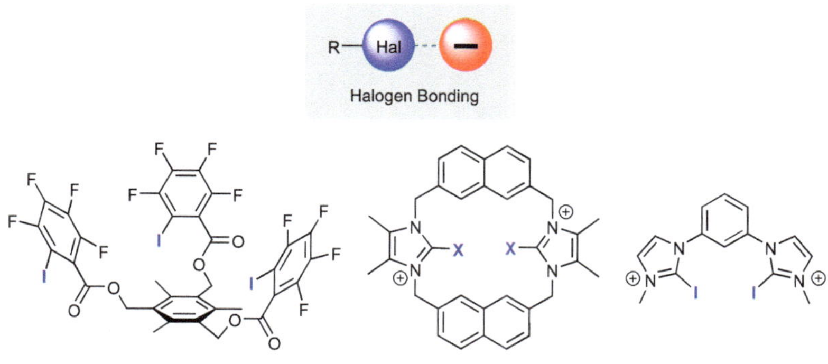

4.27 | 4.28a X = Br, 4.28b X = I | 4.29

Chart 4.12

potential advantage of excellent anion selectivity due to the very high halogen bond directionality.

4.2.6 Anion–π Interactions

Anion–π interactions are electrostatic interactions between anions and the facial surfaces of electron-poor aromatic rings.[17,18] There is extensive evidence from crystal structures and theory that this type of intermolecular interaction can make profound contributions to the stabilization of host–guest systems in the solid state or the gas phase. However, experimental proof for anion–π interactions operating in solution is rare. One of the first receptors whose anion affinity was partly attributed to anion–π interactions is compound **4.30b** (Chart 4.13).[18] While sulfonamide **4.30a** exhibits practically no affinity for anions in $CDCl_3$, **4.30b** forms weak complexes with halides in the same solvent ($K_a = 20$–$34\ M^{-1}$). The affinity difference cannot be rationalized solely by the lower pK_a value of the sulfonamide N–H group in **4.30b** and therefore is evidence for a through-space contribution of the pentafluorophenyl group to anion binding.

A similar experimental comparison of the tripodal receptors **4.31a** and **4.31b** showed that **4.31b** with electron-poor aromatic residues forms stronger complexes with halides, a result that is consistent with anion–π interactions.[18] However, the possibility of aromatic $CH\cdots X^-$ interactions makes it hard to precisely delineate the individual contributions to complex stabilization.

The recently described calix[4]pyrrole derivative **4.32** binds anions primarily by hydrogen bond formation to the NH groups analogously to the parent compound **4.15** (Chart 4.7).[72] A systematic characterization of halide complexation, using derivatives of **4.32** with different electron-withdrawing substituents on the aromatic residues flanking the binding site, demonstrated that binding becomes progressively more attractive as the

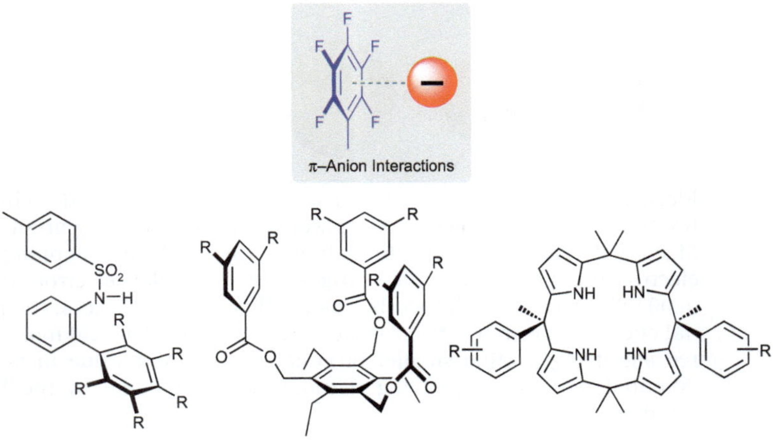

4.30a R = H
4.30b R = F

4.31a R = H
4.31b R = NO_2

4.32

Chart 4.13

number and electron-withdrawing character of the aromatic substituents increases. Reinforcement of anion binding caused by these substituents was attributed to anion–π interactions and their electrostatic nature. The contribution to increased complex stability was estimated to be in the range −2.9 to −4.2 kJ mol^{-1}.

4.3 Spherical Anions

4.3.1 Anion Properties

The four singly charged halides fluoride, chloride, bromide, and iodide mainly differ in size and electron density. Both effects are directly evident when comparing the electrostatic potential (EP) surfaces of these anions, shown in Figure 4.2. To facilitate comparison, the color code is consistent with the EP surfaces of other anions in Figures 4.6, 4.8, and 4.10. The color range from red to blue covers the EP range between −850 and −200 kcal mol^{-1}, with blue also signifying potentials above the upper limit. Figure 4.2 shows that the increase in size of the halides from fluoride to iodide results in a significant reduction of the absolute value of the negative potential, which causes iodide to be the least strongly coordinating anion in this series. The significant decrease of the hydration enthalpy from fluoride to iodide reflects this trend (Table 4.1).

Another important aspect that needs to be taken into account when comparing the halides is their different basicity (Table 4.1). While all halides are completely deprotonated in aqueous solution, basicity decreases from fluoride to iodide by ~13 orders of magnitude. The relatively high basicity of the fluoride anion becomes pertinent when the receptor contains acidic

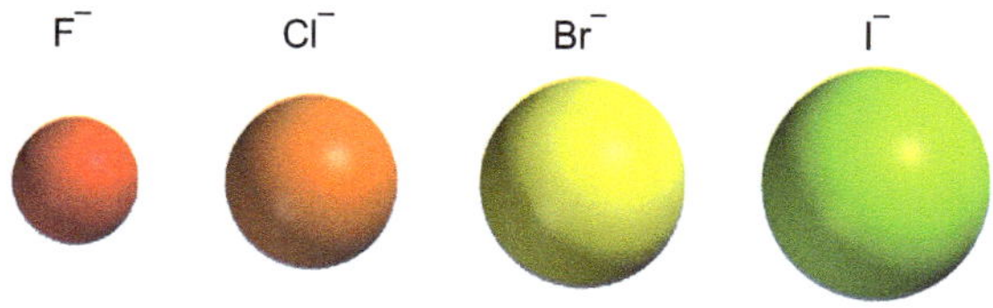

Figure 4.2 Electrostatic potential surfaces of the four halides fluoride, chloride, bromide and iodide. The surfaces were generated using Spartan 10 for Macintosh (Wavefunction, Inc., Irvine, CA, USA) by mapping AM1 electrostatic potentials onto surfaces of molecular electron density (0.002 electron Å^{-1}) followed by color-coding. In all surfaces, the potential energy values range from −850 kcal mol^{-1} to −200 kcal mol^{-1}, with red signifying a value smaller or equal to the lower value in negative potential and blue signifying a value larger or equal to the higher number in negative potential.

protons, particularly in polar aprotic organic media where deprotonation of the receptor is associated with the formation of the stable $[HF_2]^-$ anion (see Section 4.2.3).[47]

4.3.2 Recognition in Nature

All four halides are present in biological systems albeit at significantly different levels.[73] Fluoride in low concentrations has beneficial effects on the formation of dental enamel and bones. Toothpastes and table salts therefore often contain sources of fluoride such as sodium fluoride. In the USA, the UK, and Australia, water fluoridation, *i.e.* the addition of fluoride into public water supplies, is used to elicit the beneficial effects of fluoride on human health. There is, however, increasing evidence that several types of pathologies in humans, such as osteoporosis, neurological and metabolic dysfunctions, and even cancer can be caused by exposure to fluoride in too high concentrations. Direct adverse effects of fluoride anions on biochemical processes like DNA synthesis, activities of certain enzymes, and the metabolism of nutrients have also been reported, explaining the growing concern about fluoride consumption or contamination of the environment.[74]

A structurally diverse family of proteins, the so-called chloride ion channels are specifically responsible for the selective transport of chloride anions across cell membranes. Processes associated with these channel proteins include the stabilization of the membrane resting potential and the regulation of electrical excitability. Crystal structures of two chloride ion channels from *Escherichia coli* and *Salmonella typhimurium* revealed that chloride selectivity is accomplished at the narrowest part of the channels, the so-called ion filter, where two backbone NH groups and two OH groups, one from Ser107 and one from Tyr445, interact with the anion (Figure 4.3).[75]

The genetic disease cystic fibrosis is caused by malfunction of the cystic fibrosis transmembrane conductance regulator (CFTR), a chloride channel protein located in epithelial cells of the lung and other organs. This

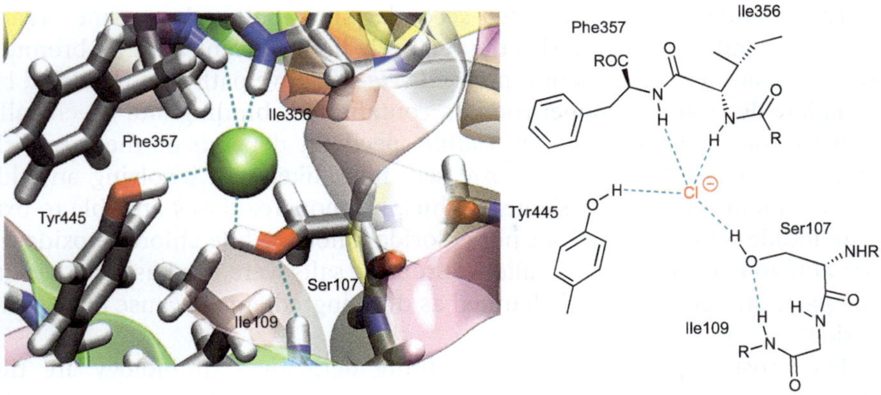

Figure 4.3 Binding arrangement of a chloride anion inside the narrowest pore region of the chloride ion channel from *Salmonella typhimurium* (left) and schematic representation of the underlying interactions (right). The picture was generated using Visual Molecular Dynamics from the structure deposited in the Research Collaboratory for Structural Bioinformatics Protein Data Bank (code 1KPL).[75]

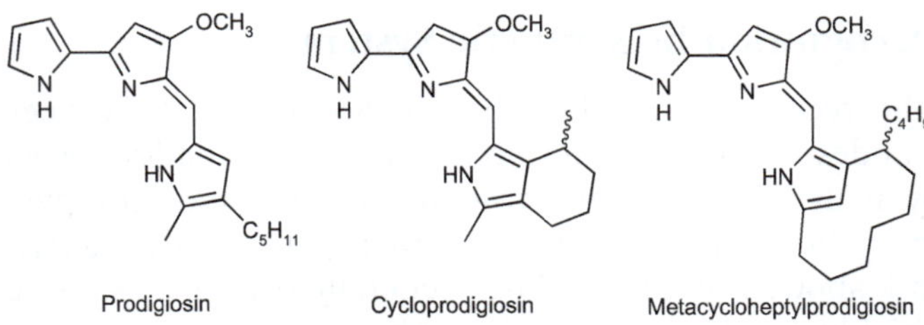

Figure 4.4 Examples of prodigiosins.

multifunctional transport protein controls the flow of water and chloride ions in and out of the cell. When CFTR does not work properly, insufficient chloride and fluid transport causes the buildup of thick mucus in the lungs leading in turn to a pronounced susceptibility towards fatal pulmonary infections. Molecular strategies that can restore or replace membrane chloride transport function may be possible therapeutic approaches to "channelopathies" like cystic fibrosis.

The best known, naturally occurring, small molecules with chloride recognition ability are the prodigiosins (Figure 4.4), a series of pyrrole alkaloids produced by micro-organisms such as *Streptomyces* and *Serratia*.[76] Prodigiosins have been pursued as lead compounds for anticancer and immunosuppression therapies. Some of their diverse biological activities, such as uncoupling of proton translocation and inhibition of vacuolar acidification, are attributed to the ability of these compounds to act as membrane symporters for H^+ and Cl^-.[77] Transport involves ion-pairing of a protonated prodigiosin molecule and the chloride counterion.

The abundance of brominated secondary metabolites in marine organisms is usually associated with the relatively high concentration of bromide salts in seawater. Bromination proceeds *via* oxidative pathways mediated by vanadate-dependent haloperoxidases containing a binding site specifically tailored for the vanadate anion. Halide specificity depends on the ease with which the respective halide is oxidized, most likely not involving an additional anion recognition element. Thus, bromoperoxidases are able to oxidize iodide and bromide, but not chloride anions, while chloroperoxidases are able to oxidize all three halides. Bromide salts were used as sedatives in the past, but are now widely banned as drug ingredients because of chronic toxicity.

The most important iodine-containing derivatives in biology are the thyroid hormones thyroxine and tri-iodothyronine, which both contain iodinated aryl rings. Biosynthesis of these compounds requires iodide; lack of iodide in the diet therefore usually results in dysfunction of the thyroid gland. To prevent the associated health problems, table salt is usually complemented by small amounts of sodium iodide. Seafood is also a good source of iodide.

4.3.3 Recognition by Synthetic Systems

Most of the receptor types described in Section 4.2 have been used for binding halides. Structural information underlying halide recognition was derived by analyzing the crystal structures of the halide complexes of acyclic,[78] macrocyclic (**4.33a–c**)[31] or macrobicyclic polyamines (**4.34**) (Chart 4.14).

This work showed that the halides generally behave as spherical charges having a highly variable coordination sphere. Anion coordination relies on N–H···X^- hydrogen bonds and N^+···X^- electrostatic interactions, sometimes with additional contributions from C–H···X^- hydrogen bonds. The number of possible N–H···X^- hydrogen bonding interactions increases with the size of the anion, and, in general, a pyramidal coordination geometry is observed with acyclic or monocyclic polyammonium receptors leaving one side of the anion exposed. This coordination geometry is largely determined by the conformation and the protonation sites of the receptor, the complex stoichiometry, and the size and electronegativity of the anion.[31] Steed and

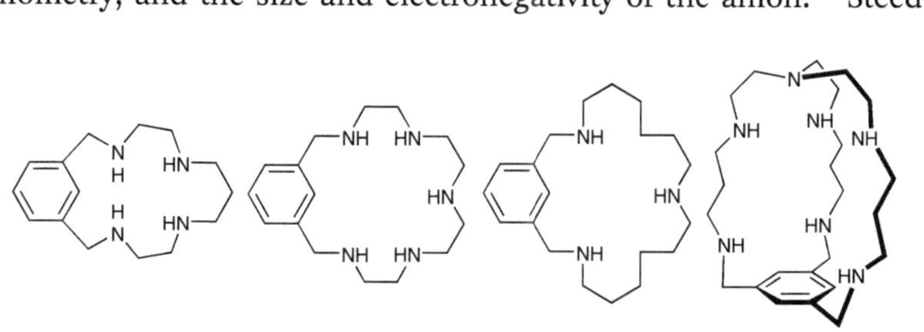

4.33a 4.33b 4.33c 4.34

Chart 4.14

coworkers have concluded that "halides should be considered as species that are able to participate in hydrogen bond networks in a variety of geometries, depending on the features of the polyprotonated environment that are to be bonded".[78] This statement is underscored by the fact that halides can be completely incorporated into the cavities of suitable receptors, such as azacryptands, or the more rigid derivative **4.34** where the encapsulated halide is surrounded by hydrogen bond donors from all sides (Chart 4.14).

An important criterion for the performance of halide receptors is binding selectivity. Not surprisingly, it is easier to achieve selective binding of fluoride than the other halides. Fluoride is the smallest halide and binds preferentially to a receptor cavity that is too small for larger anions. Azacryptand **4.3** (Chart 4.2) and the bicyclic lactam **4.10** (Chart 4.5) are good examples of this behavior. Fluoride affinity of the hexaprotonated form of **4.3** is an impressive $\log K_a$ of 10.55, while the next larger chloride anion is only bound with a $\log K_a$ of <2, resulting in a fluoride over chloride selectivity that is greater than eight orders of magnitude.[43] Fluoride affinity of **4.10** is weaker than that of **4.3** because it does not involve ion pairing, but this receptor also binds fluoride ≥ 20 times stronger than chloride.[43] A crystal structure confirmed that **4.10** incorporates the fluoride anion in the center of the macrotricyclic cavity where it interacts with six converging NH groups. This structure persists in solution, as demonstrated by ^{19}F NMR spectroscopy.

Increasing the cavity size of such macrobicyclic systems generally causes the expected shift of binding selectivity towards larger halides, but neither binding affinity nor selectivity of the corresponding receptors remain as high as those of **4.3** or **4.10**. One reason is the increased flexibility of the expanded receptors, which allows them to accommodate anions of different sizes. The other aspect to consider is the decrease in electron density from fluoride to iodide, which leads to intrinsically weaker receptor–anion interactions.

Based on these considerations, the most obvious receptor design for the weakly coordinating iodide anion involves the use of rigid macrocycles enclosing a cavity of suitable size. The rigid structure prevents the smaller anions from interacting simultaneously with all binding sites inside the cavity. Compounds **4.35**, **4.36**, **4.37** (Chart 4.15), and the bambus[6]uril **4.19** (Chart 4.9) are examples of such compounds.

Receptor **4.35** binds chloride and iodide anions in $CDCl_3$ (containing 2% DMSO-d_6), but coordination of iodide is significantly stronger.[79] Iodide complexation involves formation of 2 : 1 complexes in which the anion resides in a spacious cavity between two molecules of **4.35**, where it can efficiently interact with all six NH groups (Figure 4.5).

Similar behavior was observed for receptor **4.36**, which is structurally related to triazolophane **4.17** (Chart 4.8).[80] The cavity size of **4.36** is actually optimal for complexation of the smaller halides and, indeed, fluoride, chloride, and bromide bind to **4.36** in a 1 : 1 fashion. However, the corresponding complexes are less stable than those of **4.17** because of

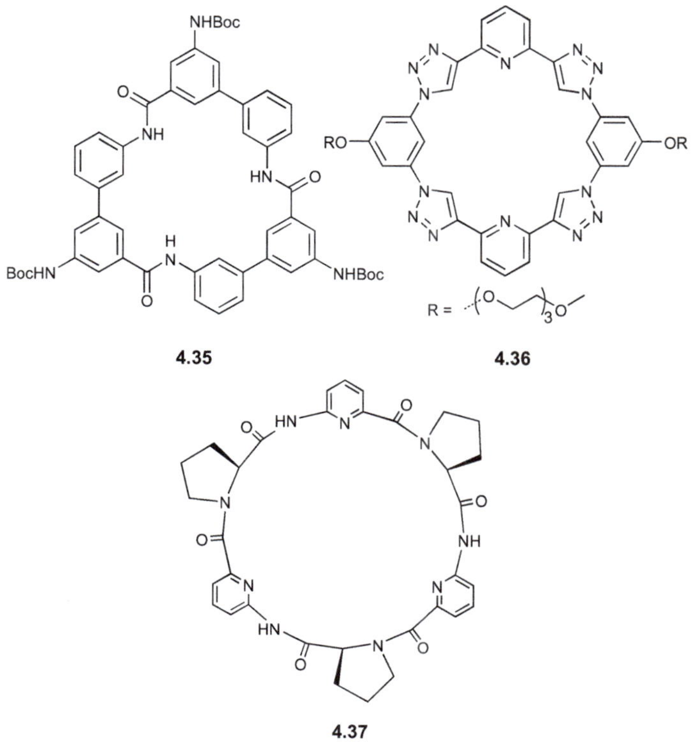

Chart 4.15

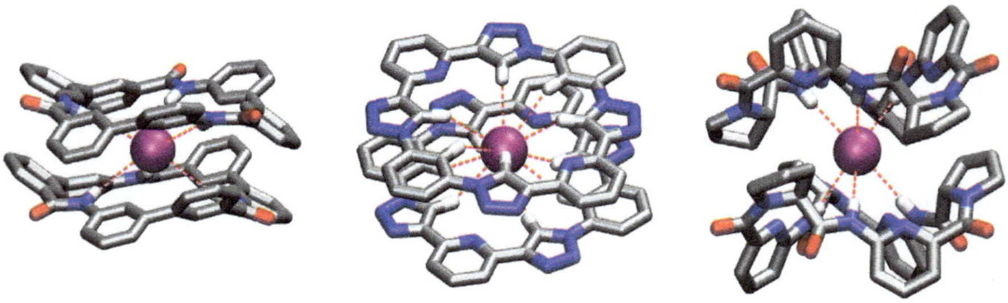

Figure 4.5 Calculated structures of the sandwich type iodide complexes of receptors **4.35** (left)[79] and **4.36**,[80] and crystal structure of the corresponding complex of **4.37** (right).[81] Calculations were performed by the author by using Spartan 10 and the Merck molecular force-field. Hydrogen atoms except those on the NH groups are omitted for reasons of clarity.

repulsive interactions between the pyridyl nitrogen lone pairs within **4.36** and the anionic guests. These destabilizing effects are attenuated in a 2 : 1 sandwich complex (Figure 4.5). The combined advantages of sandwich complex formation coupled with the inability of iodide to fit inside the cavity of **4.36** leads to a 2 : 1 complex with relatively high stability.

The cyclic hexapeptide, **4.37** containing an alternating sequence of L-proline and 6-aminopicolinic acid subunits also forms sandwich-type complexes with halides (Figure 4.5).[81] Complex stabilities increase in the order chloride < bromide < iodide with selectivity for iodide again being due to the rigidity of the peptide and the size match of the cavity between two cyclopeptide rings and the largest halide.

A notable difference in the behavior of receptors **4.35** and **4.37** concerns the medium in which complex formation takes place. Both receptors interact with iodide in DMSO, but under these conditions only weakly stable 1 : 1 complexes are formed. The stability constant for the 1 : 1 iodide complex of **4.37** amounts to 150 M^{-1} in DMSO-d_6, while iodide affinity of **4.35** was estimated to be $K_a < 10$ M^{-1}. In the case of **4.35**, reduction of solvent polarity and concomitant strengthening of the direct anion–receptor interactions leads to a 2 : 1 sandwich complex. In remarkable contrast, the propensity of cyclopeptide **4.37** to form a sandwich complex is enhanced by *increasing* solvent polarity. For example, a sandwich-type complex is formed in aqueous solvent mixtures such as water/methanol, 4 : 1 (*v*/*v*). In this case, 2 : 1 complex formation is attributed to hydrophobic interactions between the nonpolar surfaces of the two cyclopeptide rings that come in close contact (Figure 4.5).[19] The special nature of hydrophobic interactions, becoming more efficient with rising water content, explains the ability of **4.37** to bind anions in highly competitive media. Figure 4.5 suggests that other receptors might exhibit the same hydrophobic receptor stacking process, but suitably designed water-soluble derivatives are required to test this concept.

One way to circumvent the problem of poor anion binding ability in aqueous solution is to incorporate the receptor into the lipophilic core of micelles. The potential of this approach was demonstrated for fluoride recognition, which is particularly difficult in water because of the strong solvation of fluoride anions (Table 4.1). The receptor used in this work was the salophen-UO_2 complex **4.38** (Chart 4.16), known to bind hard Lewis bases by coordinative interactions in organic solvents and in the solid state. Complex **4.38** is not water soluble, but it can be solubilized in micellar solutions composed of 50 mM aqueous cetyltrimethylammonium bromide (CTAB).[82] NMR relaxivity and nuclear Overhauser effect measurements showed that under these conditions **4.38** is preferentially located at the interfaces of the CTAB micelles. Therefore, anion-binding properties of **4.38** are retained, allowing the receptor to bind to fluoride anions with an affinity of 10 800 M^{-1}. Other concepts for the development of fluoride-selective receptors and probes have been reviewed.[74,82,83]

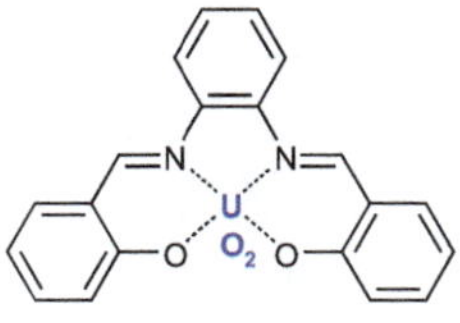

4.38

Chart 4.16

Due to the biological importance of chloride transport across bilayer membranes, significant effort has been directed towards the development of synthetic chloride transporters. This area of research has been extensively reviewed in the past,[6–13] and only a few chloride transporters are presented here.

The Davis group performed systematic investigations on the anion-binding properties of bile acid derivatives containing hydrogen bond donor sites, such as urea, thiourea, sulfonamide, carbamate, trifluoroacetamide, and isophthalamide moieties on the concave side of the steroid scaffold.[84,85] Advantages of these so-called "cholapods" are their inherent lipophilicity and the highly convergent, well preorganized arrangement of binding sites, leading to an efficient shielding of a bound anion from the surrounding medium. As a consequence, extremely high anion affinity was observed for some derivatives in organic media. For example, the chloride complex of **4.39** (Chart 4.17) has an impressive stability of $K_a = 1.8 \times 10^{11}\ M^{-1}$ in water-saturated chloroform. In collaboration with the Smith group, the cholapods were shown to act as transport carriers for chloride anions across vesicle membranes.[85] The anion transport proceeds *via* an antiport mechanism and, in general, transport efficiency correlates with chloride affinity. In a related study it was shown that cholapods also facilitate the translocation of phospholipid head-groups through bilayer membranes ("flip-flop").[6,85]

Inspired by the ability of prodigiosins (Figure 4.4) to promote the co-transport of H^+/Cl^- across cell membranes, several groups investigated whether analogs of these compounds also act as anion transporters.[10] In this context, other natural products were considered, such as tambjamine derivatives **4.40** (Chart 4.17) or entirely synthetic systems such as isophthalamides, thioureas, or tripodal trisureas.[12] Many of these carriers transport anions by an antiport mechanism involving chloride/bicarbonate or chloride/nitrate exchange. The results from numerous investigations have yielded a better understanding of the transport mechanisms, and they will facilitate future development of new carriers. Important aspects that have to

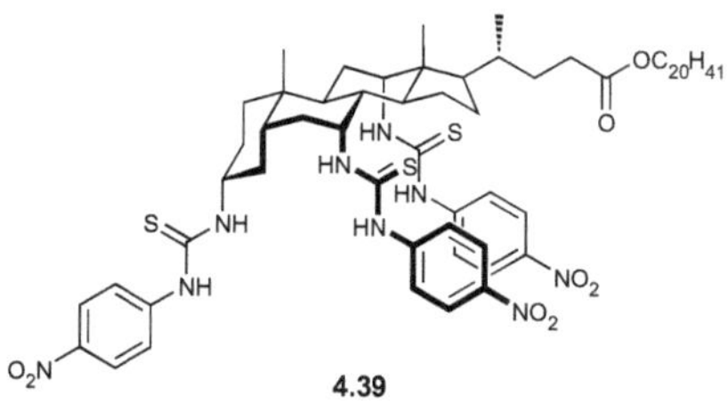

4.39

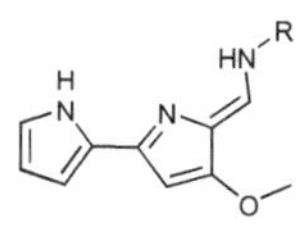

4.40

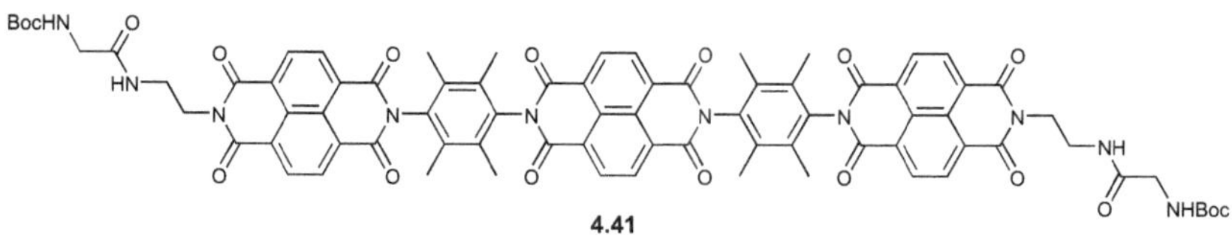

4.41

Chart 4.17

be considered in this respect are carrier lipophilicity and structural flexibility. Systematic investigations involving derivatives of tambjamine **4.40** revealed that transport properties peak when the compound has an optimal lipophilicity, usually expressed quantitatively in terms of the octanol/water partition coefficients log *P*.[12] Compounds with relatively polar structures partition inefficiently into the membrane while highly nonpolar derivatives reside in the lipophilic center of the bilayer and do not reach the chloride anions at the aqueous membrane interface. Recently, transport studies involving trisurea receptors based on a cyclohexane scaffold have shown that increased receptor flexibility leads to reduced chloride binding affinity (as expected), but also improved transport efficiency due to faster complexation and decomplexation rates.[86]

In addition to carrier mediated transport there has been a notable effort to promote anion transport across bilayer membranes by forming ion channels. Examples of synthetic channel systems are the oligo(*p*-phenylene)-*N*,*N*-naphthalenediimide (O-NDI) and oligo(*p*-phenylene)-*N*,*N*-perylenediimide rods studied by the Matile group.[11] These rigid rod-like, electron-deficient aromatic systems, termed anion-π slides, have been designed to span bilayer membranes. They mediate antiport of anions; O-NDI rod **4.41** (Chart 4.17), for example, preferentially transports chloride over fluoride, bromide, and iodide anions. The hypothesis that anion transport is guided by anion–π interactions is supported by the results of a systematic investigation correlating anion affinity with transport properties for a series of perylenediimide derivatives.

Gokel and co-workers investigated a series of amphiphilic heptapeptides of the general structure $(R_2)NCO–X–CO–(Aaa)_n–OR'$ for their ability to form chloride channels within lipid bilayers.[7] The compounds contain four subunits, namely the N-terminal anchor R, the connector chain X, the actual peptide, and the C-terminal anchor R′. A series of systematic structural studies that varied the individual subunits in these peptides revealed the mechanism of anion transport, structural aspects important for efficient transport, and the anion interacting subunits in the peptide part of these amphiphiles.

4.4 Linear Anions

4.4.1 Anion Properties

The EP surfaces of the cyanide and the azide anion are shown in Figure 4.6. The symmetrical electronic structure of the azide anion requires both terminal nitrogen atoms to possess the same negative potential, while the cyanide anion is asymmetric with the higher negative potential on the more electronegative nitrogen atom. The other important difference is anion basicity. While the azide anion is fully deprotonated at physiological pH, cyanide anions are abundant only in basic solution. At pH 7.4, the dominant species is hydrogen cyanide (Table 4.1).

Figure 4.6 Electrostatic potential surfaces of the azide (left) and the cyanide (right) anions. (See Figure 4.2 caption for description of surface generation.)

Of these two linear anions, only cyanide is found in biological systems. It is produced by cyanogenic micro-organisms and it exists as cyanohydrin groups on cyanogenic glycosides found in higher plants such as sorghum, flax, giant taro, bamboo, or cassava. Cyanogenic glycosides are also found in almonds or apricot kernels. Excessive consumption can lead to toxic concentrations of cyanide, formed by enzymatic cleavage of the cyanogenic glycosides, and cause acute and chronic health problems.[87] A particular concern is improperly prepared cassava, which is an important carbohydrate source for a large fraction of the population in South America and parts of Africa.

4.4.2 Recognition in Nature

Cyanide inhibits cytochrome *c* oxidase, which is the last enzyme in the respiratory chains of mitochondria and aerobic bacteria and catalyzes the reduction of molecular oxygen to water. Cytochrome *c* oxidase contains two copper centers, Cu_A and Cu_B, as well as two heme A subunits, *a* and a_3. In the oxidized enzyme, heme a_3 and Cu_B form a binuclear center in which oxygen and its reduced intermediates are held until the catalytic cycle is completed. Coordination of cyanide to these metal centers inhibits an essential step in the oxidative phosphorylation leading to asphyxia and eventually to death. The structure of fully reduced bovine cytochrome *c* oxidase containing a bound cyanide ion (Figure 4.7) illustrates the mode of cyanide complexation.[88] The cyanide anion coordinates to the iron center of the heme unit *via* the carbon atom and to the copper ion *via* the nitrogen atom, displacing one imidazole group. Other linear anions, including azide, behave similarly and are also highly toxic. This is the reason why sodium azide is often used as a preservative in aqueous laboratory reagents and biologic fluids.

4.4.3 Recognition by Synthetic Systems

Ideal receptors for linear anions feature an oblong shape with a cavity containing two binding sites at the apexes. Very efficient binding of azide anion was achieved using the bimetallic complexes of azacryptands, the so-called cascade complexes. For example, dicopper(II) cryptate **4.42** (Chart 4.18) was shown to interact with various anions such as N_3^-, OCN^-, SCN^-, SO_4^{2-}, $HCOO^-$, CH_3COO^-, HCO_3^-, and NO_3^-.[64] Complex formation

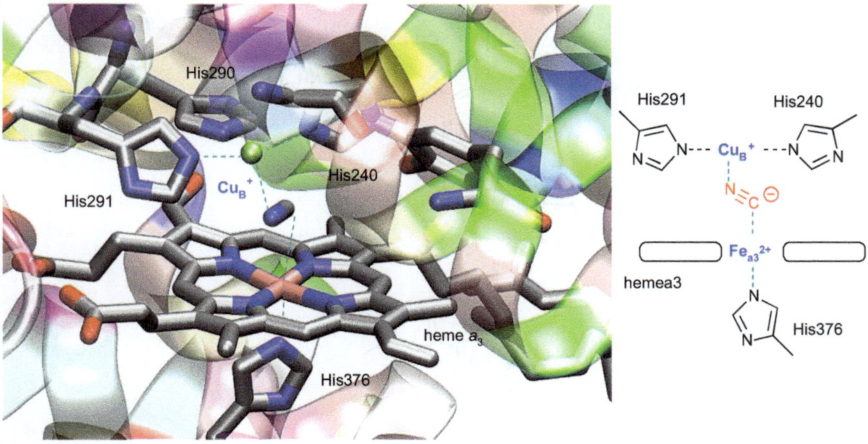

Figure 4.7 Coordination of the cyanide anion to the heme a_3 and Cu_B centers of bovine cytochrome *c* oxidase (left) and schematic representation of the underlying interactions (right). The picture was generated using Visual Molecular Dynamics from the structure deposited in the Research Collaboratory for Structural Bioinformatics Protein Data Bank (code 3AG4).[88]

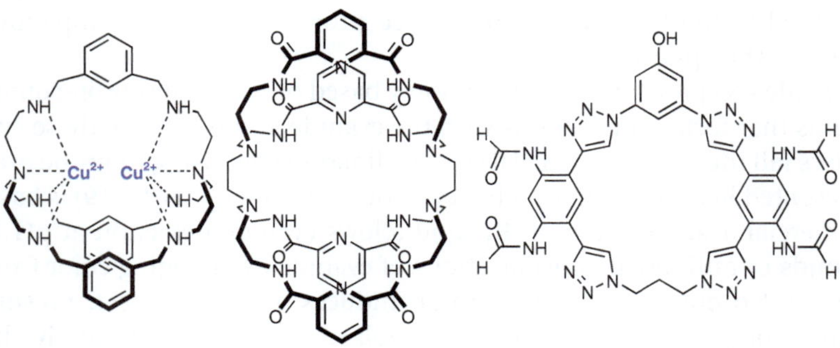

4.42 **4.43** **4.44**

Chart 4.18

can easily be detected by the color change of an aqueous solution of this host from green to blue. The most stable complex is formed with azide ($\log K_a = 4.75$) which fits almost perfectly between the two copper centers.

Extending the cavity diameter of the neutral hydrogen-bonding receptors **4.10** (Chart 4.5) and **4.17** (Chart 4.8) in one direction yields the corresponding analogs **4.43**[89] and **4.44**[90] (Chart 4.18) and shifts the binding preference from spherical fluoride to the linear $[HF_2]^-$ anion.

One of few receptors without metal centers to bind cyanide is a derivative of triazolophane **4.17** (Chart 4.8) (containing $(OCH_2CH_2)_3OCH_3$ units instead of the *tert*-butyl groups).[91] Computational investigations indicate preferential binding of the linear cyanide anion along one axis of the

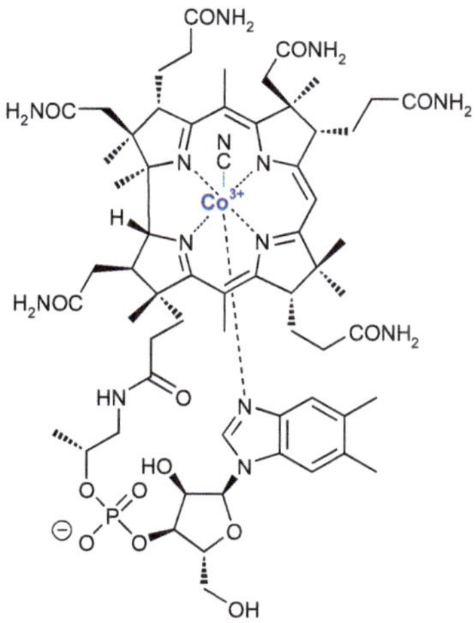

4.45

Chart 4.19

macrocyclic cavity. Binding studies in CH_2Cl_2 showed that the receptor formed 1:1 complexes with chloride or cyanide anion with exactly the same stability constants ($K_a = 2.8 \times 10^6$ M^{-1}). That cyanide is bound with the same affinity as chloride, in spite of the higher basicity of the former, was attributed to the high rigidity of the receptor and its more complementary shape for the spherical halide.

Cyanide receptors are more commonly based on metal- or boron-containing systems that form dative bonds with the cyanide anion.[87] Since these cyanide binders fall into the realm of classic coordination chemistry, only one example is presented here. The colorimetric cyanide probe **4.45** (Chart 4.19) is based on commercial available vitamin B_{12} and allows cyanide detection at pH 7.5. It responds to millimolar concentrations of cyanide by changing color from red to violet. No other investigated anion induces a similar effect, presumably because only cyanide is able to replace the intramolecularly bound dimethylbenzimidazole ligand from the metal center leading to a complex containing two coordinated cyanide anions. Spectrophotometric assays using this receptor allow detection of cyanide in blood[92] and quantification of the total cyanogenic glycosides in fresh and processed cassava products.[93]

It is worth noting that many indicators for cyanide have been developed whose mode of action involves a color change induced by chemical reaction with the anion. Although highly useful, this so-called chemodosimeter approach is beyond the scope of this chapter.[94]

4.5 Trigonal Planar Anions

4.5.1 Anion Properties

EP surfaces of the two trigonal planar anions, nitrate and carbonate, are shown in Figure 4.8. There is significantly less negative potential on the

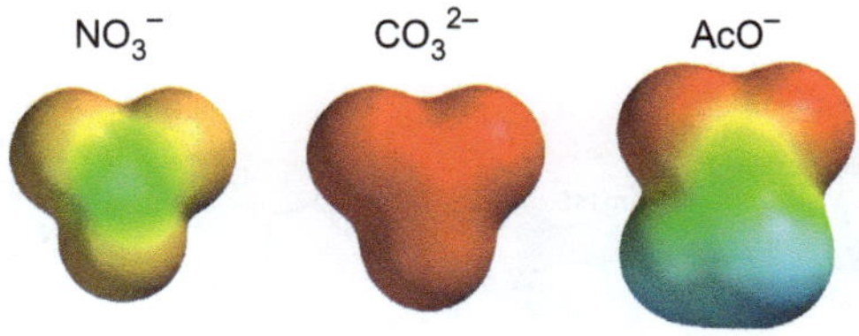

Figure 4.8 Electrostatic potential surfaces of the (left–right) nitrate, carbonate, and acetate anions. (See Figure 4.2 caption for description of surface generation.)

oxygen atoms of the nitrate anion in comparison to carbonate, due to the higher electronegativity of nitrogen in comparison to carbon and the lower charge of the nitrate anion. Nitrate is therefore more difficult to bind than carbonate. The carbonate dianion is, however, more basic and exists only at high pH. Table 4.1 shows that the predominant species at physiological pH is the hydrogencarbonate anion, HCO_3^-. Figure 4.8 also depicts the EP surface of the acetate anion as a representative of carboxylates, which are treated in this section because of their structural relationship to carbonate. Carboxylates are typically deprotonated at neutral pH and the two negatively charged oxygen atoms can interact with anion binding sites on receptors. The organic residue, which could contain additional functionalities including other carboxylate groups, provides an additional handle for binding and selective recognition.

4.5.2 Recognition in Nature

There is no major biological process involving nitrate anions; however, nitrate salts serve as an important nitrogen source in fertilizers. Contamination of water runoff from over-fertilized fields can cause eutrophication of rivers and lakes, *i.e.* an initial excess of algal blooms followed by the depletion of oxygen in the water and the breakdown of the ecosystem. Carbonate anions, generated biochemically by the enzyme carboanhydrase from CO_2, are important components of the blood buffer and of the salts that make up the endoskeleton of vertebrates. The acetyl group is the essential component of acetyl-coenzyme A. Carboxylates, in general, can be found in biological systems in the form of fatty acids or the various intermediates of the Krebs cycle.

An important binding motif for carboxylate groups in proteins, also found in the phosphate-binding protein (see Section 4.6.2), involves the formation of a salt bridge reinforced by two parallel hydrogen bonds between the carboxylate oxygen atoms and two NH groups of a protonated arginine guanidinium moiety. This mode of binding is utilized, for example, by carboxypeptidase A, the enzyme responsible for the hydrolytic cleavage of a C-terminal hydrophobic amino acid from a polypeptide chain. It is illustrated in Figure 4.9 by the crystal structure of carboxypeptidase A with bound L-*N*-(hydroxyaminocarbonyl)phenylalanine.[95]

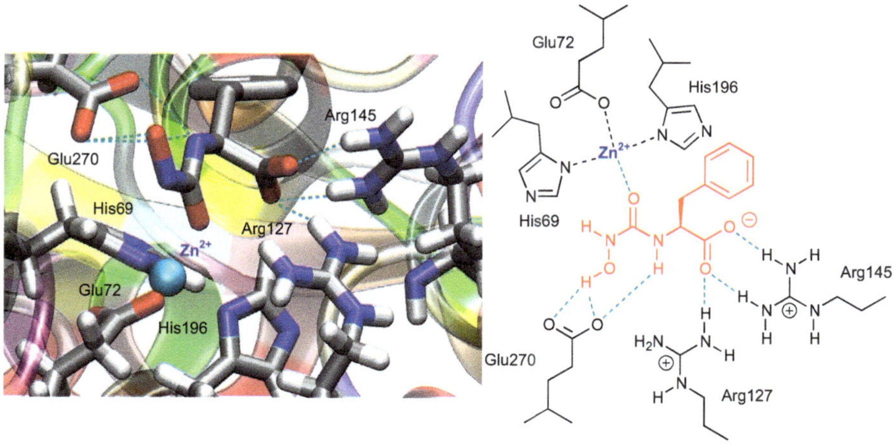

Figure 4.9 Arrangement of L-*N*-(hydroxyaminocarbonyl)phenylalanine inside the binding site of carboxypeptidase A (left) and schematic representation of the underlying interactions (right). The picture was generated using Visual Molecular Dynamics from the structure deposited in the Research Collaboratory for Structural Bioinformatics Protein Data Bank (code 1HEE).[95]

4.5.3 Recognition by Synthetic Systems

In light of the environmental importance of nitrate as one cause for freshwater eutrophication, considerable effort has been devoted to the development of nitrate-selective receptors. To achieve selective extraction from water, a receptor has to recognize weakly coordinating nitrate anions in preference to other ubiquitous anions such as chloride, sulfate, or phosphate. Information about the preferred binding geometry of nitrate anions was obtained by characterizing the complexes of receptor **4.46** (Chart 4.20).[96] This receptor binds nitrate salts of alkali metal cations as ion pairs. Crystal structures showed that the complex structure depends on a subtle interplay of steric factors, coordination to the metal cation, and hydrogen bonding with the receptor NH residues. Importantly, this work provided evidence that hydrogen bonds directed towards the oxygen lone pairs of the nitrate anion are stronger than hydrogen bonds to the π-electrons.

Computations suggested that trisurea **4.47** (Chart 4.20) should represent a promising nitrate receptor,[97] but affinity for this anion turned out to be weak ($K_a = 17\ M^{-1}$ in DMSO-d_6) and unselective. Other anions, such as acetate, cyanide, and bromide are bound significantly stronger because **4.47** is flexible enough to easily adapt to different anion shapes.

Greater success at selective nitrate recognition has been achieved by using receptors with rigidified structures. For example, the C_3 symmetrical cyclophane **4.48** (Chart 4.20) preferentially interacts with planar anions, but the nitrate complex of **4.48** with a K_a of 300 M^{-1} is less stable in CD_2Cl_2/CD_3CN, 1 : 3 (v/v) than the acetate complex ($K_a = 770\ M^{-1}$).[98] Thus, receptor selectivity is mainly governed by anion basicity.

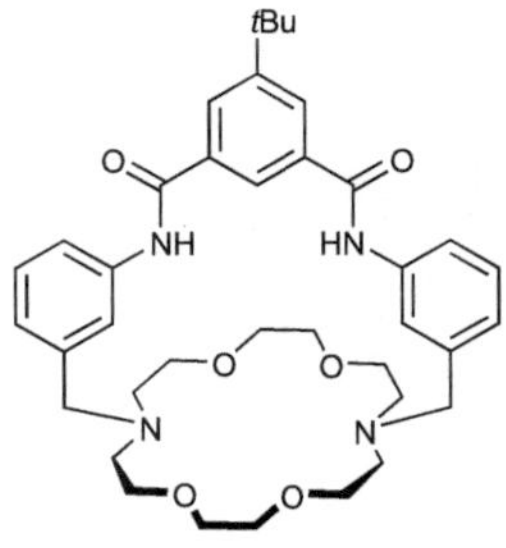

4.46

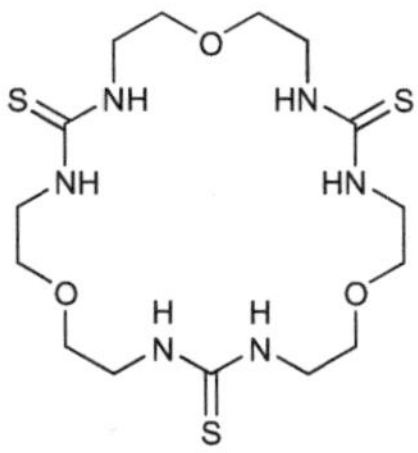

4.47

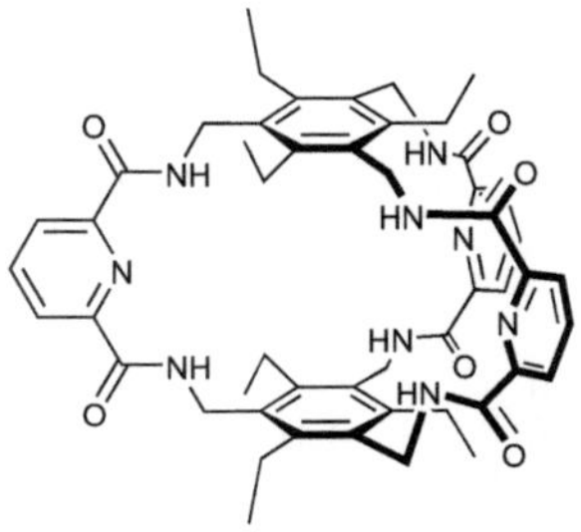

4.48

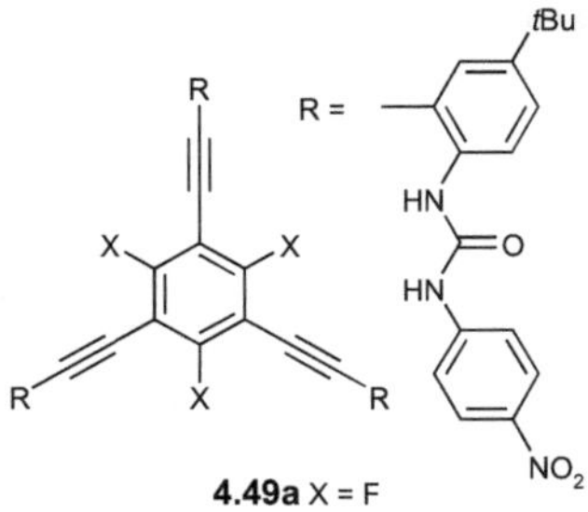

4.49a X = F
4.49b X = H

Chart 4.20

Tripodal trisurea **4.49a** (Chart 4.20) is a rare example of a nitrate-selective receptor. This compound was observed to bind to a nitrate anion in 10% DMSO-d_6/$CDCl_3$ with a K_a of 24 100 M^{-1}, while the next stable chloride complex has a K_a of only 12 200 M^{-1}. The unusually high nitrate binding constant was attributed to cooperative contributions including anion–π interactions between the central aromatic unit of **4.49a** and the planar oxoanion. This assumption was supported by the significantly weaker nitrate binding exhibited by **4.49b**, whose structure lacks fluorine substituents in the aromatic core. These results suggest that anion–π interactions might be an effective design strategy for nitrate-selective receptors.

Some carbonate receptors possess the unusual feature of being able to sequester carbon dioxide from the atmosphere. Trapping involves either non-covalent complexation of a carbonic acid species or covalent bond formation and creation of a carbamic acid derivative. Practical application of these compounds as large-scale sequestration agents seems unlikely, but the results do show how a receptor can shift a chemical equilibrium by stabilizing a reactive species. In the case of receptor **4.13a** (Chart 4.6), formation of the carbonate complex was observed by slowly evaporating a DMSO solution containing an additional equivalent of tetrabutylammonium hydroxide in an open vessel.[99] Compounds **4.50**[100] and **4.51**[101] (Chart 4.21) were shown to crystallize with carbonate after treatment with tetrabutylammonium fluoride in DMSO. The copper complex **4.52** fixes CO_2 covalently by formation of a carbamic acid at one of the aliphatic amine groups.[102]

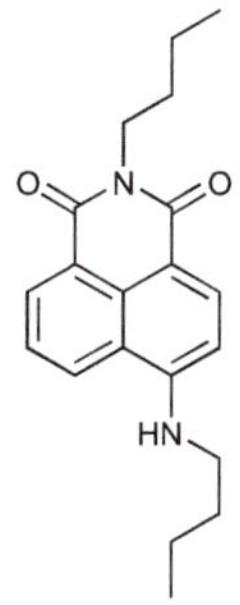

4.50

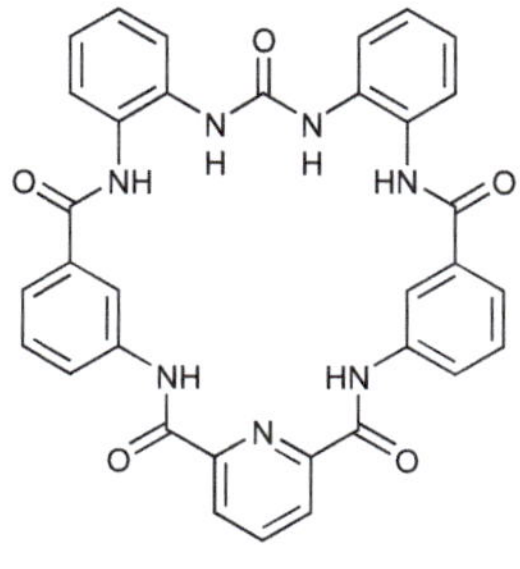

4.51

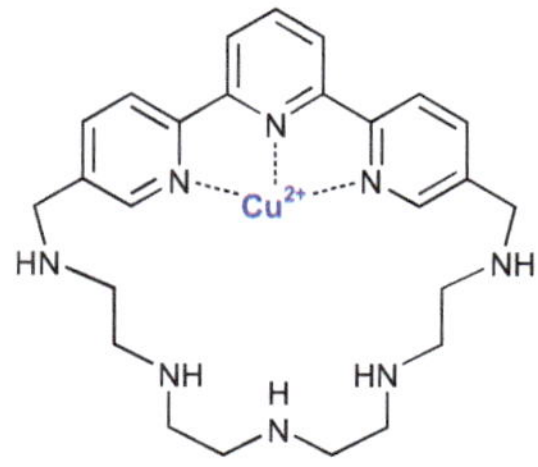

4.52

Chart 4.21

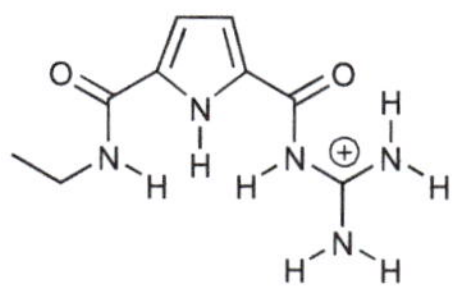

4.53

Chart 4.22

An indicator displacement assay for bicarbonate was developed based on the dicopper(II) complex **4.42** (Chart 4.18).[103] The fluorescent indicator dye, coumarin 343, is quenched when it associates with **4.42** in water buffered to pH 7. Addition of bicarbonate displaces the indicator and induces recovery of the fluorescence. The assay was used to quantify the amount of bicarbonate in mineral water.

Acetate receptors are numerous, involving almost every receptor type used in anion coordination chemistry. Examples include cyclic and bicyclic polyammonium receptors, lactams, or dimetallic transition metal complexes. In addition, urea, thiourea, guanidinium, or squaramide groups attached to an appropriate receptor scaffold represent functional units with which carboxylate groups efficiently interact *via* hydrogen bonds. Although strong acetate binding could be observed for several receptors and in different solvents, binding selectivity over other anions such as halides or phosphates was often moderate.

A highly efficient, selective, and versatile carboxylate binding motif is the 2-guanidiniocarbonyl pyrrole introduced by Schmuck in 1999.[36,104] The parent receptor **4.53** (Chart 4.22) was shown to bind acetate anion in 40% D_2O/DMSO-d_6 with an affinity of 2790 M^{-1}, ~60 times more strongly than the *N*-acetyl guanidinium ion lacking the additional pyrrole unit. The remarkably high acetate affinity was ascribed to cooperative effects of the pyrrole and amide NH groups of **4.53**. Subsequently, the same binding motif was demonstrated to be highly effective for achieving sequence-selective peptide recognition (see Chapter 8), developing enzyme mimics, and stabilizing self-assembling systems.[36,104–106]

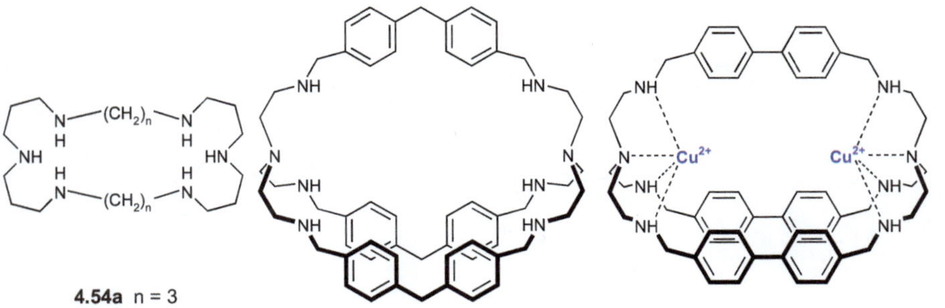

4.54a n = 3
4.54b n = 7
4.54c n = 10

4.55

4.56

Chart 4.23

Strategies to obtain synthetic receptors for dicarboxylates often involve incorporation of two carboxylate binding sites into a macrocycle or an acyclic receptor. Binding studies are conducted to find the optimal match of guest with the multivalent receptor. An example of this approach is the receptor series **4.54a–c** (Chart 4.23) with different distances between the two 1,5,9-triazanonane moieties.[32] The three receptors were screened for association with dicarboxylates $^{-}OOC(CH_2)_nCOO^{-}$ of different chain lengths. Receptor **4.54a** exhibited the highest affinity for oxalate ($n = 0$), **4.54b** for glutaric acid ($n = 3$) while longer dicarboxylic acids such pimelic acid ($n = 5$) and suberic acid ($n = 6$) were bound best by **4.54c.** Similar studies were conducted using other classes of bicyclic polyammonium receptors (**4.55** exhibits preferential binding of adipate ($n = 4$))[43] or their corresponding dimetallic transition metal complexes (**4.56** prefers to bind glutarate and adipate).[64]

Attaching three amino imidazoline moieties to a 1,3,5-triethylbenzene core yields receptors that are selective for tricarboxylates. Receptor **4.7** (Chart 4.3), for example, has been shown to bind citrate with a binding constant of 6.9×10^3 M^{-1} in D_2O, while dicarboxylates or monocarboxylates are bound with considerably reduced affinity.[107] The high selectivity of **4.7** allowed quantitative determination of the citrate content of various sports drinks by an indicator displacement assay. Structural modification of the parent receptor **4.7** by introducing boronic acids induced selectivity for dicarboxylates containing vicinal diols. The corresponding receptor **4.57** (Chart 4.24) binds tartrate with a binding constant of 5.5×10^4 M^{-1} in 25% water/methanol.[107] Malate, the corresponding dicarboxylate with one hydroxy group less, binds to **4.57** with similar strength, while succinate, the dicarboxylate without a hydroxy group, has a lower affinity by almost two orders of magnitude. The contributions of the individual binding sites to the recognition of α-hydroxycarboxylates and diols by receptors **4.7**, **4.57**, and the analog **4.58** containing two boronic acid groups has been evaluated. The results show that the receptor lacking the boronic acids is selective for tricarboxylates,[108] while the presence of the boronic acid groups induces selectivity for α-hydroxycarboxylates over simple 1,2-alkane diols. Isothermal titration calorimetry provided information about the thermodynamics of

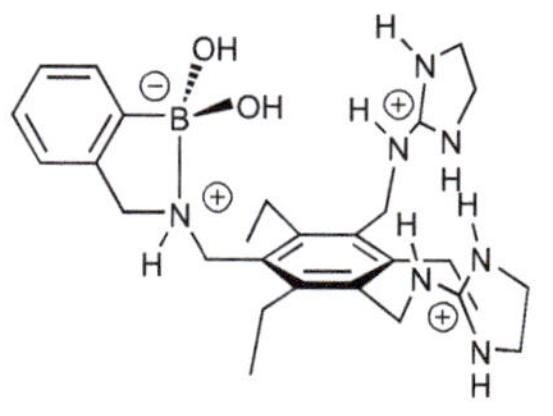

4.57

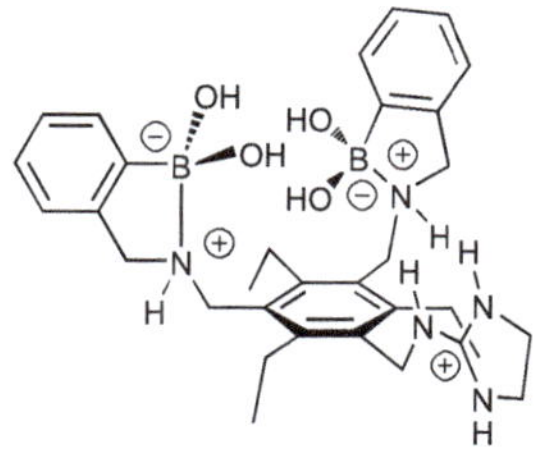

4.58

Chart 4.24

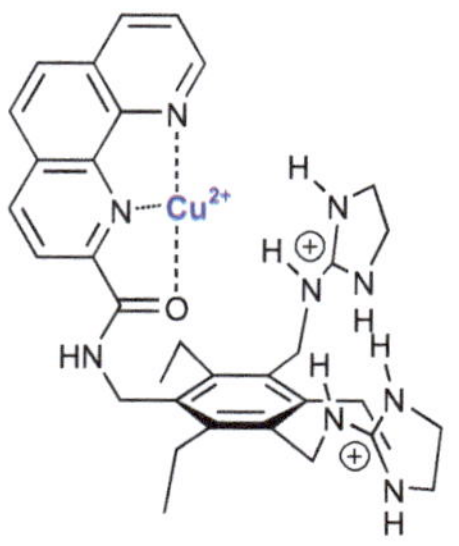

4.59

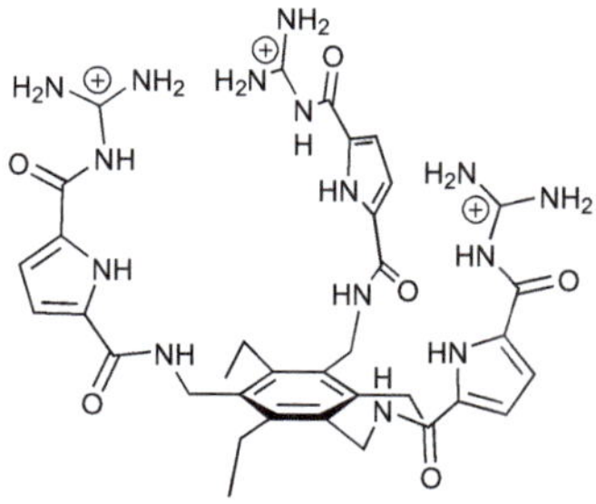

4.60

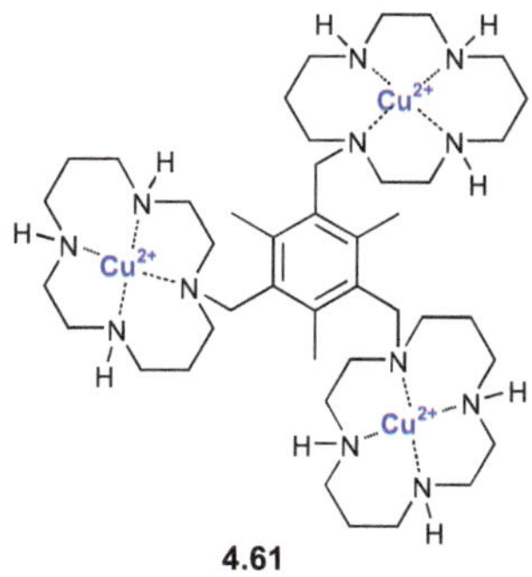

4.61

Chart 4.25

binding, and the cooperativity of the individual groups in complex formation was also elucidated. Based on this work, indicator ensembles were developed for the simultaneous quantitative estimation of the malate and the tartrate concentration in water[109] and of malate in Pinot Noir.[110] In addition, a colorimetric test strip for tartrate has been devised.[111]

Other tripodal citrate receptors are the copper(II) complex **4.59**,[112] the tris(2-guanidiniocarbonyl pyrrole) derivative **4.60**,[36] and the tris(copper(II)-cyclam) **4.61**[64] (Chart 4.25), all of which possess high affinity for citrate in highly competitive solvents including water ($\log K_a = 5.20$ for **4.60** in water and $\log K_a = 5.59$ for **4.61** in 0.01 M aqueous 4-(2-hydroxyethyl)-1-piperazineethanesulfonic acid (HEPES) buffer at pH 7) and are readily incorporated into indicator displacement assays for citrate detection.

Enantioselective recognition of α-hydroxycarboxylates such as mandelate or lactate has been achieved with chiral macrocycle **4.62** (Chart 4.26) containing a spirofluorene unit.[113] Compound **4.62** shows preferential binding of the *R*-enantiomers of mandelate and lactate in DMSO, which was attributed to unfavorable steric effects in the complexes of the corresponding *S*-enantiomers. The macrocyclic tetramine **4.63** allows enantioselective malate recognition.[114]

The structures of coordinatively unsaturated chiral lanthanide complexes, *e.g.* **4.64**, with bound acetate, lactate, citrate, or selected amino acids, have

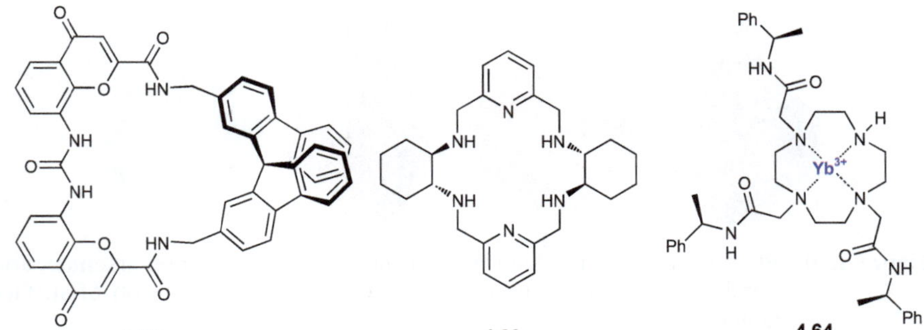

4.62 4.63 4.64

Chart 4.26

been characterized in the solid state and in solution by various spectroscopic techniques.[115] These investigations provided detailed insight into how the guest anion influences lanthanide coordination geometry.

4.6 Tetrahedral Anions

4.6.1 Anion Properties

Of the tetrahedral oxoanions, phosphate, and to a lesser extent sulfate have the highest biological relevance. Their EP surfaces along with those of the structurally related AsO_4^{3-} anion are shown in Figure 4.10. This figure clearly illustrates the pronounced negative EP of sulfate, phosphate, and arsenate, which results from the high negative charge of these anions and renders all three of them strongly coordinating. Differences in the EP surfaces only become visible when extending the color-coded EP range below -850 kcal mol^{-1}. The effects of a lower charge combined with a higher electronegativity of the central atom become evident when comparing the EP surfaces of these three anions with perchlorate (ClO_4^-), also depicted in Figure 4.10. This comparison demonstrates that perchlorate anions are weakly coordinating and therefore more difficult to bind than sulfate or phosphate anions, also illustrating why perchlorate anions are found at the opposite end of the Hofmeister series.

4.6.2 Recognition in Nature

The major difference between sulfate and phosphate anions in aqueous solution is their protonation state at physiological pH. While sulfate anions are fully deprotonated, phosphate anions exist as an almost 1 : 1 mixture of the monoprotonated HPO_4^{2-} and diprotonated $H_2PO_4^-$ forms. The fully deprotonated PO_4^{3-} anion is only present at high pH. Nature makes use of the different pK_a values of sulfate and phosphate anions to differentiate them as evidenced by the different binding modes of the sulfate-binding protein (SBP) and the phosphate-binding protein (PBP). Both bacterial

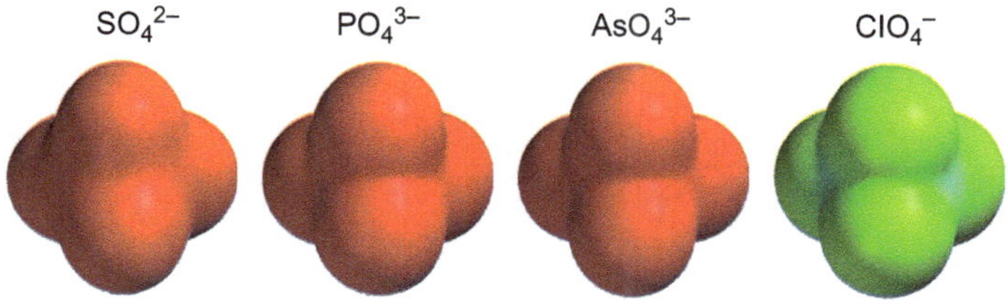

Figure 4.10 Electrostatic potential surfaces of the sulfate, phosphate, arsenate, and perchlorate anions. (See Figure 4.2 caption for description of surface generation.)

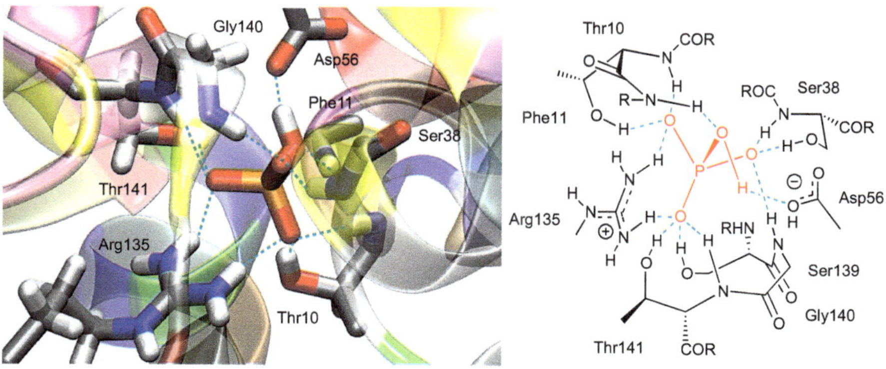

Figure 4.11 Binding arrangement of a hydrogenphosphate anion inside the binding site of the phosphate-binding protein (left) and schematic representation of the underlying interactions (right). The picture was generated using Visual Molecular Dynamics from the structure deposited in the Research Collaboratory for Structural Bioinformatics Protein Data Bank (code 1IXH).[116]

proteins are responsible for transport of the respective anions through the periplasmatic space once they have passively diffused across the outer cell membrane. The captured anions are subsequently transported across the inner membrane by specific transporter proteins.

Phosphate complexation of PBP involves incorporation of a fully desolvated hydrogenphosphate anion into a deep cleft within the protein. Here the anion is held tightly by 11 hydrogen bonds, as shown in Figure 4.11.[116] Two of these hydrogen bonds are part of a salt bridge between the guanidinium group of Arg135 and the anion. One hydrogen bond involves a hydrogen bond acceptor, namely the carboxylate group in the side chain of Asp56, which interacts with the hydrogenphosphate proton. If a fully deprotonated tetrahedral oxoanion such as sulfate is included into the binding pocket, a repulsive interaction occurs in this position rendering the sulfate complex of PBP significantly less stable.

The sulfate complex of SBP is slightly more stable than the hydrogenphosphate complex of PBP (K_d of the former is 0.12×10^{-6} M, while K_d of the

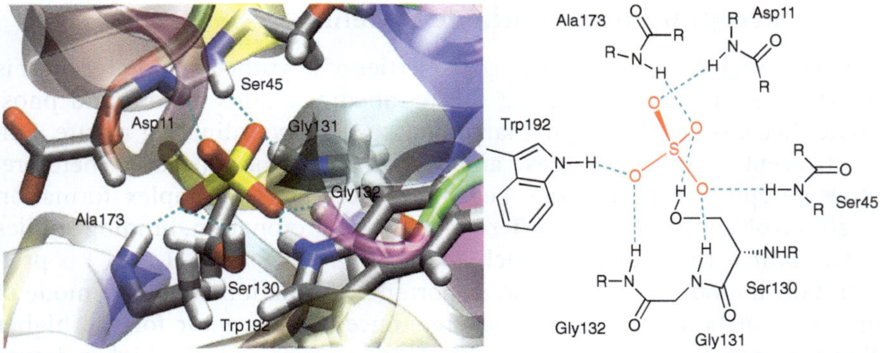

Figure 4.12 Binding arrangement of a sulfate anion inside the binding site of the sulfate-binding protein from *Salmonella typhimurium* (left) and schematic representation of the underlying interactions (right). The picture was generated using Visual Molecular Dynamics from the structure deposited in the Research Collaboratory for Structural Bioinformatics Protein Data Bank (code 1SBP).[117]

latter is 0.31×10^{-6} M), although the number of hydrogen bonds between protein and substrate is lower. In this case, there are seven hydrogen bonds, mostly between the oxygen atoms of the substrate and NH groups of the protein backbone (Figure 4.12).[117] In addition, a tryptophan NH and a serine OH group are also involved in binding, while salt bridges to positively charged residues are absent. The lack of a hydrogen bond acceptor in the active site renders SBP selective for fully deprotonated tetrahedral oxoanions. Hydrogenphosphate is, for example, bound with a dissociation constant of only 6×10^{-2} M.

The toxicity of arsenate is due to its close structural relationship with phosphate, allowing arsenate to replace phosphate in many reactions. The lower stability of arsenate esters leads to disruption of important biochemical pathways, such as oxidative phosphorylation. Organisms growing in an arsenate-rich environment have therefore devised finely tuned mechanisms to achieve selective recognition of phosphate over arsenate anions. Comparison of the crystal structures of a PBP overexpressed under phosphate-limiting conditions containing either HPO_4^{2-} or $HAsO_4^{2-}$ revealed small but significant differences in the length and distance of the crucial hydrogen bond involving the aspartate residue that acts as the hydrogen bond acceptor in the active site,[118] which result in a roughly 4500-fold higher affinity of the protein for phosphate over arsenate. These results strongly suggest that nature takes extreme care to avoid incorporation of arsenate into DNA, questioning the validity of earlier work that came to a different conclusion.[119] In contrast, chromate anions enter cells by a transport mechanism actually intended for sulfate and phosphate transport. The acute toxicity of chromate is due to the oxidation properties of chromium(VI) species.

4.6.3 Recognition by Synthetic Systems

Any screening study of the binding properties of a synthetic anion receptor is incomplete without considering the tetrahedral anions sulfate and phosphate. Because of their high charge and strongly coordinating nature, efficient receptor binding to these anions can usually be expected. Therefore, other receptor characteristics become important. As complex formation might involve anions with a different degree of protonation, binding studies must address the question, which species (*e.g.* HPO_4^{2-} or $H_2PO_4^{-}$) is preferentially bound. In this context, important information about the mode of binding can be derived from the preference of a receptor for the higher charged species (mainly electrostatic) or the species with the higher degree of protonation (mainly hydrogen-bonding). Another important criterion of a receptor is its sulfate/phosphate selectivity or selectivity for another tetrahedral anion.

Because of its inherent importance, the published work on receptors for sulfate[120] and phosphorylated substrates including phosphate anions[121,122] has been reviewed. In addition, a review is available which summarizes the approaches to recognize tetrahedral anions in general.[123] The review points out that nearly all types of anion receptors have affinity for sulfate or phosphate anions. The most commonly used receptors are polyammonium-based systems, receptors containing metal ions, or ones with pyrrole, amide, urea, guanidinium, or other hydrogen bond donors. Based on their extensive survey the authors came to the following conclusions:

- The size of the cavity available for anion binding and the overall flexibility of the receptor are the major factors in determining its selectivity. More rigid systems ensure a high fidelity in the arrangement of the hydrogen bond donors, which improves affinity and selectivity.
- Characteristic differences in receptor properties can be seen for the individual binding motifs. For example, protonated polyammonium receptors are characterized by very high association constants, even in water, and typically display a strong preference for tetrahedral oxoanions.
- A further increase in affinity and selectivity for phosphate anions is observed for polyammonium receptors containing metal centers. Phosphate selectivity of such receptors relative to chloride, sulfate, or nitrate can approach 10^4 in aqueous solution.
- Pyrrole-based anion receptors mainly rely on hydrogen bonding interactions to the pyrrole NH group in anion binding. As a consequence, selectivities are largely determined by the Lewis basicity of the anion and macrocyclic polypyrrole receptors therefore usually prefer to bind fluoride rather than phosphate. However, chloride or bromide anions are bound less strongly than phosphate because the oxygen atoms of the phosphate anion can engage in several strong hydrogen bonding interactions with a polypyrrole-based system.

In general, affinity for nitrate or sulfate is low, although it can be improved by an appropriate receptor design.

- Receptors with amide or urea groups possess high affinity for phosphate anions, although interference from acetate or chloride anions is commonly observed. Because of the higher acidity of the NH groups, thioamides or thioureas generally bind stronger to anions than the corresponding oxo-analogs.
- Guanidinium-based receptors can be designed to bind selectively either to sulfate or phosphate anions. Related receptors containing amidinium or imidazolinium moieties often prefer carboxylate or halide anions. Aligning these groups around the cavity of a suitable scaffold can also yield phosphate- or sulfate-selective systems. As in the case of ammonium-based receptors, introduction of a suitable metal ion significantly improves affinity and selectivity.

An important general feature of oxoanions is the ability to accept more hydrogen bonds from a receptor than only one per (deprotonated) oxygen atom, particularly if the receptor can surround the anion from all sides. The minimum number of hydrogen bonds found in a sulfate complex with a synthetic receptor is six, while **4.10** (Chart 4.5) forms eight hydrogen bonds with a sulfate anion.[124] Computational results indicate that the sulfate anion can engage in a maximum of 12 hydrogen bonds to a receptor, three to each oxygen atom.[125] However, very few receptors reach this number.[41] Even the naturally-occuring SBP only employs seven hydrogen bonds. One receptor specifically designed to bind a sulfate anion *via* 12 hydrogen bonds is the tetrahedral metallacage **4.65** (Chart 4.27).[126] This compound was derived by a computational approach that was based on the optimized structure of the complex between a sulfate ion and six urea units. Estimation of complex stability yielded an apparent stability constant of 6×10^6 M^{-1} for the sulfate complex of **4.65** in water, similar to the value observed for SBP. However, in contrast to the protein, which binds sulfate solely by hydrogen-bonding, electrostatic interactions between the positively charged cage and the negative sulfate anion are likely to enhance complex stability. Experiments involving other tetrahedral anions revealed the following selectivity trend for **4.65**: $PO_4^{3-} \gg CrO_4^{2-} > SO_4^{2-} > SeO_4^{2-} > MoO_4^{2-} > WO_4^{2-}$. This order was

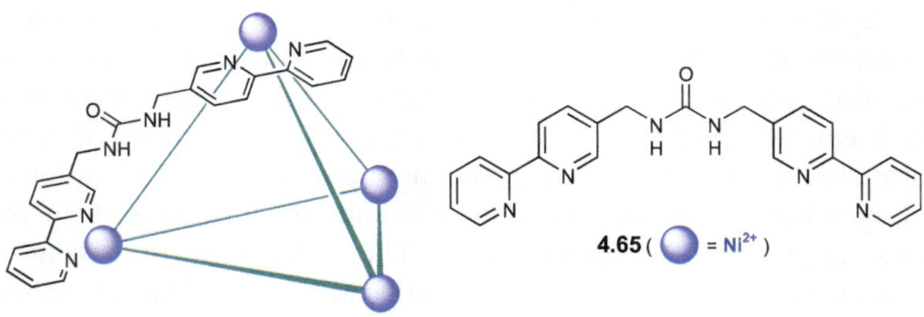

Chart 4.27

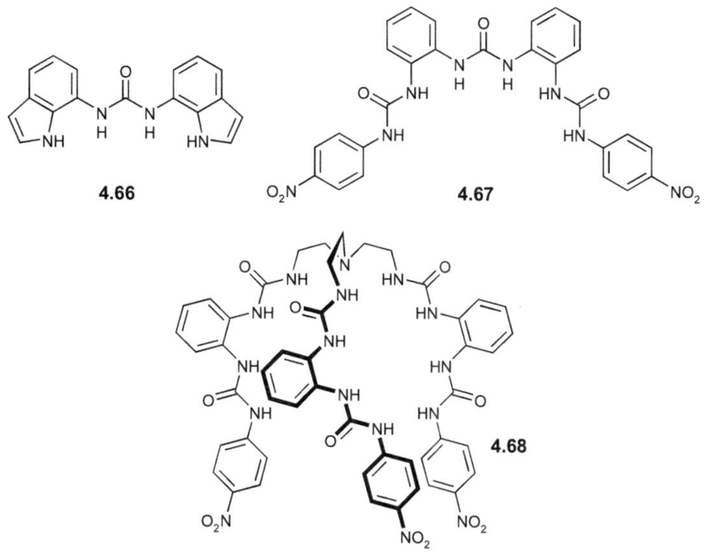

Chart 4.28

rationalized on the basis of the charge, size, hydration, basicity, and hydrogen-bond acceptor abilities of the anions.[126]

Other examples of urea-derived receptors that can coordinate to fully deprotonated tetrahedral oxoanions and engage in 12 hydrogen bonds are compounds **4.13a,b** (Chart 4.6),[125,127] **4.66**,[128] and **4.67** (Chart 4.28).[129] While receptor **4.66** binds the anion in a 3 : 1 (host–guest) fashion, receptors **4.13a,b** and **4.67** form 2 : 1 complexes, albeit with different geometries. Importantly, receptors **4.66** and **4.67** are also able to stabilize the strongly basic PO_4^{3-} anion in the solid state, and **4.67** does so even in solution.

A receptor that achieves saturation of the hydrogen-bonding pattern of fully deprotonated tetrahedral oxoanions in a 1 : 1 binding mode is the hexaurea **4.68** (Chart 4.28).[130] In the solid state, **4.68** folds around a sulfate anion with each urea group occupying an edge of a tetrahedron to interact with the anion. In solution (0.5% D_2O/DMSO-d_6), complex formation is more complicated, involving the formation of 2 : 1 guest–host complexes in the presence of an excess of sulfate anions. Formation of these higher complexes can be suppressed by increasing the water content of the solution up to 25%. This receptor also allows extraction of sulfate anions from water into $CDCl_3$.

Once the oxoanion is protonated, the maximum number of hydrogen bonds it can accept from a donor depends on the number of remaining deprotonated oxygen atoms. The protonated OH groups on the anion can act as hydrogen bonds donors contributing to complex formation if the receptor contains appropriate acceptor sites. The selectivity of receptor **4.69** (Chart 4.29) for hydrogensulfate ($K_a = 108\,000$ M^{-1}) and dihydrogenphosphate anions ($K_a = 29\,000$ M^{-1}) over acetate anions ($K_a = 18\,000$ M^{-1}) could be related to the presence of hydrogen bond acceptors in the form of the

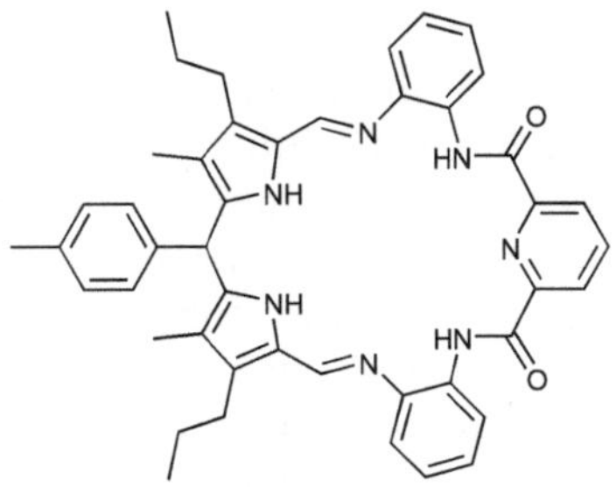

4.69

Chart 4.29

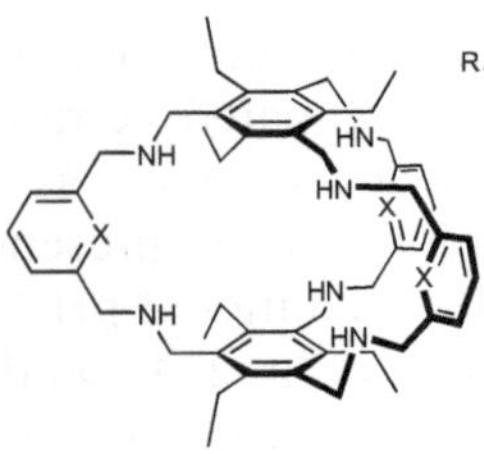

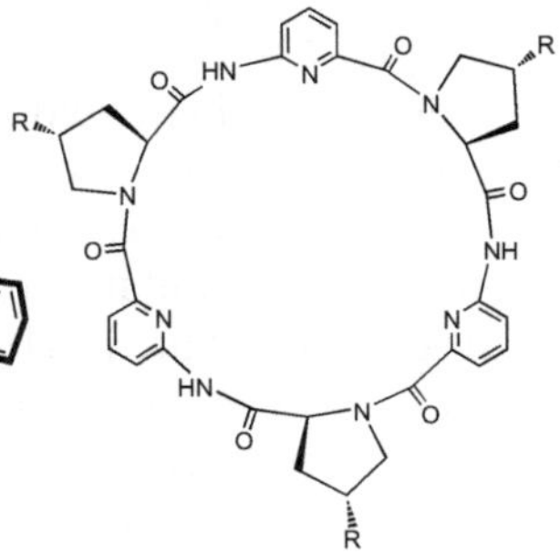

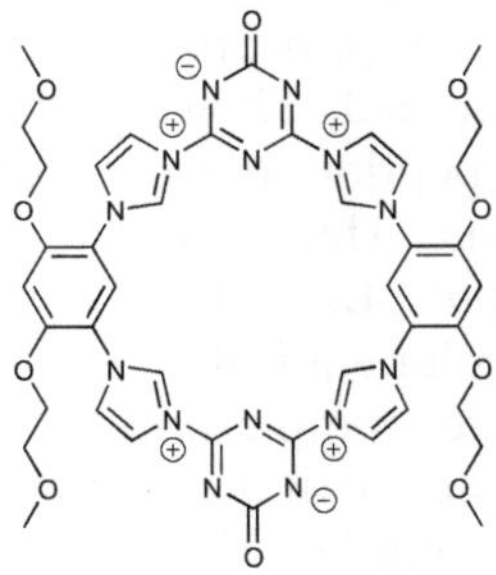

4.70a X = CH
4.70b X = N

4.71 R = 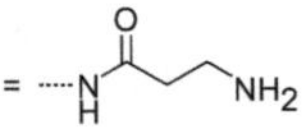

4.72

Chart 4.30

pyridine ring nitrogen atoms and imine groups within the macrocycle.[131] Unfortunately, no information is available about the affinity of **4.69** for the sulfate anion.

The different protonation states of sulfate and phosphate anions in aqueous solution at physiological pH provide an opportunity to achieve selective binding, with the binding modes of the SBP and PBP serving as guidelines. Thus, receptors arranging only hydrogen bond donors around the binding site are likely to possess selectivity for the sulfate ion whereas those that also feature hydrogen bond acceptors will preferentially bind to hydrogenphosphate or dihydrogenphosphate ions. The following examples illustrate this concept.

The relatively rigid macrobicyclic polyammonium cages **4.70a,b** (Chart 4.30) bind a variety of different anions, including halides and acetate, but possess particular selectivity for tetrahedral oxoanions.[132] Interestingly, stability of the sulfate complex of **4.70a** increases when going from the pentaprotonated form of **4.70a** to the hexaprotonated form, while the stability of the dihydrogenphosphate complex decreases in the same direction. Thus, the fully protonated form of **4.70a** mimics the behavior of the SBP, binding strongly to fully deprotonated anions while protonated anions experience repulsive interactions when bound inside the cavity containing only hydrogen bond donors. Consequently, the introduction of acceptors into the receptor cavity should have a profound impact on this selectivity. Indeed, receptor **4.70b**, containing pyridine ring nitrogen atoms as hydrogen bond acceptors, is able to stabilize the dihydrogenphosphate anion, even if all aliphatic amino groups are protonated, as evidenced by the slight increase of dihydrogenphosphate affinity when adding the sixth proton to the pentaprotonated form of this receptor.[133]

With a $\log K_a$ of 4.2, cyclopeptide **4.71** (Chart 4.30) possesses appreciable affinity for sulfate anions in aqueous acetate buffer at pH 4.8.[134] Binding involves hydrogen bonding of the anion to the peptide NH groups and to the protonated amino groups in the side chains, which fold around the anion in the complex. Interestingly, sulfate affinity of **4.71** is retained ($\log K_a = 2.4$) even in phosphate buffer at pH 7.0 where the receptor is still fully protonated, showing that the lack of hydrogen bond acceptors in the cyclopeptide prevents efficient interactions with protonated tetrahedral anions. Thus, the cyclopeptide mimics the behavior of SBP, as does receptor **4.72** which was shown to bind sulfate anions in the presence of 20 molar equivalents of dihydrogenphosphate in HEPES buffer at pH 7.0.[135] Sulfate complexation with **4.72** involves formation of a 2:1 sandwich complex with a structure related to the complexes shown in Figure 4.5. In this complex, the anion interacts *via* eight hydrogen bonds with the C2 hydrogen atoms of the imidazolium units.

As mentioned previously, a useful approach to achieve recognition of tetrahedral anions in water relies on receptors containing metal cations. Prototypes of dimetallic receptors are compounds **4.73** and **4.74** (Chart 4.31) which have high affinity for phosphate ions. With a $\log K_a$ of 8.5, the smaller receptor **4.73** preferentially interacts with diphosphate,[136] while the larger analog **4.74** possesses selectivity for triphosphate ($\log K_a = 8.0$).[137] Complex formation involves bridging of the two copper cations by a perched anion. The cavities of related macrobicyclic receptors (*e.g.* **4.42** or **4.56**) are too small to allow for efficient inclusion of tetrahedral anions.

Another binding motif for tetrahedral anions comprises three charged substituents connected to a copper-tren unit. Examples are receptors **4.24**

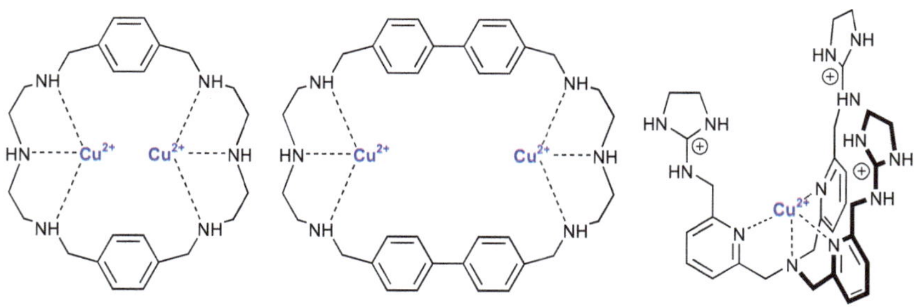

Chart 4.31

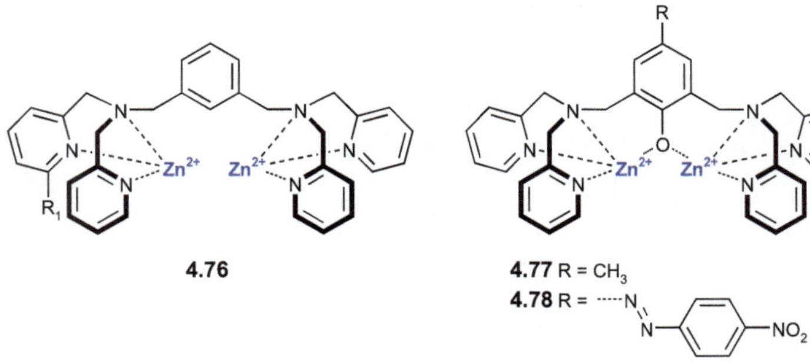

Chart 4.32

(Chart 4.11) and **4.75** (Chart 4.31), which bind to anions by a combination of coordination bonds, electrostatic interaction, and hydrogen bonding to the protonated amino imidazoline or amino moieties.[138] These receptors bind to hydrogenphosphate in 2% methanol/water (5 mM HEPES, pH 7.4) with K_a of 2.5×10^4 M^{-1} and 1.5×10^4 M^{-1}, respectively. Other tetrahedral anions tested in this study were arsenate, perrhenate, and sulfate. Arsenate binding was comparable to phosphate. Interactions of perrhenate with **4.75** were almost negligible ($K_a<100$ M^{-1}), and with **4.24** they were weaker than phosphate binding by about one order of magnitude. Sulfate was weakly bound by both receptors ($K_a<100$ M^{-1}). The high phosphate affinity and selectivity of **4.75** allowed construction of a dye displacement assay for the quantification of phosphate levels in horse serum and saliva.[139]

Receptors based on zinc(II) dipicolylamine complexes are particularly selective for phosphate anions. The dimetallic parent compound **4.76** (Chart 4.32) binds dihydrogenphosphate anions with a binding constant of 1.93×10^5 M^{-1} in 50 mM HEPES buffer at pH 7.[140] In combination with pyrocatechol violet or coumarin dye, the dizinc complex **4.77** allows colorimetric or fluorimetric phosphate detection in aqueous solution.[141,142] Anions such as sulfate, halides, carbonate, acetate, azide, or nitrate do not interfere in this assay. The azophenol-based derivative **4.78** is selective for diphosphate.[143]

Such zinc(II) dipicolylamine subunits were also attached to several other molecular scaffolds, *e.g.* cyclopeptides,[144] to develop receptors or probes for phosphate or diphosphate anions. It should also be mentioned that some of these receptors (*e.g.* **4.26** (Chart 4.11)) have been used to detect phosphorylated side chains in proteins (see Chapter 8).[145]

Motivation for the development of sulfate and phosphate receptors extends beyond the biomedical field. An important environmental application is phosphate extraction from water as a means to remove pollutants or to recycle fertilizers. Removal of sulfate from water is also a central process in the treatment of nuclear waste. The main challenge associated with extraction of these anions is the Hofmeister bias, namely the large energetic penalty for moving well hydrated ions into nonpolar media.[146] An apparent exception to this rule is the cyclo[8]pyrrole **4.79** (Chart 4.33). When dissolved

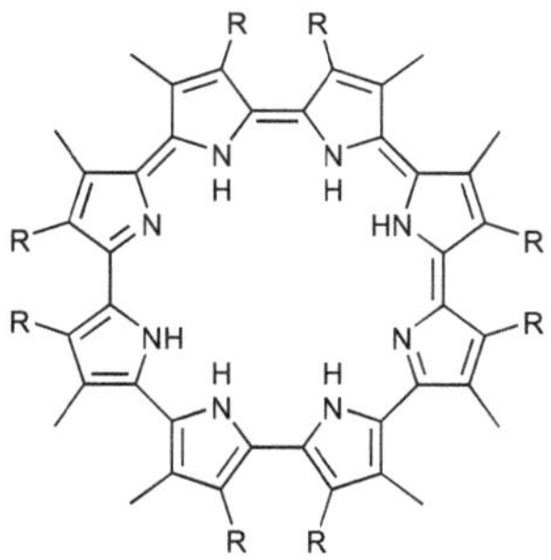

4.79
R = $C_{11}H_{23}$

Chart 4.33

in toluene, the nitrate salt of the diprotonated form of **4.79** is able to extract sulfate anions from the aqueous phase to the organic phase by anion metathesis.[147] Similarly, sulfate carriers have been developed with tren units or cyclopeptides incorporating trisureas.[148] Another promising approach to extract tetrahedral oxoanions from water involves the use of metal organic frameworks with tailored binding sites for anions.[146]

4.7 Future Directions and Challenges

Extensive work in the field of anion coordination chemistry over the years has made accessible a broad range of synthetic molecular receptors for inorganic anions. Many of these systems were prepared to elucidate the general principles of anion coordination, and the knowledge gained allowed researchers to increasingly target more applied areas, for example the selective detection of inorganic anions in their biochemical environment, the transport of anions across lipid membranes, or the extraction of anions from biochemical or environmental matrices. These applications have broad impact in technologies such as medicinal diagnostics, environmental monitoring, or wastewater treatment. To reach these goals, the following challenges need to be addressed:

- Selective anion recognition in a biochemical system or in the environment requires the anion to be bound in water, potentially in the presence of higher concentrations of competing anions. The receptors need to be sufficiently water-soluble and they need to possess high binding selectivity.
- Environmentally relevant anions have to be reliably detected at very low concentrations. As a consequence, receptor affinity needs to be in the millimolar range at least, and generally significantly better.
- Receptors for anion transport or aqueous extraction need to transfer hydrated anions into a nonpolar medium with high selectivity. Overcoming the Hofmeister bias is a major design challenge.
- To enable detection of anions in biological systems or in nature the binding event must be coupled to a suitable output signal. Technically

simple naked-eye or test-strip based assays that do not require elaborate equipment will be most convenient and will allow inexperienced users to test if water is adequately pure for consumption or agricultural use.

With the impressive development of anion coordination chemistry in recent years there is a realistic prospect that many real-world problems can be solved, particularly if concepts such as receptor immobilization, multivalency, and cooperativity are properly considered in the receptor design. The attractiveness of anion coordination chemistry lies in the fact that the basic scientific question of how to bind an anion eventually leads to the discovery of new and valuable receptor-based technologies.

References

1. M. T. Albelda, J. C. Frías, E. García-España and H.-J. Schneider, *Chem. Soc. Rev.*, 2012, **41**, 3859.
2. L. Fabbrizzi, M. Licchelli, G. Rabaioli and A. Taglietti, *Coord. Chem. Rev.*, 2000, **205**, 85.
3. C. Suksai and T. Tuntulani, *Chem. Soc. Rev.*, 2003, **32**, 192.
4. R. Martínez-Máñez and F. Sancenón, *Chem. Rev.*, 2003, **103**, 4419.
5. T. Gunnlaugsson, M. Glynn, G. M. Tocci (née Hussey), P. E. Kruger and F. M. Pfeffer, *Coord. Chem. Rev.*, 2006, **250**, 3094.
6. B. D. Smith and T. N. Lambert, *Chem. Commun.*, 2003, 2261.
7. G. W. Gokel and N. Barkey, *New J. Chem.*, 2009, **33**, 947.
8. J. T. Davis, O. Okunola and R. Quesada, *Chem. Soc. Rev.*, 2010, **39**, 3843.
9. C. J. E. Haynes and P. A. Gale, *Chem. Commun.*, 2011, **47**, 8203.
10. P. A. Gale, *Acc. Chem. Res.*, 2011, **44**, 216.
11. A. V. Jentzsch, A. Hennig, J. Mareda and S. Matile, *Acc. Chem. Res.*, 2013, **46**, 2791.
12. P. A. Gale, R. Pérez-Tomás and R. Quesada, *Acc. Chem. Res.*, 2013, **46**, 2801.
13. H. Valkenier and A. P. Davis, *Acc. Chem. Res.*, 2013, **46**, 2898.
14. Y. Hua and A. H. Flood, *Chem. Soc. Rev.*, 2010, **39**, 1262.
15. P. Metrangolo and G. Resnati, *Halogen Bonding: Fundamentals and Applications*, Springer, Heidelberg, 2010.
16. A. Priimagi, G. Cavallo, P. Metrangolo and G. Resnati, *Acc. Chem. Res.*, 2013, **46**, 2686.
17. B. L. Schottel, H. T. Chifotides and K. R. Dunbar, *Chem. Soc. Rev.*, 2008, **37**, 68.
18. A. Frontera, P. Gamez, M. Mascal, T. J. Mooibroek and J. Reedijk, *Angew. Chem.*, 2011, **123**, 9736; *Angew. Chem. Int. Ed.*, 2011, **50**, 9564.
19. S. Kubik, *Chem. Soc. Rev.*, 2010, **39**, 3648.
20. E. A. Kataev and C. Müller, *Tetrahedron*, 2014, **70**, 137.
21. Y. Marcus, *Ion Properties*, Marcel Dekker, New York, 1997.
22. P. Lo Nostro and B. W. Ninham, *Chem. Rev.*, 2012, **112**, 2286.

23. F. P. Schmidtchen, *Chem. Soc. Rev.*, 2010, **39**, 3916.
24. F. Biedermann, W. M. Nau and H.-J. Schneider, *Angew. Chem.*, 2014, **126**, 11338; *Angew. Chem., Int. Ed.*, 2014, **53**, 11158.
25. H.-J. Schneider, *Supramolecular Systems in Biomedical Fields*, RSC, Cambridge, 2013.
26. A. Bianchi, K. Bowman-James and E. García-España, *Supramolecular Chemistry of Anions*, Wiley-VCH, New York, 1997.
27. J. L. Sessler, P. A. Gale and W.-S. Cho, *Anion Receptor Chemistry*, RSC, Cambridge, 2006.
28. K. Bowman-James, A. Bianchi and E. García-España, *Anion Coordination Chemistry*, Wiley-VCH, Weinheim, 2011.
29. For the most recent review in this series, see: P. A. Gale, N. Busschaert, C. J. E. Haynes, L. E. Karagiannidis and I. L. Kirby, *Chem. Soc. Rev.*, 2014, **43**, 205.
30. C. H. Park and H. E. Simmons, *J. Am. Chem. Soc.*, 1968, **90**, 2431.
31. C. A. Ilioudis and J. W. Steed, *J. Supramol. Chem.*, 2001, **1**, 165.
32. J. M. Llinares, D. Powell and K. Bowman-James, *Coord. Chem. Rev.*, 2003, **240**, 57.
33. E. García-España, P. Díaz, J. M. Llinares and A. Bianchi, *Coord. Chem. Rev.*, 2006, **250**, 2952.
34. P. Mateus, N. Bernier and R. Delgado, *Coord. Chem. Rev.*, 2010, **254**, 1726.
35. M. D. Best, S. L. Tobey and E. V. Anslyn, *Coord. Chem. Rev.*, 2003, **240**, 3.
36. C. Schmuck, *Coord. Chem. Rev.*, 2006, **250**, 3053.
37. P. Blondeau, M. Segura, R. Pérez-Fernández and J. de Mendoza, *Chem. Soc. Rev.*, 2007, **36**, 198.
38. F. P. Schmidtchen and G. Müller, *J. Chem. Soc., Chem. Commun.*, 1984, 1115.
39. F. P. Schmidtchen, *Chem. Ber.*, 1981, **114**, 597.
40. C. R. Bondy and S. J. Loeb, *Coord. Chem. Rev.*, 2003, **240**, 77.
41. S. O. Kang, R. A. Begum and K. Bowman-James, *Angew. Chem.*, 2006, **118**, 8048; *Angew. Chem., Int. Ed.*, 2006, **45**, 7882.
42. S. O. Kang, M. A. Hossain and K. Bowman-James, *Coord. Chem. Rev.*, 2006, **250**, 3038.
43. S. O. Kang, J. M. Llinares, V. W. Day and K. Bowman-James, *Chem. Soc. Rev.*, 2010, **39**, 3980.
44. Z. Zhang and P. R. Schreiner, *Chem. Soc. Rev.*, 2009, **38**, 1187.
45. V. Amendola, L. Fabbrizzi and L. Mosca, *Chem. Soc. Rev.*, 2010, **39**, 3889.
46. V. Amendola, G. Bergamaschi, M. Boiocchi, L. Fabbrizzi and M. Milani, *Chem. – Eur. J.*, 2010, **16**, 4368.
47. V. Amendola, D. Esteban-Gómez, L. Fabbrizzi and M. Licchelli, *Acc. Chem. Res.*, 2006, **39**, 343.
48. J. L. Sessler and J. M. Davis, *Acc. Chem. Res.*, 2001, **34**, 989.
49. J. L. Sessler, S. Camiolo and P. A. Gale, *Coord. Chem. Rev.*, 2003, **240**, 17.

50. P. A. Gale, P. Anzenbacher Jr. and J. L. Sessler, *Coord. Chem. Rev.*, 2001, **222**, 57.
51. C.-H. Lee, H. Miyaji, D.-W. Yoon and J. L. Sessler, *Chem. Commun.*, 2008, 24.
52. T. G. Levitskaia, M. Marquez, J. L. Sessler, J. A. Shriver, T. Vercouter and B. A. Moyer, *Chem. Commun.*, 2003, 2248.
53. P. A. Gale, *Chem. Commun.*, 2008, 4525.
54. H. Juwarker, J.-m. Suk and K.-S. Jeong, *Chem. Soc. Rev.*, 2009, **38**, 3316.
55. S. Lee, C.-H. Chen and A. H. Flood, *Nat. Chem.*, 2013, **5**, 704.
56. J. Svec, M. Dusek, K. Fejfarova, P. Stacko, P. Klán, A. E. Kaifer, W. Li, E. Hudeckova and V. Sindelar, *Chem. – Eur. J.*, 2011, **17**, 5605.
57. E. Galbraith and T. D. James, *Chem. Soc. Rev.*, 2010, **39**, 3831.
58. C. R. Wade, A. E. J. Broomsgrove, S. Aldridge and F. P. Gabbaï, *Chem. Rev.*, 2010, **110**, 3958.
59. M. Hirai and F. P. Gabbaï, *Chem. Sci.*, 2014, **5**, 1886.
60. V. Amendola and L. Fabbrizzi, *Chem. Commun.*, 2009, 513.
61. D. J. Mercer and S. J. Loeb, *Chem. Soc. Rev.*, 2010, **39**, 3612.
62. P. D. Beer and E. J. Hayes, *Coord. Chem. Rev.*, 2003, **240**, 167.
63. E. J. O'Neil and B. D. Smith, *Coord. Chem. Rev.*, 2006, **250**, 3068.
64. L. Fabbrizzi and A. Poggi, *Chem. Soc. Rev.*, 2013, **42**, 1681.
65. S. Shinoda and H. Tsukube, *Analyst*, 2011, **136**, 431.
66. C. Lincheneau, F. Stomeo, S. Comby and T. Gunnlaugsson, *Aust. J. Chem.*, 2011, **64**, 1315.
67. P. Metrangolo, F. Meyer, T. Pilati, G. Resnati and G. Terraneo, *Angew. Chem.*, 2008, **120**, 6206; *Angew. Chem., Int. Ed.*, 2008, **47**, 6114.
68. G. Cavallo, P. Metrangolo, T. Pilati, G. Resnati, M. Sansotera and G. Terraneo, *Chem. Soc. Rev.*, 2010, **39**, 3772.
69. M. G. Sarwar, B. Dragisic, S. Sagoo and M. S. Taylor, *Angew. Chem.*, 2010, **122**, 1718; *Angew. Chem. Int., Ed.*, 2010, **49**, 1674.
70. F. Zapata, A. Caballero, N. G. White, T. D. W. Claridge, P. J. Costa, V. Félix and P. D. Beer, *J. Am. Chem. Soc.*, 2012, **134**, 11533.
71. S. M. Walter, F. Kniep, L. Rout, F. P. Schmidtchen, E. Herdtweck and S. M. Huber, *J. Am. Chem. Soc.*, 2012, **134**, 8507.
72. L. Adriaenssens, G. Gil-Ramírez, A. Frontera, D. Quiñonero, E. C. Escudero-Adán and P. Ballester, *J. Am. Chem. Soc.*, 2014, **136**, 3208.
73. K. L. Kirk, *Biochemistry of the Elemental Halogens and Inorganic Halides*, Plenum Press, New York, 1991.
74. M. Cametti and K. Rissanen, *Chem. Soc. Rev.*, 2013, **42**, 2016.
75. R. Dutzler, E. B. Campbell, M. Cadene, B. T. Chait and R. MacKinnon, *Nature*, 2002, **415**, 287.
76. A. Fürstner, *Angew. Chem.*, 2003, **115**, 3706; *Angew. Chem., Int. Ed.*, 2003, **42**, 3582.
77. J. T. Davis, *Top. Heterocycl. Chem.*, 2010, **24**, 145.
78. C. A. Ilioudis, K. S. B. Hancock, D. G. Georganopoulou and J. W. Steed, *New J. Chem.*, 2000, **24**, 787.
79. K. Choi and A. D. Hamilton, *J. Am. Chem. Soc.*, 2003, **125**, 10241.

80. Y. Li, M. Pink, J. A. Karty and A. H. Flood, *J. Am. Chem. Soc.*, 2008, **130**, 17293.
81. S. Kubik, R. Goddard, R. Kirchner, D. Nolting and J. Seidel, *Angew. Chem.*, 2001, **113**, 2722; *Angew. Chem., Int. Ed.*, 2001, **40**, 2648.
82. M. Cametti and K. Rissanen, *Chem. Commun.*, 2009, 2809.
83. Y. Zhou, J. F. Zhang and J. Yoon, *Chem. Rev.*, 2014, **114**, 5511.
84. A. P. Davis and J.-B. Joos, *Coord. Chem. Rev.*, 2003, **240**, 143.
85. A. P. Davis, *Coord. Chem. Rev.*, 2006, **250**, 2939.
86. J. A. Cooper, S. T. G. Street and A. P. Davis, *Angew. Chem.*, 2014, **126**, 5715; *Angew. Chem., Int. Ed.*, 2014, **53**, 5609.
87. F. H. Zelder and C. Männel-Croisé, *Chimia*, 2009, **63**, 58.
88. K. Muramoto, K. Ohta, K. Shinzawa-Itoh, K. Kanda, M. Taniguchi, H. Nabekura, E. Yamashita, T. Tsukihara and S. Yoshikawa, *Proc. Natl. Acad. Sci. U. S. A.*, 2010, **107**, 7740.
89. S. O. Kang, D. Powell, V. W. Day and K. Bowman-James, *Angew. Chem.*, 2006, **118**, 1955; *Angew. Chem., Int. Ed.*, 2006, **45**, 1921.
90. R. O. Ramabhadran, Y. Liu, Y. Hua, M. Ciardi, A. H. Flood and K. Raghavachari, *J. Am. Chem. Soc.*, 2014, **136**, 5078.
91. R. O. Ramabhadran, Y. Hua, Y.-j. Li, A. H. Flood and K. Raghavachari, *Chem. – Eur. J.*, 2011, **17**, 9123.
92. C. Männel-Croisé and F. Zelder, *Anal. Methods*, 2012, **4**, 2632.
93. L. D. Tivana, J. Da Cruz Francisco, F. Zelder, B. Bergenståhl and P. Dejmek, *Food Chem.*, 2014, **158**, 20.
94. K. Kaur, R. Saini, A. Kumar, V. Luxami, N. Kaur, P. Singh and S. Kumar, *Coord. Chem. Rev.*, 2012, **256**, 1992.
95. J. H. Cho, D. H. Kim, S. J. Chung, N.-C. Ha, B.-H. Oh and K. Y. Choi, *Bioorg. Med. Chem.*, 2002, **10**, 2015.
96. J. M. Mahoney, K. A. Stucker, H. Jiang, I. Carmichael, N. R. Brinkmann, A. M. Beatty, B. C. Noll and B. D. Smith, *J. Am. Chem. Soc.*, 2005, **127**, 2922.
97. R. Herges, A. Dikmans, U. Jana, F. Köhler, P. G. Jones, I. Dix, T. Fricke and B. König, *Eur. J. Org. Chem.*, 2002, 3004.
98. A. P. Bisson, V. M. Lynch, M.-K. C. Monahan and E. V. Anslyn, *Angew. Chem.*, 1997, **109**, 2435; *Angew. Chem., Int. Ed.*, 1997, **36**, 2340.
99. I. Ravikumar and P. Ghosh, *Chem. Commun.*, 2010, **46**, 1082.
100. T. Gunnlaugsson, P. E. Kruger, P. Jensen, F. M. Pfeffer and G. M. Hussey, *Tetrahedron Lett.*, 2003, **44**, 8909.
101. S. J. Brooks, P. A. Gale and M. E. Light, *Chem. Commun.*, 2006, 4344.
102. E. García-España, P. Gaviña, J. Latorre, C. Soriano and B. Verdejo, *J. Am. Chem. Soc.*, 2004, **126**, 5082.
103. L. Fabbrizzi, A. Leone and A. Taglietti, *Angew. Chem.*, 2001, **113**, 3156; *Angew. Chem., Int. Ed.*, 2001, **40**, 3066.
104. C. Schmuck and L. Geiger, *Curr. Org. Chem.*, 2003, **7**, 1485.
105. C. Schmuck and P. Wich, *Top. Curr. Chem.*, 2007, **277**, 3.
106. T. H. Rehm and C. Schmuck, *Chem. Soc. Rev.*, 2010, **39**, 3597.

107. S. L. Wiskur, H. Ait-Haddou, J. J. Lavigne and E. V. Anslyn, *Acc. Chem. Res.*, 2001, **34**, 963.
108. S. L. Wiskur, J. J. Lavigne, A. Metzger, S. L. Tobey, V. Lynch and E. V. Anslyn, *Chem. – Eur. J.*, 2004, **10**, 3792.
109. S. L. Wiskur, P. N. Floriano, E. V. Anslyn and J. T. McDevitt, *Angew. Chem.*, 2003, **115**, 2116; *Angew. Chem., Int. Ed.*, 2003, **42**, 2070.
110. A. M. Piątek, Y. J. Bomble, S. L. Wiskur and E. V. Anslyn, *J. Am. Chem. Soc.*, 2004, **126**, 6072.
111. B. T. Nguyen, S. L. Wiskur and E. V. Anslyn, *Org. Lett.*, 2004, **6**, 2499.
112. L. A. Cabell, M. D. Best, J. J. Lavigne, S. E. Schneider, D. M. Perreault, M.-K. Monahan and E. V. Anslyn, *J. Chem. Soc., Perkin Trans. 2*, 2001, 315.
113. A. Tejeda, A. I. Oliva, L. Simón, M. Grande, M. C. Caballero and J. R. Morán, *Tetrahedron Lett.*, 2000, **41**, 4563.
114. A. González-Álvarez, I. Alfonso, P. Díaz, E. García-España and V. Gotor, *Chem. Commun.*, 2006, 1227.
115. R. S. Dickins, S. Aime, A. A. Batsanov, A. Beeby, M. Botta, J. I. Bruce, J. A. K. Howard, C. S. Love, D. Parker, R. D. Peacock and H. Puschmann, *J. Am. Chem. Soc.*, 2002, **124**, 12697.
116. Z. Wang, H. Luecke, N. Yao and F. A. Quiocho, *Nat. Struct. Biol.*, 1997, **4**, 519.
117. J. W. Pflugrath and F. A. Quiocho, *J. Mol. Biol.*, 1988, **200**, 163.
118. M. Elias, A. Wellner, K. Goldin-Azulay, E. Chabriere, J. A. Vorholt, T. J. Erb and D. S. Tawfik, *Nature*, 2012, **491**, 134.
119. F. Wolfe-Simon, J. Switzer Blum, T. R. Kulp, G. W. Gordon, S. E. Hoeft, J. Pett-Ridge, J. F. Stolz, S. M. Webb, P. K. Weber, P. C. W. Davies, A. D. Anbar and R. S. Oremland, *Science*, 2011, **332**, 1163.
120. I. Ravikumar and P. Ghosh, *Chem. Soc. Rev.*, 2012, **41**, 3077.
121. C. Bazzicalupi, A. Bencini and V. Lippolis, *Chem. Soc. Rev.*, 2010, **39**, 3709.
122. A. E. Hargrove, S. Nieto, T. Zhang, J. L. Sessler and E. V. Anslyn, *Chem. Rev.*, 2011, **111**, 6603.
123. E. A. Katayev, Y. A. Ustynyuk and J. L. Sessler, *Coord. Chem. Rev.*, 2006, **250**, 3004.
124. S. O. Kang, M. A. Hossain, D. Powell and K. Bowman-James, *Chem. Commun.*, 2005, 328.
125. R. Custelcean, B. A. Moyer and B. P. Hay, *Chem. Commun.*, 2005, 5971.
126. R. Custelcean, P. V. Bonnesen, N. C. Duncan, X. Zhang, L. A. Watson, G. Van Berkel, W. B. Parson and B. P. Hay, *J. Am. Chem. Soc.*, 2012, **134**, 8525.
127. R. Dutta and P. Ghosh, *Chem. Commun.*, 2014, **50**, 10538.
128. C. Caltagirone, J. R. Hiscock, M. B. Hursthouse, M. E. Light and P. A. Gale, *Chem. – Eur. J.*, 2008, **14**, 10236.
129. C. Jia, B. Wu, S. Li, Z. Yang, Q. Zhao, J. Liang, Q.-S. Li and X.-J. Yang, *Chem. Commun.*, 2010, **46**, 5376.

130. C. Jia, B. Wu, S. Li, X. Huang, Q. Zhao, Q.-S. Li and X.-J. Yang, *Angew. Chem.*, 2011, **123**, 506; *Angew. Chem., Int. Ed.*, 2011, **50**, 486.
131. J. L. Sessler, E. Katayev, G. D. Pantos, P. Scherbakov, M. D. Reshetova, V. N. Khrustalev, V. M. Lynch and Y. A. Ustynyuk, *J. Am. Chem. Soc.*, 2005, **127**, 11442.
132. P. Mateus, R. Delgado, P. Brandão, S. Carvalho and V. Félix, *Org. Biomol. Chem.*, 2009, **7**, 4661.
133. P. Mateus, R. Delgado, P. Brandão and V. Félix, *J. Org. Chem.*, 2009, **74**, 8638.
134. A. Schaly, R. Belda, E. García-España and S. Kubik, *Org. Lett.*, 2013, **15**, 6238.
135. H. Zhou, Y. Zhao, G. Gao, S. Li, J. Lan and J. You, *J. Am. Chem. Soc.*, 2013, **135**, 14908.
136. L. Fabbrizzi, N. Marcotte, F. Stomeo and A. Taglietti, *Angew. Chem.*, 2002, **114**, 3965; *Angew. Chem., Int. Ed.*, 2002, **41**, 3811.
137. N. Marcotte and A. Taglietti, *Supramol. Chem.*, 2003, **15**, 617.
138. S. L. Tobey, B. D. Jones and E. V. Anslyn, *J. Am. Chem. Soc.*, 2003, **125**, 4026.
139. S. L. Tobey and E. V. Anslyn, *Org. Lett.*, 2003, **5**, 2029.
140. J. A. Drewry, S. Fletcher, H. Hassan and P. T. Gunning, *Org. Biomol. Chem.*, 2009, **7**, 5074.
141. M. S. Han and D. H. Kim, *Angew. Chem.*, 2002, **114**, 3963; *Angew. Chem., Int. Ed.*, 2002, **41**, 3809.
142. R. G. Hanshaw, S. M. Hilkert, H. Jiang and B. D. Smith, *Tetrahedron Lett.*, 2004, **45**, 8721.
143. D. H. Lee, J. H. Im, S. U. Son, Y. K. Chung and J.-I. Hong, *J. Am. Chem. Soc.*, 2003, **125**, 7752.
144. R. B. P. Elmes and K. A. Jolliffe, *Chem. Commun.*, 2015, **51**, 4951.
145. T. Sakamoto, A. Ojida and I. Hamachi, *Chem. Commun.*, 2009, 141.
146. R. Custelcean and B. A. Moyer, *Eur. J. Inorg. Chem.*, 2007, 1321.
147. L. R. Eller, M. Stępień, C. J. Fowler, J. T. Lee, J. L. Sessler and B. A. Moyer, *J. Am. Chem. Soc.*, 2007, **129**, 11020.
148. N. Busschaert, L. E. Karagiannidis, M. Wenzel, C. F. E. Haynes, N. J. Wells, P. G. Young, D. Makuc, J. Plavec, K. A. Jolliffe and P. A. Gale, *Chem. Sci.*, 2014, **5**, 1118.

CHAPTER 5

Synthetic Receptors for Carbohydrates

ALEXANDER DRAGANOV,[a] DANZHU WANG,[a] JALISA HOLMES,[a] KAILI JI,[a] BINGHE WANG,*[a] SCOTT K. SHAW[b] AND BRADLEY D. SMITH*[b]

[a] Department of Chemistry and Center for Diagnostics and Therapeutics, Georgia State University, Atlanta, Georgia 30303, USA; [b] Department of Chemistry and Biochemistry, 236 Nieuwland Science Hall, University of Notre Dame, IN 46556, USA

*Email: wang@gsu.edu; smith.115@nd.edu

5.1 Introduction

Figure 5.1 shows a schematic summary of glycoconjugate biosynthesis and cell surface recognition.[1] Monosaccharides in the extracellular milieu are selectively transported inside mammalian cells and converted into molecular building blocks (typically nucleoside sugars). Intracellular trafficking processes move the building blocks into secretory compartments where they are assembled into oligosaccharides and attached to lipids or proteins. These glycolipid and glycoprotein conjugates are expressed on the cell surface, where they serve as targeting ligands for receptors that are located on other cells or pathogens. The molecular recognition of cell surface glycoconjugates (sometimes called glycans) plays important functional roles in biological and pathological processes, including cancer development and metastasis, viral and bacterial infections, stem cell differentiation, cell–cell adhesion, and egg fertilization.[2,3] Because of the importance of

Monographs in Supramolecular Chemistry No. 14
Synthetic Receptors for Biomolecules: Design Principles and Applications
Edited by Bradley D. Smith

Published by the Royal Society of Chemistry, www.rsc.org

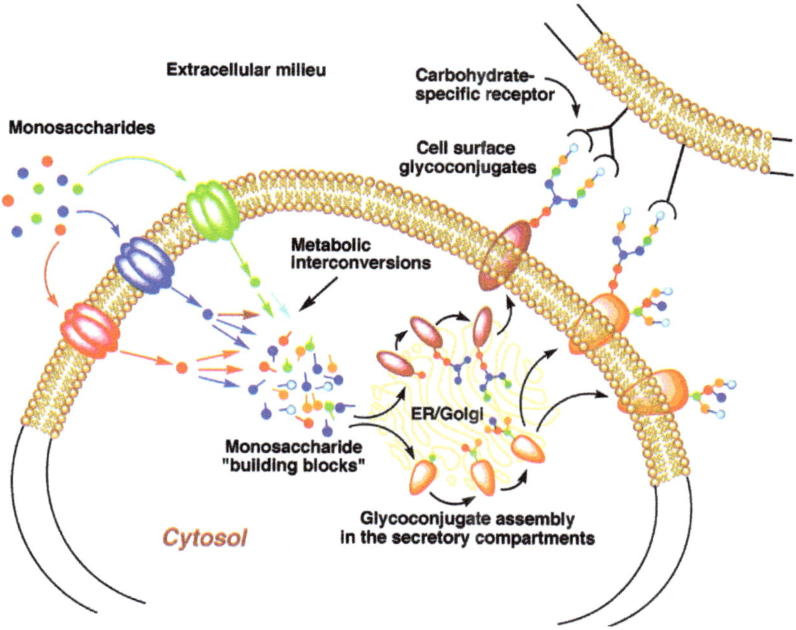

Figure 5.1 Summary of glycoconjugate biosynthesis and cell surface recognition. Synthetic receptors are needed to recognize the different carbohydrate structures for various diagnostic or therapeutic applications. (From *Science*, 2001, **291**, 2357. Reprinted with permission from AAAS.)[1]

carbohydrates and carbohydrate recognition in physiological and pathological processes, researchers are developing chemical tools to inhibit or control such processes.[3] The focus of this chapter is on the design and development of synthetic receptors that have an ability to selectively target the various types of carbohydrate structures, including monosaccharides, oligosaccharides, and glycoconjugates. The different receptor categories are described, along with their binding properties and how they can be utilized for diagnostic or therapeutic applications.

5.2 Chemical Structures and Physical Properties of Carbohydrates

Shown in Figure 5.2 are the four mutarotational isomers of glucose, the most abundant organic molecule on the planet. Like all the common monosaccharides, glucose is highly water soluble and uncharged at neutral pH. The structure shown in the box is the β-pyranose isomer. The equatorial hydroxyl groups are located around the periphery of the molecule and the axial CH residues project up and down to create hydrophobic faces. The β-pyranose structure is in spontaneous equilibrium with the other three isomers. This dynamic isomerization process is called mutarotation, a

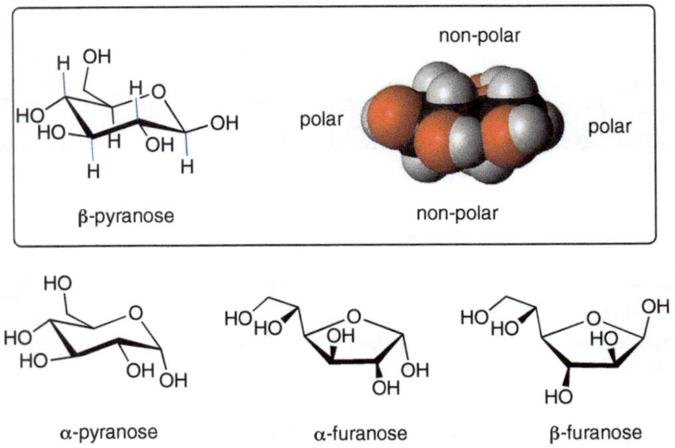

Figure 5.2 The four mutarotational isomers of glucose and a model of the β-pyranose structure illustrating the different patches of molecular polarity and potential for intramolecular OH···O hydrogen bonding.

property exhibited by all reducing sugars and a complicating feature of carbohydrate molecular recognition.

The high number of stereogenic centers in a monosaccharide means that there are a large numbers of possible isomeric structures and linkage points. Furthermore, certain hydroxyl groups can be modified and converted into amines, acylated amines, sulfates, or carboxylates. The major monosaccharides in the bloodstream of healthy humans are glucose, glucosamine, and fructose, with concentrations of 3.5–5.5, 4.2–6.2, and 0.03–0.3 mM, respectively.[4] The structural complexity increases rapidly as the number of saccharide units increases. For example, >10 million tetrasaccharides can be assembled from just nine monoscaccharides. Compiled in Figure 5.3 are the most abundant monosaccharide units, which are used as building blocks to create mammalian oligosaccharide structures.[4] The large number of different glycoconjugates on the surface of a cell form a dense coating called collectively the glycocalyx, and the over- or under-expression of a specific glycoconjugate is often a disease-specific biomarker and thus a valuable molecular target for biomedical imaging or drug targeting.

From the perspective of a molecular receptor there is very little difference between the exterior of a sugar molecule and a small cluster of water molecules (Figure 5.4).[5] Favorable carbohydrate binding to a receptor cavity has to overcome the energetic penalty for desolvating both the receptor and the carbohydrate guest. Furthermore, the molecular model in Figure 5.2 highlights the likelihood of intramolecular hydrogen bonding, which weakens the ability of a carbohydrate to form intermolecular hydrogen bonds.[6] As a result, intermolecular hydrogen bond strengths in a carbohydrate:receptor complex are usually weaker than a typical hydrogen bond. However, the

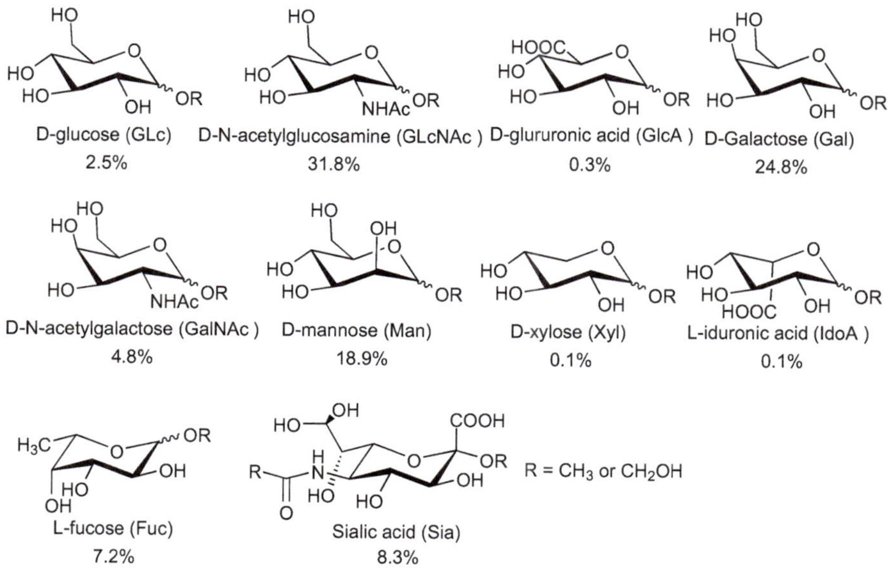

Figure 5.3 Structures of mammalian monosaccharide units and their abundance, as evaluated from a statistical analysis of a glycan databank of mammalian oligosaccharide structures.
(Reprinted from *Coord. Chem. Rev.*, 2014, **268**, 22, © 2014, with permission from Elsevier.)[4]

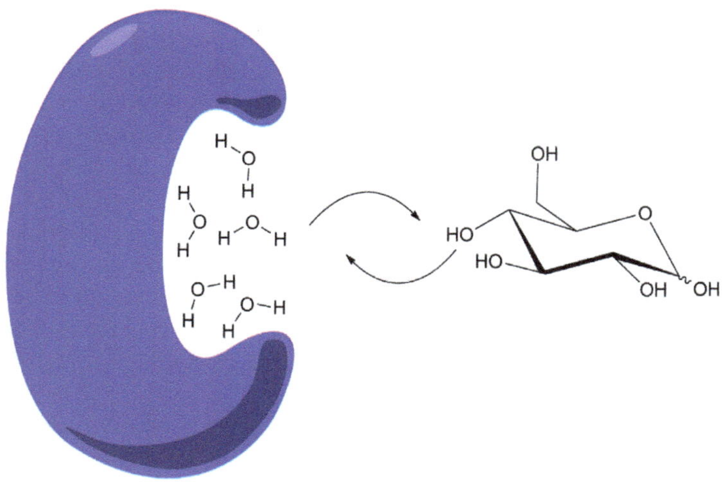

Figure 5.4 The challenge of carbohydrate recognition. The receptor must recognize a sugar molecule whose exterior is similar to a cluster of water molecules. (Adapted from ref. 5 with permission from the Royal Society of Chemistry.)

patchwork structure of monosaccharides, like β-glucopyranose, with non-polar axial faces and a polar equator provides a clue to how binding can be achieved using receptors with a complementary cavity surface. Effective

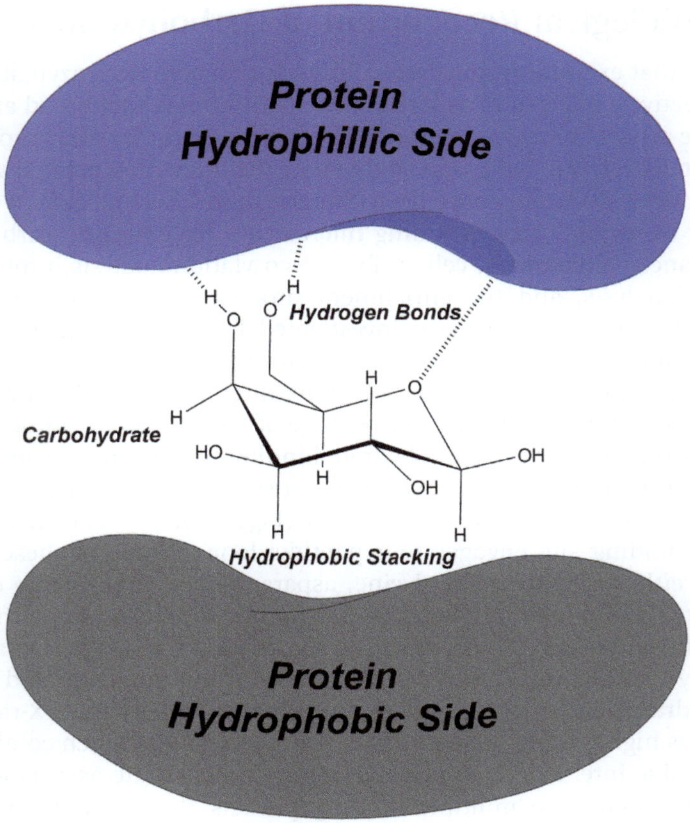

Figure 5.5 Effective receptor:carbohydrate association is driven by a synergistic network of polar and non-polar interactions.
(Adapted from *Anal. Bioanal. Chem.*, 2012, **402**, 3161, © 2012, Springer-Verlag, with kind permission from Springer Science and Business Media.)[7]

receptor/carbohydrate association is achieved by a synergistic network of polar and non-polar interactions (Figure 5.5).[7]

Another reversible interaction that is effective in water is coordination of carbohydrate oxygen atoms to a Lewis acid. With biological receptors, the Lewis acid is often a Ca^{2+} cation and the reversible coordination bonds not only increase affinity, but introduce binding selectivity, based on the carbohydrate stereochemistry. The effectiveness of Lewis acid coordination for sugar recognition has been exploited for practical applications. For example, the higher affinity of Ca^{2+} for fructose over glucose is the basis for the industrial separation process that makes high fructose corn syrup. As described below, Lewis acid coordination is an effective strategy that is often used by researchers trying to design synthetic receptors. Before the discussion of synthetic receptors, it is helpful to examine in more detail how nature achieves carbohydrate recognition.

5.3 Biological Recognition of Carbohydrates

Proteins that exhibit carbohydrate affinity but do not have enzymatic activity are collectively referred to as lectins. The definition is very broad and covers a diverse array of protein structures. The legume lectin family is probably the most extensively studied group, and to date the three-dimensional structures of well over 1000 lectins have been characterized.[8] Lectins can be used in a number of technologies including microarrays for detecting carbohydrate-based cancer biomarkers, cell surface glycosylation analysis, protein glycosylation analysis, and the enrichment process for proteomic studies.[9,10] In addition, lectins have been investigated for tumor targeting, antiviral therapy, anti-tumor therapy, and drug delivery.[11]

Generally speaking, lectin–carbohydrate interactions tend to be relatively weak, with binding constants in the range of 10^3–10^4 M^{-1}.[12] Although there is wide structural variety, carbohydrate binding typically occurs in a shallow pocket on the lectin surface and involves two major components: hydrogen bonding and hydrophobic interactions (Figure 5.5).[13] The actual carbohydrate binding site engages four peptidyl loops. Three of these typically contain either aspartic acid, glycine, asparagine, or an aromatic or hydrophobic residue, while the fourth loop is highly variable and determines the specificity of the lectin.[14] Lectins may also contain one or two metal cations (typically Ca^{2+} or sometimes Mn^{2+}) about 4 Å apart and located near the carbohydrate-binding site.[15] Shown in Figure 5.6 are two X-ray crystal structures highlighting the network of hydrogen bonds, which combine with hydrophobic interactions to produce lectin–carbohydrate association.

Lectins often have multiple binding pockets that allow them to have higher affinities for oligosaccharides over monosaccharides, largely due to multivalent interactions between binding partners.[16] Multivalent binding is also seen with glycoconjugates having multiple sugars on their surfaces. For example, there are 13 mannose residues on the surface of the protein carboxypeptidase Y and the association constant with the lectin concanavalin A is 2.7×10^6 M^{-1}, which is much higher than the binding constant with free mannose.[17] Some classes of lectins, such as galectins, use protein self-assembly to achieve multivalent binding of cell surface glycans (Figure 5.7).[18]

In addition to lectins, other protein systems can recognize carbohydrate targets. For example, antibodies are developed for carbohydrate antigen recognition.[19] However, carbohydrate antibodies often have low affinity and have weak specificity.[19] There are many enzymes that can bind monosaccharides and catalyze a range of oxidation and cleavage reactions. Moreover, glycosidases and hydrolases can recognize polysaccharides with special structural features. The protein recognition domain found in a carbohydrate-active site is called the carbohydrate-binding module. There are currently 64 families of CBM based on their amino acid sequence similarity.[20] The adenosine triphosphate (ATP)-binding-cassette (ABC) transporters, which are found in every living cell, use the energy of ATP binding and hydrolysis to transport molecules across membranes.

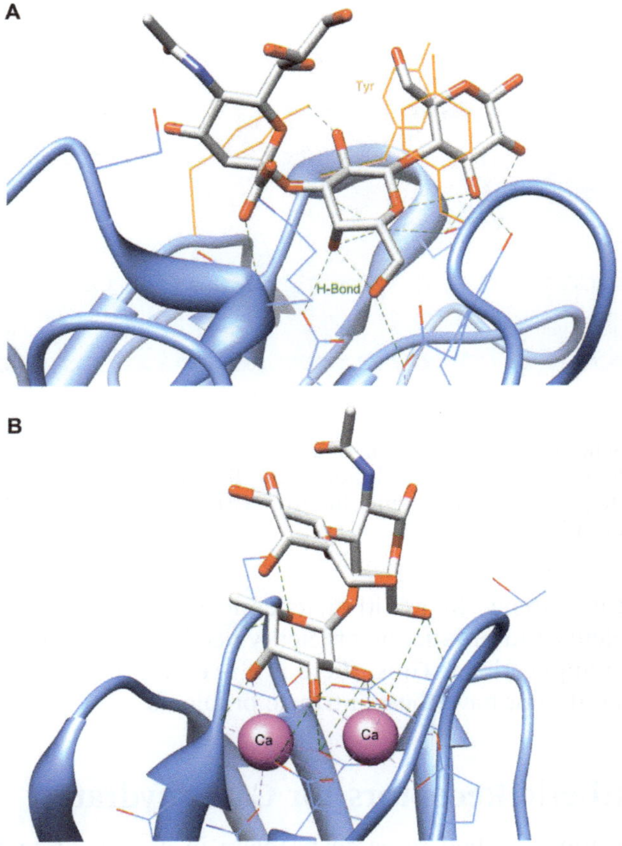

Figure 5.6 Examples of shallow lectin binding sites. (A) *Maackia amurensis* leukoagglutinin complexed with sialyllactose (NeuAcα2-3Galβ1-4Glc) (Protein Data Bank (PDB) code 1DBN). Note the network of hydrogen bonds and the hydrophobic contacts with three tyrosine side chains (orange). (B) *Pseudomonas aeruginosa* lectin B complexed with Lewis a (Le^a: Galβ1-3[Fucα1-4]GlcNAc) (PDB code 1W8H) with coordination of two calcium ions. (Adapted from ref. 13 with permission from the Royal Society of Chemistry.)[13]

ABC transporters are responsible for the export of a wide variety of cell-surface glycoconjugates in bacteria.[21] In contrast, the glucose transporter (GLUT) family of transmembrane proteins observed in most mammalian cells mediate the energy-independent transport of glucose and related hexoses. Binding of a hexose causes a conformational change in the protein, which releases the hexose on the other side of the membrane in a form of facilitated diffusion.[22]

Overall, lectins and antibodies are very useful tools for biomedical research and they have many potential therapeutic applications.[10] However, these natural "receptors" often have the drawback of broad substrate specificity, low affinity, and sometimes toxicity, which limit their applications in

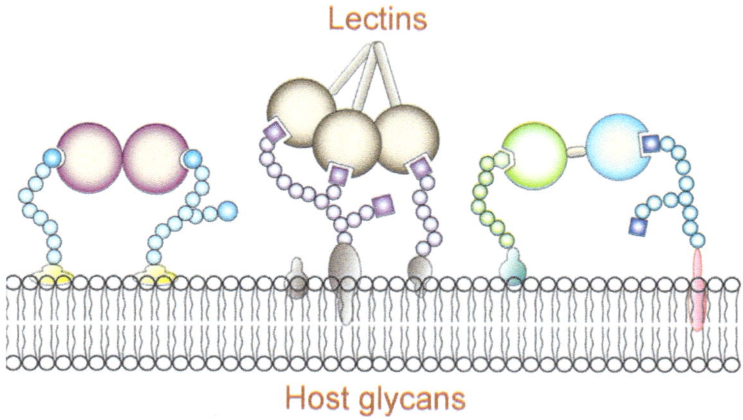

Figure 5.7 Schematic representation of different ways that lectins can self-assemble to achieve multivalent binding to cell surface glycans.
(Reproduced with permission from *Front. Immunol.*, 2012, **3**, 199 © 2012 Frontiers Media.)[18]

the therapeutic arena. In addition, chemical and biochemical stability, immunogenicity, and production efficiency need to be considered for large-scale human applications. Thus, there is a need for artificial carbohydrate receptors that do not have these intrinsic problems.

5.4 Synthetic Receptors for Carbohydrates

To date, the most studied research problem in carbohydrate detection is sensing of glucose levels in the bloodstream of diabetic patients.[23] There is still no cure for diabetes; however, proper control of glucose concentration can help tremendously reduce complications and prolong life. A critical step in controlling blood glucose level is appropriate monitoring. The ideal solution is a continuous, real-time, and non-invasive monitoring method. Potentially, there are several ways to approach the problem, such as direct spectroscopic methods and enzyme-based sensing devices.[24] In principle, chemosensors could be extremely effective if they could be packed inside miniaturized devices that report glucose levels in real time.[25] Synthetic receptors for other monosaccharide structures, as well as oligosaccharides, are expected to be very useful for various sensing, delivery, and triggered release applications (see Chapter 1).[26]

The field of glycobiology is expanding very quickly and there is increasing awareness that cell surface glycans are important biomarkers of disease. Thus, there is an increasing need for targeting agents and separation systems that can be incorporated into diagnostic and therapeutic devices. The first step in each of these applications is production of a synthetic carbohydrate receptor or "synthetic lectin". The emphasis of this chapter is on synthetic receptors that operate effectively under biomedical conditions, and this typically means aqueous solution. The following summary of synthetic receptors is separated into three molecular classes: boronic acids, organic macrocycle and cleft-shaped molecules, and metal coordination complexes.

5.4.1 Boronic Acid Receptors for Carbohydrates

Boronic acids are part of a broader group of boron containing compounds that are known to form covalent complexes with a range of anions, biomolecules, and enzymes. Boronic acid receptors for carbohydrates are sometimes referred to as "borolectins". Since the topic has been reviewed extensively over the last few years,[2,4,27,28] the goal here is to provide a concise overview of the fundamental concepts.

Boronic acids are Lewis acids and the ionization of phenylboronic acid in aqueous solution is represented in Figure 5.8. Boronic acids form reversible covalent complexes with 1,2- and 1,3-diols in aqueous solution, and due to the chelate effect, the pK_a for the diol-boronic acid equilibrium is usually lower than the pK_a for the boronic acid ionization alone (although there are exceptions).[29] The magnitude of the difference depends greatly on the structure of the diol. Typically, five-membered boronate rings are more stable than six-membered rings. Furthermore, if the diol is structurally incorporated into a cyclic molecule such as a sugar, then complexation is much stronger with *cis*-1,2-diols than with *trans*-1,2-diols. Shown in Figure 5.9 are representative stability constants for the reversible association of borate with saccharide diols to form borate esters.[4]

These structural factors explain why boronic acids, (a) tend to bind monosaccharides in the furanose form more tightly than the pyranose form, and (b) bind with monosaccharides in the following order of association constants: glucose < galactose < fructose.[30] As a rule, sugar complexation with aryl boronic acids is favored by electron-withdrawing substituents on the aryl ring and when the pH is above the pK_a of the boronic acid.[31,32] The pH dependency of the binding equilibrium is an important factor that must be considered during receptor design, and may present a reproducibility issue if there is a natural pH fluctuation in the sample. If the goal is to bind carbohydrates at physiological pH, the receptor's chemical structure may have to be fine-tuned for optimal performance.

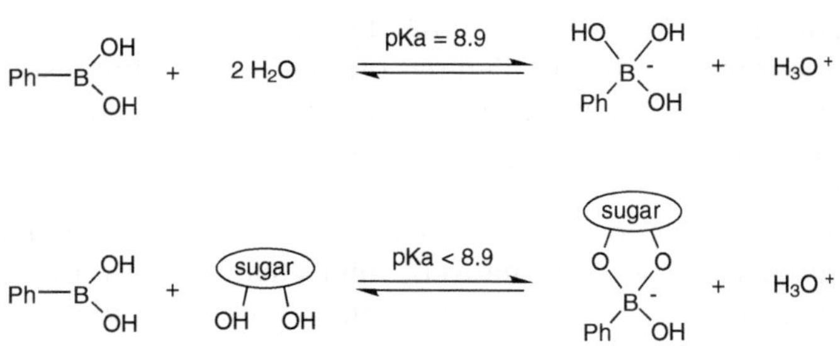

Figure 5.8 Ionization equilibria for phenyl boronic acid.

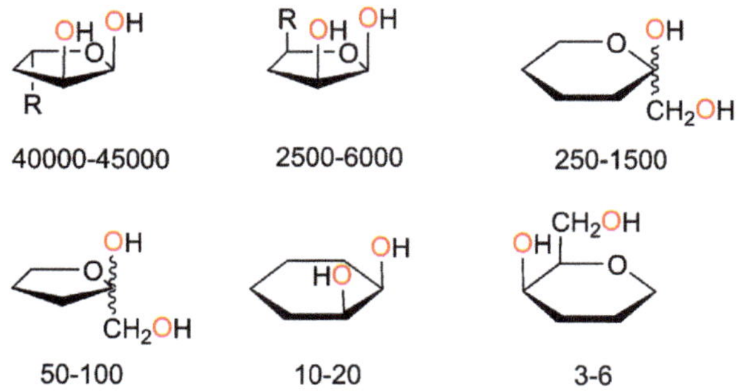

Figure 5.9 Diols that have the highest affinity for borate. The numbers under the structures denote the range of stability constants observed in M^{-1}. (Reprinted from *Coord. Chem. Rev.*, 2014, **268**, 22, © 2014, with permission from Elsevier.)[4]

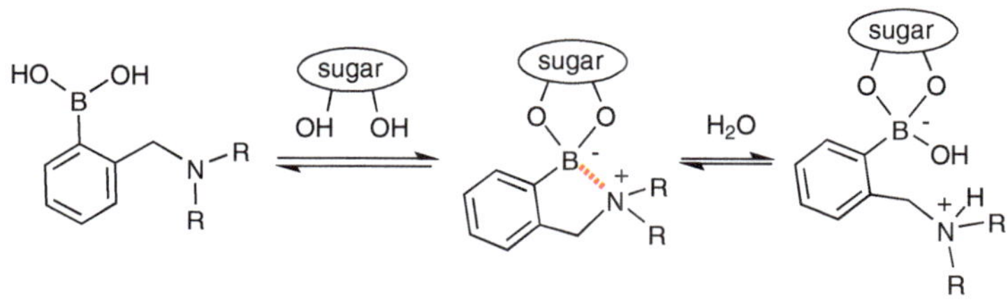

Figure 5.10 Sugar complexation by a monoboronic acid receptor with proximal nitrogen.

Shown in Figure 5.10 is an important structural modification, which promotes sugar binding at neutral pH. Boronic acids with an amino group in a 1,5-relationship produce stabilized boronate complexes due to formation of an internal boron–nitrogen bond. Under certain conditions there is evidence for a hydrated structure with a protonated amine, but in either case, the proximal nitrogen leads to enhanced sugar binding and a valuable mechanism for sugar sensing.

The reversible association of boronic acids with carbohydrates has been known for many decades, and has been utilized extensively for separations.[33] Affinity chromatography systems use immobilized boronic acids for selective retention of eluted sugars.[34] Acrylamide hydrogels have been developed with appended boronic acids for high-performance electrophoresis.[35] Boronic acids have been incorporated into molecularly imprinted polymers, self-assembled monolayers, and liquid membrane transport systems have been devised with boronic acids as membrane transport carriers.[35] Boronic acids have also been incorporated into many types of advanced materials for controlled release,[36] or alternatively, molecular devices that use sugar binding as a trigger to activate a subsequent supramolecular event such as membrane fusion.[37] But most of the early work on boronic acids for carbohydrate binding has focused on sensing or imaging.

Various optical sensing strategies have been examined, including methods that produce changes in fluorescence, absorption, and circular dichroism. The desirable properties for these optical boronic acid reporters include large spectroscopic changes upon binding—preferably in a ratiometric fashion, high water solubility, and easy structural modification. Shown in Figure 5.11 is a selection of boronic acids that have been used in optical sensing assays. Boronic acids that do not contain a chromophore can be incorporated into indicator displacement assays (see Chapter 1), involving carbohydrate displacement of a colored or fluorescent dye (indicator) from the receptor. This assay is easy to construct because of the availability of commercially available dyes with diol-like motifs that

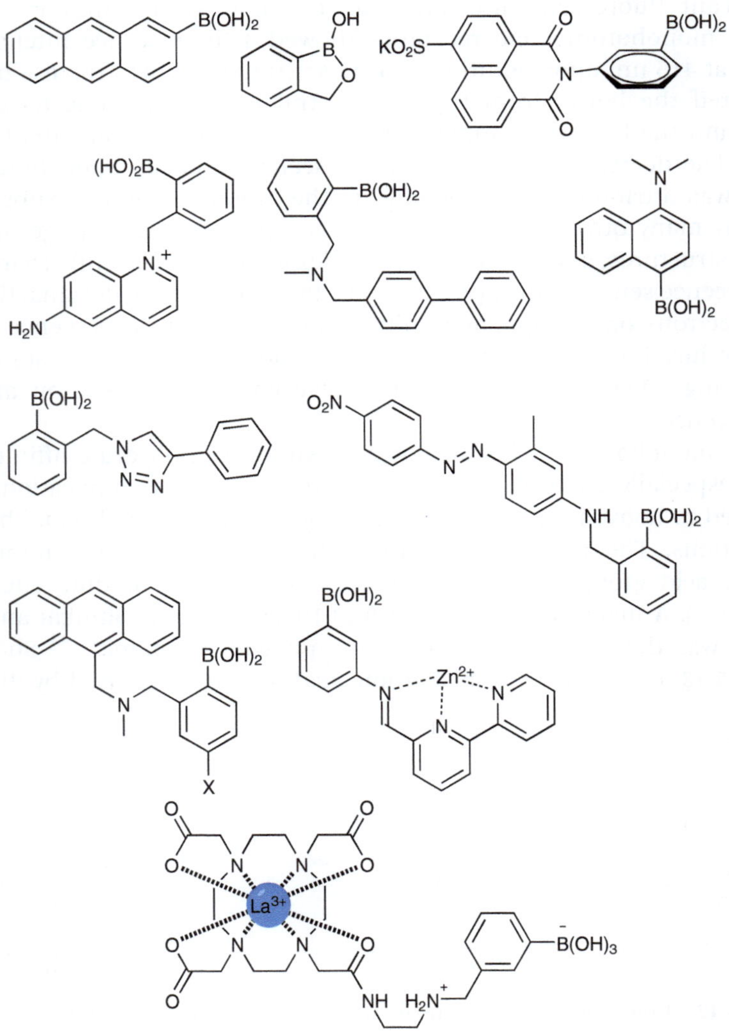

Figure 5.11 Selection of monoboronic acid receptors.

associate with a boronic acid and produce a signal change (*e.g.*, alizarin red-S). A limitation of this type of assay is that it works in batch-wise fashion and is not suitable for longitudinal sensing. This latter application is best achieved using a receptor that incorporates a chromophore into its structure. Furthermore, the boronic acid binding must produce an observable change in the chromophore signal. An example of a colorimetric sensor is shown in Figure 5.12. The receptor is an azo dye and sugar binding to the boronic acid leads to a color change from red to yellow that is easily observed with the naked eye.[38,39]

The versatility and high sensitivity of fluorescence sensing has motivated a major effort to develop fluorescent boronic acids for sugar sensor. In 1992, Yoon and Czarnik reported that anthracene boronic acids exhibited significant fluorescent intensity changes upon sugar binding.[40] These simple monoboronic acid receptors showed a fluorescence intensity decrease at 416 nm upon addition of a sugar. At neutral pH, fructose binding converted the boron from trigonal neutral form to anionic tetrahedral form, and the fructose complex was observed to have an apparent pK_a of ~5.9. The decreased anthracene fluorescence intensity upon binding of sugar was attributed to quenching by the anionic complex. Subsequent work by many other research groups has explored a wide range of fluorescent structures, and discovered examples with a more useful "switch-on" fluorescence sensing effect (Figure 5.11). In many of these designs, the lone pair electrons on the appended nitrogen can quench fluorescence due to photoinduced electron transfer to the fluorophore excited state. The quenching effect is diminished upon saccharide complexation and the fluorescence is restored.

Most monoboronic acids do not have strong carbohydrate affinity, and this is especially the case for glucose. Therefore, in many applications there is a need to boost affinity and selectivity by introducing additional binding interactions. The most obvious approach is to increase the number of boronic acid groups and thus produce a multivalent binding effect (see Chapter 2). A major advance in the mid-1990s, made by Shinkai and coworkers, was the development of bisboronic acid receptors.[41–43] Shown in Figure 5.13 are examples of bisboronic acid receptors reported by different

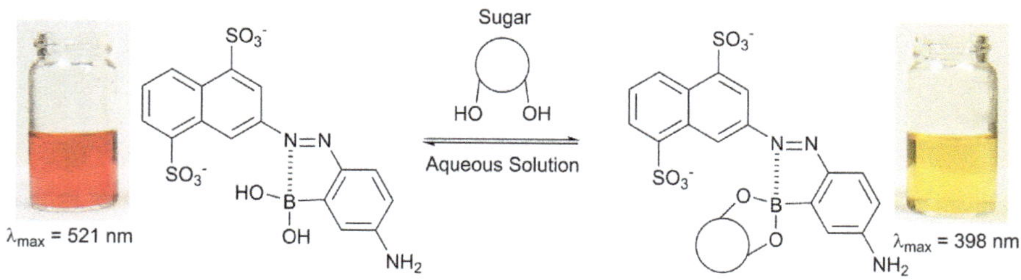

Figure 5.12 Colorimetric sugar sensing using a boronic acid receptor. (Reprinted from *Bioorg. Med. Chem. Lett.*, 2007, **17**, 3789, © 2007, with permission from Elsevier.)[38]

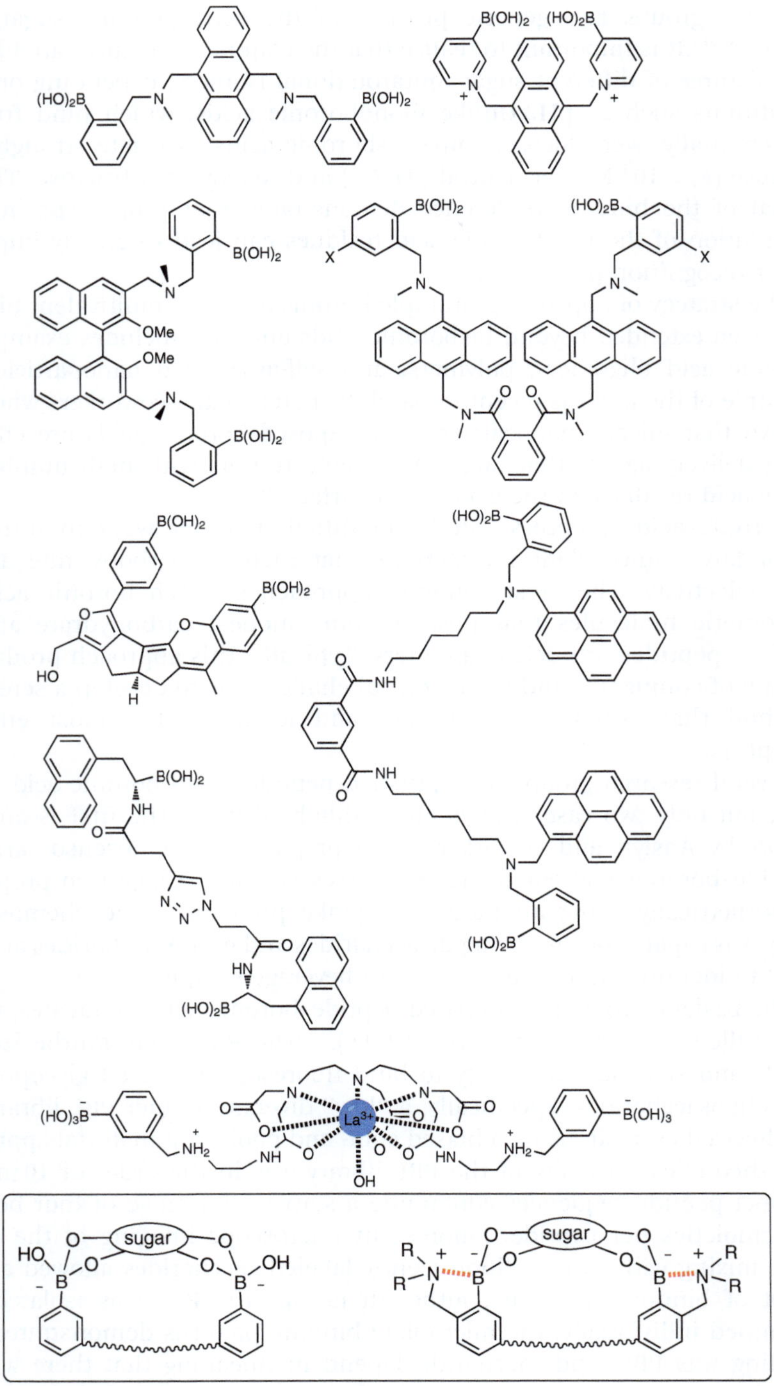

Figure 5.13 Selection of bisboronic acid receptors.

research groups and generic pictures of the corresponding sugar complexes.[44,45] It is important to realize that the complex structures are likely to be mixtures of different sugar mutarotational isomers, depending on local conditions such as pH. Unlike monoboronic acids, which bind fructose preferentially over glucose, some bisboronic acids associate strongly with glucose ($K_a > 10^3$ M^{-1} at neutral pH 7.8) and weakly with fructose. This reversal of the binding preference demonstrated that proper spacing and orientation of the two boronic acid residues can lead to greatly improved sugar recognition properties.

The strategy of appending multiple boronic acids for multivalent binding has been extended beyond bisboronic acids and now includes examples of boronic acid oligomers, polymers, and self-assembled nanoparticles. An example of the latter is recent research by Kataoka and coworkers who have shown that micelle nanocarriers with exposed boronic acids are effective drug delivery agents that target melanoma tumors with high numbers of sialic acid residues on the cancer cell surface.[46]

Boronic acid receptors can be modified in other ways to introduce secondary non-covalent interactions that increase carbohydrate affinity and selectivity. The most common approaches attach boronic acids to oligomeric molecules that possess some modest carbohydrate affinity, such as peptides or nucleic aptamers. Typically, this approach produces a library of conjugates and the technical challenge is to develop a screening method that evaluates the library and identifies the most effective receptors.

Several research groups have pursued peptide-based boronic acid receptors, but only two case studies are highlighted here. The first is an early report by Anslyn and coworkers, who prepared a chemosensor array of peptide–boronic acid conjugates and assessed their recognition properties colorimetrically using an indicator uptake protocol.[47] The chemosensor array was capable of classifying disaccharides and monosaccharides and was used to identify specific saccharides in beverage samples.

The Lavigne group also prepared peptide–boronic acid conjugates, which they called peptide boronolectins (PBLs).[48] The PBLs were synthesized on beads and screened for ability to bind fluorescently-labeled glycoproteins and oligosaccharides. Specifically, a "low" diversity 12-mer PBL library was synthesized on resin using a biased split- and pool-combinatorial approach. The theoretical diversity of the PBL library was in the order of 10 million distinct peptide sequences containing a statistical average of four boronic acid moieties per peptide. Fluorescent microscope imaging of the beads after mixing with various fluorescently labeled saccharides allowed assessment of binding selectivity. Although no specific PBL was isolated and examined individually, the microplate binding patterns demonstrated that binding was PBL- and saccharide-dependent (meaning that there was selectivity among the different PBLs in the pool). This type of pattern recognition strategy has the potential for development as a diagnostic assay for the presence of glycosylated biomarkers.

Another way to improve boronic acid recognition ability is to attach them to nucleic acid aptamers. The properties of aptamers and the synthesis and selection technology are described in Chapter 2. Suffice to state here that aptamers are oligonucleotides that fold into three-dimensional structures and create binding sites.[49,50] A large library of oligonucleotides can be generated (over 10^{12} sequences), and established methods are used for selecting sequences with affinity for a suitable target binding ability. Typically, the aptamers selected for carbohydrate binding are weak binders.[51] For example, RNA aptamers for glucose, mannose, and galactose have low affinities with dissociation constants in the range between 10^{-4} and 10^{-5} M.[52] To circumvent this deficiency, Wang and coworkers produced aptamer–boronic acid conjugates.[53] They found that boronic acid-modified thymidine triphosphate (B-TTP) with different linkers could be incorporated into the DNA chain by polymerase chain reaction and that the modified DNA underwent intrinsic fluorescent changes upon carbohydrate binding.[54] Furthermore, B-TTP aptamers were isolated against fibrinogen with K_d values in the nanomolar range.[55] The aptamers did not bind with good reproducibility, possibly due to the heterogeneous nature of fibrinogen glycosylation, the sensitivity to buffer, and the easy denaturation of fibrinogen. However, the results show that covalent incorporation of a boronic acid moiety into an aptamer sequence increases the likelihood of producing a high affinity for carbohydrate receptor. Most recently, an elegant non-covalent version of this conjugation strategy was reported by Yang and coworkers. They developed an aptamer that forms a high-affinity ternary complex with glucose bound to a bisboronic acid receptor.[56]

While most boronic acid sensing systems use optical reporter groups it is worth noting that alternative detection methods have been developed. As indicated in Figure 5.8, association of a boronic acid with a sugar leads to a drop in pH which can be readily detected by various technologies. In addition, the increased boronic acid ionization can cause changes in turbidity or electrochemical response.[57] One interesting new idea is the development of molecular probes that are boronic acid-substituted shape shifting bullvalenes. The probes bear a ^{13}C label and are employed in nuclear magnetic resonance (NMR) sensor arrays for polyhydroxylated compounds such as carbohydrates, flavanols, and sialic acids.[58]

In recent years, researchers have begun to move on from solution state sensing of mono- and disaccharides to other challenging tasks in carbohydrate recognition; one of these is molecular imaging of glycoconjugates in biomedical samples. These glycoconjugates often serve as biomarkers for diseases such as cancer, thus their detection is of great importance. Wang and coworkers have developed fluorescent bisboronic acid receptors that can selectively bind sialyl Lewis X (sLe^X), which is a tumor-associated biomarker.[59–62] Specifically, cell surface sLe^X is known to promote cellular adhesion in embryotic stages,[63] and overexpression of sLe^X has been associated with the development of a number of cancers.[64,65] The fluorescent bisboronic acid receptors were tested against three different cancer cell lines

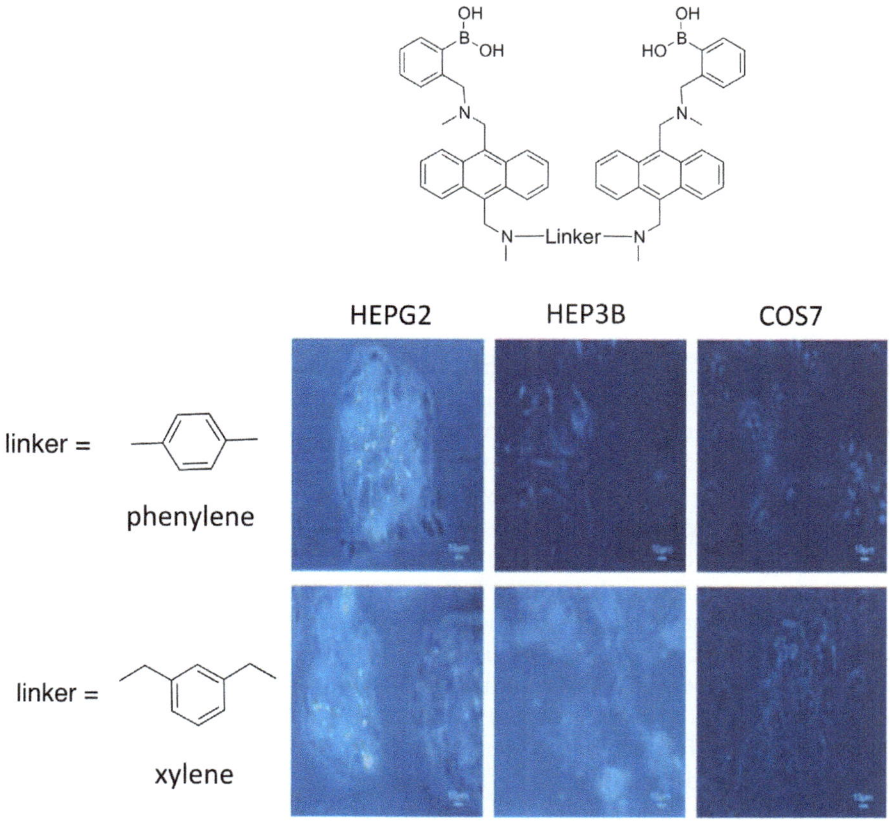

Figure 5.14 HEPG2 cells express only sLe^{X}, HEP3B cells express only Lewis^{Y}, and COS7 cells do not express either antigen. The micrographs show cell-selective staining with 5 μM of fluorescent bisboronic acid receptors (ex 370 nm, em 426 nm).
(Reprinted from *Bioorg. Med. Chem. Lett.*, 2002, **12**, 2175, © 2002, with permission from Elsevier.)[59]

HEPG2 (sLe^{X}-expression only), HEP3B (Lewis^{Y} only), and COS7 (no sLe^{X} or Lewis^{Y} expression). The fluorescence micrographs in Figure 5.14 show that the intensity of the cell staining depended on the structure of the bisboronic acid receptor. The receptor with the shorter phenylene spacer only stained the cell line expressing sLe^{X} (HEPG2), the and receptor with the longer xylene spacer stained the cell lines expressing Lewis^{Y} (HEP3B) or sLe^{X} (HEPG2). None of the receptors showed affinity for COS7 cell lines, providing proof that these compounds are sLe^{X} and Lewis^{Y} selective.

Further work modified the structure of sLe^{X} -selective bisboronic acid by conjugating a boron-dipyrromethene (BODIPY) fluorophore to create a fluorescent imaging probe that enabled tumor imaging *in vivo*.[66] The fluorescent probe was injected into mice with implanted tumors, which overexpress sLe^{X}. After a 24 h washout period, it was found that the

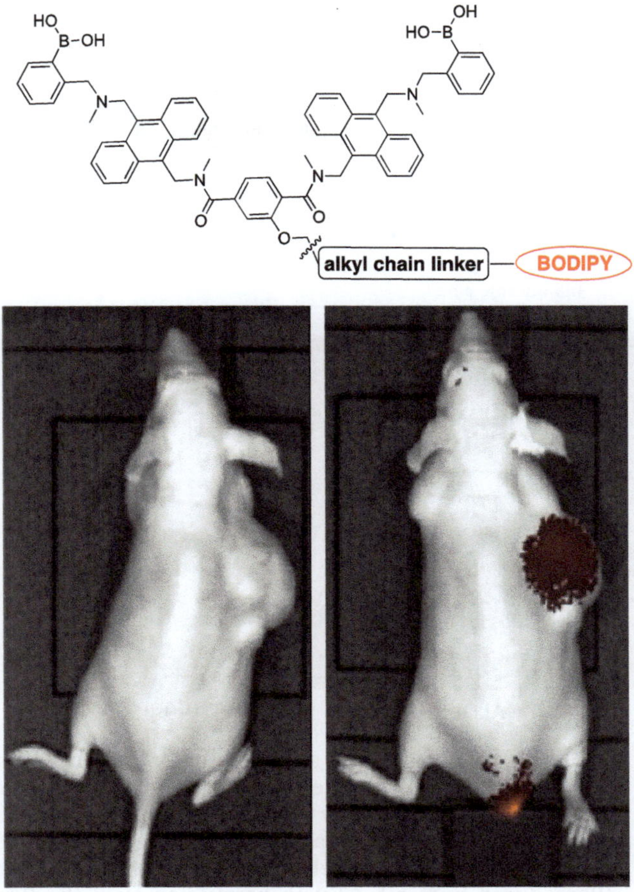

Figure 5.15 Structure of fluorescent probe and fluorescent images of a mouse bearing a tumor that overexpresses sLe^x. The left image is the mouse before probe injection and the right image is the mouse at 24 h after tail vein injection of the probe, showing accumulation at the tumor site. (Reprinted from *Bioorg. Med. Chem. Lett.*, 2013, **23**, 6307, © 2013, with permission from Elsevier.)[66]

fluorescent probe was concentrated in the tumor area, allowing ready detection of the tumor (Figure 5.15).[66] This work demonstrates the tremendous potential of boronic acid probes for *in vivo* optical imaging of carbohydrate biomarkers.

Wang and coworkers have also used bisboronic compounds in combination with matrix-assisted laser desorption ionization (MALDI)-mass spectrometry (MS) technology for histological imaging of cancer tissues that express sLe^X.[62] Figure 5.16 shows a bisboronic acid that was modified by linking to a matrix-free mass spectrometric tag (a trityl group) through a thio-ether, which can be cleaved by the MALDI laser, giving rise to a trityl carbocation detectable by the mass spectrometric analyzer. The tag was

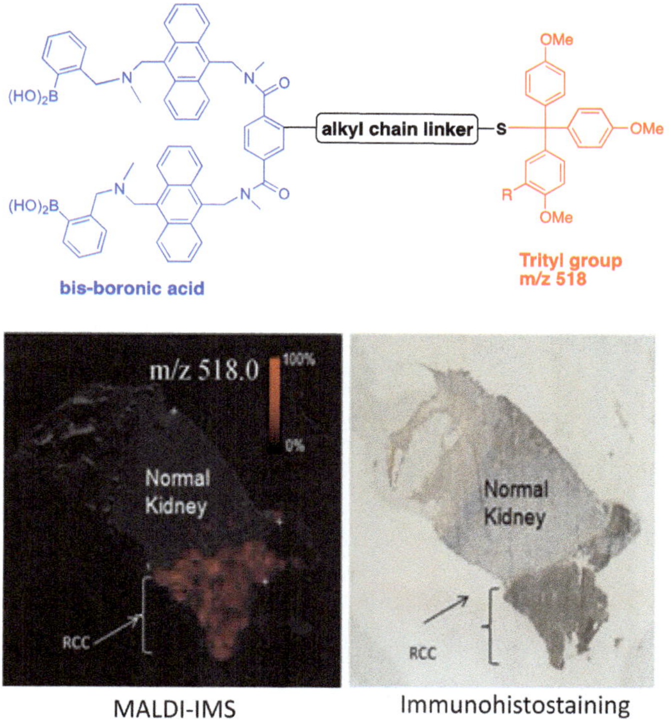

Figure 5.16 Top: structure of bisboronic acid receptor for use as a mass spectroscopic tag. Bottom: matrix-assisted laser desorption ionization (MALDI)-imaging mass spectrometry (IMS) and immunohistostaining images of kidney tissue showing overlap in the tumor region (renal cell carcinoma (RCC)) and not in normal cell areas.
(Reproduced from ref. 62 with permission from the Royal Society of Chemistry.)[62]

tested using a histology section of renal tumor tissue that contains an area of normal healthy cells and a region of cancer cells. After incubation of the bisboronic acid probe with the tissue, MALDI-MS imaging showed that the *m/z* 518 fragment from the probe corresponded to the same tumorigenic region that was visualized by standard immunostaining assay (Figure 5.16). The combination of synthetic molecule boronic acid receptors to target biomarkers, such as sLe^X, and analytical methods, such as MALDI-MS, is a promising approach for developing novel cancer diagnostics.

5.4.2 Organic Macrocycle and Cleft-shaped Molecules

The past two decades has seen steady development of biomimetic receptors that use weak non-covalent interactions for carbohydrate binding. Initially, most association systems were studied in organic solvent because the non-covalent interactions are stronger.[4,67,68] However, increasing success has led to demonstrations of effective binding in competitive solvents, including water.[4,67–69] One aim of this research is to illuminate the factors controlling carbohydrate binding by biological receptors and a long-term goal is to eventually produce receptors that have practical applications.

As discussed earlier, biological receptors associate with carbohydrates by forming a network of hydrogen bonds and "hydrophobic" interactions.[70–75] One common strategy for developing synthetic receptors that mimic protein binding pockets is to employ macrocyclic or cleft-shaped structures with appropriate sites for non-covalent interactions. A few key factors need to be considered when using this method to create synthetic carbohydrate receptors; (a) the size of the cavity of the receptor should be large enough to accommodate the carbohydrate target; (b) the receptor structure should be rigid enough to prevent intramolecular interactions that would block access to the cavity; and (c) the structural functionality of the receptor should have the desired mix of polar and non-polar interactions.[76]

Shown in Figure 5.17 is a small collection of organic macrocycle and cleft-shaped structures that have been reported to bind to mono- or disaccharides in water or aqueous/organic solvents.[4,13,67,68,76–80] In addition, polymeric systems have also been developed to recognize naturally occuring polysaccharides.[81] The binding events have been studied using various spectroscopic techniques, such as NMR spectroscopy, circular dichroism, isothermal titration calorimetry (ITC), and absorption spectroscopy. It is

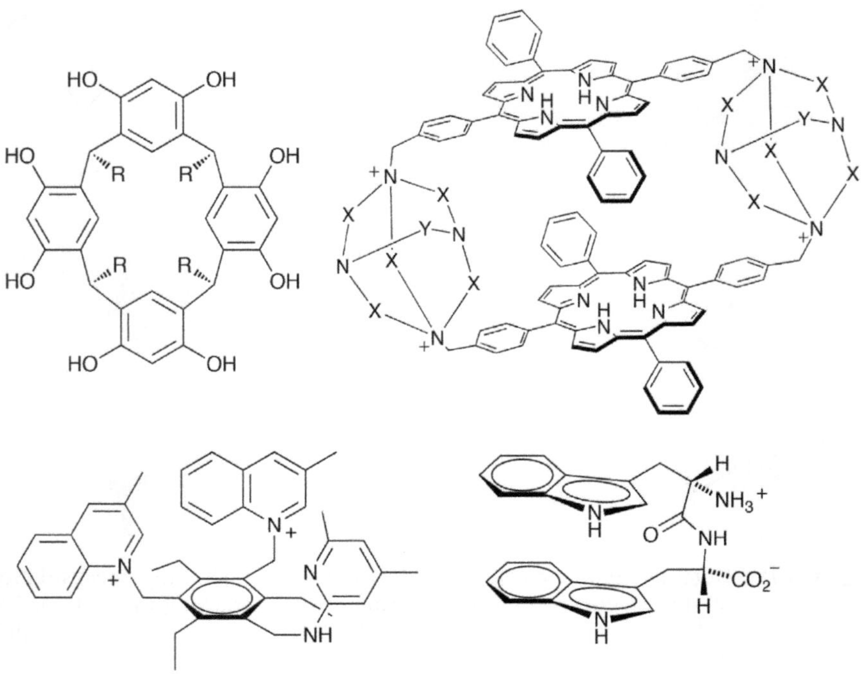

Figure 5.17 Organic receptors for carbohydrates.

worth noting that interpretation of these studies in water can sometimes be complicated due to receptor self-aggregation.

To date, the most effective organic receptors for carbohydrates are the organic cage molecules of Davis and coworkers, which are designed to fully encapsulate the guest saccharide (Figure 5.18). These cage molecules are referred to as "temple receptors" because they have an aromatic roof and floor that can form CH–π interactions and polar side-columns that can form hydrogen bonds. The prototype version in Figure 5.18 showed impressive binding of O-linked β-*N*-acetylglucosamine (GlcNAcβ-OMe) with a K_a of 630 M^{-1} in water.[76,82] The host/guest structures were determined by detailed NMR methods and the binding constants confirmed by ITC. Furthermore, ITC showed that binding was primarily enthalpy-driven. The next-generation design included fluorescent anthracene components and exhibited relatively good D-glucose binding affinity of ~55 M^{-1} and a selectivity of 50 : 1 over D-galactose or D-mannose.[83] The selectivity for guest structure with all-equatorial substituents supports the proposed picture of guest encapsulation inside the cage. The fluorescence response suggests a path towards

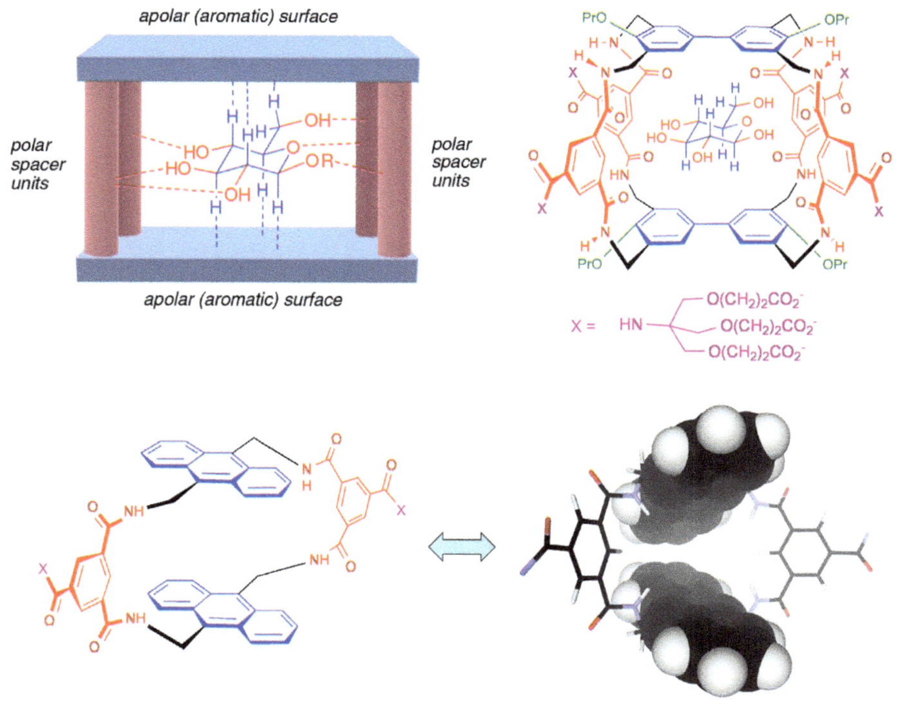

Figure 5.18 Organic "temple receptors" for carbohydrates, showing encapsulation of glucose guest and representation of CH–π stacking interaction. Top right: prototype design. Bottom: fluorescent next-generation receptor with anthracene components.
(Reprinted by permission from Macmillan Publishers Ltd: *Nat. Chem.*, 2012, **4**, 718, © 2012.)[83]

further development as sugar sensor. Together, the "temple receptors" illustrate how rational design and synthetic organic chemistry can be combined to produce lectin mimics that have millimolar carbohydrate affinities that are close to the natural protein systems.

5.4.3 Metal Cation Coordination Complexes

Synthetic receptors that incorporate metal centers are well-known to bind anions and electron-rich guests in aqueous solution (Chapter 4). The strength and directionality of metal coordination bonds leads to useful affinities and guest binding selectivities. In the case of carbohydrate binding, synthetic receptors with metal cations as binding sites can be considered as mimics of calcium containing lectins. To date, only a few carbohydrate binding studies have been reported. One illustrative example is the indicator displacement assay described in Figure 5.19.[68] The assay employed an admixture of *N*-methyl anthranilic acid as the indicator dye coordinated to a simple Pd(II) coordination complex, and screening revealed that the dye fluorescence emission response was dependent on carbohydrate identity.[84]

More recent studies have examined the carbohydrate recognition properties of dinuclear metal complexes.[85] The example shown in Figure 5.20, with Mn(II) or Mg(II) cations, resembles the active site of xylose–glucose

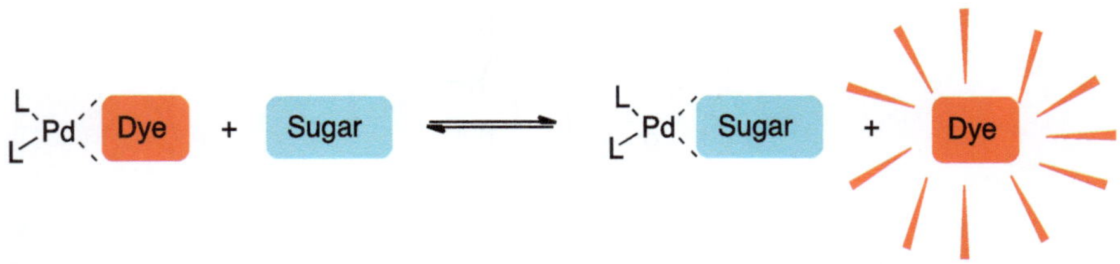

Figure 5.19 Indicator displacement assay showing a fluorescence increase due to displacement of fluorescent dye by carbohydrate.
(Reproduced from ref. 84 with permission from the Royal Society of Chemistry.)

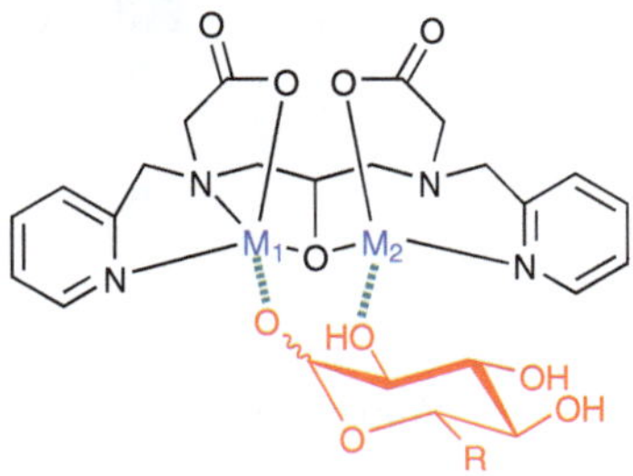

Figure 5.20 Structural mimic of the xylose–glucose isomerase binding site and proposed structure of D-glucose complex.
(Reprinted from *Carbohydr. Res.*, 2014, **384**, 87, © 2014, with permission from Elsevier.)[86]

isomerases. Fluorescence titration experiments showed selective binding to D-glucose and D-xylose in alkaline solution with binding constants of 190 M^{-1} and 1600 M^{-1}, respectively.[86] Obviously the requirement for strongly basic pH is a practical limitation, but the work is an interesting mimic of the active site of sugar metabolizing enzymes. The design of synthetic enzyme mimics that catalyze important carbohydrate reactions is likely to become an emerging future research topic.

The remarkable photonic properties of lanthanide coordination complexes make them attractive candidates for development as synthetic receptors for sensing applications. They appear to be well suited for binding to the sialic acids that are involved in many biological and pathological phenomena. This was demonstrated by a recent report of several lanthanide-macrocyclic polyazacarboxylate hexadentate complexes with europium, dysprosium, or terbium as the metal centers.[87] Shown in Figure 5.21 is a 221% emission enhancement produced by binding of *N*-acetylneuraminic acid (Neu5Ac) to a Dy(III) complex (L) at pH 10.2. In comparison, association with neutral monosaccharides was significantly weaker. NMR studies suggested that Neu5Ac binding involved: (a) a singular coordination of the acetamide oxygen of the monosaccharide to the metal center, and (b) hydrogen bonding between a hydroxide on the lanthanide and a saccharide glycerol unit. This study and related work using lanthanide based

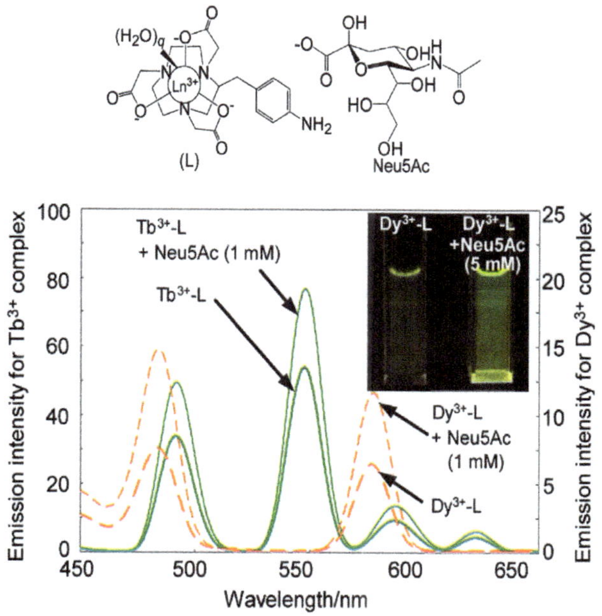

Figure 5.21 Structures of the lanthanide-macrocyclic polyazacarboxylate hexadentate complex (L) and *N*-acetylneuraminic acid (Neu5Ac). Emission enhancement upon binding.
(Reprinted with permission from *Inorg. Chem.*, 2013, **52**, 6239. © 2013 American Chemical Society.)[87]

receptors[88] provides a foundation for new approaches towards sensing and imaging of cancer biomarkers.

5.5 Future Directions and Challenges

It is surprising that the intense research investment over the past two decades on synthetic chemosensors for glucose has not yet produced a commercial device. It is possible that this may never happen, because alternative glucose monitoring methods based on less invasive technologies may be more competitive in the market place. Whatever the clinical outcome, there is still an unmet need for new carbohydrate sensors that operate in biomedical and environmental research settings. There is a general consensus that the field of glycobiology is in its infancy and continued expansion is expected for some time. New carbohydrate recognition technologies will improve purification and diagnostic screening of biomedical samples. These technologies require chemically well-defined and robust immobilized affinity surfaces and microarrays. In addition, there is a need for new classes of targeted carbohydrate probes that can effectively stain histopathology samples or act as injectable agents for molecular imaging of disease in living subjects.

Carbohydrate recognition in water is very challenging and the field needs new classes of synthetic receptors with improved binding affinity and selectivity. In many cases, the step-wise approach of rational design, synthesis, and evaluation is too slow. It is likely that new receptors will be discovered by the inherently faster method of library synthesis and high-throughput screening. Promising carbohydrate receptors have recently been produced by Ravoo and coworkers using "dynamic combinatorial chemistry",[89] a strategy that self-assembles templated receptors under thermodynamic control.[90] Separate batches of 12 peptides containing thiol groups were allowed to undergo reversible disulfide bond formation in the presence of an unreactive carbohydrate template in aqueous solution. The equilibrating systems produced macrocyclic peptides with impressive affinities for the specific monosaccaride and disaccharide template.[91] Further success is anticipated using this innovative method.

To date, most applications of synthetic carbohydrate receptors have been in imaging and sensing. Looking to the future, it is likely that there will be advances in targeted delivery and therapeutics. This can be achieved by simply substituting the reporter groups used for imaging with other types of cargo that have therapeutic activity. A promising example is the recent work of Raines and coworkers, who showed that covalent attachment of about seven boronic acids to the enzyme RNase A produced a bioconjugate that could bind to the glycocalyx of mammalian cells and subsequently enter the cytosol. Once inside the cell the enzymatic action of the RNase induced cell death.[92] A subsequent study by Perez and coworkers showed that a tau protein conjugate with multiple boronic acids is able to enter non-neural or neural cultured cells lacking tau protein.[93] In addition, boronated-tau accumulated in the hippocampal neurons of tau knockout mice after intracranial injection *in vivo*. These findings suggest that boronated proteins may have generally enhanced cell-surface binding and cytosol delivery capabilities. This is an exciting possibility, since it is relatively straightforward to incorporate 4-boronophenylalanine into the genetic code of *Escherichia coli* and create proteins bearing this unnatural boronated amino acid.[94]

It is important to remember that most lectins have only modest affinities for their carbohydrate targets, and thus it is likely that significant biological activity can be obtained by incorporating currently known synthetic lectins into appropriately designed bioconjugates.

Acknowledgements

We are grateful for funding support from the National Institutes of Health (GM059078 to BDS and DK55062, CA88343, NO1-CO-27184, CA113917, CA123329, GM086925, GM084933, and CA159567 to BW).

References

1. C. R. Bertozzi and L. L. Kiessling, *Science*, 2001, **291**, 2357.
2. *Carbohydrate Recognition: Biological Problems, Methods, and Applications*, ed. B. Wang and G.-J. Boons, Wiley, Hoboken, 2011.
3. *Chemosensors: Principles, Strategies, and Applications*, ed. B. Wang and E. V. Anslyn, Wiley, Hoboken, 2011.
4. J. A. Peters, *Coord. Chem. Rev.*, 2014, **268**, 1.
5. A. P. Davis, *Org. Biomol. Chem.*, 2009, **7**, 3629.
6. M. M. Deshmukh, L. J. Libero and S. R. Gadre, *J. Phys. Chem. A*, 2008, **112**, 10.
7. X. Q. Zeng, C. A. S. Andrade, M. D. L. Oliveira and X. L. Sun, *Anal. Bioanal. Chem.*, 2012, **402**, 3161.
8. S. Perez, A. Sarkar, C. Breton, S. Drouillard, A. Rivet and A. Imberty, *Glyco3D: A Portal for Structural Glycoscience*, 2013, http://glyco3d.cermav.cnrs.fr.
9. H.-J. Gabius, S. André, J. Jiménez-Barbero, A. Romero and D. Solís, *Trend Biomed. Sci.*, 2011, **36**, 298.
10. H. Ghazarian, B. Idoni and S. B. Oppenheimer, *Acta Histochem.*, 2011, **113**, 236.
11. C. Bies, C. M. Lehr and J. F. Woodley, *Adv. Drug Delivery Rev.*, 2004, **56**, 425.
12. E. J. Toone, *Curr. Opin. Struct. Biol.*, 1994, **4**, 719.
13. J. Arnaud, A. Audfray and A. Imberty, *Chem. Soc. Rev.*, 2013, **42**, 4798.
14. N. M. Young and R. P. Oomen, *J. Mol. Biol.*, 1992, **228**, 924.
15. H. Lis and N. Sharon, *Chem. Rev.*, 1998, **98**, 637.
16. D. K. Tarun and C. F. Brewer, *Adv. Carbohydr. Chem. Biochem.*, 2010, **63**, 15.

17. K. Lebed, A. J. Kulik, L. Forro and M. Lekka, *J. Colloid Interface Sci.*, 2006, **299**, 41.
18. G. R. Vasta, H. Ahmed, M. Nita-Lazar, A. Banerjee, M. Pasek, S. Shridhar, P. Guha and J. A. Fernandez-Robledo, *Front. Immunol.*, 2012, **3**, 199.
19. G. Cunto-Amestry, T. K. Dam, P. Luo, B. Monzavi-Karbassi, C. F. Brewer, T. C. Van Cott and T. Kieber-Emmons, *J. Biol. Chem.*, 2001, **276**, 9.
20. B. L. Cantarel, P. M. Coutinho, C. Rancurel, T. Bernard, V. Lombard and B. Henrissat, *Nucleic Acids Res.*, 2009, **37**, 6.
21. L. Cuthbertson, M. S. Kimber and C. Whitfield, *PNAS*, 2007, **104**, 6.
22. B. Thorens and M. Mueckler, *Am. J. Physiol. Endocrinol. Metab.*, 2010, **298**, E141.
23. G. Danaei, M. M. Finucane, Y. Lu, G. M. Singh, M. J. Cowan, C. J. Paciorek, J. K. Lin, F. Farzadfar, Y. H. Khang, G. A. Stevens, M. Rao, M. K. Ali, L. M. Riley, C. A. Robinson and M. Ezzati, *Lancet*, 2011, **378**, 31.
24. K. R. Ervin and E. J. Kiser, *Diabetes Technol. Ther.*, 1999, **1**, 3.
25. Y. J. Heo and S. Takeuchi, *Adv. Healthcare Mater.*, 2013, **2**, 43.
26. A. Matsumoto, K. Kataoka and Y. Miyahara, *Polym. J.*, 2014, **46**, 483.
27. T. D. James, M. D. Phillips and S. Shinkai, *Boronic Acids in Saccharide Recognition*, RSC, Cambridge, 2006.
28. X. Yang, Y. Cheng, S. Jin and B. Wang, in *Artificial Receptors for Chemical Sensors*, ed. V. M. Mirsky and Yatsimirsky, Wiley-VCH Verlag GmbH & Co., Weinheim, 2010, pp. 169.
29. J. Yan, G. Springsteen and B. Wang, *Tetrahedron*, 2004, **60**, 11205.
30. J. C. Norrild and H. Eggert, *J. Am. Chem. Soc.*, 1995, **117**, 1479.
31. J. Yan, H. Fang and B. H. Wang, *Med. Res. Rev.*, 2005, **25**, 490.
32. R. Pizer and L. Babcock, *Inorg. Chem.*, 1977, **16**, 1677.
33. R. Nishiyabu, Y. Kubo, T. D. James and J. S. Fossey, *Chem. Commun.*, 2011, **47**, 1106.
34. Q. Zhang, N. Tang, J. W. Brock, H. M. Mottaz, J. M. Ames, J. W. Baynes, R. D. Smith and T. O. Metz, *J. Proteome Res.*, 2007, **6**, 2323.
35. Y. Guan and Y. Zhang, *Chem. Soc. Rev.*, 2013, **42**, 8106.
36. R. J. Ma and L. Q. Shi, *Polym. Chem.*, 2014, **5**, 1503.
37. Z. Y. Zhang and B. D. Smith, *J. Am. Chem. Soc.*, 1998, **120**, 7141.
38. Y. Egawa, R. Gotoh, S. Niina and J. Anzai, *Bioorg. Med. Chem. Lett.*, 2007, **17**, 3789.
39. Y. Egawa, R. Miki and T. Seki, *Materials*, 2014, 7, 1201.
40. J. Yoon and A. W. Czarnik, *J. Am. Chem. Soc.*, 1992, **114**, 5874.
41. T. D. James, K. R. A. Samankumara Sandanayake and S. Shinkai, *Chem. Commun.*, 1994, 477.
42. T. D. James, K. R. A. Samankumara Sandanayake, R. Iguchi and S. Shinkai, *J. Am. Chem. Soc.*, 1995, **117**, 8982.
43. T. D. James, H. Shinmori and S. Shinkai, *Chem. Commun.*, 1997, 71.
44. G. Kaur, H. Fang, X. Gao, H. Li and B. Wang, *Tetrahedron*, 2006, **62**, 2583.
45. T. D. James, K. Sandanayake and S. Shinkai, *Angew. Chem., Int. Ed. Engl.*, 1994, **33**, 2207.

46. S. Deshayes, H. Cabral, T. Ishii, Y. Miura, S. Kobayashi, T. Yamashita, A. Matsumoto, Y. Miyahara, N. Nishiyama and K. Kataoka, *J. Am. Chem. Soc.*, 2013, **135**, 15501.
47. N. Y. Edwards, T. W. Sager, J. T. McDevitt and E. V. Anslyn, *J. Am. Chem. Soc.*, 2007, **129**, 13575.
48. Y. Zou, D. L. Broughton, K. L. Bicker, P. R. Thompson and J. J. Lavigne, *ChemBioChem*, 2007, **8**, 2048.
49. W. Sun, L. P. Du and M. Y. Li, *Curr. Pharm. Des.*, 2010, **16**, 2269.
50. T. Hayashi, H. Oshima, T. Mashima, T. Nagata, M. Katahira and M. Kinoshita, *Nucleic Acids Res.*, 2014.
51. O. Oyelaran, Q. Li, D. Farnsworth and J. C Gildersleeve, *J. Proteome Res.*, 2010, **8**, 10.
52. F. Kleinjung, S. Klussman, V. A. Erdmann, F. W. Scheller, J. P. Fürste and F. F. Bier, *Anal. Chem.*, 1998, **70**, 4.
53. Y. Cheng, C. Dai, H. Peng, S. Zheng, S. Jin and B. Wang, *Chem. - Asian J.*, 2011, **6**, 6.
54. X. Yang, C. Dai, A. D. C. Molina and B. Wang, *Chem. Commun.*, 2010, **46**, 3.
55. M. Li, N. Lin, Z. Huang, L. Du, C. Altier, H. Fang and B. Wang, *J. Am. Chem. Soc.*, 2008, **130**, 3.
56. K. A. Yang, M. Barbu, M. Halim, P. Pallavi, B. Kim, D. M. Kolpashchikov, S. Pecic, S. Taylor, T. S. Worgall and M. N. Stojanovic, *Nat. Chem.*, 2014, **6**, 1003.
57. G. Vancoillie, S. Pelz, E. Holder and R. Hoogenboom, *Polym. Chem.*, 2012, **3**, 1726.
58. J. F. Teichert, D. Mazunin and J. W. Bode, *J. Am. Chem. Soc.*, 2013, **135**, 11314.
59. W. Q. Yang, S. H. Gao, X. M. Gao, V. V. R. Karnati, W. J. Ni, B. H. Wang, W. B. Hooks, J. Carson and B. Weston, *Bioorg. Med. Chem. Lett.*, 2002, **12**, 2175.
60. V. V. Karnati, X. M. Gao, S. H. Gao, W. Q. Yang, W. J. Ni, S. Sankar and B. H. Wang, *Bioorg. Med. Chem. Lett.*, 2002, **12**, 3373.
61. W. Yang, H. Fan, S. Gao, X. Gao, W. Ni, V. Karnati, W. B. Hooks, J. Carson, B. Weston and B. Wang, *Chem. Biol.*, 2004, **11**, 439.
62. C. F. Dai, L. H. Cazares, L. F. Wang, Y. Chu, S. M. L. Wang, D. A. Troyer, O. J. Semmes, R. R. Drake and B. H. Wang, *Chem. Commun.*, 2011, **47**, 10338.
63. G. F. Clark, S. Oehninger, M. S. Patankar, R. Koistinen, A. Dell, H. R. Morris, H. Koistinen and M. Seppala, *Hum. Reprod.*, 1996, **11**, 467.
64. S. Iida, H. Tsuiji, Y. Nemoto, Y. Sano, M. A. Reddish and T. Irimura, *Oncol. Res.*, 1998, **10**, 407.
65. C. L. Ye, K. Kiriyama, C. Mistuoka, R. Kannagi, K. Ito, T. Watanabe, K. Kondo, S. Akiyama and H. Takagi, *Int. J. Cancer*, 1995, **61**, 455.
66. Y. Chu, D. Wang, K. Wang, Z. L. Liu, B. Weston and B. Wang, *Bioorg. Med. Chem. Lett.*, 2013, **23**, 6307.
67. M. Mazik, *RSC Adv.*, 2012, **2**, 2630.

68. S. Jin, Y. Cheng, S. Reid, M. Li and B. Wang, *Med. Res. Rev.*, 2010, **30**, 171.
69. S. Kubik, *Angew. Chem., Int. Ed.*, 2009, **48**, 1722.
70. J. C. Morales, D. Zurita and S. Penades, *J. Org. Chem.*, 1998, **63**, 9212.
71. J. Jimenez-Barbero, E. Junquera, M. Martinpastor, S. Sharma, C. Vicent and S. Penades, *J. Am. Chem. Soc.*, 1995, **117**, 11198.
72. S. Vandenbussche, D. Diaz, M. Carmen Fernandez-Alonso, W. Pan, S. P. Vincent, G. Cuevas, F. Javier Canada, J. Jimenez-Barbero and K. Bartik, *Chem. – Eur. J.*, 2008, **14**, 7570.
73. G. Terraneo, D. Potenza, A. Canales, J. Jimenez-Barbero, K. K. Baldridge and A. Bernardi, *J. Am. Chem. Soc.*, 2007, **129**, 2890.
74. M. I. Chavez, C. Andreu, P. Vidal, N. Aboitiz, F. Freire, P. Groves, J. L. Asensio, G. Asensio, M. Muraki, F. J. Canada and J. Jimenez-Barbero, *Chem. – Eur. J.*, 2005, **11**, 7060.
75. M. D. Fernandez, F. J. Canada, J. Jimenez-Barbero and G. Cuevas, *J. Am. Chem. Soc.*, 2005, **127**, 7379.
76. D. B. Walker, G. Joshi and A. P. Davis, *Cell. Mol. Life Sci.*, 2009, **66**, 3177.
77. K. Kobayashi, Y. Asakawa, Y. Kato and Y. Aoyama, *J. Am. Chem. Soc.*, 1992, **114**, 10307.
78. T. Reenberg, N. Nyberg, J. O. Duus, J. L. J. van Dongen and M. Meldal, *Eur. J. Org. Chem.*, 2007, 5003.
79. V. Kral, O. Rusin and F. P. Schmidtchen, *Org. Lett.*, 2001, **3**, 873.
80. C. Geffert, M. Kuschel and M. Mazik, *J. Org. Chem.*, 2013, **78**, 292.
81. G. Fukuhara and Y. Inoue, *J. Am. Chem. Soc.*, 2011, **133**, 768.
82. Y. Ferrand, E. Klein, N. P. Barwell, M. P. Crump, J. Jimenez-Barbero, C. Vicent, G.-J. Boons, S. Ingale and A. P. Davis, *Angew. Chem., Int. Ed.*, 2009, **48**, 1775.
83. C. Ke, H. Destecroix, M. P. Crump and A. P. Davis, *Nat. Chem.*, 2012, **4**, 718.
84. V. Janowski and K. Severin, *Chem. Commun.*, 2011, **47**, 8521.
85. S. Striegler and M. G. Gichinga, *Chem. Commun.*, 2008, 5930.
86. A. Patra and M. Bera, *Carbohydr. Res.*, 2014, **384**, 87.
87. K. Ouchi, S. Saito and M. Shibukawa, *Inorg. Chem.*, 2013, **52**, 6239.
88. O. Alpturk, O. Rusin, S. O. Fakayode, W. H. Wang, J. O. Escobedo, I. M. Warner, W. E. Crowe, V. Kral, J. M. Pruet and R. M. Strongin, *Proc. Natl. Acad. Sci. U. S. A.*, 2006, **103**, 9756.
89. P. T. Corbett, J. Leclaire, L. Vial, K. R. West, J.-L. Wietor, J. K. M. Sanders and S. Otto, *Chem. Rev.*, 2006, **106**, 3652.
90. M. Rauschenberg, S. Bomke, U. Karst and B. J. Ravoo, *Angew. Chem., Int. Ed.*, 2010, **49**, 7340.
91. M. Rauschenberg, S. Bandaru, M. P. Waller and B. J. Ravoo, *Chem. – Eur. J.*, 2014, **20**, 2770.
92. G. A. Ellis, M. J. Palte and R. T. Raines, *J. Am. Chem. Soc.*, 2012, **134**, 3631.
93. M. Perez, R. Cuadros, N. Pallas-Bazarra, C. Garcia, E. Langa, J. Jurado-Arjona, F. Hernandez and J. Avila, *J. Alzheimers Dis.*, 2014, **40**, 143.
94. E. Brustad, M. L. Bushey, J. W. Lee, D. Groff, W. Liu and P. G. Schultz, *Angew. Chem., Int. Ed.*, 2008, **47**, 8220.

CHAPTER 6

Synthetic Receptors for Nucleosides and Nucleotides

JINROK OH,[a] HYUN-WOO RHEE[b] AND JONG-IN HONG*[a]

[a] Department of Chemistry, Seoul National University, Gwanak-ro 1, Gwanak-gu, Seoul 151-747, Korea; [b] Department of Chemistry, School of Natural Science, Ulsan National Institute of Science and Technology (UNIST), UNIST-gil 50, Ulsan 689-798, Korea
*Email: jihong@snu.ac.kr

6.1 Physical Properties and Biological Functions of Nucleosides and Nucleotides

Nucleotide is the common name for molecules that are composed of the following three parts: nucleobase, sugar, and phosphate, while a *nucleoside* lacks the phosphate group. The generic structures of nucleosides and nucleotides are depicted in Scheme 6.1. These molecules are substrates for many enzymes, such as nucleic acid polymerases, kinases, oxidoreductases, and others. Nucleotides are the building blocks of nucleic acids and are incorporated into nucleic acid sequences by polymerases in the process of transcription. Nucleotides also serve to carry chemical energy within cells for biomolecule transformation and metabolism. Furthermore, they are involved in cellular signal transduction pathways, and participate in enzymatic reactions as important cofactors.

Among nucleosides/nucleotides, adenosine triphosphate (ATP) is the most ubiquitous for several reasons. It acts as the energy currency of biosystems, serves as a phosphate group donor, and is the precursor of the

Monographs in Supramolecular Chemistry No. 14
Synthetic Receptors for Biomolecules: Design Principles and Applications
Edited by Bradley D. Smith

Published by the Royal Society of Chemistry, www.rsc.org

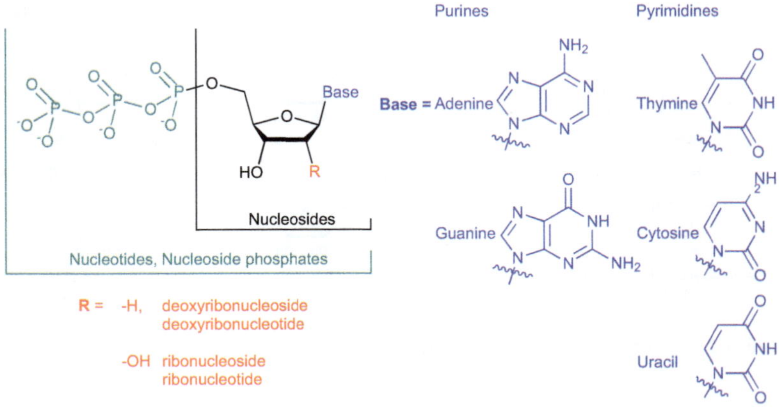

Scheme 6.1 General structures of nucleosides and nucleotides.

second-messenger molecule cyclic adenosine 3′,5′-monophosphate (cAMP). Other nucleotides, such as guanosine triphosphate (GTP), thymidine triphosphate (TTP), uridine triphosphate (UTP), and cytidine triphosphate (CTP) all have their own specific roles such as activators of partner biomolecules. Additionally, diphosphates, monophosphates and nucleosides are all biologically significant and generally related to the strict regulation of enzymatic reactions.

GTP and guanosine diphosphate (GDP) also work as signaling molecules in guanosine nucleotide-binding proteins (G-proteins). Generally, GTP binding induces an activated G-protein state, while GDP binding turns off the protein activity. G-proteins transmit environmental cellular signals with their associated transmembrane proteins, called G-protein-coupled receptors (GPCRs). A range of GPCR proteins detect various extracellular signals such as light, odor, or hormone molecules. Upon external stimulation, a GPCR changes its conformation and the cytoplasmic domain called the guanine nucleotide exchange factor (GEF) becomes activated. Next, this activated GEF causes the nearby G-protein to release the bound GDP, after which the empty binding site of the G-protein associates with cellular GTP. As a consequence, the external stimulus initiates a signal transduction pathway through a successive series of protein activation events. This pathway enables G-proteins to regulate many biological mechanisms, including cellular motility and development.[1]

Other nucleotides, such as nicotinamide adenine dinucleotide (NAD^+) and flavin adenine dinucleotide (FAD) are very important in cellular redox chemistry, because they contain a redox-active functional group, either nicotinamide or isoalloxazine, which can be reversibly reduced and oxidized (Scheme 6.2). They act as cofactors for redox enzymes involved in the electron transport chain (ETC) or dehydrogenases. In particular, eosinophils (white blood cells) contain a high concentration of FADs, which are responsible for the generation of hydrogen peroxide or superoxides.

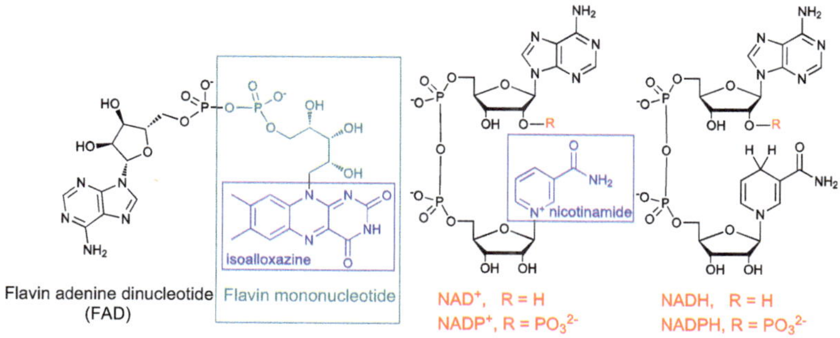

Scheme 6.2 Structures of redox nucleotides.

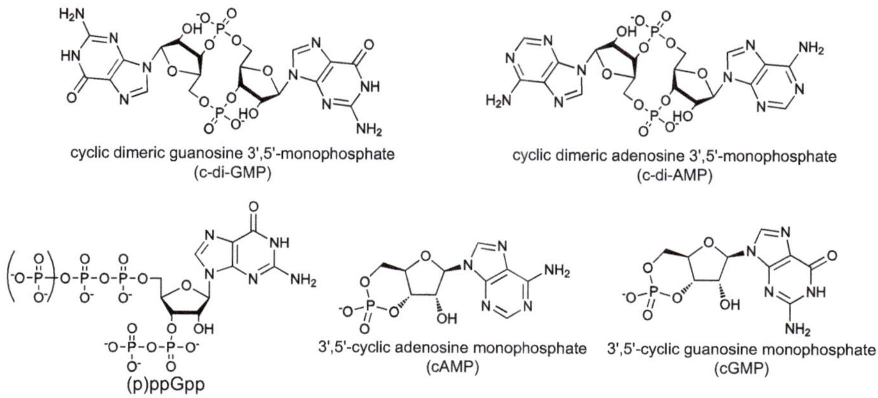

Scheme 6.3 Structures of nucleotide-based signaling molecules.

There are also cyclic forms of nucleotides, such as cAMP and cyclic guanosine 3′,5′-monophosphate (cGMP), which function as second-messenger molecules. Furthermore, in bacterial cells, as well as mammalian cells, multi-phosphorylated signaling molecules called alarmones are primarily in charge of stringent responses in situations such as starvation or heat shock. Several nucleotides act as alarmones, including guanosine 3′,5′-bispyrophosphate (ppGpp), cyclic dimeric guanosine 3′,5′-monophosphate (c-di-GMP), cAMP, and diadenosine 5′,5‴-P1,P4-tetraphosphate (ApppppA). Some of these molecules (*e.g.* ApppppA) are found in both bacterial and mammalian cells and others have so far only been observed in bacterial cells. Both prokaryotes and eukaryotes utilize linear and cyclic nucleotides to regulate diverse cellular processes in response to extracellular cues. In bacteria, cyclic nucleotides such as c-di-GMP, c-di-AMP, cGMP, and cAMP, as well as linear nucleotides such as ppGpp and guanosine 3′-diphosphate, 5′-triphosphate (pppGpp) have emerged as important second messengers involved in the regulation of processes that regulate virulence factor production or biofilm formation (Scheme 6.3).

The biology community regards the discovery of these signaling nucleotides as landmark events. In 1957, Earl Sutherland determined that cAMP mediated the hyperglycemic effects of epinephrine and glucagon and was awarded the Nobel Prize for this work. A few years later, in 1963, cGMP was extracted from rat urine and was also found to exist in prokaryotes. In 1970, Cashel and Gallant first reported the existence of ppGpp and pppGpp in *Escherichia coli*, and it was later shown that the concentrations of ppGpp and pppGpp increased during stress. It was another decade before the next nucleotide second messenger was discovered in bacteria. In 1987, Benziman discovered that c-di-GMP acted as an allosteric modulator of cellulose synthase in *Gluconacetobacter xylinus*, where c-di-GMP is associated with biofilm formation as well as virulence factor production in a good number of clinically important bacteria. As a result, there is now immense interest in c-di-GMP signaling. Almost two decades after Benziman's seminal discovery, Hopfner reported that bacteria also use c-di-AMP for signaling. Details of these nucleotides are well summarized in a recent review by Sintim *et al.*[2] However, the physiology of these cyclic nucleotides and other nucleotides is not clearly understood, in part because there is a lack of methods to selectively detect them in living cells.

Over the years, many synthetic analogs of nucleotides and nucleosides have been prepared and found to exhibit pharmaceutical activity. For example, many anti-viral drugs, so-called nucleoside reverse transcriptase inhibitors (NRTIs) have been developed to disrupt viral replication through the incorporation of an artificial nucleoside into a viral genome. One of the successful NRTIs for anti-viral treatment of human immunodeficiency virus (HIV) is azidothymidine (AZT). The natural thymidine contains a 3′-OH group, while AZT has a 3′-azido group in place of –OH. Therefore, when AZT is incorporated into a viral genome by viral reverse transcriptase, the 3′-azido group of AZT cannot make additional 5′-3′ phosphodiester bonds with other nucleotides, which is absolutely crucial to extend the DNA chain. This directly results in early termination of the viral genome, and consequently, a fatal effect on the viral reproduction cycle.[3] Although AZT was the first US Food and Drug Administration (FDA)-approved NRTI, it was shown to cause significant side effects, including cardiomyopathy, anemia, and neutropenia. These side effects were due to the incorporation of AZT into the patient's genomic DNA through human DNA polymerase, which results in imperfect DNA extension and replication during cellular reproduction. Other NRTIs have also been developed as anti-retroviral drugs: didanosine is an FDA-approved inosine analog; lamivudine is a cytidine analog; and abacavir is a guanosine analog molecule (Scheme 6.4). These drugs are currently used for treating viral diseases and cancers caused by viral infections.[4]

6.2 Biological Recognition of Nucleosides and Nucleotides

In nature, enzymes and nucleic acids have developed very selective ways to bind specific nucleotides in the milieu of other nucleotides with very similar

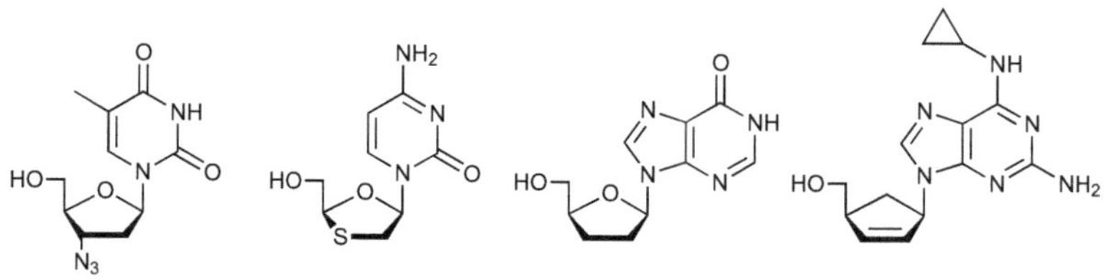

Scheme 6.4 Structures of selected nucleoside reverse transcriptase inhibitors (NRTIs).

structures. The biopolymers achieve this by creating a nucleotide binding pocket that is lined with functional groups that form non-covalent interactions. This section describes several extraordinary examples of natural receptors which recognize specific nucleotides.

Protein kinase A (PKA) belongs to a family of enzymes, known as cAMP-dependent protein kinases, whose activity depends on cellular levels of cAMP. The binding of cAMP is very selective and sensitive ($K_d \sim 10$ nM). The cAMP signaling pathway is one of the most extensively studied topics in cellular biology. Variations in intracellular cAMP levels have direct effect on diverse cellular mechanisms, such as the metabolism of gene expression where PKA functions like an airport control tower. PKA is a holo-tetrameric protein composed of two regulatory and two catalytic subunits. Association between cAMP and PKA triggers the dissociation of the catalytic subunits from the regulatory subunits. Next, the two dissociated catalytic subunits phosphorylate the cytoplasmic substrates, which then diffuse into the nucleus, where they direct some of the long-term effects of cAMP in the regulation of gene expression. The regulatory subunit of PKA contains two tandem highly conserved cyclic nucleotide-binding domains (CNB-A and -B). The co-crystal structure of the regulatory subunit of PKA with cAMP reveals a network of electrostatic and π–π stacking interactions in the binding cavity (Figure 6.1).[5] The phosphate and the ribose ring of cAMP interact with CNB through several hydrogen bonds and electrostatic contacts. The conserved residues (Glu200 and Glu324) interact with the 2′-OH of the ribose ring, while Arg209 and Arg333 residues interact with the phosphate group. The adenine ring of cAMP stacks against the aromatic side chains of Trp260 and Tyr371.

In biological systems, redox enzymes such as oxidoreductases contain redox-responsible nucleotides as prosthetic groups. These enzymes have evolved to have extremely high association constants, because partially oxidized or reduced nucleotide molecules can generate harmful products such as superoxides or peroxides. Flavoenzymes are oxidoreductases that catalyze a wide range of biological processes through electron transfer reactions using FAD or flavin mononucleotide (FMN) as a prosthetic group. Thus, the biological recognition ability of flavoenzymes is of high interest. Watt and coworkers have elucidated many detailed crystal structures of FMN-bound

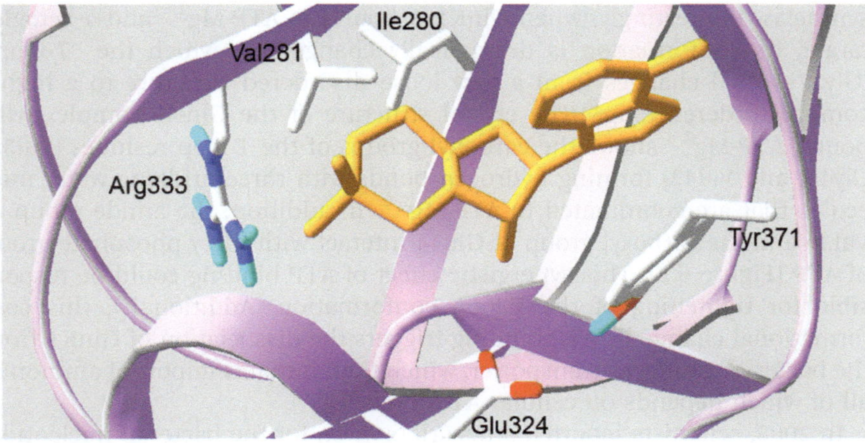

Figure 6.1 Cyclic adenosine monophosphate recognition by a protein kinase A regulatory subunit (Protein Data Bank ID: 1RGS).

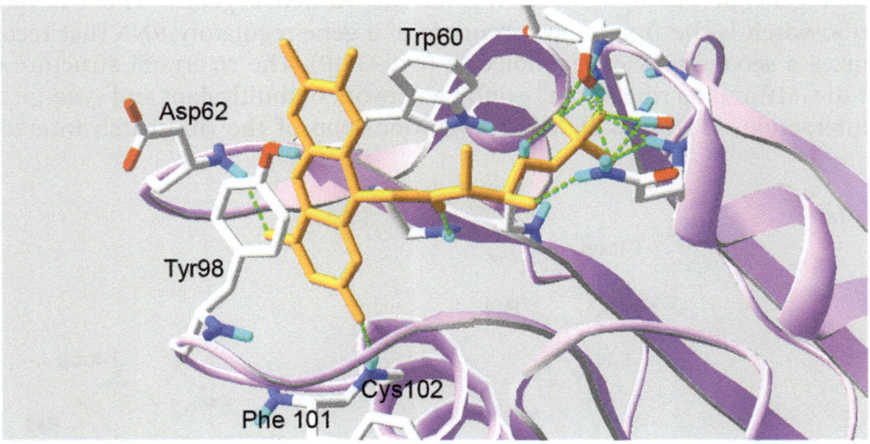

Figure 6.2 Flavin mononucleotide recognition by Flavodoxin (Protein Data Bank ID: 3FX2).

flavoenzymes.[6] The entire molecule of FMN forms extensive hydrogen bonding networks (Figure 6.2). The FMN isoalloxazine group forms a hydrogen-bond network with amide groups of the surrounding two segments of the enzyme, whereas the ribityl and phosphate groups interact with both amide and hydroxyl groups of the amino acid residues (Thr and Ser). The isoalloxazine also stacks against several aromatic residues (Trp60 and Tyr98).

Another biological example of selective nucleotide recognition is found in GlnK1, a bacterial ATP-binding protein that is a member of the PII family. This protein is a homo-trimeric intracellular protein from *Methanocaldococcus jannaschii*, which controls ammonia transport related to glutamine synthesis. Interestingly, when GlnK1 is bound to ATP-Mg^{2+} and α-ketoglutarate, its conformation is dramatically changed, in which the "T-loop" (Gly37–Val53) changes from a very loose disordered structure to a highly compact ordered loop.[7] The crystal structure of the GlnK1 complex with bound ATP-Mg^{2+} shows the carbonyl groups of the T-loop residues (Val38, Gly40, and Val42) forming hydrogen bonds with three bridging water molecules that are coordinated to ATP-Mg^{2+}. In addition, the amide group of Val38 and the carboxyl group of Glu44 interact with the γ-phosphate group of ATP (Figure 6.3). This synergistic effect of ATP binding could be responsible for tightening of the T-loop conformation. Additionally, this conformational change by ATP binding triggers the dissociation of GlnK1 from the bacterial ammonia transporter, which then initiates import of ammonia, all of which depends on cellular ATP levels.

In 2008, scientists reported a new riboswitch for the bacterial nucleotide, an important bacterial signal molecule that regulates survival processes ranging from biofilm formation to virulence gene expression.[8] Interestingly, the c-di-GMP riboswitch is transcribed from the upstream sequence of the tfoX gene of *Vibrio cholerae*, one of the virulence genes. The c-di-GMP riboswitch is the first known example of a gene-regulatory RNA that recognizes a second-messenger molecule (c-di-GMP). The co-crystal structure of c-di-GMP and its riboswitch[9] exhibit a network of multivalent and synergistic interactions (Figure 6.4). The recognition loop of the riboswitch interacts

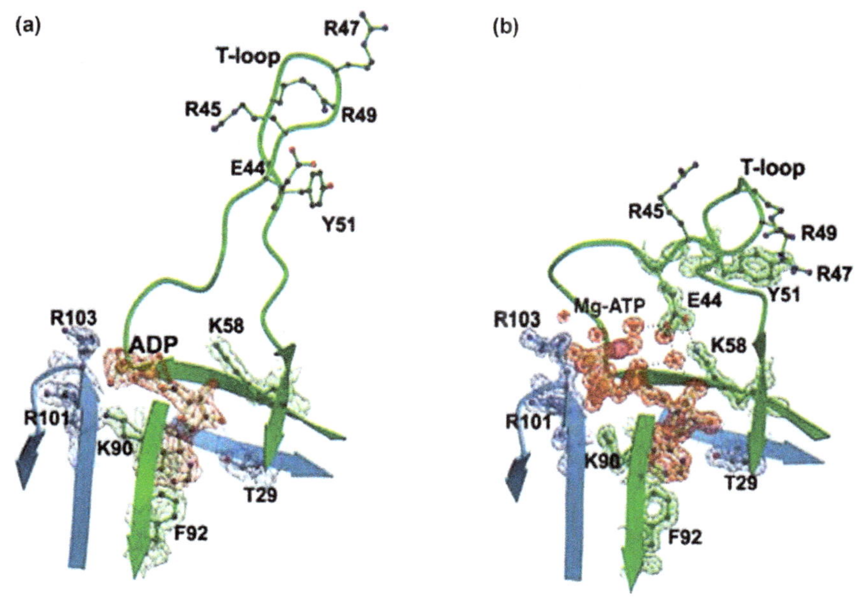

Figure 6.3 Structural perturbations near the adenosine triphosphate (ATP) binding site of GlnK1 before (a) and after (b) ATP binding.
Adapted with permission from ref. 7.

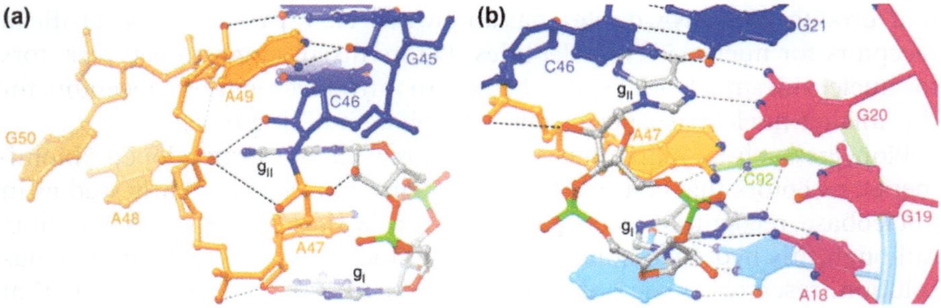

Figure 6.4 Cyclic-di-GMP recognition by type I c-di-GMP riboswitch in two different aspects.
Reprinted by permission from Macmillan Publishers Ltd: *Nat. Struct. Mol. Biol.*, ref. 9, © 2009.

with each guanine base through a combination of hydrogen bonds with the Watson–Crick and Hoogsteen edges (see Figure 7.2 in Chapter 7) and aromatic stacking interactions. These synergistic binding events result in high specificity and subnanomolar affinity, which is the tightest binding reported for a riboswitch–small molecule pair.

6.3 Synthetic Receptors for Nucleosides and Nucleotides

6.3.1 Classical Approaches Based on Weak Interactions

In the 1950s, the beautiful crystal structure of the DNA double helix sparked academic interest in the underlying non-covalent interactions. In the 1960s, Rich and coworkers investigated the hydrogen bonding of purine and pyrimidine bases in non-aqueous solvents that eliminated aromatic stacking interactions.[10] Their study clearly showed that hydrogen bonding between adenine and thymine (or uracil) derivatives and between guanine and cytosine derivatives was highly selective, confirming the importance of electronic and geometric complementarity in the structure of the double-stranded nucleic acids. In addition to hydrogen bonding interactions, aromatic–aromatic stacking between adjacent sets of base pairs is an important feature of the helical nucleic acid structure. In 1996, Kool and coworkers studied the energetics of these stacking interactions separately from hydrogen bonding interactions by introducing a natural or non-natural nucleobase into a base-paired duplex.[11] They found that the strength of the aromatic stacking interaction depends on the structure and polarity of the aromatic groups. Unlike non-natural nucleobases with a nonpolar structure and large surface area, natural nucleobases are less effective in forming stacking interactions. The relatively weak stacking in the natural system might be needed because DNA helix unwinding is necessary for DNA replication and transcription. The desire to better understand the basic forces that produce the DNA duplex has motivated chemists to design artificial receptors for nucleosides/nucleotides. Before discussing artificial receptors for nucleosides/nucleotides, it is helpful to introduce synthetic receptors for the smaller nucleobase, sugar, and phosphate building blocks.

Pioneers such as Julius Rebek and Andrew Hamilton developed "biomimetic" receptors that can selectively recognize nucleobases. Early studies in nucleobase recognition focused on the utilization of hydrogen bonding. Although this hydrogen-bond-based recognition is not effective in aqueous solution, researchers were able to identify the basic binding elements that govern selective recognition. The new insight led to synthetic receptors with much higher specificity than the natural analogs.

6.3.1.1 Nucleobase Recognition with Hydrogen-bonding Receptors

Adenine is a simple purine with a primary aromatic amine group, making the Watson–Crick face and Hoogsteen face quite similar with respect to hydrogen bonding environments, each of which provides one hydrogen bonding donor and one hydrogen bonding acceptor. This structural feature makes it easier to construct synthetic receptors for adenine than for the other nucleobases. In 1987 Rebek and coworkers designed receptor **6.1** to recognize both Watson–Crick and Hoogsteen faces simultaneously, while the receptor naphthyl linker interacted through aromatic stacking interactions (Scheme 6.5).[12] Later, it was found that there are attractive secondary interactions between the imide carbonyl groups of host **6.1** and the polarized C–H groups (C^8–H and C^2–H) of 9-ethyladenine. This pre-organized host molecule (**6.1**) showed high affinity towards 9-ethyladenine ($K_a = 1.1 \times 10^4$ M^{-1} in 30% CD_3CN in $CDCl_3$) and acted as a synthetic transport carrier of adenosine derivatives across a liquid organic membrane.[13] Due to its structural simplicity and versatility, receptor **6.1** became the starting point for numerous studies related to the recognition of adenosine derivatives.

In the late 1980s and early 1990s, Hamilton[14] and Rebek[15] developed synthetic receptors for thymine derivatives through triple hydrogen bonding interactions, using 2,6-bis(acylamino)pyridine or 2,4-diaminotriazine

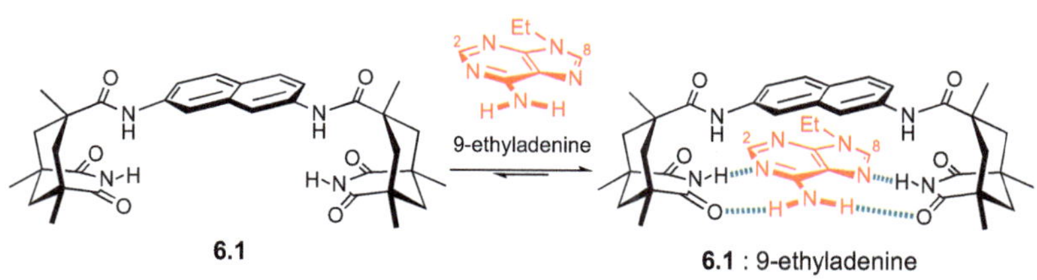

Scheme 6.5 Molecular recognition of **6.1** for 9-ethyladenine.
Reprinted with permission from ref. 12. © 1987 American Chemical Society.

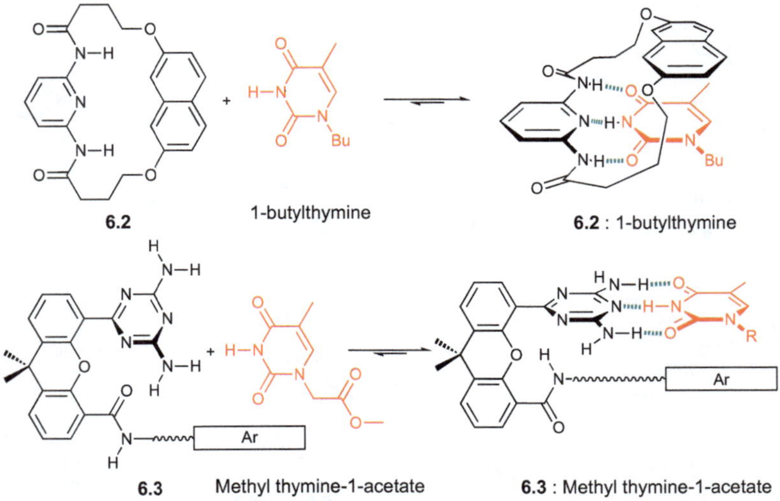

Scheme 6.6 Synthetic receptors for thymine derivatives.
Reprinted with permission from ref. 14 and 15. © 1987 American Chemical Society and © 1991 American Chemical Society.

(Scheme 6.6). To achieve high affinity toward nucleobases they incorporated π–π stacking interactions to produce the molecular hinge **6.2** and xenthene-derived receptor **6.3**. Both achieved relatively high association constants ($\sim 10^3$ M^{-1}) for thymine derivatives in organic solvents.

Although all these synthetic receptors were initially designed for the recognition of thymine, their recognition ability and functions toward flavins were also studied by Yano,[16–18] and Rotello.[19,20] Since thymine and flavins have the imide group in common, it is not surprising that synthetic receptors for thymine exhibited similar affinity toward flavin species. Interestingly, the reduction potential of flavins could be modulated by the series of synthetic receptors **6.6a–e** (Scheme 6.7). Hydrogen bonding interactions with the flavin imide were found to lower the reduction potential of flavins, whereas enhanced π–π stacking raised it. The results were rationalized in the following way: H-bonding with the imide group decreases flavin electron density leading to a "lowest unoccupied molecular orbital (LUMO)-lowering effect". In contrast, π–π stacking interactions increase electron density and form a partial negative charge around flavin species, thus effectively creating a "LUMO-raising effect".

The level of gene expression is regulated not only at the transcription level but also by the presence of interfering RNA, which blocks translation. Therefore, methods that strongly stabilize RNA/DNA duplexes are expected to regulate levels of gene expression leading to control of cellular metabolism. Nowadays, this strategy is known as the "antisense approach", which inhibits translation by complementary binding to mRNA. Guanine has the greatest number of hydrogen bonding sites for complexation. Therefore, it is

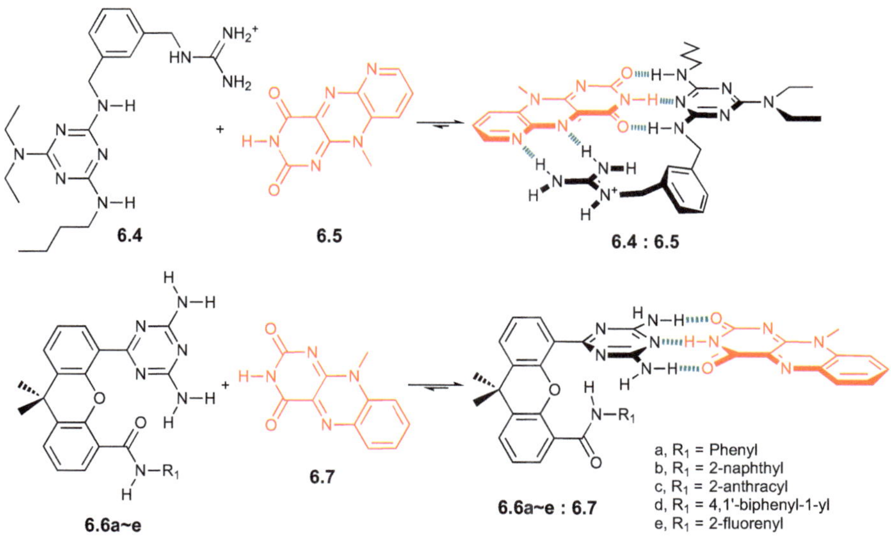

Scheme 6.7 Synthetic receptors for isoalloxazine derivatives.
Reproduced from ref. 18 with permission from the Royal Society of Chemistry and reprinted with permission from ref. 19. © 1997 American Chemical Society.

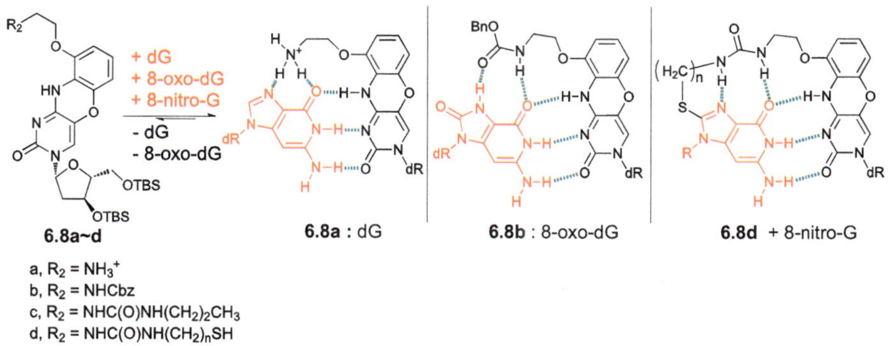

Scheme 6.8 Synthetic receptors for guanosine derivatives.
Reprinted with permission from ref. 21, 22 and 25. © 1998 American Chemical Society. © 2007 WILEY-VCH Verlag GmbH & Co. KGaA, Weinheim. © 2014 American Chemical Society.

an attractive target for the antisense approach and synthetic receptors for guanine derivatives (G-clamps) have been extensively studied.

G-clamp **6.8a**, reported in 1998 by Matteucci, was the first example of the specific recognition of guanosine using hydrogen bonding interactions (Scheme 6.8).[21] By incorporating a single G-clamp into a 10-mer DNA strand, the stability of the DNA duplex was greatly enhanced, which was demonstrated by the increase in the melting temperature. With regard to the development of a guanosine receptor, the electron-rich nature of the guanine moiety was also utilized for the development of fluorescent probes for guanine derivatives, because binding of electron-rich guanine generally decreases the fluorescence emission of probes.

Almost 10 years later, Sasaki and coworkers developed a series of G-clamp variants, called 8-oxoG-clamps, for selective fluorescence detection of 8-oxo-deoxyguanosine (8-oxo-dG), by modifying the free amine moiety of **6.8a** (Scheme 6.8).[22] 8-Oxo-dG is formed by the oxidation of guanosine, and is regarded as the inducer of G:C to T:A mutation. The cellular concentration of 8-oxo-dG is believed to have a direct relation to oxidative stress levels within some diseases.[23] A simple variation of the G-clamp structure (for example, benzyloxycarbonyl (Cbz) protection of the pendent primary amine of **6.8a**) made a large difference in the recognition of different guanosine derivatives. While **6.8a** binds deoxyguanosine ($K_a = 7.1 \times 10^5$ M^{-1}) and 8-oxo-dG ($K_a = 7.3 \times 10^5$ M^{-1}) in $CHCl_3$ buffered at pH 7 with similar affinity, **6.8b** specifically recognizes 8-oxo-dG in $CHCl_3$ buffered at pH 7. To measure its binding ability toward nucleosides, the researchers took advantage of the fluorescent nature of **6.8b** and its strong quenching by 8-oxo-dG, while other nucleosides such as guanosine had no quenching effect. This extraordinary difference was attributed to a single mismatch in the desirable hydrogen bonding pattern. Additionally, ^1H nuclear magnetic resonance (NMR) analysis clearly indicated the formation of highly ordered multiple hydrogen bonds in the complex of **6.8b** and 8-oxo-dG. The binding constant between 8-oxo-dG and **6.8b** was estimated to be 2.3×10^6 M^{-1} in $CHCl_3$ (0.1% dimethyl sulfoxide (DMSO)) buffered with organic acid and base. Since a hydrogen-bonding-only recognition system works poorly in aqueous media, the use of a detergent was required to successfully detect 8-oxo-dG in aqueous media.

Another oxidized guanosine derivative, 8-nitroguanosine, is formed by the reaction of intracellular reactive nitrogen species (RNS) and guanosine. 8-Nitroguanosine is then converted into 8-thioguanosine by reacting with hydrogen sulfide.[24] Taking advantage of the transformation of 8-nitroguanosine to 8-thioguanosine and a previously reported guanosine recognition system, Sasaki developed another interesting receptor for the covalent capture of 8-nitroguanosine.[25] To satisfy the hydrogen-bonding pattern with 8-nitroguanosine, a urea unit was incorporated instead of the pendent amine moiety of **6.8a**. The association constant between 8-nitroguanosine and urea-functionalized host **6.8c** was found to be 3.6×10^6 M^{-1} in chloroform; 8-nitroguanosine formed stronger complexes with **6.8c** than with **6.8b** (0.9×10^6 M^{-1}) due to a difference in a single hydrogen bonding. Finally, a thiol group was introduced into the terminal alkyl chain of **6.8c** to covalently "grasp" 8-nitroguanosine by a nucleophilic aromatic substitution reaction, releasing nitrite. Several nitroG-Grasps were synthesized by manipulating the length of the alkyl chain and varying the urea/carbamate connector. With these receptors, a systematic study of the reaction kinetics was performed and efficient reactivity could be achieved with **6.8d**, which contains both a urea and a propylthiol group.

6.3.1.2 Phosphate Recognition with Cationic Receptors

Phosphates, diphosphates, and triphosphates are essential elements of nucleotides. Positively charged groups were envisioned as a complementary binding unit for the anionic phosphate group of nucleotides. In this approach, mainly nitrogen-containing cationic species were employed, such as ammonium, imidazolium, and guanidinium. In particular, the structural features of the guanidinium–phosphate interaction attracted attention of supramolecular chemists. Hargrove *et al.* have published an excellent review of the recognition of phosphate-containing biomolecules.[26]

In 1991, Rebek and colleagues reported a synthetic receptor targeting dinucleotide, ApA (Scheme 6.9).[27] They synthesized a receptor **6.9** composed of two adenine-recognition moieties connected to each side of a bicyclic guanidinium ion. In the design strategy, all the possible intermolecular interactions such as hydrogen bonding, π–π stacking, and electrostatic interactions were utilized for the recognition of ApA by **6.9**. Although the association constant between ApA and **6.9** was not evaluated, it is likely to be $>10^6$ M^{-1} in nonpolar organic solvents, based on the previously reported energetics of binding.[12,28] With the use of receptor **6.9**, ApA dinucleotide could be efficiently extracted into organic solvents, and its plausible binding mode with **6.9** was elucidated by various NMR techniques, including the nuclear Overhauser effect (NOE) and rotating-frame Overhauser effect spectroscopy, as the complex **6.9**:ApA.

In 1994, Rebek and coworkers developed **6.10** as a hydrophilic variant of **6.9** to explore its recognition ability toward cAMP through multiple intermolecular interactions in aqueous media (Scheme 6.10).[29] Incorporating several hydroxyl groups into the receptor structure significantly increased its solubility in water. The newly synthesized receptor **6.10** was able to recognize

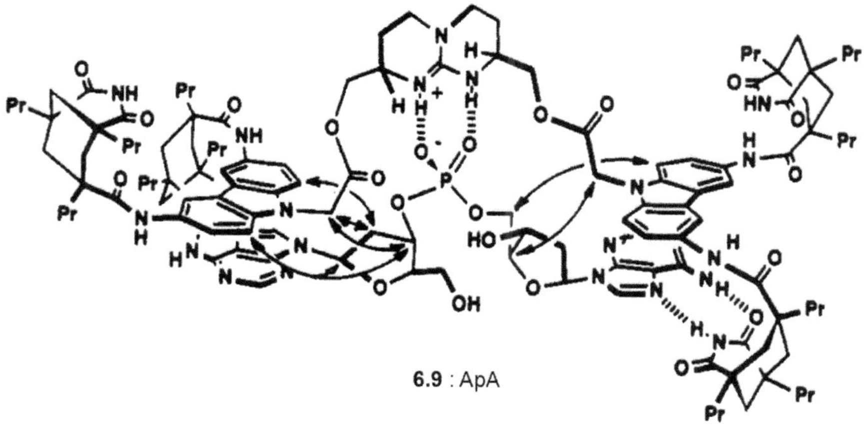

Scheme 6.9 Molecular recognition of **6.9** for a dinucleotide ApA. Arrows represent intermolecular nuclear Overhauser effect contacts.
Adapted with permission from ref. 27. © 1991 American Chemical Society.

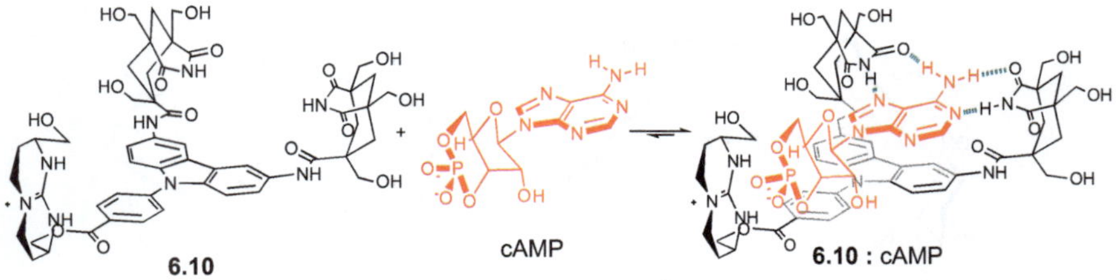

Scheme 6.10 Molecular recognition of **6.10** for cAMP.
Reprinted with permission from ref. 29. © 1994 American Chemical Society.

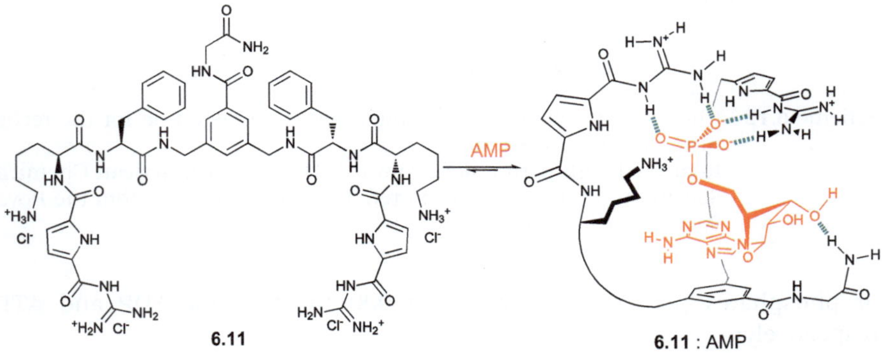

Scheme 6.11 Chemical structure of **6.11** and schematic illustration for the recognition of AMP by **6.11**.
Reprinted with permission from ref. 30. © 2011 WILEY-VCH Verlag GmbH & Co. KGaA, Weinheim.

cAMP with a binding affinity of 600 M^{-1} in buffered aqueous media (pH 6.0), which was three- to fourfold stronger than binding with adenosine or 9-ethyladenine. Also, they quantitatively measured the guanidinium–phosphate interaction in buffered aqueous media, which was ~0.3–0.6 kcal mol^{-1}, whereas the overall binding energy between **6.10** and cAMP was ~3 kcal mol^{-1}.

Schmuck developed a molecular tweezer **6.11** for nucleotides, consisting of two symmetric peptide arms and an *m*-phenylene linker (Scheme 6.11).[30] The two peptide arms contain (1) lysine and guanidinocarbonylpyrrole, both of which are positively charged in neutral conditions, for hydrogen bonding and electrostatic interactions with the nucleotide phosphate moiety, and (2) phenylalanine, which forms a cleft with a *m*-phenylene linker, and enables a π–π stacking interaction with the nucleobase. The structure of the associated complex was elucidated by a battery of NMR studies and molecular dynamics simulations. The highly pre-organized receptor showed an exceptionally high association constant of 7.6×10^5 M^{-1} for AMP in 100% aqueous media (pH = 7.0), with a decrease in affinity for every increase in the number

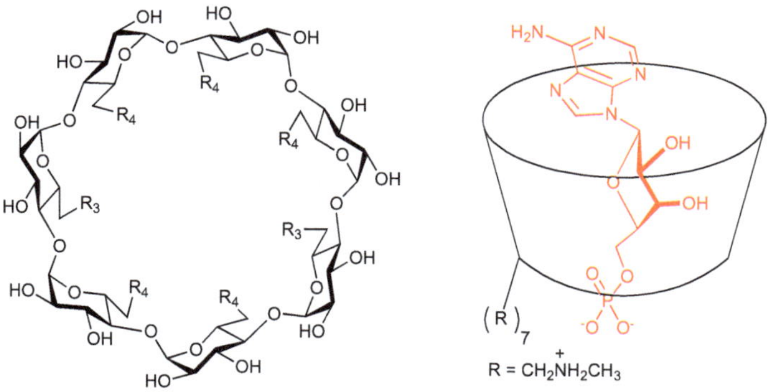

6.12a~c a, R_3 = -NHCH$_3$, R_4 = OH

b, R_3 = R_4 = NHCH$_3$

c, R_3 = R_4 = NHCH$_2$CH$_2$OH

6.12b : AMP

Scheme 6.12 Chemical structures of **6.12** and schematic illustration for the recognition of AMP by **6.12b**.
Reprinted with permission from ref. 31, © 1994 American Chemical Society, and reproduced from ref. 32 with permission from the Royal Society of Chemistry.

of phosphates ($K_a = 3.3 \times 10^5$ M^{-1} and 0.8×10^5 M^{-1} for ADP and ATP, respectively).

From a supramolecular point of view, a nucleotide is an amphiphilic molecule, which contains both hydrophobic and hydrophilic components. Thus, there have been extensive studies of nucleotide recognition using container molecules. Cyclodextrins have been most successfully utilized as host molecules. Cyclodextrins are cyclic oligosaccharides, composed of α-D-glucopyranosides with α-1,4 glycosidic bonds, making a jar-like structure with a slightly hydrophobic cavity and polar rim at the openings.

In 1994, Schneider and Eliseev investigated modified cyclodextrins for the recognition of nucleotides;[31] the hydroxyl groups at the lower rim of β-cyclodextrin were substituted with methylamines, which provided positive charge at physiological pH. A heptacationic cyclodextrin derivative **6.12b** showed a binding constant of 3×10^6 M^{-1} with ATP in neutral water (Scheme 6.12). Systematic studies on the recognition ability of **6.12a** or **b** for various types of nucleotides identified the contribution of each binding interaction to the thermodynamic stability of the complex. The stabilization of the binding complex was primarily attributed to the electrostatic interaction between ammonium and phosphate, followed by hydrogen bonding interactions with hydroxyl groups on the toroidal rims of β-cyclodextrin, and least of all, the hydrophobic interaction between the cavity and nucleobase. This assumption was in accordance with the structure of the complex **6.12b**:AMP that was predicted from NMR experiments (NOE) and computational study (Chemistry at Harvard Macromolecular Mechanics (CHARMM) molecular simulation). The cyclodextrin cavity was filled with the entire ribose ring and part of the nucleobase, and association was driven by strong electrostatic interactions of the phosphate with the pendent methylammonium groups on the cyclodextrin rim.

Darcy and coworkers modified the structure of Schneider's cyclodextrin by attaching a hydroxyethylamine moiety on the smaller opening to make **6.12c** and evaluated its ability for ATP recognition.[32] This simple modification produced a remarkably increased binding affinity towards ATP, of 6×10^9 M^{-1}, which is 2000-fold higher than with **6.12b**. The protonated hydroxyethylamino side chains not only served as a complementary binding site for phosphates, but they also formed an extended hydrophobic cavity which partially excluded water molecules and reinforced the electrostatic interaction between ammonium and phosphate.

An example of ATP transport using an amphiphilic host was reported by Regen and coworkers.[33] They synthesized a "molecular umbrella" **6.13** which is composed of spermidine, cholic acid, and arginine, to produce a "facial amphiphile" with a phosphate recognition unit at a central scaffold. Within the hydrophobic environment of a lipid bilayer, **6.13** was expected to mask the hydrophilicity of a polar ATP guest. Receptor **6.13** was tested as a transporter for ATP through a phospholipid bilayer membrane. In the absence of **6.13**, <1% of ATP was transported out of phospholipid vesicles after 48 h, while 81% of ATP was transported in the presence of **6.13** (Scheme 6.13).

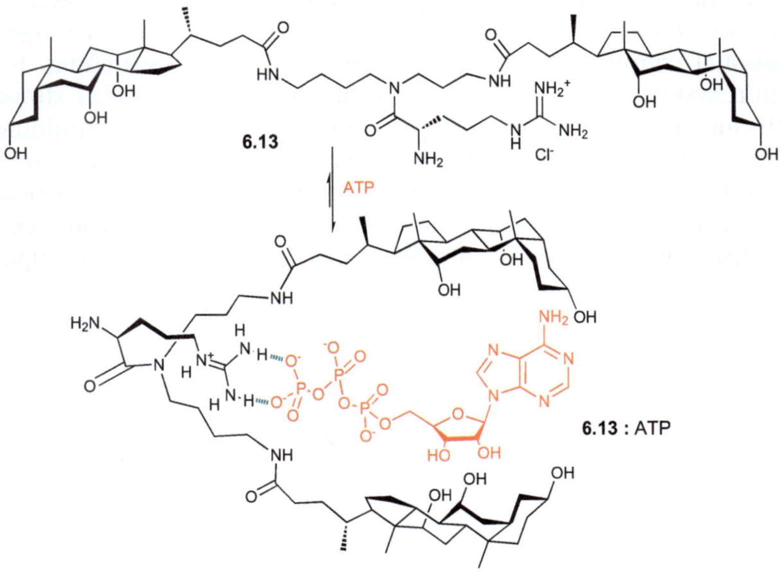

Scheme 6.13 Chemical structure of **6.13** and proposed structure of the complex **6.13**:ATP in nonpolar media.
Reprinted with permission from ref. 33. © 2003 American Chemical Society.

6.3.1.3 Ribonucleoside Recognition with Boronic Acid Receptors

In 1996, Shinkai and colleagues reported a synthetic receptor **6.14** for nucleosides with three distinct functional groups:[34] (1) boronic acid for reversible covalent binding of a *cis* diol (see Chapter 5 for detailed summary of boronic acid receptors for carbohydrates), (2) positively charged pyridinium for stabilization of an anionic boron center, and (3) azobenzene derivative for a chromogenic response. Based on Corey-Pauling-Koltun (CPK) space-filling molecular models, the azophenyl system of **6.14** was expected to stack with an adjacent adenosine guest. Based on titration experiments, adenosine showed the highest affinity towards **6.14** ($6.16 \times 10^4\ M^{-1}$ in buffered aqueous media, pH 10.5) with the greatest absorbance change among tested ribonucleosides (Scheme 6.14). There was almost no association between **6.14** and 2′-deoxyadenosine, consistent with covalent recognition of the ribose diol by the boronic acid.

Another example of a five-carbon sugar recognition of nucleosides was reported by Clayden and coworkers.[35] They prepared a peptide-based dynamic helical molecule (**6.15**) containing NMR-active nuclei (^{13}C) and boronic ester, each positioned at the C-terminus and N-terminus, respectively (Scheme 6.15). In a methanolic solvent system, receptor **6.15** rapidly switches its helical conformations ($M \rightleftarrows P$) on the NMR time scale. But when a chiral *cis* diol binds the boronic ester unit of **6.15**, the dynamic helical conformation is restricted by the chirality of bound substrates. The conformational information from a boronate–diol binding site is transferred to the C-terminus, where a spectroscopic reporter ^{13}C nucleus is located >2 nm away, and characteristic ^{13}C resonance signals indicate long-range conformational communication by ligand-induced helical switching. In this dynamic system, they found that all the purine nucleosides drove the equilibrium to form the *M* helix preferentially, whilst the pyrimidine nucleosides did not show any obvious trend. This finding is counterintuitive, because the same chirality of ribofuranose in the purine and pyrimidine nucleosides should induce the same helical structure upon complexation. While insertion of the solvent into the dative B–N bond plays an important

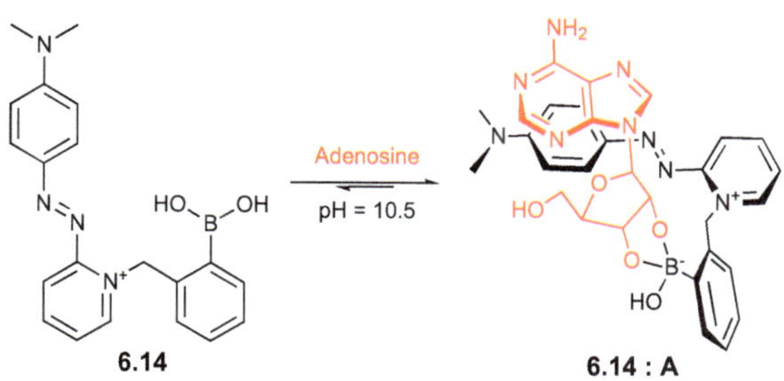

Scheme 6.14 Molecular recognition of **6.14** for adenosine. Reproduced with permission from ref. 34. © 1996 Japan Science and Technology Information Aggregator, Electronic.

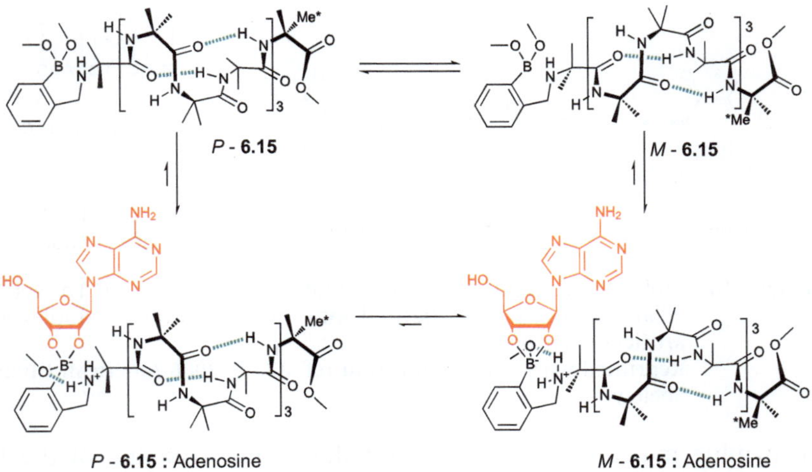

Scheme 6.15 Molecular recognition of **6.15** for adenosine and recognition-driven helicity equilibria.
Reprinted by permission from Macmillan Publishers Ltd: *Nat. Chem.*, ref. 35, © 2013.

role in rigidifying the receptor–substrate complex and thus induces asymmetry into the helical structure, this unusual induction of helicity suggests that secondary hydrogen bonding interactions between the nonbonding part of the substrate and the helical receptor must also play a crucial role in the preferential folding of the bound structure.

6.3.2 Metal Coordination-assisted Recognition

In general, synthetic receptors based on H-bonds and/or π–π stacking interactions exhibit relatively low nucleotide binding affinity in aqueous solution. In contrast, receptors containing metal cation are effective synthetic receptors for nucleotides in water. Transition metal coordination complexes such as Zn(II)-dipicolylamine (Zn-DPA) and Zn(II)-cyclen (Zn-cyclen) associate strongly with nucleotide phosphates. In addition, mercury, zinc, and cadmium complexes are useful for nucleobase recognition.

6.3.2.1 Nucleobase Recognition with Various Metal Complexes

(a) Mercury(II)–Thymine. Since the first report on the reversible reaction between thymine and mercuric chloride,[36] the imide–mercury reaction has become critical for the recognition of thymine. In 2006, Ono and coworkers found that the melting temperature of a T–T mismatched DNA duplex was greatly increased by the presence of mercury ions (Scheme 6.16).[37] The 2:1 binding stoichiometry of thymine:mercury was confirmed from circular dichroism and NMR titration experiments. This system is particularly interesting because the insertion of mercury between

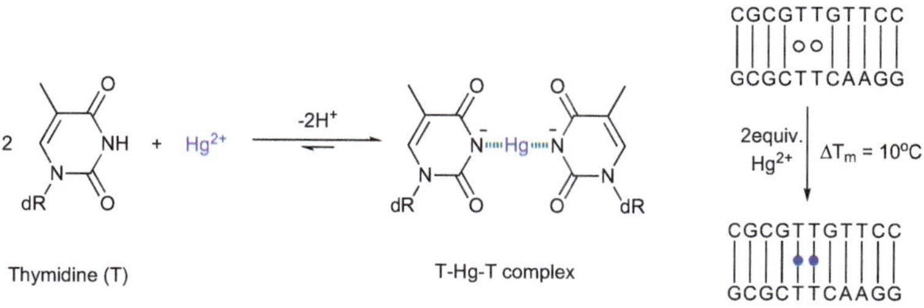

Scheme 6.16 Molecular recognition of thymidine for mercury(II) ion. The termolecular complex was introduced into the synthetic DNA duplex as an artificial base pair.
Reprinted with permission from ref. 37. © 2006 American Chemical Society.

two thymine moieties does not disturb the overall structure of the DNA duplex. The T–Hg–T complex has been incorporated into numerous molecular devices. For example, Mirkin and coworkers developed a colorimetric detection system for mercury(II) ions using the T–Hg–T interaction.[38]

(b) Zinc(II)–Thymidine. Using a simple Zn–cyclen complex, Kimura and colleagues achieved a high binding constant of up to 4×10^5 M^{-1} with thymidine in aqueous media.[39] NMR titration experiments revealed that the imide proton had disappeared, indicating deprotonation of imide NH. Moreover, a crystal structure of the complex between AZT and Zn–cyclen clearly showed a deprotonated imide N coordinated to the zinc center and the complex was further stabilized through additional hydrogen bonding interactions between imide carbonyls and cyclen NHs. This is possible because NHs in Zn–cyclen become acidic enough to form hydrogen bonds with the imide carbonyls whose basicity becomes even stronger upon imide N–Zn coordination. The unique presentation of the imide moiety in thymine (or uracil) means that it can be selectively recognized by Zn–cyclen. Kimura and coworkers further developed a lipophilic Zn–cyclen derivative **6.16b** and showed that it could extract all imide-bearing nucleosides from water to an organic solvent such as chloroform.[40] Exploiting the pH-dependent nature of imide:Zn–cyclen complex formation, they demonstrated liquid membrane transport of thymidine from high aqueous pH to low aqueous pH (Scheme 6.17).

(c) Zinc(II)–Guanine. In 1994, Kimura and coworkers prepared acridine-pendent Zn–cyclen and examined its interaction with nucleosides.[41] Among tested nucleosides, guanosine showed moderate affinity ($K_a \sim 10^4$ M^{-1}) toward **6.16c** in aqueous media, which was not as strong as **6.16c**:thymidine complex (Scheme 6.18). The binding affinity exhibited a pH dependency in the range of 6–9, where the protonation state of bound guanosine was changed. Since Zn–cyclen complexes without a pendent aromatic

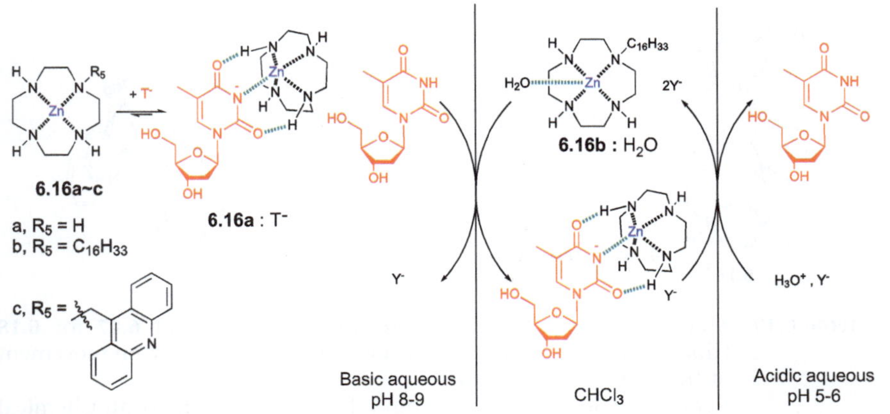

Scheme 6.17 Molecular recognition of **6.16** for thymidine. The lipophilic **6.16b** was employed as a synthetic thymidine carrier across $CHCl_3$.
Reprinted with permission from ref. 40. © 1998 American Chemical Society.

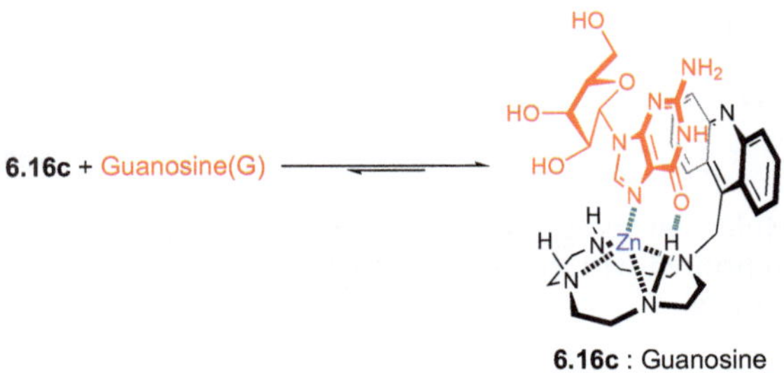

Scheme 6.18 Molecular recognition of **6.16c** for guanosine.
Reprinted with permission from ref. 41. © 1994 American Chemical Society.

moiety have almost no affinity with guanosine, it is clear that the pendent acridine contributes to the enhanced binding affinity through π–π stacking with the guanine surface, which was supported by an X-ray structure analysis.

(d) Zinc(ii)–Flavin. Zn–cyclen is also a good binding unit for the imide moiety within flavins and König and coworkers have exploited this feature to develop flavin probes.[42] Photoinduced electron transfer (PeT) was utilized as a sensing tool for the detection of riboflavin tetraacetate **6.18** in neutral aqueous media by a Zn–cyclen receptor with an appended phenothiazine as the electron donor (Scheme 6.19). The phenothiazine–isoalloxazine dyad was chosen because their electronic levels were suitable for PeT induction in an aqueous system. The researchers successfully

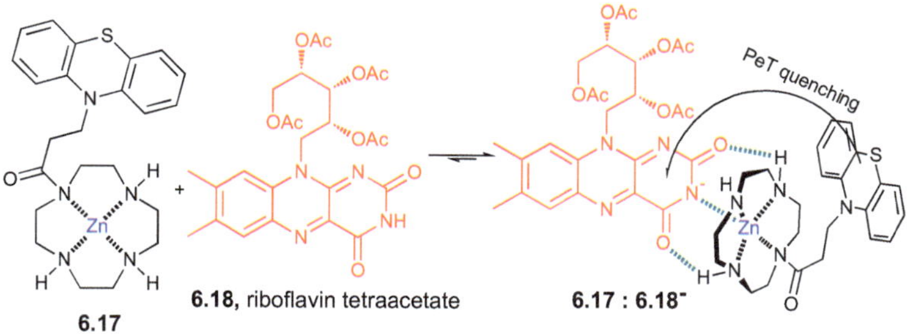

Scheme 6.19 Molecular recognition and detection mechanism of **6.17** for **6.18** (riboflavin tetraacetate). Half-arrow represents the electron movement at the excited state of flavin.
Reprinted with permission from ref. 42. © 1999 American Chemical Society.

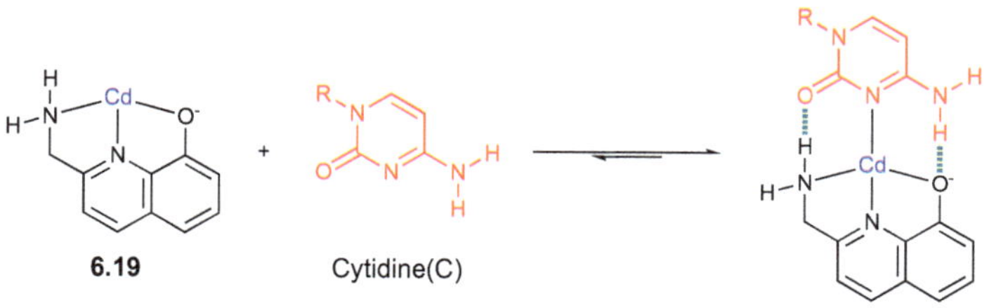

Scheme 6.20 Molecular recognition of **6.19** for cytidine.
Reprinted with permission from ref. 43. © 2002 American Chemical Society.

monitored static quenching of the isoalloxazine fluorescence emission upon complexation of **6.17** and **6.18**, and the responsible PeT mechanism was verified with the help of various spectroscopic tools, such as transient absorption spectroscopy. In addition, an association constant between **6.17** and **6.18** was estimated to be $7.9 \times 10^5\ M^{-1}$, which is similar to the association constant previously reported by Kimura.

(e) **Cadmium(II)–Cytidine.** In 2002, Chin reported a synthetic receptor that selectively binds cytidine.[43] This receptor **6.19** is a cadmium complex of 2-aminomethyl-8-hydroxyquinoline, and was designed to resemble a natural G–C hydrogen bonding network, substituting one hydrogen atom for a metal ion. The **6.19:**cytidine pair has an association constant of 117 M^{-1} in DMSO, which is 30-fold higher than that of a natural G–C base pair ($\sim$4 M^{-1}). The selectivity for cytidine is remarkably high with almost negligible affinity for the other nucleobases (<2 M^{-1}). Interestingly, replacing Cd(II) (covalent radius 1.48 Å) with Zn(II) (1.25 Å) eliminated binding because the ligand and cytidine were brought too close for favorable interactions (Scheme 6.20).

6.3.2.2 *Phosphate Recognition with Zinc(II) Complexes*

(a) Zinc(II)–Dipicolylamine Complex. Among the known metal containing receptors for phosphates, Zn-DPA is one of the most successful. In addition, Co(III)- and Cu(II) complexes with dipicolylamine derivatives act as artificial catalysts for the hydrolysis of phosphodiester bonds of ribonucleotides.[44,45] Inspired by the structures of Zn-DPA/phosphate complexes, both Kim and Hong independently developed optical "sensors", which could selectively detect phosphate anions in aqueous solution. Later, Hong and coworkers found that synthetic receptor **6.20** exhibited an enormously high binding affinity toward pyrophosphate (PP_i) ($K \sim 6.6 \times 10^8$ M^{-1}) and 1000 times selectivity over $HPO_4{}^{2-}$ in 100% aqueous media (Scheme 6.21).[46] The binding mode was fully elucidated by an X-ray crystallographic analysis. A crystal structure of the PPi complex revealed that the two sets of oxygen anions on each P are coordinated to two zinc centers of **6.20** to generate the doubly hexacoordinated zinc complex (**6.20**:PPi). Subsequent work has shown that bis(Zn-DPA) receptors have high affinity for the pyrophosphate group that is common to all nucleoside triphosphates.

(b) Zinc(II)–Cyclen Complex. Like Zn-DPA complexes, Zn–cyclen complexes have been utilized often as phosphate receptors. For example, Kimura and colleagues prepared a Zn–cyclen derivative **6.21** with a 2-pyridyl group directly attached to an amine N of cyclen. In **6.21**, the 2-aminopyridine unit acts as a fluorescent reporter unit, and the fluorescence is emitted by two distinct mechanisms; twisted intramolecular charge transfer (TICT) and locally excited state (LE).[47] A crystal structure of **6.21** revealed a strong dative bond between pyridine N and zinc center (2.17 Å) by which the TICT mechanism was activated. The researchers hypothesized

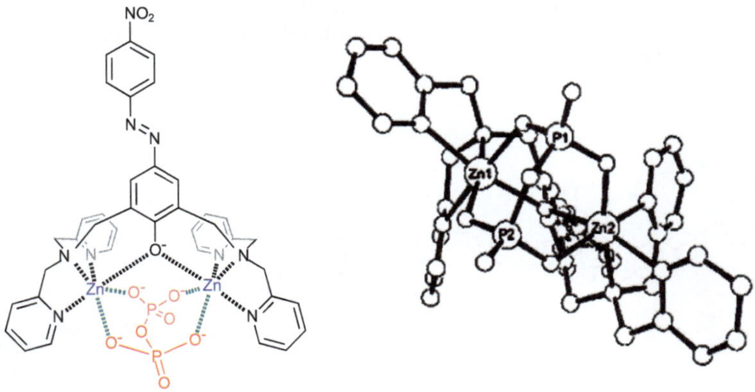

6.20 : pyrophosphate (PPi)

Scheme 6.21 Molecular recognition of **6.20** for pyrophosphate (PPi) and the crystal structure of the complex **6.20**:PPi.

that coordination of pyridine N to the zinc center could be disturbed by binding of an external anionic substrate, leading to a change in the emission mechanism from TICT to LE. Indeed, a ratiometric spectral change was observed in titration experiments with various anionic substrates including carboxylate, imide anion, and phosphate (Scheme 6.22). Interestingly, there was no relationship between association constants and TICT/LE ratios, suggesting that the spectral change depends not only on the basicity but also on the size of substrate molecules.

6.3.2.3 Selective Receptors for Nucleotides using Metal Complexes

As mentioned in previous sections, Kimura and colleagues studied interactions of Zn–cyclen with phosphate, thymine, isoalloxazine, and guanine. They later studied the recognition of thymidine nucleotides[48] and thymidylylpoly(phosphothymidines) (Scheme 6.23).[49] For this purpose, they prepared multinuclear Zinc(II) complexes bridged by *m*- or *p*-phenylenes. For recognition of thymidylylpoly(phosphothymidines) they found that log(K_{app})

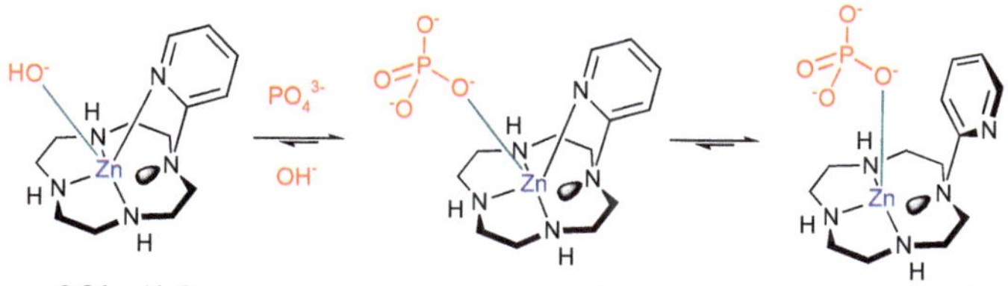

6.21 : H_2O
TICT

6.21 : PO_4^{3-}
TICT

6.21 : PO_4^{3-}
LE

Scheme 6.22 Molecular recognition of **6.21** for phosphate and its detection mechanism through conformational equilibria.
Reprinted with permission from ref. 47. © 2004 American Chemical Society.

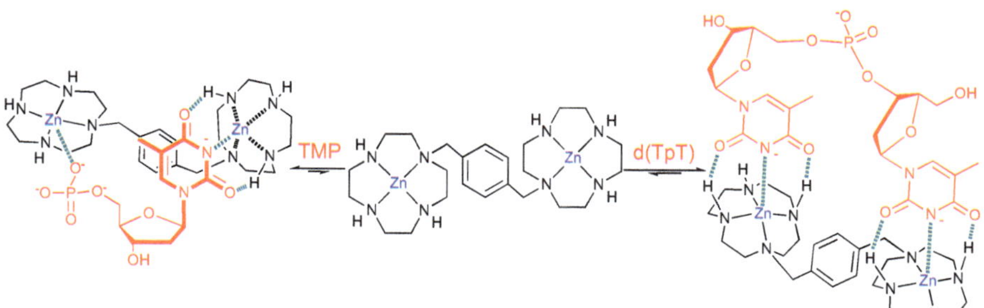

6.22 : TMP$^-$

6.22

6.22 : d(TpT)$^{2-}$

Scheme 6.23 Differential recognition of **6.22** for TMP and d(TpT).
Reprinted with permission from ref. 48 and 49. © 2000 American Chemical Society and © 1999 WILEY-VCH Verlag GmbH, Weinheim, Fed. Rep. of Germany.

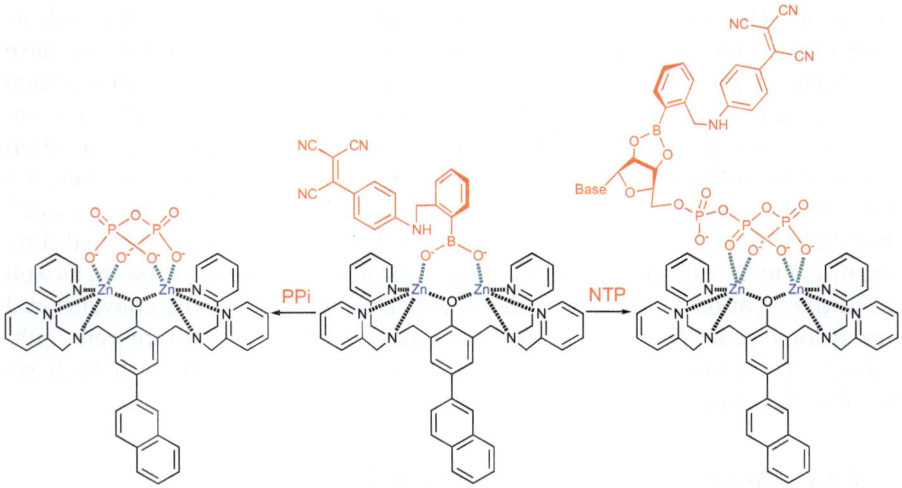

6.23 : PPi **6.23** : **BA** **6.23** : NTP-**BA**

Scheme 6.24 Differential recognition of **6.23:BA** complex for pyrophosphate (PPi) and nucleotide triphosphate (NTP).
Reprinted with permission from ref. 50. © 2011 American Chemical Society.

was proportional to the number of T:Zn–cyclen complexes within the system; with values of 3.2, 6.23, and 9.2 for T:Zn, TpT:Zn_2, and TpTpT:Zn_3, respectively. The association constants between thymidine nucleotides (TXP) and receptor **6.22** were estimated to be ≥1000 times higher than that for dT alone.

An important practical goal is to develop fluorescent receptors that can selectively detect a small amount of pyrophosphate in the presence of a large amount of nucleotide triphosphates (NTPs), because many important biochemical reactions within cells involve release of PPi in the presence of NTPs and other phosphorylated biomolecules. For selective sensing of PPi over NTPs, Hong and colleagues designed an ensemble system which comprises a fluorescent bis(Zn-DPA) receptor **6.23** and boronic acid **BA** as a receptor quencher. The two components, **6.23** and **BA**, recognize PPi and the *cis* diol of ribose, respectively (Scheme 6.24).[50] When PPi is added to the **6.23:BA** ensemble, PPi displaces **BA** from **6.23**, concomitantly turning on fluorescence. However, when NTP is added to the **6.23:BA** ensemble, the displaced **BA** re-attaches to ribose *cis* diol to produce the termolecular complex **6.23**:NTP-**BA** whose fluorescence is quenched as before (**6.23:BA** ensemble). Using the ability of **6.23** to distinguish between PPi and NTPs, the researchers were able to monitor the RNA polymerase reaction by measuring the amount of PPi released during the enzymatic reaction.

6.3.3 Selective Fluorescent Probes for Nucleotides

The development of nucleotide receptors is an offshoot from anion receptor research in the 1990s. At that time, much attention was given to synthetic receptors which could discriminate phosphate from other anions, such as fluoride, acetate, and citrate.[51] Since then supramolecular chemists have synthesized synthetic receptors which are not only able to discriminate among various nucleotides but are also capable of measuring the concentration of a certain nucleotide using fluorescence. As such, they are often referred to as *fluorescent sensors*. Several highly selective fluorescent sensors for specific nucleotides have been developed and a recent review by Yoon[52] highlights new developments in this area. They are useful for the real-time monitoring of test tube enzymatic reactions of nucleotide conversion, as well as in the creation of label-free assays, which do not require the conventional hazardous radioisotope-labeled substrates or expensive fluorescent substrates. The following sections describe fluorescent sensors with high selectivity for specific nucleotides.

6.3.3.1 Selective Fluorescent Probes for NTPs

(a) ATP. Hamachi and colleagues developed a xanthene-based Zn-DPA complex **6.24** that has selectivity for nucleotides and pyrophosphate over other anions. The "turn-on" fluorescent sensing mechanism of **6.24** is unique. The nonfluorescent deconjugated state of **6.24** was formed by an unprecedented nucleophilic attack of water which was bound to two Zn(II) ions. Then, binding of the triphosphate group of ATP to the bis(Zn-DPA) moiety of **6.24** resulted in the recovery of the conjugated xanthene structure which showed a strong fluorescence emission ($K_a = 1.3 \times 10^6$ M^{-1}, $F/F_o = 33$) (Scheme 6.25).[53] Because ATP is the most abundant nucleotide in living cells, this probe enabled fluorescence imaging of intracellular ATP particulate stores in living cells. The work is a pioneering example of intracellular ATP imaging in living cells using a synthetic receptor. The xanthene-based probe also has strong binding affinity and similar fluorescence response for other nucleotides (for GTP: $K_a = 1.7 \times 10^6$ M^{-1}, $F/F_o = 34$) and pyrophosphate ($K_a = 4.0 \times 10^7$ M^{-1}, $F/F_o = 55$),[53] which means that the fluorescent response could be generated from not only the

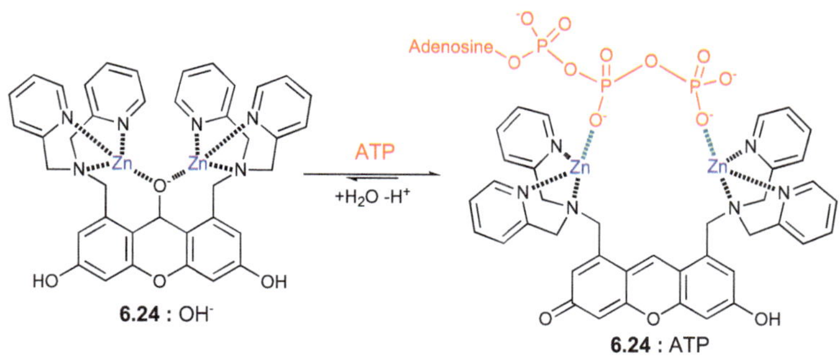

Scheme 6.25 Molecular recognition and detection mechanism of **6.24** for ATP. Reprinted with permission from ref. 53. © 2008 American Chemical Society.

ATP store but also from other nucleotides or pyrophosphate. Therefore, the result was further verified with an imaging assay using quinacrine, a frequently used probe for detecting ATP stores. Later work modified the probe structure and developed Förster resonance energy transfer (FRET)-based sensors for ratiometric visualization of the ATP levels inside living cells as well as real-time fluorescence monitoring of enzymatic reactions involving adenosine polyphosphates.[54] A positively charged pyronin ring was attached as a cellular compartment targeting unit to produce a probe which could measure the ATP concentrations in specific organelles of a living cell.[55] Furthermore, introduction of a lipid anchor into **6.24** enabled selective detection of extracellular release of nucleoside polyphosphates on the plasma membrane surface.[55]

In 2009, Yoon and colleagues reported a pyrene-conjugated imidazolium compound **6.25** for selective sensing of ATP (Scheme 6.26).[56] Since probe **6.25** has two pyrene units, it exhibits a pyrene excimer emission in aqueous media due to a π–π stacked conformation. When ATP was introduced, **6.25** showed ratiometric behavior in which pyrene monomeric emission was increased whereas excimeric emission was decreased. This behavior was observed only for ATP, whereas pyrophosphate or other nucleotides such as GTP showed no obvious enhancement in monomer emission. As a result, a large fluorescence intensity ratio for ATP (I_{375}/I_{487}) enabled measurement of intracellular ATP levels and monitoring the enzymatic reaction of apyrase. The researchers postulated that only the ATP adenine can intercalate between the two pyrene moieties to produce the pyrene monomer fluorescence.

(b) GTP. In 2012, Kim and colleagues reported a cyclophane **6.26** containing imidazolium and anthracene,[57] which was synthesized in only a single step, and showed superb selectivity toward GTP among tested analytes (Scheme 6.27). The association constant with GTP was estimated to be 1×10^6 M^{-1} in 100% aqueous solution at pH 7.4, which is as high as

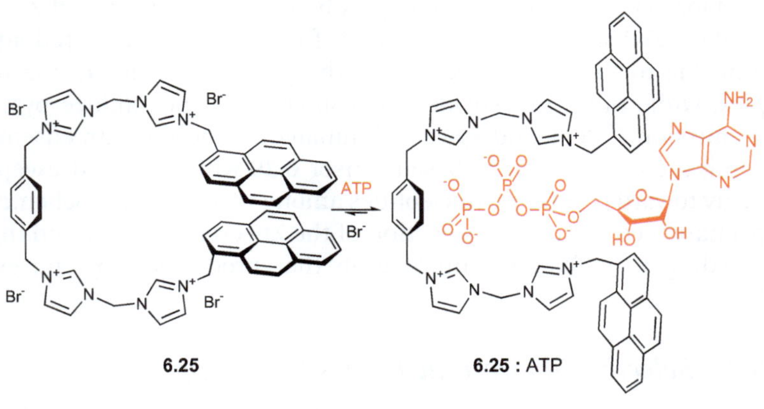

Scheme 6.26 Molecular recognition of **6.25** for ATP.
Reprinted with permission from ref. 56. © 2009 American Chemical Society.

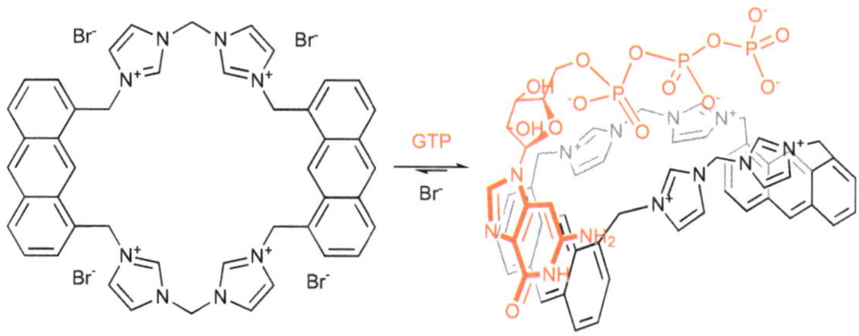

Scheme 6.27 Molecular recognition of **6.26** for GTP.
Reproduced from ref. 57 with permission from the Royal Society of Chemistry.

that with bis(Zn-DPA) complexes. The different responses of **6.26** towards ATP and GTP are presumably due to the nature and geometry of aromatic interactions. They proposed that guanine–anthracene undergoes face-to-face stacking interaction to produce excimer emission, while the adenine–anthracene pair is likely to be stabilized by C–H–π interactions between the C-2 hydrogen of adenine and the π surface of anthracene.

(c) TTP. Because bis(Zn-DPA) itself has strong affinity toward a triphosphate group (K_a ~10^6 M^{-1}), it is challenging to make a selective probe for a single type of nucleotide using Zn-DPA. In order to tackle this problem, Hong and colleagues combined a thymine-binding unit (Zn–cyclen) and a diphosphate-binding Zn-DPA unit to make a unique ensemble system for the selective sensing of TTP over other nucleotides.[58] The structures of **6.27** and **6.28** have appended energy donor and acceptor units that enable intermolecular FRET (Scheme 6.28). Assembly of the ensemble system as a TTP complex produced high energy transfer efficiency (91%) compared to other nucleotides (~75%).

Ensemble systems based on pyrene are an effective way to create a ratiometric fluorescent sensor. A 2:1 complex between pyrene-labelled Zn–cyclen and TTP or UTP showed intense excimer fluorescence in buffered aqueous solution due to the close proximity of the pyrene moieties in the ternary complex, while other nucleoside triphosphates did not induce pyrene excimer emission.[59] Zeng and Spiccia eliminated the need for an ensemble by designing the elegant pyrene-based sensor **6.29** which showed exceptional selectivity toward thymidine nucleotides among various NTPs (Scheme 6.29). The primary recognition is association of the Zn–cyclen moiety and thymine base leading to a folded conformation that produces a pyrene excimer band.[60]

6.3.3.2 *Selective Fluorescent Probes for ppGpp*

Guanosine 3′-diphosphate-5′-di(tri)phosphate, (p)ppGpp, is a bacterial alarmone that generates a stringent response under harsh conditions such as nutritional deprivation. When bacteria are placed in stress circumstances

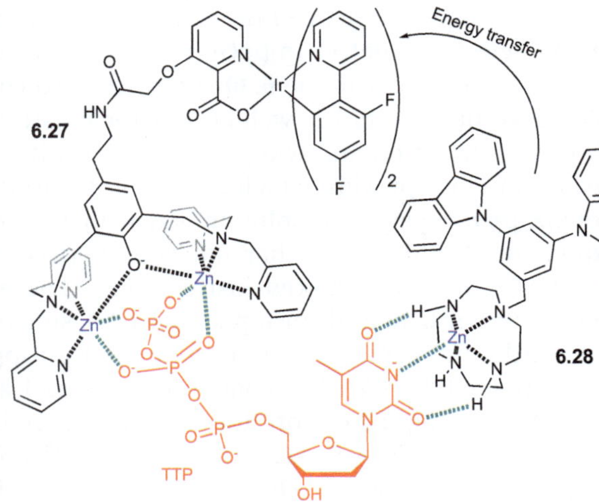

Scheme 6.28 Molecular recognition and detection mechanism of **6.27–6.28** ensemble system for thymidine triphosphate (TTP). The arrow represents an energy transfer process within the **6.27**-TTP-**6.28** termolecular complex. Reprinted with permission from ref. 58. Copyright © 2008 WILEY-VCH Verlag GmbH & Co. KGaA, Weinheim.

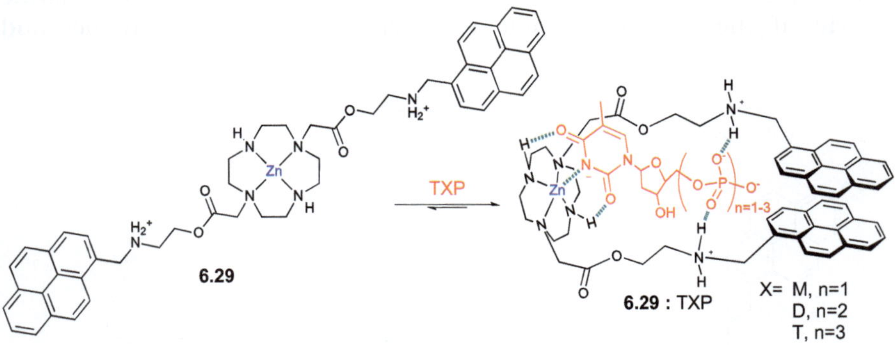

Scheme 6.29 Molecular recognition of **6.29** for thymidine phosphates (TXPs). The complexation brings two pyrene moieties into close proximity. Reproduced from ref. 59 with permission from the Royal Society of Chemistry.

such as amino acid starvation or heat shock, (p)ppGpp is swiftly synthesized from ATP and GDP by RelA on the bacterial ribosomal complex. (p)ppGpp is synthesized by the transfer of pyrophosphate from ATP to the 3′-OH of GDP (or GTP). (p)ppGpp influences many bacterial survival mechanisms, and this facilitates bacteria persistence under stress conditions. Furthermore, (p)ppGpp acts as a global regulator of gene expression in microorganisms. Therefore, precise and real-time detection of (p)ppGpp should lead to a deeper understanding of bacterial physiology.

In 2008, Hong and colleagues developed the first fluorescent receptor for (p)ppGpp, which can selectively detect (p)ppGpp in the presence of other nucleotides in water.[61] Fluorescent probe **6.30** comprises two components: pyrene and bis(Zn-DPA). Bis(Zn-DPA) is well-known for its strong binding affinity to pyrophosphate in water, and two pyrene units exhibit unique excimer emission when in close proximity (Scheme 6.30). In titration experiments, addition of ppGpp to an aqueous solution of **6.30** increased excimeric emission approximately 17-fold while other nucleotides (ATP, GTP, CTP, TTP, UTP, cAMP, cGMP, and PPi) showed just a monomeric emission at 380 nm. A Job plot between **6.30** and ppGpp showed 2:1 stoichiometry in water, as expected. Furthermore, **6.30** was used to monitor ppGpp synthesis in the presence of extracted *E. coli* ribosomal complexes and also ppGpp degradation by enzymes. Using **6.30**, Jeon and Chung and colleagues observed real-time ppGpp hydrolysis by SpoT, which was identified for the first time as a metazoan ppGpp hydrolase.[62] Other researchers found two novel point mutations of the RelA protein in clinical *Staphylococcus aureus*, which caused an up-regulation in the ppGpp concentration. This up-regulated ppGpp resulted in reduced linezolid susceptibility and switched on the stringent response to promote persistent infection.[63] Because of its strong binding affinity toward oligophosphates, **6.30** has been used in other applications as an adenylyl cyclase activity assay probe[64] and a flavin probe.[65]

Fluorescent Zn-DPA receptors can be used for measuring real-time enzyme activity if the receptor is able to discriminate between product and

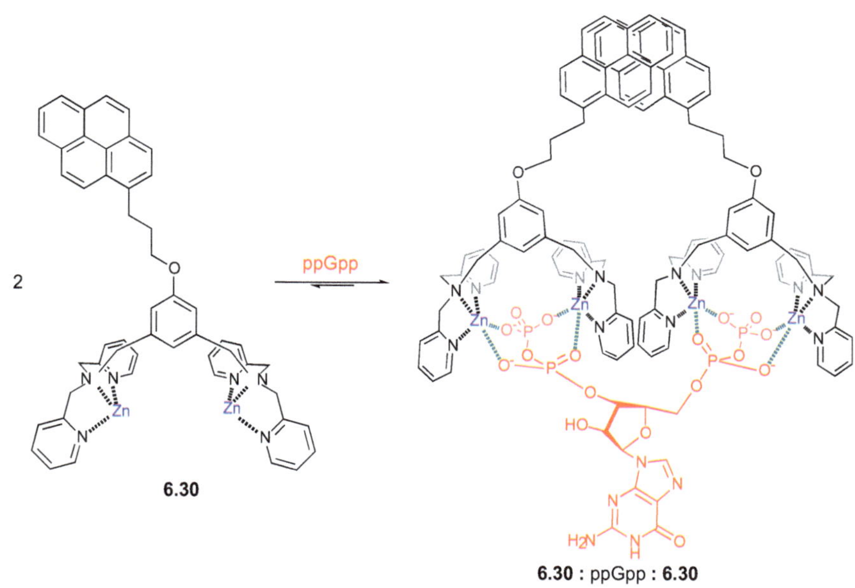

Scheme 6.30 Molecular recognition and detection mechanism of **6.30** for ppGpp. Reprinted with permission from ref. 61. © 2008 American Chemical Society.

substrates. For example, the Hamachi group was able to successfully develop a real-time glycosyltransferase assay using a simple anthracene/bis(Zn-DPA) conjugate receptor,[66] because this receptor has a greater binding affinity toward the product molecule uridine 5′-diphosphate (UDP, $K_a = 3.8 \times 10^5$ M^{-1} in HEPES buffer) over the starting material, UDP-galactose (glycosyl donor, $K_a = 4.4 \times 10^3$ M^{-1} in HEPES buffer), and other molecules (*e.g.* starting glycosyl acceptor) in the enzyme reaction mixture. Promiscuous binding of Zn-DPA receptors can be circumvented in a test-tube reaction using purified enzymes. However, it can be problematic in a cell lysate enzyme activity assay or cell imaging applications. These types of applications require more selective probes.

6.3.3.3 *Selective Fluorescent Probes for Flavin Nucleotides*

Flavins are important nucleosides/nucleotides in biological systems because of their numerous roles in redox reactions of metabolic processes. They exist in three major forms as coenzymes in nature, flavin adenine dinucleotide (FAD), flavin mononucleotide (FMN) and riboflavin (vitamin B_2). With different molecular structures, they have different roles in biological systems. For example, FMN performs a major role as a prosthetic group of oxidoreductases such as nicotinamide adenine dinucleotide (NADH) dehydrogenase, while FAD plays a key role in the mitochondrial ETC as an electron transfer mediator. In human white blood cells, FAD also plays an important role in redox reactions to generate hydrogen peroxide or superoxides that can damage a parasitic organism. The essential part of flavins in these redox reactions is the isoalloxazine ring, which exhibits a strong green fluorescence emission in its oxidized form. Because all the flavins have the same fluorophore, isoalloxazine, they exhibit similar emission spectra in water, although their intrinsic fluorescence quantum yields vary with substituents.

As mentioned previously, König *et al.* utilized a Zn(II)–cyclen/phenothiazine conjugate for the recognition and sensing of riboflavin tetraacetate through its specific binding with the imide moiety of isoalloxazine. This binding resulted in fluorescence quenching of the isoalloxazine ring through PeT, and thus one cannot discriminate between FMN and FAD in living cells through the use of Zn(II)–cyclen.[42]

In 2007, Hong and colleagues devised an extremely simple sensing assay that can discriminate FAD from FMN by monitoring changes in fluorescence.[67] In aqueous media, FAD showed greatly decreased fluorescence quantum yield because adenine and isoalloxazine stack strongly enough to quench the fluorescence of isoalloxazine. However, when bis(Zn-DPA) complex **6.31** bound FAD through Zn coordination to the diphosphate of FAD, the stacked conformation between adenine and isoalloxazine of FAD was disrupted, which resulted in the recovery of the fluorescence of isoalloxazine (Scheme 6.31). Complexation of FAD with **6.31** ($K_a = 1.11 \times 10^5$ M^{-1}) induced a seven-fold enhancement in fluorescence

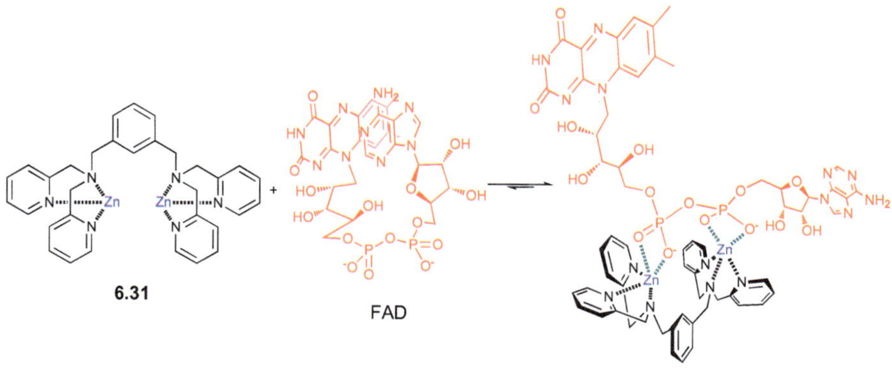

Scheme 6.31 Molecular recognition and detection mechanism of **6.31** for FAD. Reprinted with permission from ref. 67. © 2007 American Chemical Society.

compared to FAD itself in water. Additionally, **6.31** was utilized for the detection of FAD that is present in granules of eosinophils. Incubation of eosinophils with **6.31** resulted in bright greenish fluorescence that was detected using fluorescence microscopy. Furthermore, the tandem use of **6.31** in combination with flow cytometry served as an immunological diagnosis tool for eosinophilia. It was also found that the fluorescence intensity of neutrophils with toxic granules was greatly enhanced in the presence of **6.31**.[68] Because the toxic granules in neutrophils are a marker for an acute bacterial infection, **6.31** enables a simple diagnosis method.

It was later revealed that **6.31** not only binds to the diphosphate of FAD but also catalyzes the lysis reaction of FAD into riboflavin 4′,5′-cyclic phosphate (cFMN) and AMP under ambient conditions. The catalytic efficiency of artificial enzyme, **6.31**, was compared with the natural enzyme (FMN cyclase) that catalyzes this transformation. The k_{cat}/K_m of **6.31** was determined to be 36 nM^{-1} min^{-1}, whereas that of the natural enzyme was reported to be 5.6–32.9 μM^{-1} min^{-1}, depending on the coordinating transition metal ions (Co^{2+} or Mn^{2+}). An electron-rich phenothiazine group was introduced into the receptor structure to produce **6.32**, which could induce PeT with the isoalloxazine unit in the flavin. As expected, FMN showed decreased fluorescence upon binding with **6.32**, whereas the fluorescence of riboflavin was hardly perturbed (Scheme 6.32). As a result, **6.32** enabled the first discriminative detection of FAD, FMN, and riboflavin. This discriminating ability is quite useful for determining the concentration of flavins in a given enzymatic mixture of a flavin conversion reaction. Sensor **6.32** was successfully applied in real-time fluorescent assays of various enzymatic reactions related to flavin conversions such as riboflavin kinase, alkaline phosphatase, and FAD synthetase.[69] However, it is noteworthy to mention that **6.32** is not a selective receptor for either FAD or FMN. It can also bind phosphorylated biomolecules such as phosphorylated peptides or proteins

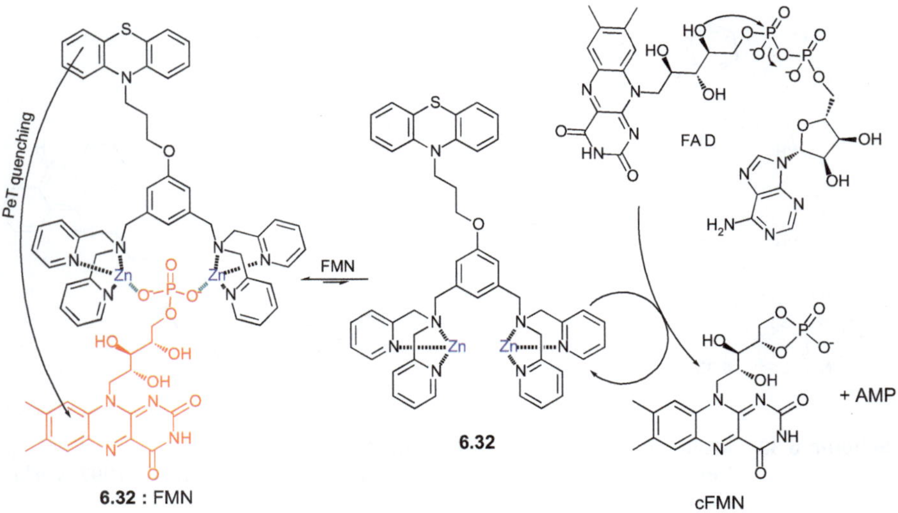

Scheme 6.32 Differential recognition mechanism of **6.32** for FAD and FMN. Reprinted with permission from ref. 69. © 2009 American Chemical Society.

and it was utilized in real-time monitoring of kinase activity, screening of kinase inhibitors, and cancer diagnosis based on abnormal kinase activity observed in patients' samples.[70]

6.3.3.4 Selective Fluorescent Probes for Nicotinamide Dinucleotides

Similarly to flavin nucleotides, nicotinamide nucleotides are also important biomolecules in cellular redox chemistry. Nicotinamide nucleotides exist in four different structures, namely NAD^+, NADH, $NADP^+$, and NADPH, depending on the redox state and the presence/absence of phosphate, where P and H denote an additional phosphate group at the 2′-position of adenosine, and a hydride at the 4-position of nicotinamide, respectively. Due to their significant roles as cellular redox mediators and their structural characteristics, nicotinamide nucleotides have received much attention from supramolecular chemists.

In 1997, Lehn and Fenniri reported a series of synthetic receptors for nicotinamide nucleotides, based on oxa-azacrown ethers with pendent acridines (Scheme 6.33).[71] Azacrown ethers are positively charged at neutral conditions and interact electrostatically with phosphates, while appended acridines are expected to change their intrinsic fluorescence *via* interactions with the aromatic components of analytes. In general, NADPH showed a three-fold higher binding constant than $NADP^+$ ($K_a = 3\times10^8$ M^{-1} and 5×10^5 M^{-1} for NADPH and $NADP^+$ in aqueous buffer at pH 7.6, respectively). The weaker interaction of $NADP^+$ with **6.33** might originate from

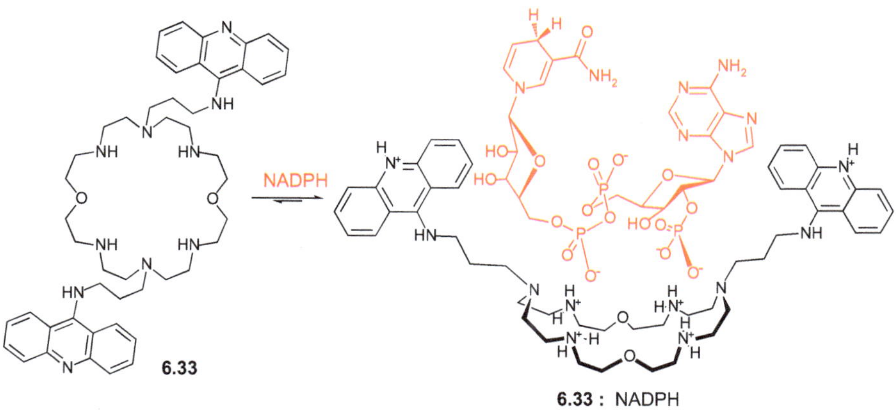

Scheme 6.33 Molecular recognition of **6.33** for NADPH.
Reprinted with permission from ref. 71. Copyright © 1997 Verlag GmbH & Co. KGaA, Weinheim.

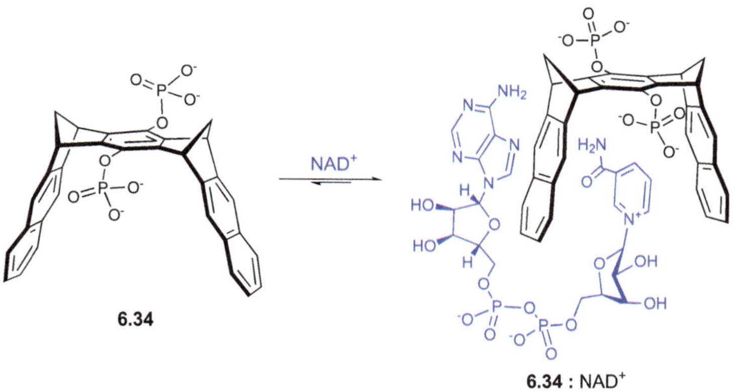

Scheme 6.34 Molecular recognition of **6.34** for NAD^+.
Reprinted with permission from ref. 72. © 2009 John Wiley & Sons, Ltd.

two aspects: (1) lower electrostatic attraction of **6.33** to $NADP^+$ (three negative charges) than NADPH (four negative charges), and (2) repulsion of the two positively charged aromatic moieties, nicotiniumamide and acridinium.

Polkowska *et al.* reported a U-shaped "molecular clip" **6.34** that showed highly selective affinity toward NAD^+ in buffered aqueous solution (Scheme 6.34).[72] The designed receptor **6.34** is composed of benzene and two naphthalenes linked together through a bicyclo[2,2,1]heptane structure, which forms an electron-rich hydrophobic cavity with extended aromatic surfaces. Therefore, in the presence of NAD^+, either the nicotinamide or adenine moiety is expected to fill the clip cavity through π–π stacking interactions. NMR experiments showed a 1:1 binding stoichiometry between **6.34** and NAD^+. But which aromatic moiety was inside the cavity?

A combination of NMR studies and *ab initio* calculations concluded that nicotinamide predominantly fills the cavity. This is expected, because the positively charged nicotinamide interacts more effectively than neutral adenine with the concave, electron-rich π system of **6.34**. The association constant between **6.34** and NAD^+ was determined to be $\sim 5\times10^3\ M^{-1}$ in buffered aqueous solution at neural pH by NMR and fluorescence titrations. Although this is weaker association than the recognition systems mentioned previously, the clip system offers insight into the inclusion of NAD^+ within the aromatic cavities of redox enzymes. Furthermore, appropriately designed molecular clips can be used to inhibit NAD-mediated enzymes (see Chapter 9).

Inspired by the recognition of ppGpp by **6.30**, Hong and Oh explored the discriminatory ability of **6.30** for NAD^+ and $NADP^+$ (Scheme 6.35).[65] It was expected that (1) the low-lying LUMO of nicotinamide would perturb the fluorescence of pyrene by oxidative PeT quenching from the excited pyrene to NAD^+ and (2) the existence of the extra phosphate group in $NADP^+$ would lead to excimer formation, as in the case of ppGpp complexation. As expected, a decrease in pyrene monomeric emission was observed with the addition of either NAD^+ or $NADP^+$. However, enhancement of pyrene excimer emission was observed only for $NADP^+$. The apparent association constant was estimated to be $4.4\times10^4\ M^{-1}$ and $3.7\times10^5\ M^{-1}$ for NAD^+ and $NADP^+$, respectively. The excimeric emission of pyrene was maintained even when a large amount of $NADP^+$ was added, whereas a higher concentration of ppGpp had a diminishing effect in the emission intensity. They proposed a cooperative binding effect due to the decreased number of negative charges on the $NADP^+$.

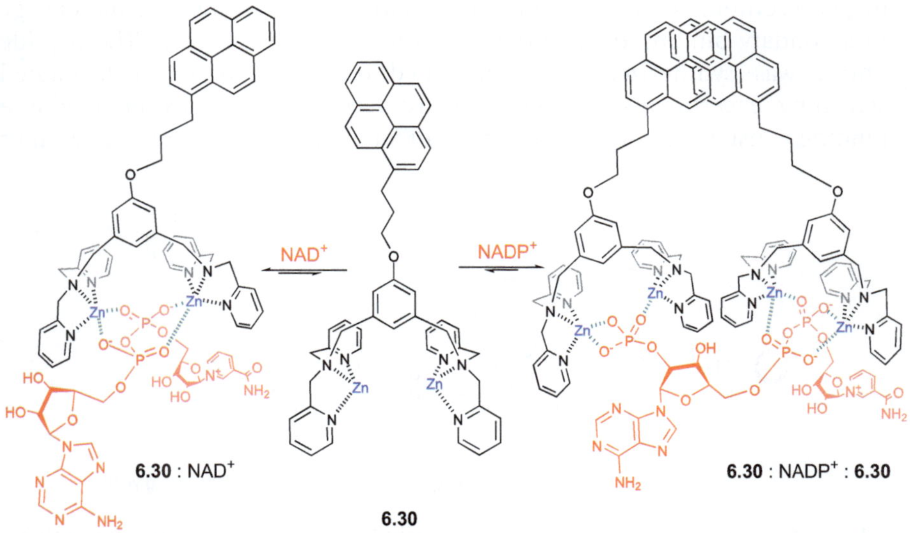

Scheme 6.35 Differential recognition of **6.30** for NAD^+ and $NADP^+$. Reprinted with permission from ref. 65. © 2013 American Chemical Society.

6.3.4 Other Approaches for Highly Selective Nucleoside/Nucleotide Receptors

It is a formidable task to design a synthetic receptor that will target a specific nucleotide with high affinity and selectivity over thousands of other phosphorylated biomolecules in a cell. If the goal is to produce a fluorescent probe, then the molecular design must also include a mechanism that signals when analyte binding has occurred. Unfortunately, current molecular modeling techniques are often not sufficiently accurate for *de novo* rational design of a high-performance synthetic receptor. In fact, serendipity is still one of the major factors controlling the discovery of new receptors and fluorescent probes. Several researchers have responded to this reality by inventing research algorithms that accelerate the pace of synthetic receptor production and sensor discovery. In this section, four approaches will be introduced: peptide libraries, diversity-oriented fluorophore libraries, molecularly imprinted polymers, and hybrid biosensors.

6.3.4.1 *Peptide (Library)-based Receptors*

Perhaps the most efficient way of developing a selective synthetic receptor is to start with a useful binding scaffold and create a diverse library of synthetic analogs for screening.[73] This method was successfully employed by Anslyn and coworkers in a project to produce an ATP-selective receptor **6.35** (Scheme 6.36).[74] A rigid hexasubstituted benzene scaffold was created with two guanidinium linkers providing a pre-organized binding cavity for the triphosphate unit of ATP. The linkers were extended with amino acid residues to give a combinatorial library of receptors with peptide arms that may engage in secondary binding interactions with the nucleobase of ATP. The peptide library was synthesized and maintained on solid phase for facilitated screening experiments. An ATP-selective receptor was discovered with a binding constant of 3.4×10^3 M^{-1} in HEPES buffer. Furthermore, the receptor

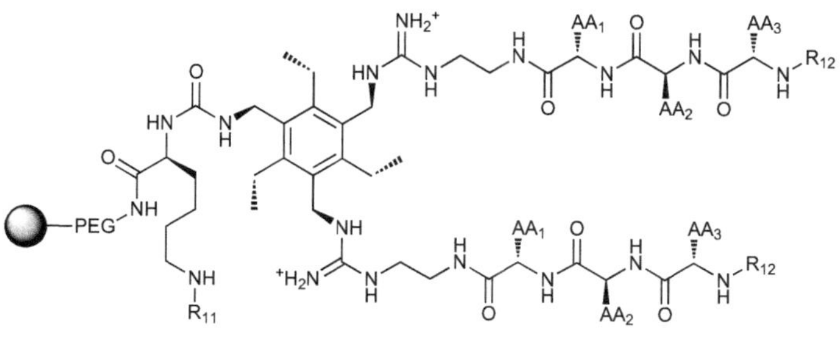

6.35

Scheme 6.36 General structure of the resin-bound sensor library for anions. AA: amino acid; R: fluorophore.
Reprinted with permission from ref. 74. © 2000 American Chemical Society.

exhibited high selectivity for ATP over the structurally similar nucleotides AMP and GTP. Later, this system was expanded into an array of randomly chosen receptors that exhibited differential recognition of nucleotides.[75] Principal component analysis was used to identify the array recognition patterns and the system successfully discriminated ATP, GTP, and AMP.

The Waters group has developed a monomeric β-hairpin peptide scaffold for molecular recognition of various targets including ATP.[76] They designed the β-sheet peptide **6.36** composed of 12 amino-acid residues with Asn-Gly dipeptide as a turn motif (Scheme 6.37). The side arms included aromatic residues (Trp) for π–π stacking interactions with the nucleobase and positively charged residues (Arg, Lys) for electrostatic attractions with the triphosphate unit. By monitoring the fluorescence decrease of a peptidyl tryptophan residue, the association constant with ATP was estimated to be $5.8\times10^3\ M^{-1}$ in buffered aqueous solution. NMR titration experiments revealed that the diagonally positioned Trp–Trp pair forms a binding cleft for intercalation of the nucleobase. Since the binding is driven by electrostatic attraction, the association constant was greatly reduced by the presence of inorganic salt ($21\ M^{-1}$ in 500 mM NaCl, 10 mM acetate buffer), and similarly decreased for AMP with less negative charge ($K_a = 109\ M^{-1}$ in 10 mM acetate buffer).

6.3.4.2 *Diversity-oriented Fluorophore Libraries*

The diversity-oriented fluorophore library approach (DOFLA) is a promising method for rapid discovery of selective fluorescent probes. DOFLA has led to new fluorescent receptors for various biomolecules. For example, a selective fluorescent probe for GTP was developed using DOFLA.[77] Chang and

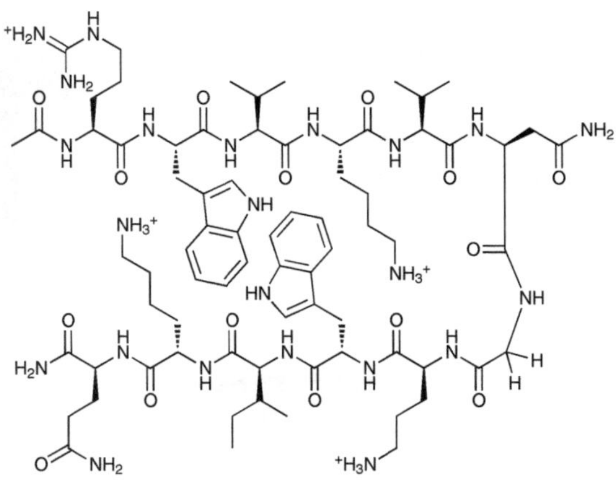

6.36

Scheme 6.37 Molecular structure of **6.36**.
Reprinted with permission from ref. 76. © 2003 American Chemical Society.

coworkers synthesized a 96-member combinatorial library of benzimidazolium-based compounds using solid-phase synthesis. In that library, they found that compound **6.37**, later dubbed GTP green, showed a selective fluorescent response to GTP. When excited at 480 nm, an approximately 80-fold fluorescence increase in emission at 540 nm was observed only for GTP, in comparison to small changes induced by all other nucleotides (Figure 6.5). The study demonstrates the power of DOFLA for generating selective fluorescent probes for nucleotides.

Recently, Hong and colleagues developed selective synthetic receptors for nucleotides using diversity-oriented combinatorial synthesis.[78] More than 250 fluorescent compounds, each comprising a metal ion chelator and styryl-based fluorophore, were synthesized *via* a Knoevenagel condensation between picolinium (or quinolinium) derivatives and benzaldehyde of metal-coordinated chelators. Among these fluorescent probes, **6.38**, whose structure included a mercury(II)-coordinated azathia crown ether and a styrylpyridinium fluorophore, showed an ultraselective fluorescence response toward dTTP (2′-deoxythymidine 5′-triphosphate) among various other nucleotides (dATP, dCTP, dGTP, dTTP, ATP, ADP, and AMP) and PPi in aqueous HEPES buffer solution at pH 7.4 (Figure 6.6). Further studies showed that imide-containing uridine triphosphate (UTP, uridine 5′-triphosphate) was also detected with a dissociation constant of $K_d = 1.48$ μM. Other compounds with the same mercury(II)-coordinated azathiocrown ether (anion receptor unit), but different fluorophore did not show the same selectivity toward dTTP. The result indicates that probe selectivity is not controlled by the anion binding unit alone, but by the whole molecular structure of **6.38** including the fluorescent signaling unit.

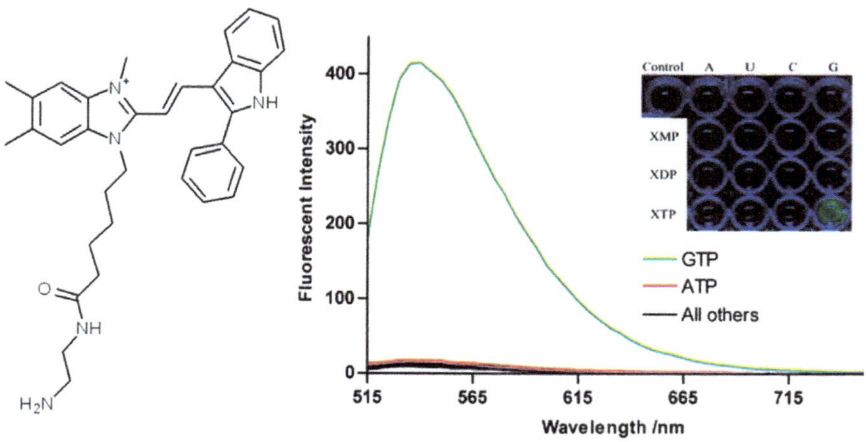

6.37, GTP green

Figure 6.5 Structure of **6.37** and its fluorescent behavior upon addition of various analytes.
Adapted with permission from ref. 77. © 2006 American Chemical Society.

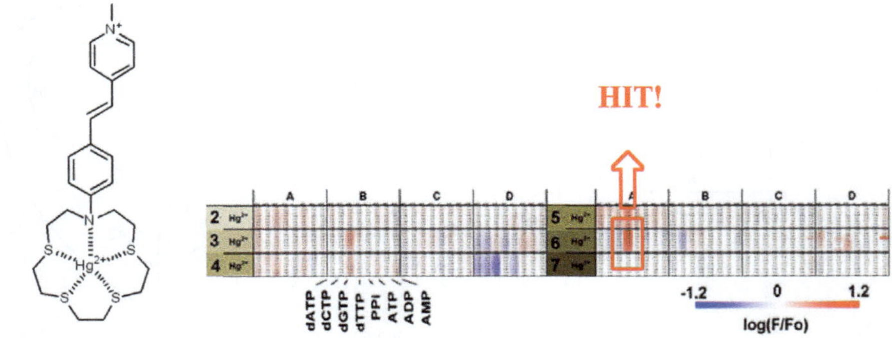

6.38, TTP selective probe

Figure 6.6 Structure of **6.38** and heat map obtained from library, especially mercury (II)-complexes.
Reprinted with permission from ref. 78. © 2013 American Chemical Society.

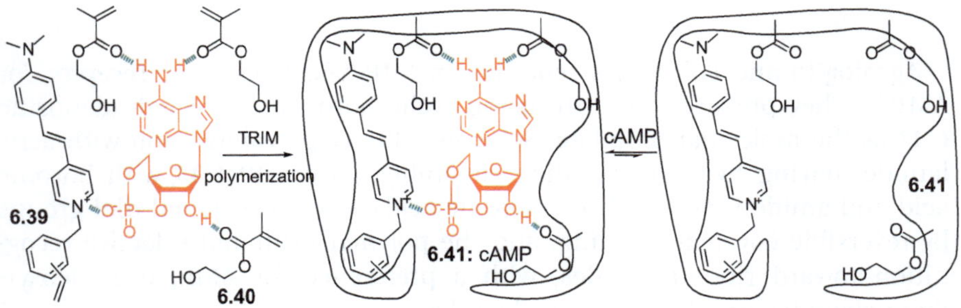

Scheme 6.38 Synthesis and molecular recognition of molecularly imprinted polymer **6.41** for cAMP.
Reprinted with permission from ref. 79. © 1998 American Chemical Society.

6.3.4.3 *Molecularly Imprinted Polymers*

The molecularly imprinted polymer (MIP) technique is a conceptually powerful tool for generating synthetic receptors that can bind any specific substrate. The literature contains several demonstrations of MIP-derived synthetic receptors for nucleosides and nucleotides.

In 1998, Powell and colleagues reported an MIP for the fluorescent detection of cAMP.[79] The MIPs were prepared by crosslinking the vinyl-functionalized monomers **6.39** and **6.40** using trimethylolpropane trimethacrylate (TRIM) in the presence of various molecular templates such as cAMP (Scheme 6.38). Monomer **6.39** has both a positive charge and is fluorescent in nature, thus it was expected to recognize the negative charge of cAMP with fluorescent signal changes, while **6.40** provided a three-dimensional hydrogen bonding network for interactions with the cAMP. The resulting MIP **6.41** exhibited high affinity (3.5×10^5 M^{-1}) toward cAMP in an aqueous solvent system, with no obvious binding of structurally similar cGMP.

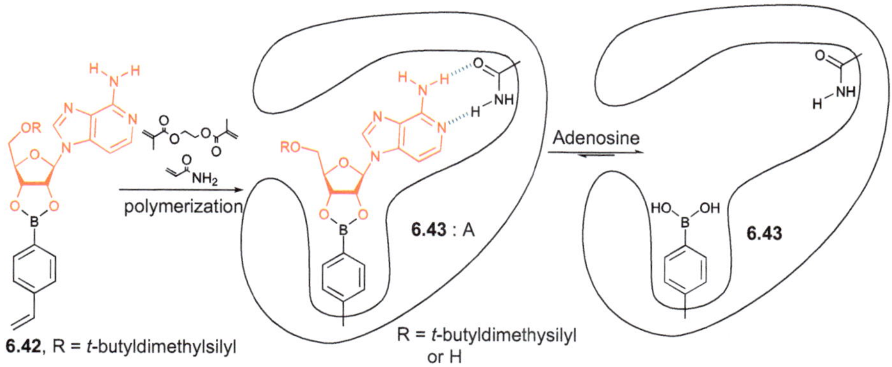

6.42, R = *t*-butyldimethylsilyl

Scheme 6.39 Synthesis and molecular recognition of molecularly imprinted polymer **6.43** for adenosine.
Reprinted from *Analytica Chimica Acta*, ref. 80, © 2008, with permission from Elsevier.

Agrofoglio and colleagues reported an MIP-based synthetic receptor for AMP.[80] They prepared the MIP using cyclic boronate-protected adenosine **6.42** as the molecular template (Scheme 6.39). Copolymerization with acrylamide, having hydrogen-bonding capability, gave MIP **6.43** with boronic acid and amide functionalities. Boronic acids recognize vicinal diol groups by reversible covalent bonding, and the polymer exhibited selective recognition toward ribonucleotides, with a preference for AMP, over deoxyribonucleotides or other cyclic nucleotides.

6.3.4.4 Hybrid Biosensors

There are several classes of natural receptors for nucleosides and nucleotides that bind with high affinity and selectivity. However, these natural receptors cannot be directly used for selective "detection", because they lack a suitable reporting mechanism. One way to establish a "detection system" is to embed a fluorophore into the receptor binding domain, with the expectation that the fluorescence signal will be modulated by guest binding. Here we describe a few examples of how researchers have converted natural nucleotide receptors into fluorescent hybrid biosensors.

In 2000, Tsien and Taylor successfully demonstrated selective fluorescent biosensing for cAMP in a living cell.[81,82] They utilized the cAMP recognition-induced dissociation between the catalytic and regulatory domains of PKA. FRET was established by genetically engineering both the regulatory and catalytic domains to have a chimeric structure with cyan fluorescent protein (CFP) and yellow fluorescent protein (YFP), respectively. Cells with engineered proteins showed fluorescent behavior with ~5% FRET efficiency. When a cAMP-raising agent such as forskolin was added, disruption in FRET efficiency was observed, indicating a rise in intracellular cAMP (Figure 6.7).

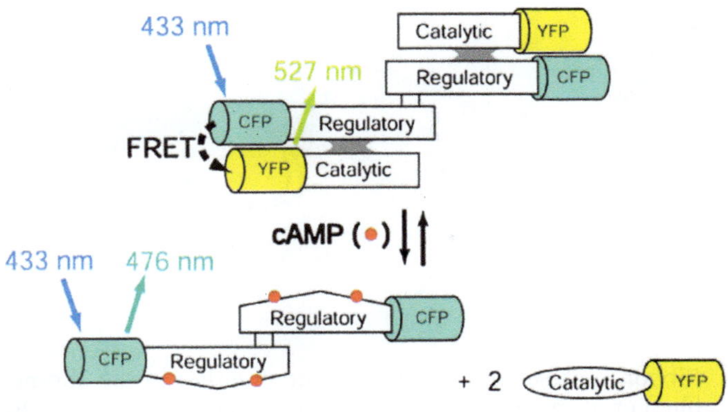

Figure 6.7 Schematic representation of fluorescent cAMP sensing with chimeric protein kinase A domains.
Adapted from ref. 82 with permission from the Royal Society of Chemistry.

Another example, which utilizes a natural recognition system, was reported by Miller and colleagues.[83] To measure the concentration of cyclic-di-GMP in an individual cell, they used a genetically engineered protein composed of a PilZ domain and two fluorescent protein subunits. The working principle is as follows: recognition of cyclic-di-GMP by a PilZ domain induces a conformational change that results in the alteration of the FRET efficiency. Although the FRET efficiency was not greatly altered, it was enough to detect an unusual distribution of cyclic-di-GMP between swarmer cells and stalker cells during their asymmetric division process.

Genetically engineered proteins have been designed to measure the intracellular ratio of ATP and ADP, which reflects the energy state of a cell.[84,85] Yellen and colleagues used both a GlnK1 domain and circularly permuted monomeric Venus (cpmVenus) to construct an ATP and ADP responsive biosensor. Binding of ATP or ADP rigidifies the conformation of GlnK1 in a different degree, which in turn enhances the fluorescence intensity of the conjugated fluorescent protein, cpmVenus (Figure 6.8). Also, because the GlnK1 domain binds ATP or ADP so tightly, the binding site of GlnK1 is presumably saturated under the cellular concentration, thus exhibiting a ratio-dependent emission profile regardless of the concentration of ATP or ADP. Using this chimeric protein, the researchers were able to detect changes in the intracellular ratio of ATP and ADP in the presence of external stimuli.

Inspired by the way nature uses segments of RNA to bind small molecules (*e.g.* riboswitches), scientists have developed ways of producing unnatural nucleic acid sequences, called aptamers, for selective binding of small molecules. The sequences are discovered by a powerful *in vitro* selection process termed systematic evolution of ligands by exponential enrichment (SELEX). Nowadays, SELEX is used not only to prepare a receptor but also to

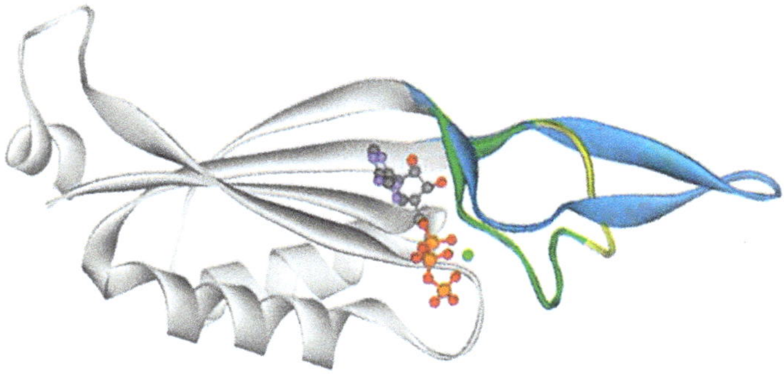

Figure 6.8 Ribbon representation of GlnK protein either in the absence (gray and blue) or in the presence (gray, green, yellow) of ATP-Mg^{2+}. Blue: flexible T loop; yellow: cpFP inserted region.
Reprinted by permission from Macmillan Publishers Ltd: *Nat. Methods*, ref. 84, © 2009.

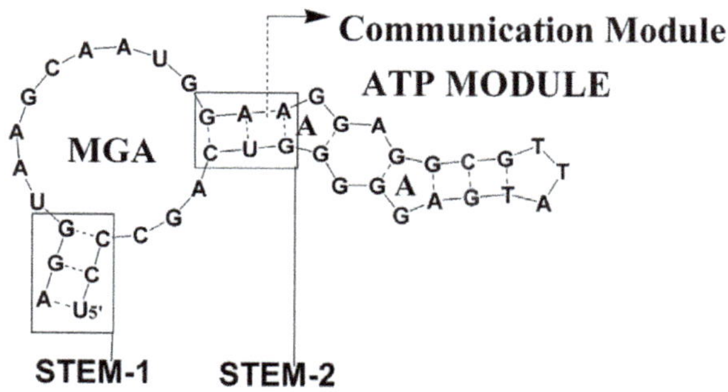

Figure 6.9 Structure of fluorescent ATP-sensing aptamer. MGA: malachite green binding site; ATP module: ATP binding site.
Adapted with permission from ref. 89. © 2004 American Chemical Society.

embed a function. In 2003, Tsien and colleagues reported that an aptamer for the dye malachite green (MG) can dramatically enhance fluorescence quantum yields (2360-fold) of bound MG. This provides a fluorescent detection method for monitoring the actual binding event at the binding site of aptameric sensors incorporating the MG aptamer.[86]

Szostak and colleagues isolated selective aptamers for ATP ($K_d = 4.8$ μM)[87] and GTP ($K_d = 75$ nM),[88] respectively. To produce a highly selective optical probe, they combined the ATP aptamer as a "recognition domain" and Tsien's MG aptamer as a "signaling domain" (Figure 6.9).[89] These two binding sites were linked together through a short communication module which transduced the ATP binding event into stabilization of bound MG, leading to increased fluorescence of bound MG. This chimeric aptamer sensor could detect 10 μM of ATP with exceptional selectivity.

In 2008, Tan and coworkers developed another effective ATP recognition system using the same ATP aptamer, but a different strategy called a molecular beacon.[90] They used a quencher-conjugated short DNA sequence that is complementary to a fluorophore-conjugated aptamer. Upon ATP binding, the proximal fluorophore and quencher units are forced apart, which switches on fluorescence by disrupting the quenching process (Figure 6.10).

Recently, Morii and colleagues demonstrated fluorescent nucleotide recognition using a ribonucleopeptide (RNP) complex (a stable complex of RNA and a peptide) (Figure 6.11).[91] Rev-responsive element (RRE) is a well-known RNA segment that tightly binds Rev peptide. This RNP complex was chosen as a model system for a tailored receptor (RNA)–reporter (protein) conjugate. To construct a receptor library without losing binding capability, a randomized nucleotide region was introduced near RRE. From this library, they found RNPs that showed selective binding affinities toward ATP and GTP. Using these RNP pairs, they could synthesize covalently conjugated RNA–peptide pairs for specific binding of ATP and covalently conjugated RNA–peptide pairs for specific binding of GTP. In addition, they attached various fluorophores, such as 5-carboxyfluorescein or pyrene to each RNP complex to measure the ATP or GTP levels by monitoring the fluorescence intensity enhancement. As expected, each of these sensors showed an orthogonal

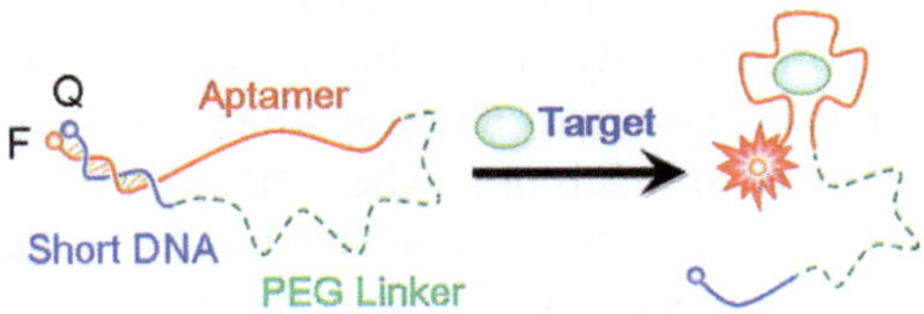

Figure 6.10 Schematic representation for the fluorescent ATP sensing with a molecular beacon strategy. F: fluorophore; Q: quencher.
Adapted with permission from ref. 90. © 2008 American Chemical Society.

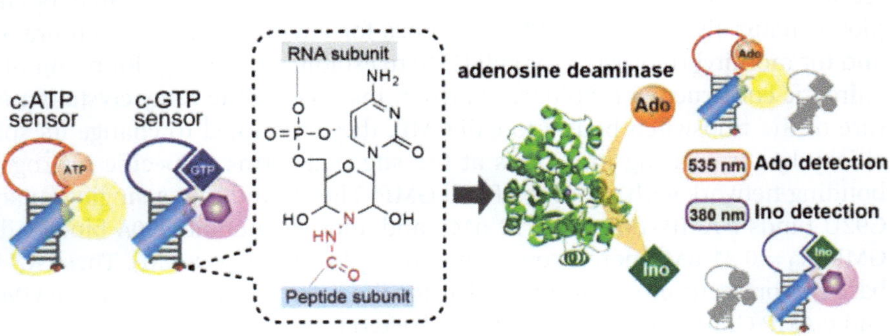

Figure 6.11 Schematic representation for the fluorescent sensing of either ATP or GTP with corresponding ribonucleopeptides and their utilization in an *in vitro* enzyme assay.
Adapted with permission from ref. 91. © 2013 American Chemical Society.

fluorescence response toward ATP ($K_d = 6$ μM) and GTP ($K_d = 7$ μM). These selective binding properties were successfully applied to real-time monitoring of *in vitro* enzymatic conversion of adenosine into inosine by adenosine deaminase.

In 2011, Jaffrey and coworkers reported several RNA aptamers which strongly bind to 4-hydroxybenzylidene imidazolinone derivatives (HBIs), small molecule analogs of the fluorophore within green fluorescent protein (GFP).[92] They termed an aptamer complex with 3,5-difluoro-4-hydroxybenzylidene imidazolinone (DFHBI) as Spinach, because the aptamer not only binds to DFHBI but also turns on the green fluorescence of DFHBI, which is greatly decreased in its free form. Additionally, Spinach showed greater photostability than fluorescein or enhanced GFP, which is crucial for long-time imaging of live cells. In fact, its superior ability as a tag for small RNA molecules was successfully demonstrated by monitoring the intracellular distribution of Spinach-fused 5S ribosomal RNA. The structure and turn-on mechanism of Spinach has been studied in detail. Spinach contains three stem loops encircling a large central loop which provides a binding site for HBIs. Mutagenesis revealed that one of the three stem loops plays critical roles in fluorescence enhancement, to which small molecule-binding aptamers are inserted for the development of RNA-based metabolite probes.[93] These RNA-based metabolite probes are composed of two tandem binding sites for metabolite and DFHBI connected by a transducer sequence where the occupation state of binding site is transmitted. The transducer sequence was designed to reinforce the structure of Spinach only when the metabolite binding site was filled. As a result, RNA-based metabolite sensors for *S*-adenosylmethionine, ADP and GTP were developed. In addition, these chimeric RNAs were successfully used to monitor metabolite dynamics in live cells.

The above strategy seems quite general and can be utilized to design fluorescent probes for essentially any biomolecular target. In 2013, Hammond and coworkers reported RNA-based fluorescent probes for bacterial second messengers, c-di-GMP and c-AMP-GMP.[94] The researchers employed naturally occurring GEMM-I (genes for environment, for membranes and for motility) riboswitch for c-di-GMP to establish the recognition unit of a chimeric construct with Spinach. Further, inspired by the X-ray crystal structure of the riboswitch bound to c-di-GMP, they attempted to change its specificity by introducing mutations at the site that forms a specific hydrogen bonding network with guanine of c-di-GMP. They found that a single mutant C92U binds c-AMP-GMP ($K_d = 19$ nM), and another mutant G20A binds c-di-GMP ($K_d = 0.21$ nM) more strongly than the wild type ($K_d = 8$ nM). These RNA-based fluorescent probes were applied for the selective detection of c-di-GMP and c-AMP-GMP production in a living bacterial cell.

An aptazyme (aptamer + ribozyme) is a ribozyme which performs catalytic functions only in the presence of analyte molecule as an allosteric modulator. In 2005, Ellington and colleagues developed a unique ATP sensing system using aptazyme in combination with a highly efficient amplification system called rolling circle amplification (RCA).[95] In order to establish an

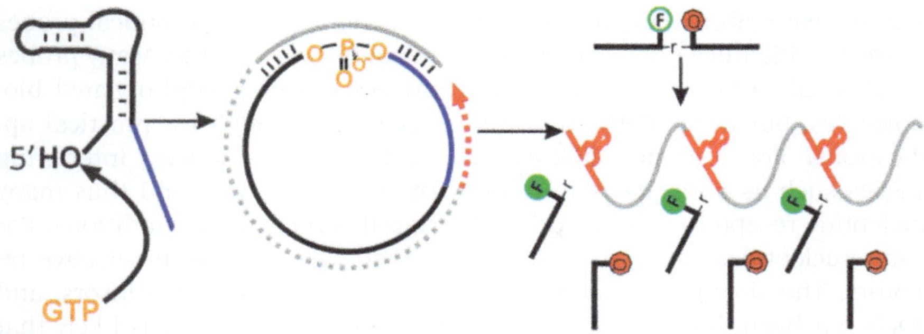

Figure 6.12 Schematic representation for the fluorescent GTP sensing with GTP-specific deoxyribozyme Dk2.
Adapted with permission from ref. 96. © 2013 American Chemical Society.

ATP sensing system, an ATP-dependent aptazyme was selected to function as DNA ligase. A DNA substrate was designed to include an iodo leaving group (electrophile) at the 5′ end and thiophosphate moiety (nucleophile) at the 3′ end, which could be intramolecularly ligated by the action of ATP-dependent aptazyme, giving circularized DNA. The resultant circular DNA served as a template for RCA in the presence of DNA polymerase. The fluorescence signal was obtained by hybridization with fluorophore-labeled complementary oligonucleotides. This system was proven to detect as little as 1 μM ATP. No obvious fluorescence signal was observed upon exposure to 10 mM of other nucleotides such as UTP, GTP and CTP because the amplification was initiated by the ATP binding-induced conformational change.

Li and coworkers reported a doubly-amplifying GTP sensing system using two deoxyribozymes (Figure 6.12).[96] The first deoxyribozyme they used was Dk2, which is highly sensitive and selective to GTP. Dk2 is known to self-phosphorylate at its 5′ end only in the presence of GTP, which not only acts as an allosteric modulator but also as a phosphate donor. Self-phosphorylated Dk2 can be cyclized by DNA ligase to give circular Dk2 that can serve as a template for the first amplification process, RCA. They appended an antisense DNA sequence to the 3′ end of Dk2, which codes the second deoxyribozyme, a sequence-selective RNA lyase. With this combination of modified Dk2 and RCA, they were able to amplify a low concentration of GTP to a large amount of RNA lyase. Finally, an increased fluorescence signal was produced by RNA lyase catalysis of a suitable fluorescent RNA substrate. Using this double-amplifying system, as little as 4 μM GTP was selectively detected without any inconvenient purification step. This system is one of the most successful model systems that utilize the catalytic activity of natural deoxyribozymes as a sensing and amplifying system for biomolecules.

6.4 Challenges and Future Directions

Over the past 10 years there has been a boom in the development of synthetic receptors, especially for phosphates and nucleotides. However, despite this intensive effort there are very few synthetic receptors or optical probes for nucleoside/nucleotide sensing that are completely specific. Many probes are claimed to be selective for certain nucleotides or phosphorylated biomolecules, but in actuality they are not selective enough for practical applications. For example, biological samples contain weakly interfering species such as phosphates in more than 1000-fold excess and thus many nucleotide receptors do not work in living cells under these conditions. For some nucleotides, such as CTP, there are virtually no known selective receptors. The slow progress towards practically useful probes, sensors, and labels has been discussed and analyzed.[97] Looking forward, it is likely that new highly selective receptors and probes can be developed through the following strategies and their combination:

(1) hybrid biosensing approach using biological units, which already have excellent selectivity;
(2) combinatorial chemistry approaches for diversity generation; and
(3) computer modeling approaches that design highly selective receptors through computational calculations.

The authors expect to see many new receptors and probes for nucleosides and nucleotides being developed in the upcoming decades, with high translational impact in healthcare and biomedical research.

References

1. R. T. Dorsam and J. S. Gutkind, *Nat. Rev. Cancer*, 2007, 7, 79.
2. D. Kalia, G. Merey, S. Nakayama, Y. Zheng, J. Zhou, Y. Luo, M. Guo, B. T. Roembke and H. O. Sintim, *Chem. Soc. Rev.*, 2013, **42**, 305.
3. R. Yarchoan, H. Mitsuya, C. E. Myers and S. Broder, *N. Engl. J. Med.*, 1989, **321**, 726.
4. L. P. Jordheim, D. Durantel, F. Zoulim and C. Dumontet, *Nat. Rev. Drug Discovery*, 2013, **12**, 447.
5. Y. Su, W. R. G. Dostmann, F. W. Herberg, K. Durick, N. H. Xuong, L. Teneyck, S. S. Taylor and K. I. Varughese, *Science*, 1995, **269**, 807.
6. W. Watt, A. Tulinsky, R. P. Swenson and K. D. Watenpaugh, *J. Mol. Biol.*, 1991, **218**, 195.
7. O. Yildiz, C. Kalthoff, S. Raunser and W. Kuhlbrandt, *EMBO J.*, 2007, **26**, 589.
8. N. Sudarsan, E. R. Lee, Z. Weinberg, R. H. Moy, J. N. Kim, K. H. Link and R. R. Breaker, *Science*, 2008, **321**, 411.
9. N. Kulshina, N. J. Baird and A. R. Ferre-D'Amare, *Nat. Struct. Mol. Biol.*, 2009, **16**, 1212.
10. Y. Kyogoku, R. C. Lord and A. Rich, *Proc. Natl. Acad. Sci. U. S. A.*, 1967, **57**, 250.
11. K. M. Guckian, B. A. Schweitzer, R. X. F. Ren, C. J. Sheils, P. L. Paris, D. C. Tahmassebi and E. T. Kool, *J. Am. Chem. Soc.*, 1996, **118**, 8182.

12. J. Rebek, B. Askew, P. Ballester, C. Buhr, S. Jones, D. Nemeth and K. Williams, *J. Am. Chem. Soc.*, 1987, **109**, 5033.
13. T. Benzing, T. Tjivikua, J. Wolfe and J. Rebek, *Science*, 1988, **242**, 266.
14. A. D. Hamilton and D. Vanengen, *J. Am. Chem. Soc.*, 1987, **109**, 5035.
15. T. K. Park, J. Schroeder and J. Rebek, *J. Am. Chem. Soc.*, 1991, **113**, 5125.
16. N. Tamura, K. Mitsui, T. Nabeshima and Y. Yano, *J. Chem. Soc., Perkin Trans. 2*, 1994, 2229.
17. N. Tamura, T. Kajiki, T. Nabeshima and Y. Yano, *J. Chem. Soc., Chem. Commun.*, 1994, 2583.
18. T. Kajiki, H. Moriya, S. Kondo, T. Nabeshima and Y. Yano, *Chem. Commun.*, 1998, 2727.
19. E. C. Breinlinger and V. M. Rotello, *J. Am. Chem. Soc.*, 1997, **119**, 1165.
20. R. Deans, G. Cooke and V. M. Rotello, *J. Org. Chem.*, 1997, **62**, 836.
21. K. Y. Lin and M. D. Matteucci, *J. Am. Chem. Soc.*, 1998, **120**, 8531.
22. O. Nakagawa, S. Ono, Z. Li, A. Tsujimoto and S. Sasaki, *Angew. Chem., Int. Ed.*, 2007, **46**, 4500.
23. S. Loft and H. E. Poulsen, *J. Mol. Med.*, 1996, **74**, 297.
24. K. A. Ahmed, T. Sawa and T. Akaike, *Amino Acids*, 2011, **41**, 123.
25. Y. Fuchi and S. Sasaki, *Org. Lett.*, 2014, **16**, 1760.
26. A. E. Hargrove, S. Nieto, T. Z. Zhang, J. L. Sessler and E. V. Anslyn, *Chem. Rev.*, 2011, **111**, 6603.
27. A. Galan, J. Demendoza, C. Toiron, M. Bruix, G. Deslongchamps and J. Rebek, *J. Am. Chem. Soc.*, 1991, **113**, 9424.
28. F. P. Schmidtchen, *Tetrahedron Lett.*, 1989, **30**, 4493.
29. Y. Kato, M. M. Conn and J. Rebek, *J. Am. Chem. Soc.*, 1994, **116**, 3279.
30. H. Y. Kuchelmeister and C. Schmuck, *Chem. – Eur. J.*, 2011, **17**, 5311.
31. A. V. Eliseev and H. J. Schneider, *J. Am. Chem. Soc.*, 1994, **116**, 6081.
32. P. Schwinte, R. Darcy and F. O'Keeffe, *J. Chem. Soc., Perkin Trans. 2*, 1998, 805.
33. V. Janout, B. W. Jing, I. V. Staina and S. L. Regen, *J. Am. Chem. Soc.*, 2003, **125**, 4436.
34. M. Takeuchi, M. Taguchi, H. Shinmori and S. Shinkai, *Bull. Chem. Soc. Jpn.*, 1996, **69**, 2613.
35. R. A. Brown, V. Diemer, S. J. Webb and J. Clayden, *Nat. Chem.*, 2013, **5**, 853.
36. S. Katz, *J. Am. Chem. Soc.*, 1952, **74**, 2238.
37. Y. Miyake, H. Togashi, M. Tashiro, H. Yamaguchi, S. Oda, M. Kudo, Y. Tanaka, Y. Kondo, R. Sawa, T. Fujimoto, T. Machinami and A. Ono, *J. Am. Chem. Soc.*, 2006, **128**, 2172.
38. J. S. Lee, M. S. Han and C. A. Mirkin, *Angew. Chem., Int. Ed.*, 2007, **46**, 4093.
39. M. Shionoya, E. Kimura and M. Shiro, *J. Am. Chem. Soc.*, 1993, **115**, 6730.
40. S. Aoki, Y. Honda and E. Kimura, *J. Am. Chem. Soc.*, 1998, **120**, 10018.
41. M. Shionoya, T. Ikeda, E. Kimura and M. Shiro, *J. Am. Chem. Soc.*, 1994, **116**, 3848.

42. B. König, M. Pelka, H. Zieg, T. Ritter, H. Bouas-Laurent, R. Bonneau and J. P. Desvergne, *J. Am. Chem. Soc.*, 1999, **121**, 1681.
43. F. Mancin and J. Chin, *J. Am. Chem. Soc.*, 2002, **124**, 10946.
44. J. S. Seo, N. D. Sung, R. C. Hynes and J. Chin, *Inorg. Chem.*, 1996, **35**, 7472.
45. M. Komiyama, S. Kina, K. Matsumura, J. Sumaoka, S. Tobey, V. M. Lynch and E. Anslyn, *J. Am. Chem. Soc.*, 2002, **124**, 13731.
46. D. H. Lee, J. H. Im, S. U. Son, Y. K. Chung and J. I. Hong, *J. Am. Chem. Soc.*, 2003, **125**, 7752.
47. S. Aoki, D. Kagata, M. Shiro, K. Takeda and E. Kimura, *J. Am. Chem. Soc.*, 2004, **126**, 13377.
48. S. Aoki and E. Kimura, *J. Am. Chem. Soc.*, 2000, **122**, 4542.
49. E. Kimura, M. Kikuchi, H. Kitamura and T. Koike, *Chem. - Eur. J.*, 1999, **5**, 3113.
50. J. H. Lee, A. R. Jeong, J. H. Jung, C. M. Park and J. I. Hong, *J. Org. Chem.*, 2011, **76**, 417.
51. D. H. Lee, K. H. Lee and J. I. Hong, *Org. Lett.*, 2001, **3**, 5.
52. Y. Zhou, Z. Xu and J. Yoon, *Chem. Soc. Rev.*, 2011, **40**, 2222.
53. A. Ojida, I. Takashima, T. Kohira, H. Nonaka and I. Hamachi, *J. Am. Chem. Soc.*, 2008, **130**, 12095.
54. Y. Kurishita, T. Kohira, A. Ojida and I. Hamachi, *J. Am. Chem. Soc.*, 2010, **132**, 13290.
55. Y. Kurishita, T. Kohira, A. Ojida and I. Hamachi, *J. Am. Chem. Soc.*, 2012, **134**, 18779.
56. Z. Xu, N. J. Singh, J. Lim, J. Pan, H. N. Kim, S. Park, K. S. Kim and J. Yoon, *J. Am. Chem. Soc.*, 2009, **131**, 15528.
57. N. Ahmed, B. Shirinfar, I. S. Youn, A. Bist, V. Suresh and K. S. Kim, *Chem. Commun.*, 2012, **48**, 2662.
58. T. H. Kwon, H. J. Kim and J. I. Hong, *Chem. - Eur. J.*, 2008, **14**, 9613.
59. F. Schmidt, S. Stadlbauer and B. König, *Dalton Trans.*, 2010, **39**, 7250.
60. Z. H. Zeng and L. Spiccia, *Chem. - Eur. J.*, 2009, **15**, 12941.
61. H. W. Rhee, C. R. Lee, S. H. Cho, M. R. Song, M. Cashel, H. E. Choy, Y. J. Seok and J. I. Hong, *J. Am. Chem. Soc.*, 2008, **130**, 784.
62. D. Sun, G. Lee, J. H. Lee, H. Y. Kim, H. W. Rhee, S. Y. Park, K. J. Kim, Y. Kim, B. Y. Kim, J. I. Hong, C. Park, H. E. Choy, J. H. Kim, Y. H. Jeon and J. Chung, *Nat. Struct. Mol. Biol.*, 2010, **17**, 1188.
63. W. Gao, K. Chua, J. K. Davies, H. J. Newton, T. Seemann, P. F. Harrison, N. E. Holmes, H. W. Rhee, J. I. Hong, E. L. Hartland, T. P. Stinear and B. P. Howden, *PLoS Pathog.*, 2010, **6**, e1000944.
64. H. W. Rhee, K. S. Kim, P. L. Han and J. I. Hong, *Bioorg. Med. Chem. Lett.*, 2010, **20**, 1145.
65. J. Oh and J. I. Hong, *Org. Lett.*, 2013, **15**, 1210.
66. J. Wongkongkatep, Y. Miyahara, A. Ojida and I. Hamachi, *Angew. Chem., Int. Ed.*, 2006, **45**, 665.
67. H. W. Rhee, H. Y. Choi, K. Han and J. I. Hong, *J. Am. Chem. Soc.*, 2007, **129**, 4524.

68. Y. S. Kim, H. H. Park, H. W. Rhee, J. I. Hong and K. Han, *Ann. Clin. Lab. Sci.*, 2009, **39**, 114.
69. H. W. Rhee, S. J. Choi, S. H. Yoo, Y. O. Jang, H. H. Park, R. M. Pinto, J. C. Cameselle, F. J. Sandova, S. Roje, K. Han, D. S. Chung, J. Suh and J. I. Hong, *J. Am. Chem. Soc.*, 2009, **131**, 10107.
70. H. W. Rhee, S. H. Lee, I. S. Shin, S. J. Choi, H. H. Park, K. Han, T. H. Park and J. I. Hong, *Angew. Chem., Int. Ed.*, 2010, **49**, 4919.
71. H. Fenniri, M. W. Hosseini and J. M. Lehn, *Helv. Chim. Acta*, 1997, **80**, 786.
72. J. Polkowska, F. Bastkowski, T. Schrader, F. G. Klarner, J. Zienau, F. Koziol and C. Ochsenfeld, *J. Phys. Org. Chem.*, 2009, **22**, 779.
73. R. Boyce, G. Li, H. P. Nestler, T. Suenaga and W. C. Still, *J. Am. Chem. Soc.*, 1994, **116**, 7955.
74. S. E. Schneider, S. N. O'Neil and E. V. Anslyn, *J. Am. Chem. Soc.*, 2000, **122**, 542.
75. S. C. McCleskey, M. J. Griffin, S. E. Schneider, J. T. McDevitt and E. V. Anslyn, *J. Am. Chem. Soc.*, 2003, **125**, 1114.
76. S. M. Butterfield and M. L. Waters, *J. Am. Chem. Soc.*, 2003, **125**, 9580.
77. S. L. Wang and Y. T. Chang, *J. Am. Chem. Soc.*, 2006, **128**, 10380.
78. H. W. Rhee, S. W. Lee, J. S. Lee, Y. T. Chang and J. I. Hong, *ACS Comb. Sci.*, 2013, **15**, 483.
79. P. Turkewitsch, B. Wandelt, G. D. Darling and W. S. Powell, *Anal. Chem.*, 1998, **70**, 2025.
80. F. Breton, R. Delepee, D. Jegourel, D. Deville-Bonne and L. A. Agrofoglio, *Anal. Chim. Acta*, 2008, **616**, 222.
81. M. Zaccolo, F. De Giorgi, C. Y. Cho, L. X. Feng, T. Knapp, P. A. Negulescu, S. S. Taylor, R. Y. Tsien and T. Pozzan, *Nat. Cell Biol.*, 2000, **2**, 25.
82. R. Y. Tsien, *Integr. Biol.*, 2010, **2**, 77.
83. M. Christen, H. D. Kulasekara, B. Christen, B. R. Kulasekara, L. R. Hoffman and S. I. Miller, *Science*, 2010, **328**, 1295.
84. J. Berg, Y. P. Hung and G. Yellen, *Nat. Methods*, 2009, **6**, 161.
85. M. Tantama, J. R. Martinez-Francois, R. Mongeon and G. Yellen, *Nat. Commun.*, 2013, **4**.
86. J. R. Babendure, S. R. Adams and R. Y. Tsien, *J. Am. Chem. Soc.*, 2003, **125**, 14716.
87. P. L. Sazani, R. Larralde and J. W. Szostak, *J. Am. Chem. Soc.*, 2004, **126**, 8370.
88. J. M. Carothers, J. H. Davis, J. J. Chou and J. W. Szostak, *RNA*, 2006, **12**, 567.
89. M. N. Stojanovic and D. M. Kolpashchikov, *J. Am. Chem. Soc.*, 2004, **126**, 9266.
90. Z. W. Tang, P. Mallikaratchy, R. H. Yang, Y. M. Kim, Z. Zhu, H. Wang and W. H. Tan, *J. Am. Chem. Soc.*, 2008, **130**, 11268.
91. S. Nakano, M. Fukuda, T. Tamura, R. Sakaguchi, E. Nakata and T. Morii, *J. Am. Chem. Soc.*, 2013, **135**, 3465.
92. J. S. Paige, K. Y. Wu and S. R. Jaffrey, *Science*, 2011, **333**, 642.

93. J. S. Paige, T. Nguyen-Duc, W. J. Song and S. R. Jaffrey, *Science*, 2012, **335**, 1194.
94. C. A. Kellenberger, S. C. Wilson, J. Sales-Lee and M. C. Hammond, *J. Am. Chem. Soc.*, 2013, **135**, 4906.
95. E. J. Cho, L. T. Yang, M. Levy and A. D. Ellington, *J. Am. Chem. Soc.*, 2005, **127**, 2022.
96. S. A. McManus and Y. F. Li, *J. Am. Chem. Soc.*, 2013, **135**, 7181.
97. O. S. Wolfbeis, *Angew. Chem., Int. Ed.*, 2013, **52**, 9864.

CHAPTER 7

Synthetic Receptors for Oligonucleotides and Nucleic Acids

BRITTANY S. MORGAN AND AMANDA E. HARGROVE*

[a]Department of Chemistry, 124 Science Drive, Duke University, Box 90346, Durham, NC 27708-0346, USA
*Email: amanda.hargrove@duke.edu

7.1 Chemical Structures and Physical Properties of Oligonucleotides and Nucleic Acids

7.1.1 Properties of Oligonucleotides and Nucleic Acids

The primary structures of deoxyribonucleic acid (DNA) and ribonucleic acid (RNA) are comprised of pentose sugars, phosphate units, and nucleobases.[1] The critical difference between the oligonucleotides lies in the pentose sugar, where the deoxyribose of DNA lacks the 2′-hydroxyl group (Figure 7.1A). Phosphate groups connect the pentose sugars between respective 5′- and 3′-hydroxyl groups, forming the negatively charged sugar–phosphate backbone. Each nucleobase is connected to the backbone through the 1′-position of the pentose (Figure 7.1B). The nucleobases are aromatic, nitrogen-containing heterocycles that serve as excellent recognition motifs for π-stacking and hydrogen bonding interactions. Three nucleobases, cytosine (C), guanine (G), and adenine (A), are common to both

Monographs in Supramolecular Chemistry No. 14
Synthetic Receptors for Biomolecules: Design Principles and Applications
Edited by Bradley D. Smith

Published by the Royal Society of Chemistry, www.rsc.org

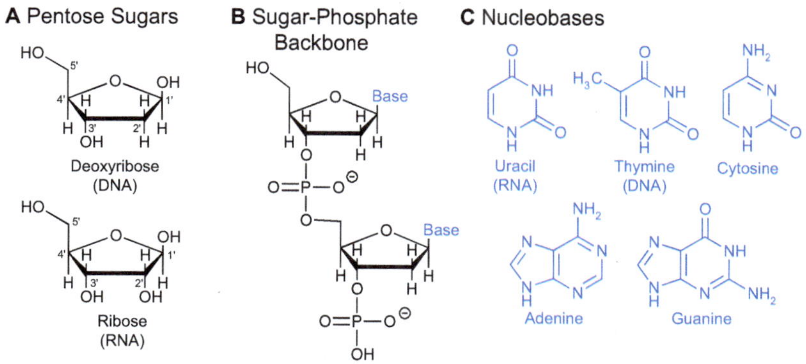

Figure 7.1 Building blocks of oligonucleotides. (A) The ribose sugars of RNA and DNA. The deoxyribose lacks a 2′-hydroxyl group. (B) The connectivity of the sugar–phosphate backbone. (C) The nucleobases of DNA and RNA. Thymine and uracil are unique to DNA and RNA, respectively.

oligonucleotides, while thymine (T) and uracil (U) are unique to DNA and RNA, respectively (Figure 7.1C).

The specific pairing of nucleobases is the most widely recognized element of oligonucleotide structure and is attributed to distinctive hydrogen-bond accepting and donating patterns between the pairs. The most common form of base pairing is termed Watson–Crick and is the structural basis of the B-form DNA double helix. While the two purine–pyrimidine pairs, A–T/U or G–C, have very similar shape and size, the pairs have slightly different stabilities due to the number of hydrogen bonding interactions (two or three, respectively) (Figure 7.2A). Other hydrogen bonding patterns are less common but nonetheless are important interactions for molecular recognition. For example, rotation around the glycosidic bond of a purine in a Watson–Crick base pair forms a non-canonical Hoogsteen base pair (Figure 7.2B). Fewer than 1% of base pairs in DNA are Hoogsteen, yet they serve as unique recognition sites for DNA transcription factors.[2] Many other non-canonical base pairs have been characterized in RNA, including wobble base pairs, sheared base pairs, and base triples, which are often involved in RNA–protein interactions.[3] For instance, HIV-1 Rev-response element (RRE) contains an imino G–A that is recognized by an asparagine residue of the protein binding partner, Rev.[4]

In addition to the hydrogen-bonding edges, nucleobases have planar, polarizable surfaces that favor highly stabilizing base stacking interactions.[5,6] In the B-form DNA double helix, each nucleobase is rotated 34° relative to adjacent nucleobases to maximize the overlap of the partial charges. In addition, the helical rise is approximately 3.3 Å, which is the sum of the van der Waals radii for carbon and nitrogen at 1.7 Å and 1.6 Å, respectively. RNA double helices and some single-stranded regions are also oriented in stable secondary structures to optimize base stacking interactions. In cellular conditions, base stacking is energetically favorable,

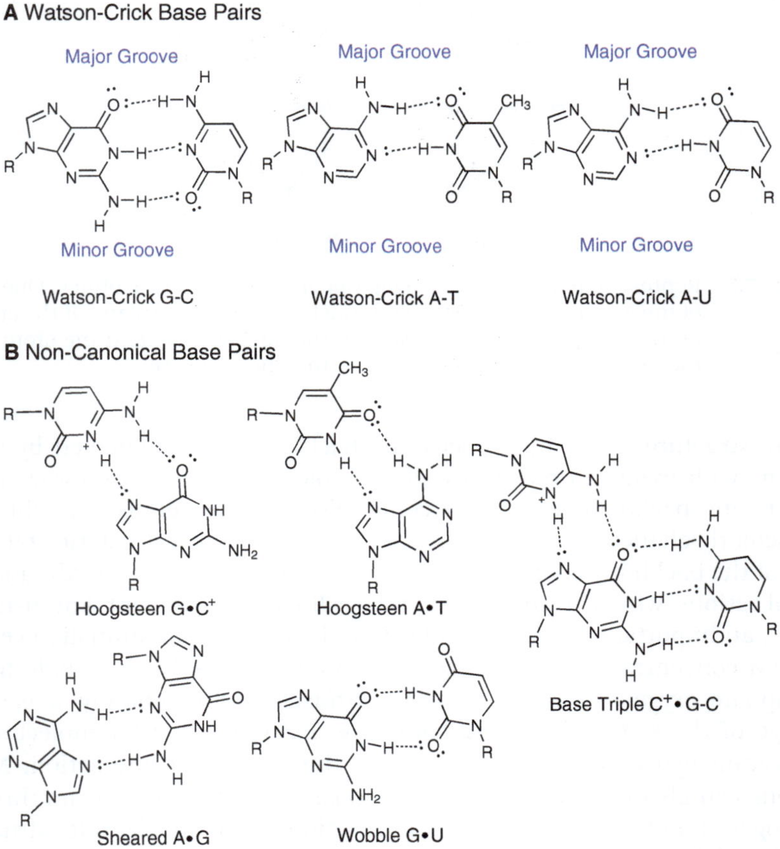

Figure 7.2 Base pairs in oligonucleotides. (A) Canonical Watson–Crick base pairs found in DNA and RNA. (B) Representative non-canonical base pairs. R represents the nucleobase–sugar connection.

cooperative, and driven by the hydrophobic effect. The contribution of the hydrophobic effect is largest with the purine bases, A and G, due to the greater surface area.

The different ribose sugars in DNA and RNA lead to many of the unique structural differences between the oligonucleotides. The ribose sugar adopts a half-boat conformer where one of the carbon atoms is located outside the plane of the pentose ring. Depending upon the position relative to the 5′-carbon, the pucker can be on the same side, leading to an *endo* sugar pucker, or on the opposite side of the plane, leading to an *exo* sugar pucker. Most sugar puckers in nucleic acids are *endo*, with the 2′-and 3′-pucker common to DNA and RNA, respectively (Figure 7.3). The structure of the ribose sugar is also a significant contributor to the relative chemical stability of DNA and RNA. In RNA, the 2′-hydroxyl residue on the ribose promotes self-cleavage of the neighboring phosphodiester bond, whereas the 2′-deoxyribose sugar of DNA promotes nucleobase hydrolysis and the creation of abasic sites.

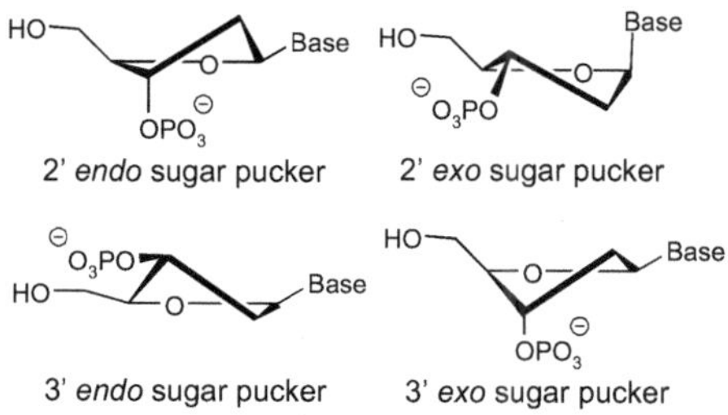

Figure 7.3 Representative examples of sugar puckers in oligonucleotides. One atom of the 5-membered ribose ring is puckered out of the plane. If the atom is on the same side as the 5′-carbon, the pucker is *endo*. If the atom is on the opposite side of the 5′-carbon, the pucker is *exo*.

The structure of the oligonucleotide backbone is also affected by interactions with metal cations and water.[7,8] Folding of the polyanionic sugar–phosphate backbone must overcome electrostatic repulsion, which is particularly challenging for globular RNA structures. Electrostatic stabilization of the backbone conformation can be achieved through coordination of metal cations, which are abundant in the cellular environment. For instance, Mg^{2+}, an important divalent metal cation, is present in mammalian cells at 10 mM concentration.[9] Metal cations can either directly coordinate to the phosphate oxygen atoms or associate diffusely and essentially screen the charge of the sugar–phosphate backbone. Exchanging water molecules in the cation hydration shell can mediate this screening interaction. Metal cations can also interact with the nucleobases and pentose sugars through hydrogen bonding and cation–π interactions. Oligonucleotide structure can be highly sensitive to the identity of the metal cation and the ionic strength of the environment. For example, oligonucleotides can form G-quadruplexes, a tetrad of guanine nucleobases, when stabilized by the monovalent cations Na^+ and K^+, but not by divalent cations. Furthermore, high salt concentrations can induce a B-form, right-handed DNA double helix to adopt a Z-form, left-handed conformation, altering the oligonucleotide recognition surface.

7.1.2 Secondary and Tertiary Structure of DNA

Despite the similarities in primary structure, the secondary structures of DNA and RNA are quite diverse. In cells, DNA primarily adopts a helical structure involving two complementary, antiparallel strands. Due to the hydrophobic effect, the nucleobases are paired within the interior of the helix and the sugar–phosphate backbone is exposed to the aqueous environment. Although base pairing is essential to form the double helix, the energetic contribution to double helix stabilization is minimal. Instead, nucleobase stacking is the major contributor to helical stability.[5] The base-paired, antiparallel strands form two grooves of unequal size: the wide,

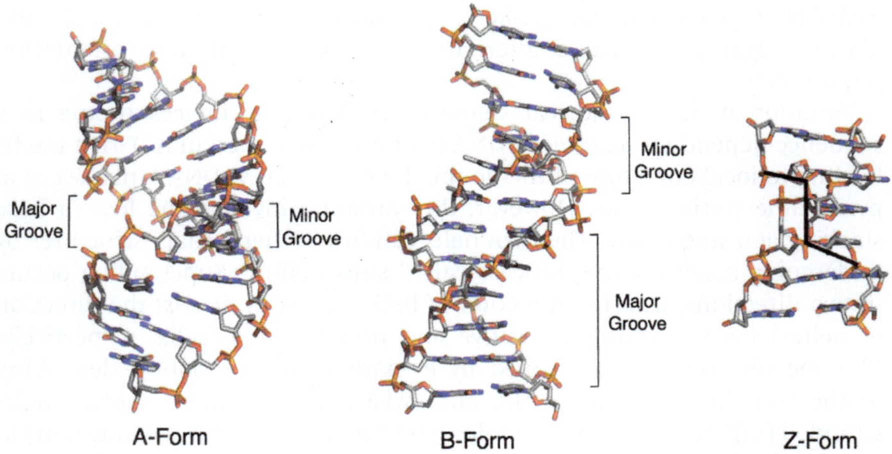

Figure 7.4 Oligonucleotides can adopt various double helix forms with altered recognition surfaces. The A-form helix is the major conformer of RNA and a minor conformer of DNA with a deep major groove and shallow minor groove (Protein Data Bank (PDB): 2D47). The B-form helix is the major conformer of DNA with a wide major groove and narrow minor groove (PDB: 1BNA). Z-form is a minor conformer of DNA with a zigzag backbone structure (PDB: 3P4J).

Table 7.1 Properties of DNA double helices.[11]

Helical form	A	B	Z
Helical sense	Right	Right	Left
Helical shape	Broadest	Intermediate	Narrowest
Helix diameter	25.5 Å	23.7 Å	18.4 Å
Base pairs per turn	11	10.4	12
Helix pitch	25.3 Å	35.4 Å	45.6 Å
Helix rise	2.3 Å	3.4 Å	3.8 Å
Base tilt	19°	1°	−9°
Base orientation	Anti	Anti	C-Anti/G-Syn

accessible major groove and the narrow, less accessible minor grove (Figure 7.4). Unique to the minor groove is an extensive network of water molecules and diffusely bound monovalent cations called the water spine.[10]

Depending upon the water and salt concentration, DNA can adopt three helical forms, A-form, B-form, or Z-form. Each helical form has unique structural features (Table 7.1)[11] and altered recognition surfaces. At physiological conditions, B-form is the dominant structure of DNA, characterized by a long, narrow, right-handed helix with a 2′-*endo* sugar pucker. Comparatively, A-form DNA is characterized by a shorter, wider, right-handed helix with a 3′-*endo* sugar pucker. In addition, the structure has altered groove accessibility with a deeper major groove and shallower minor groove. Finally, Z-form DNA is characterized by an extended, narrow, left-handed helix. The Z-form is sequence dependent and occurs in regions with CpG repeats. The zigzag-like pattern occurs due to alternating 2′- and 3′-*endo* sugar puckers and alternating syn *versus* anti base orientation (Figure 7.4).

Variation of the cylindrical shape of DNA double helices occurs in a sequence-dependent manner.[12] DNA bending is observed in A–T-rich tracts, leading to local curvature of the double helix. Similarly, DNA kinks occur at pyrimidine–purine steps. However, this variation involves the loss of base stacking at a single step. The DNA helix can form other tertiary structures by undergoing a self-twisting process called supercoiling. Supercoiling occurs in two directions, *with* the direction of helical twist or *against* the direction of helical twist, forming a positive and negative superhelix, respectively. The specific topology is defined by a mathematical equation describing writhe (W), linking number (L), and twist (T).[13] Similarly, short, single-stranded (ss)DNA segments can fold upon themselves to form stable hairpin structures. The hairpin is comprised of a complementary, base-paired stem closed by an external loop of unpaired nucleotides. In complementary regions of DNA, hairpin structures can compete with duplex formation, and such stabilized hairpins have been associated with disease.[14]

In another unique topology, four guanine nucleobases can associate through G–G Hoogsteen base pairs, forming a G-quadruplex. The structure is planar and stabilized by a monovalent cation such as Na^+ or K^+ (Figure 7.5A). G-quadruplexes can be intrastrand or interstand and are found in a parallel or antiparallel fashion. Amongst other functions, G-quadruplexes may control telomeric length and stability. In addition, cytosine-rich regions can form i-motifs at slightly acidic or neutral pH by forming $C–C^+$ hemiprotonated base pairs. Two duplexes of antiparallel strands intercalate, forming alternating $C–C^+$ pairs (Figure 7.5B). Although there are many sequences that could form i-motifs *in vivo*, the structure has only been directly observed *in vitro*.

In addition to the physiologically relevant structures, DNA can be programmed to self-assemble into unique, multidimensional structures in a method termed "DNA origami". In this technique, oligonucleotides are designed to form duplexes with single-strand overhangs called sticky ends. The sticky ends of two different duplexes are complementary, driving the assembly of multiple duplexes into a single structure. The technology has been used to construct DNA nanostructures having various functions, from molecular computing to drug delivery.

7.1.3 Secondary and Tertiary Structure of RNA

In contrast to double-stranded (ds) genomic DNA, RNA is most often found as a single-stranded oligonucleotide, leading to the great diversity observed in RNA secondary and tertiary structures. Folding of RNA is similar to that of proteins, in that primary sequence dictates secondary structure, which leads to the formation of tertiary structure. The initial folding of RNA into a defined secondary structure is dependent upon optimum base pairing and

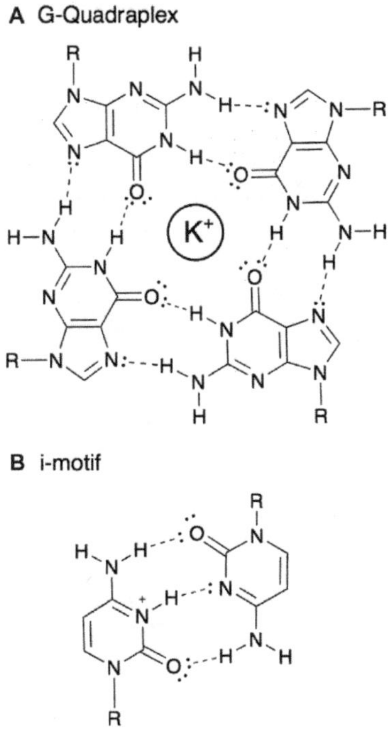

Figure 7.5 Unique DNA topologies. (A) A G-quadruplex forms between four Hoogsteen-paired guanine bases, which are stabilized by a central monovalent cation. (B) An i-motif forms between a hemiprotonated cytosine–cytosine pair at slightly acidic or neutral pH.

minimization of phosphate repulsion in the anionic backbone. Likewise, the formation of tertiary structure is highly dependent upon ion concentration, which further stabilizes backbone interactions in compact structures. Higher-order RNA structures are typically dynamic, flexible and possess multiple low-energy conformations.

Folded structures of RNA are stabilized by base pairing, which forms helical stem regions. While 2′-*endo* sugar puckers can be formed in flexible areas, helical structures are restricted to 3′-*endo* puckers due to the electrostatic repulsion between the 2′-hydroxyl and 3′-phosphate in the 2′-*endo* conformer. Due to the sugar pucker limitations, RNA mainly adopts an A-form helical structure. In contrast to DNA, RNA helices contain many more non-canonical base pairs. One of the most common is the G:U wobble base pair, which is evolutionarily conserved in the acceptor stem of transfer RNA (tRNA).[15] In addition, oligonucleotides are enriched with naturally modified nucleobases that can be incorporated into helical structures. Select modifications of DNA nucleobases can be found in genomic DNA, with methylation being the most common.[16] At the same time, RNA is enriched with >100 types of post-transcriptionally modified bases.[17] Some modifications involve simple chemical transformations, such as the bond isomerization of uridine

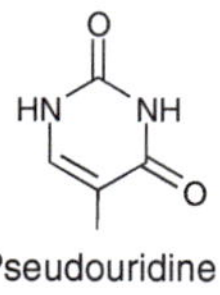

Figure 7.6 RNA is enriched with modified nucleobases. Pseudouridine is one of the most abundant.[17]

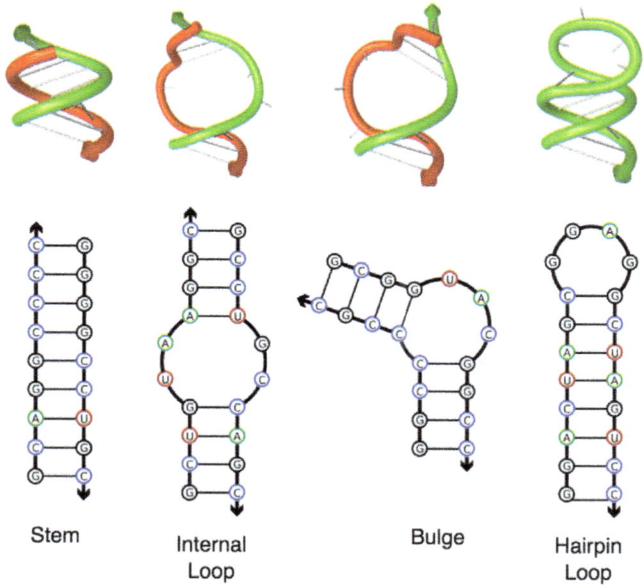

Figure 7.7 Example secondary structures of RNA. Designed using NUPACK software (www.nupack.org).

to pseudouridine (Figure 7.6), but other modifications require multiple enzymes.

When RNA folds upon itself to form stem regions, some nucleotides remain unpaired and form non-helical secondary structures. There are three common types: the bulge, internal loop, and hairpin (Figure 7.7). Bulge structures occur when nucleotides are excluded from base pairing on a single strand of the antiparallel duplex. Similarly, internal loops occur when nucleotides are excluded from base pairing on both strands. The loops can be symmetrical or asymmetrical, depending upon the number of base pairs excluded on each strand. Lastly, apical loops or hairpin structures are formed when RNA folds sharply upon itself. One of the best-characterized hairpins is the GNRA tetraloop (N: any nucleotide; R: any purine).[18] In this particular structure, the guanine stacks with the bases of the stem, and the three other nucleotides are flipped out into solution. This structure creates a unique recognition element found in many RNA–protein interactions (Figure 7.8), including the *box*B RNA and N-protein complex, which plays a role in regulating transcription in phage

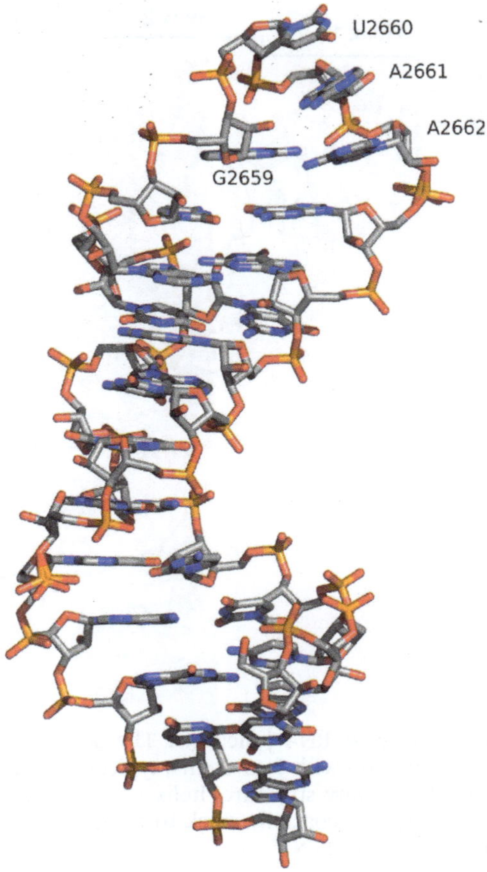

Figure 7.8 A crystal structure of a GUAA hairpin tetraloop, mimicking the sarcin/ricin loop in the 23 S ribosomal RNA of *Escherichia coli*. G2659 is stacked with the flanking stem region. U2660, A2661, A2662 are flipped-out and stacked upon each other (Protein Data Bank: 1MSY).

development. Similar to DNA, RNA is rich in other unique topologies and can also form G-quadruplexes and i-motif structures.

In addition to the varied secondary structures of RNA, additional complexity arises from the plethora of diverse tertiary forms. One of the most basic tertiary structures involves the coaxial stacking of helices, where the terminal nucleobase of an adjacent helix stacks to form one colinear helix. Coaxial stacking defines the overall globular shape of RNA. Most coaxial stacking occurs at RNA junctions, branching points of multiple secondary structures (Figure 7.9). A single junction can join two or more helices. For instance, RNase P contains a junction of 18 helices.[5] The ionic strength and linker length between helices define the topological landscape of junctions and the resulting tertiary structure.

Many other tertiary structures are prevalent in RNA, usually forming long-range base pairs. One of the more common structures is the pseudoknot,

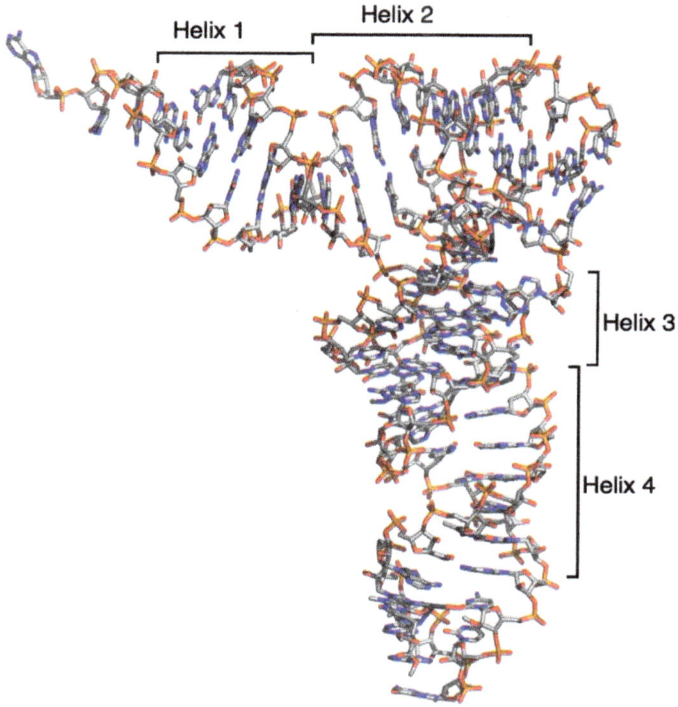

Figure 7.9 Coaxial stacking at RNA junctions. The secondary structure of tRNA contains a junction with four stems, typically represented as a four-leaf clover. In the tertiary structure, helix 1 and helix 2 coaxially stack and helix 3 and helix 4 coaxially stack to form the L-shaped configuration (Protein Data Bank: 1TN1).

where the unpaired nucleobases of a hairpin structure base pair with a single stranded region at the opposite end of the stem. The resulting structure is two helices, coaxially stacked (Figure 7.10A). Similarly, a kissing loop forms through long-range base pairing of two hairpin structures and can form with as few as two base pairs (Figure 7.10B). Other tertiary structures prevalent in RNA include the adenosine platform, base triples, and the A-minor motif, all of which are driven by the formation of canonical or non-conical base pairs. In addition to base pairing, the sugar–phosphate backbone can form unique tertiary structures. One example is the ribose zipper, which forms when two RNA strands run antiparallel to each other, forming specific hydrogen bonds between the ribose and nucleobases accessed through the minor groove.

7.1.4 Fundamentals of Oligonucleotide Recognition

The non-covalent interactions critical to oligonucleotide structure include aromatic stacking of the nucleobases, electrostatic interactions with the phosphate backbone, and a variety of hydrogen bonding interactions with the bases, ribose units, and backbone. Illustrations of these interactions

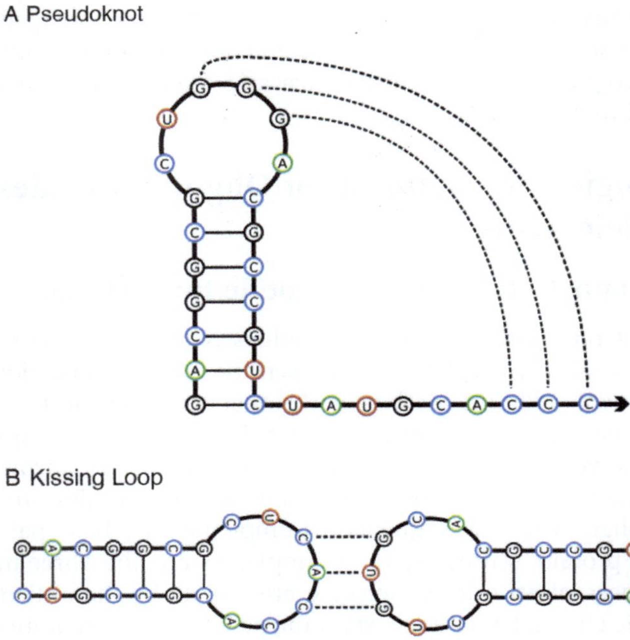

Figure 7.10 Example tertiary structures of RNA. (A) Pseudoknot structures occur when the unpaired nucleobases of a hairpin motif base pair with a single-stranded region on the opposing side of the stem. (B) A kissing loop occurs when the nucleobases of two hairpin structures base pair. Figures generated using RNAfold (http://rna.tbi.univie.ac.at/).

appear repeatedly throughout the chapter. For example, aromatic stacking interactions are observed in nucleic acid complexes with protein aromatic-rich β-strand clefts and also in complexes with intercalating small molecules. Binding affinity is often driven by hydrophobic effects with specificity determined by the electrostatic surface. Electrostatic interactions with the anionic phosphate backbone can also contribute a great deal to affinity but, as expected, often limit selectivity. In contrast, hydrogen bonding can impart significant specificity to oligonucleotide binding interactions with limited contribution to binding free energy in water. An important feature of biologically-relevant B-form DNA is its structural uniformity. The constant arrangement of hydrogen-bond donors and acceptors in the major and minor groove (Figure 7.2), as well as their consistent dimensions, renders the grooves attractive sites for specific binding of small molecules, oligonucleotides, and proteins. Furthermore, much of this recognition is programmable, allowing the *de novo* design of DNA-targeted ligands. Structural studies have revealed that shape recognition can play a large role in oligonucleotide complexes, both in helical and more complex secondary and tertiary structures. It is important to note that bridging water and salts can mediate interactions between oligonucleotides and ligands. As with all complex biomolecular recognition processes, a combination of non-covalent interactions must be employed to optimize both affinity and specificity for sequence and structure. Multidomain interactions are often critical in the biological recognition of oligonucleotides and offer a powerful but often underutilized tool in ligand design.

7.2 Biological Recognition of Oligonucleotides and Nucleic Acids

7.2.1 Oligonucleotide–Oligonucleotide Interactions

The molecular recognition of oligonucleotides is central to the transfer of biological information. Replication, transcription, and translation rely on specific oligonucleotide interactions with other oligonucleotides, proteins, and, in some cases, small molecules. Given the high fidelity and programmability of the Watson–Crick base pairing system, it is not surprising that the most predictable recognition systems for oligonucleotides are the complementary oligonucleotide sequences. Complementary base pairing is at the heart of protein synthesis, for example, where the three-nucleotide codon sequences of an mRNA molecule are "read" by the anticodon of a tRNA molecule charged with the next amino acid to be incorporated into the peptide chain (Figure 7.11A). In the context of genomes, however, complementarity of approximately 20 nucleotides is needed for unique binding of single-stranded sequences. Silencing RNAs (siRNAs) and microRNAs (miRNAs), for example, target mRNA through complementary annealing of 20–25 nucleotides. The respective duplex formed is then degraded, effectively reducing the level of that mRNA and thus its cognate protein (Figure 7.11B). In addition to canonical duplex base-pairing interactions, oligonucleotides can selectively form triple-helix structures through Hoogsteen-type interactions of a third nucleic acid strand with the major groove of duplex nucleic acids (Figure 7.11C). Intramolecular DNA triplex formation in the genome generally occurs in long tracts of homopurine or homopyrimidine bases and has been implicated in gene regulation as well as disease.[19] Single-stranded RNA can also form triple helices, both with dsDNA and dsRNA. As an example of intermolecular triplex formation, RNA viruses have been observed to employ this motif to protect polyadenylated tails of viral RNA from degradation (Figure 7.11C).[20]

7.2.2 DNA–Protein Interactions

Despite the centrality of oligonucleotide–oligonucleotide interactions, it is worth emphasizing that nearly all biological processes depend on specific nucleic acid–protein interactions. Polymerases involved in both replication and transcription, for example, depend on dynamic nucleic acid–protein binding events. While polymerases are, by necessity, promiscuous between sequences, many specific binding events are involved in the *control* of these processes. Commonly, protein interactions with genomic DNA are initiated

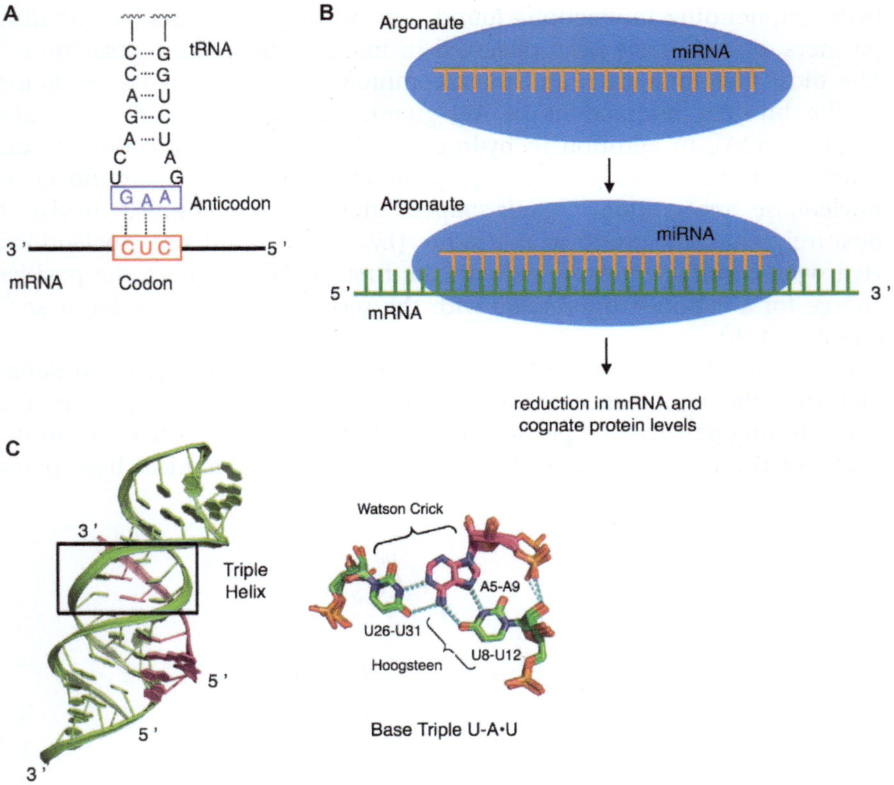

Figure 7.11 Oligonucleotide:oligonucleotide recognition. (A) In protein synthesis, the anticodon of tRNA base pairs with the codon of an mRNA, leading to the incorporation of an amino acid into a growing peptide chain. The third position can be recognized through Watson–Crick or wobble base pairing (latter shown). (B) miRNAs are 20–25-nucleotide RNA segments that form complexes with the Argonaute protein. The Argonaute–miRNA complex recognizes an mRNA through base pairing, which ultimately leads to degradation of the mRNA and reduction in cognate protein levels. (C) The polyadenylated tail of Kaposi's sarcoma-associated herpes virus forms a triplex structure with a U-rich internal loop, forming a U-A · U base triple. The triplex structure is proposed to stabilize the RNA.
From *Science*, 2010, **330**, 1244. Reprinted with permission from AAAS.[20]

by non-specific binding based on charge–charge interactions between positively charged protein residues and the DNA backbone.[21] Rapid sampling of potential binding sites leads to specific recognition. On average, protein–DNA binding interfaces are composed of 24 amino acids and 12 nucleotides and involve several types of non-covalent interactions.[12] Historically, specificity has been attributed to nucleobase–amino acid hydrogen-bonding interactions. A recent survey of possible nucleobase–amino acid hydrogen-bonding interactions found that of the 32 potential 1:1 binding partners, only 17 have been observed in nucleic acid–protein structures.[22] The most discriminating, and most common, recognition pairs are found in the bivalent interactions of Arg–guanine and Asn/Gln–adenine pairs (Figure 7.12A). In addition to hydrogen bonding, C–H···O, cation–π, and water-mediated interactions can play an important role in amino acid–nucleobase recognition.[23] Stair-shaped motifs, for example, involve a positively charged amino acid residue (Lys or Arg) and two nucleotides. Hydrogen-bonding interactions with the first nucleotide align the positive charge for a cation–π interaction with the second, often a guanine base.[24] (Figure 7.12B)

Beyond the direct amino acid–nucleobase binding modes, recent work has identified the significance of the tertiary structure of both the protein and DNA binding partners. Of particular importance are local differences in the shape of the DNA.[12] Despite the isosteric nature of canonical base pairs,

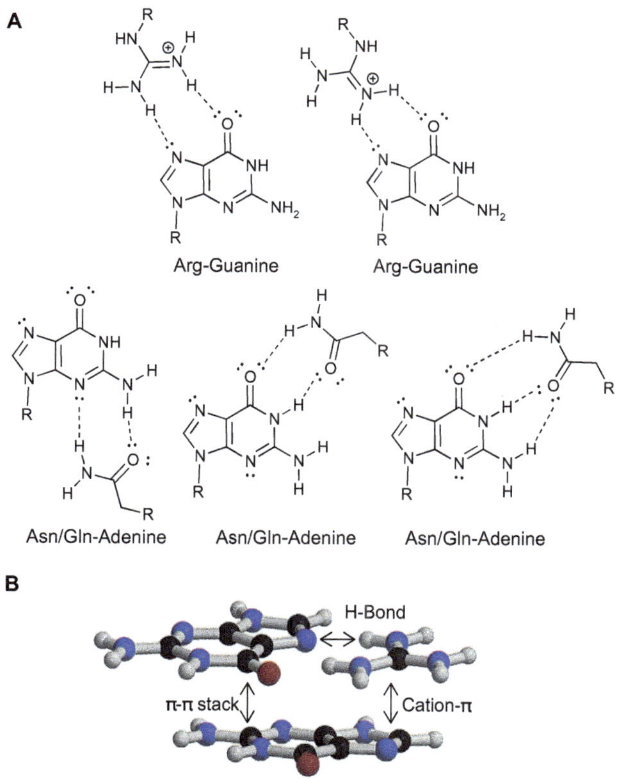

Figure 7.12 Amino acid:nucleobase interactions. (A) Amino acid–nucleobase pairs interact through a variety of hydrogen-bonding arrangements. (B) Amino acid–nucleobase pairs interact through π–π and cation–π interactions.
Reprinted with permission from *J. Phys. Chem. A.*, 2003, **107**, 6249.

variations in the chemical and conformational signature of each set as well as differences in the environment created by neighboring bases can lead to local and global changes in DNA structure. Different dinucleotide steps, for example, can lead to bending or kinking in B-DNA structures, often in association with poly-A-tracts (Figure 7.13A and B). The bending of short DNA sequences enables the DNA to wrap around histone proteins for packaging in the nucleosome. DNA kinks arise at sites with weak base-stacking interactions, such as CpG dinucleotides, where additional flexibility allows intercalation of hydrophobic protein residues. In addition, minor groove narrowing can focus the negative electrostatic potential of the minor groove relative to the major groove and thus be specifically recognized by positively charged arginine side chains (Figure 7.13C,D). Lysine–minor groove interactions are less common. More global conformational changes, such as the switch to A-form or Z-form DNA, also play an important role in specificity. For example, the widened minor groove of A-form DNA, often found in GC-rich sequences and the TATA box, allows additional interactions with

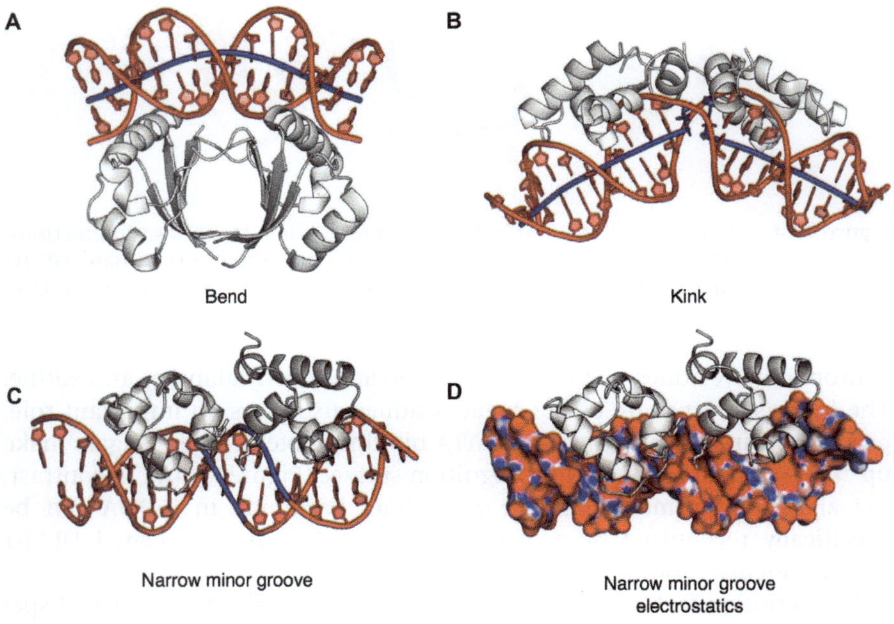

Figure 7.13 Alterations in helical DNA structure recognized by proteins. (A) Bending of a poly-A tract is recognized by human papillomavirus-18 E2. (B) Weak CpG step allows intercalation of leucine residues in the Lac repressor leading to a kink in the helical axis. (C) Example of minor groove narrowing recognized by arginine residues of the Phage 343 repressor. (D) Electrostatic potential of DNA in (C) demonstrating increased negative potential in the minor groove.
Reproduced with permission of Annual Review of Biochemistry, Volume 79, by Annual Reviews, http://www.annualreviews.org. Copyright © 2010, Annual Reviews.[12]

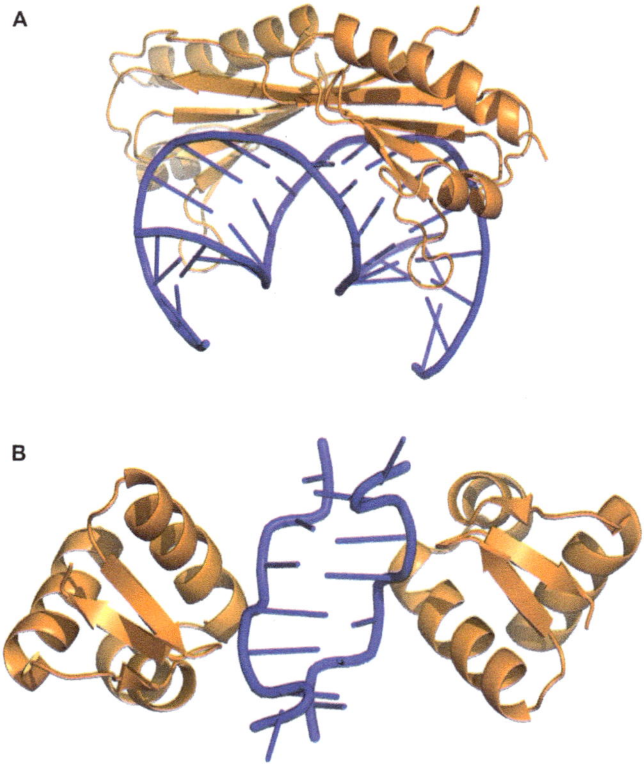

Figure 7.14 Alternative DNA–protein binding interactions. (A) Ribose interactions: complex of TATA-binding protein with DNA (Protein Data Bank (PDB): 1CDW). (B) Backbone interactions: complex of DLM-1 and Z-DNA (PDB: 2HEO).

hydrophobic residues such as alanine, leucine, phenylalanine, and valine. The increased exposure of the sugar residues also plays an important role, particularly in the binding of the TATA-binding protein, where sugars make up >50% of the protein–DNA recognition surface (Figure 7.14A). In contrast, the zigzag conformation of the phosphate backbone in Z-DNA can be specifically recognized by proteins such as the tumor-associated DLM-1 protein (Figure 7.14B).

Transcription factors are some of the best-characterized examples of specific nucleic acid-binding proteins, interacting with discrete genomic DNA sequences using a variety of structural motifs.[25,26] While control of transcription is generally achieved through combinatorial interactions in a multiprotein complex, the DNA-binding domains of transcription factors can act independently, and even in isolation, relative to the other protein domains. Common classes of DNA binding domains include zinc finger domains, helix-turn-helix (HTH) domains, and basic-leucine zipper (bZIP) domains. Zinc-finger domains are comprised of ~30 amino acid residues in the form of a short α-helix, two antiparallel β-strands and a Zn^{2+} atom coordinated to cysteine and histidine residues. Recognition occurs at the tip of the helix, with each finger coding for two to three nucleotides through binding in the major groove. In nuclear hormone receptor transcription factors, two zinc fingers simultaneously bind the major groove to recognize six DNA residues (Figure 7.15A).[27] A larger DNA-binding footprint is achieved through receptor dimerization, which can be facilitated by ligand binding, protein–protein interactions, and possibly changes in DNA conformation.

Another common DNA-binding motif is the HTH motif, in which two α-helices are connected through a short peptide linker.[12] Generally, one helix achieves recognition through hydrogen bonding and hydrophobic interactions with exposed bases in the major groove of the target DNA sequence while the other helix participates in non-specific stabilizing interactions. Furthermore, HTH units are often subdomains of larger helix bundles, including the three-helix bundle of homeodomains. Homeodomain binding to the major groove is facilitated by DNA bending and often includes narrow minor groove interactions with an arginine outside the HTH. Combination of the HTH with an antiparallel β-sheet forms the common winged helix-turn-helix (wHTH) domain found in protein families such as the forkhead and E26 transformation-specific (ETS) transcription factors and often incorporates interactions between the "wing" and the minor groove. In ETS transcription factors, the wHTH domain binds ~12–15 base pairs and achieves recognition largely through the third helix (H3), which interacts with the major groove of the consensus 5′-GGA(A/T)-3′ (Figure 7.15B).[28,29] Specifically, two conserved arginine residues interact with the two guanines of the consensus sequence, while additional base-specific interactions involve residues in the "turn" and "wing" subdomains. However, no direct base interactions are observed outside of the ETS consensus sequence, suggesting that additional specificity arises from sequence-specific backbone distortions in the DNA backbone. Major groove interactions with α-helices are also key components of the "leucine zipper" motifs commonly found in the bZIP protein family, where hydrophobic leucine residues arranged on one side of a helix interact with a matching pattern on a second, and the neighboring positively charged regions interact with DNA.

The striking atomic model of the interferon-β "enhanceosome", in which at least eight DNA-binding proteins bind cooperatively over a conserved 55-base-pair region of DNA, exemplifies the interplay of multiple proteins on a single region of the genome (Figure 7.15C).[12,30] The characterized proteins include the interferon regulatory factors (IRF) family, specifically IRF-3 and IRF-7, which use a mixed wHTH DNA binding domain consisting of four antiparallel β-sheets, three α-helices and three long loops to interact with both the major and minor grooves of a 12-base-pair sequence. Unique narrow minor groove interactions include hydrogen bonding to a spine of hydration and the insertion of histidines rather than the more common arginine residues. Four IRF proteins are neighbored by a nuclear factor κB heterodimer of RelA and p50 that relies on a critical histidine–guanine contact in addition to the conserved arginine and glutamate interactions

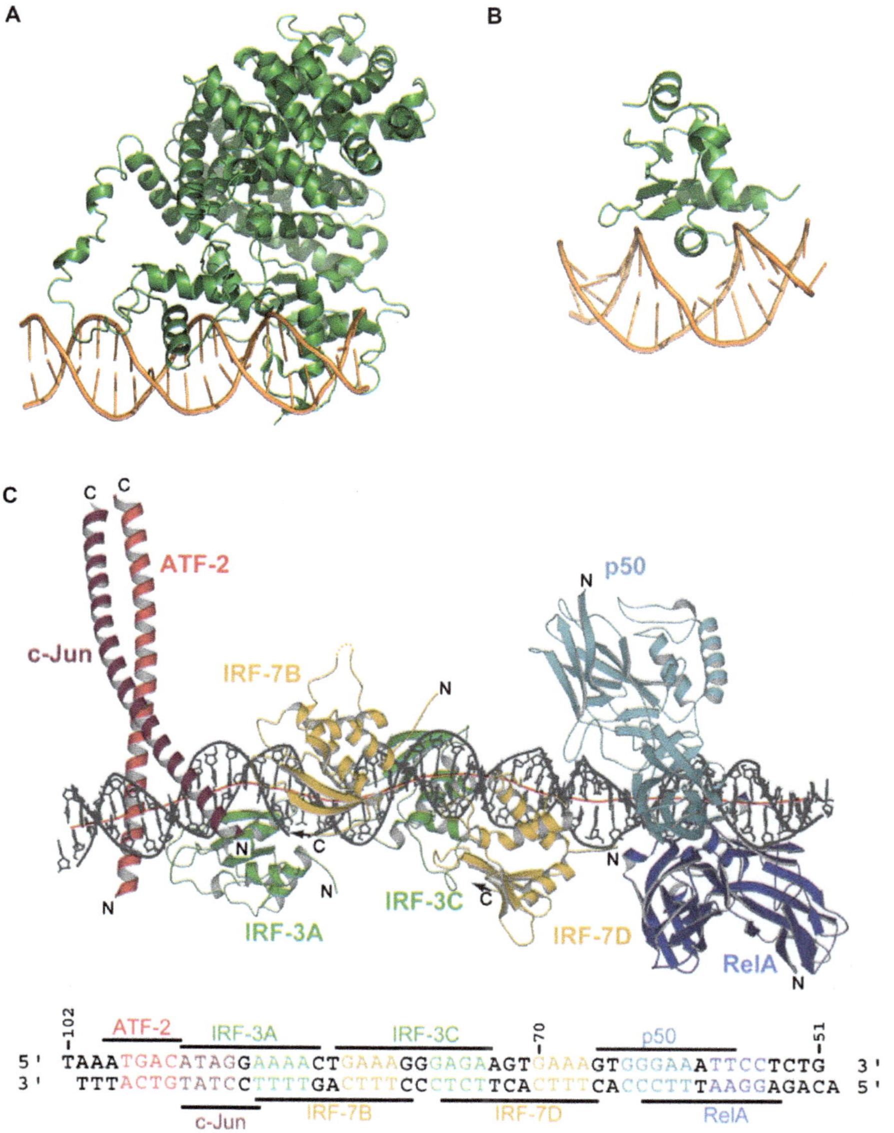

Figure 7.15 Transcription factor motifs. (A) Two zinc finger domains of retinoic acid receptor bind the major groove (Protein Data Bank (PDB): 3DZY). (B) Winged helix-turn-helix domain in E26 transformation-specific transcription factors span 12–15 DNA base pairs (PDB: 1K79). (C) Composite of X-ray crystal structures of protein:DNA interactions in the interferon-β enhancesome. Each of the eight proteins recognizes four to six base pairs to cover a total of 55 base pairs.
Reprinted from *Cell*, 2007, **129**, 1111, © 2007, with permission from Elsevier.[30]

with the guanine core. The IRF proteins are flanked on the opposite side by an Atf-2/c-Jun heterodimer formed through bZIP domains. In combination, these proteins cover >72% of the solvent accessible surface area of the enhancer DNA and form a densely packed structure, despite a lack of significant protein–protein interactions. At the same time, *in vitro* gel shift experiments revealed that individual proteins do not bind the enhancer region, suggesting that local alterations in the DNA structure contribute more strongly to the observed cooperativity, although additional factors are likely to play a role in *in vivo* signaling.

Single-stranded DNA and its protein binding partners play a critical role in many cellular processes, including DNA replication, recombination, and repair.[31] In addition to the nucleobase and backbone interactions discussed for dsDNA, ssDNA–protein interactions often incorporate hydrophobic interactions. Many of the more common ssDNA-binding motifs are also involved in RNA binding, and will thus be discussed in the following section.

7.2.3 RNA–Protein Interactions

In contrast to DNA, RNA structure is variable and highly dynamic, and RNA is rarely found without protein binding partners.[32] Specificity and affinity are achieved through combinatorial arrangements of only a handful of unique RNA-recognition domains. Such modular arrangements can be used to organize the target RNA for a specific function or, *vice versa*, for RNA to organize its cognate protein. Not surprisingly, RNA and DNA nucleobase and backbone units form many similar bonding interactions with amino acids, and there is overlap in structural domains as well. For example, transcription factor TFIIIA contains nine zinc fingers, seven of which bind to the major groove of dsRNA, while three can be used to interact with two loop structures of 5S ribosomal RNA.

Other binding modules, however, are specific to RNA and, in some cases, ssDNA. By far the most common RNA-binding domain is simply termed the RNA-recognition motif (RRM) (Figure 7.16A).[33] The RRM domain has a split α/β-topology comprised of two helices packed against a four-strand antiparallel β-sheet. RNA binding is most commonly observed on the surface of the β-sheet through three conserved residues, namely a positively charged Arg or Lys that forms a salt bridge with the phosphate backbone and two aromatic residues that participate in stacking interactions. In some structures, a third aromatic residue interacts with the sugar rings of the nucleotide. Each RRM contains two sets of binding residues, located on the central two strands of the β-sheet. While this basic motif only imparts dinucleotide binding, RRM domains are often surrounded by less conserved structural elements that expand the recognition sequence. Another ubiquitous example, the K-homology domain (KH), contains a cleft formed by a conserved GXXG loop and flanking helices, β-strands, and variable loops. In contrast to the RRM, this four-nucleotide recognition domain does not

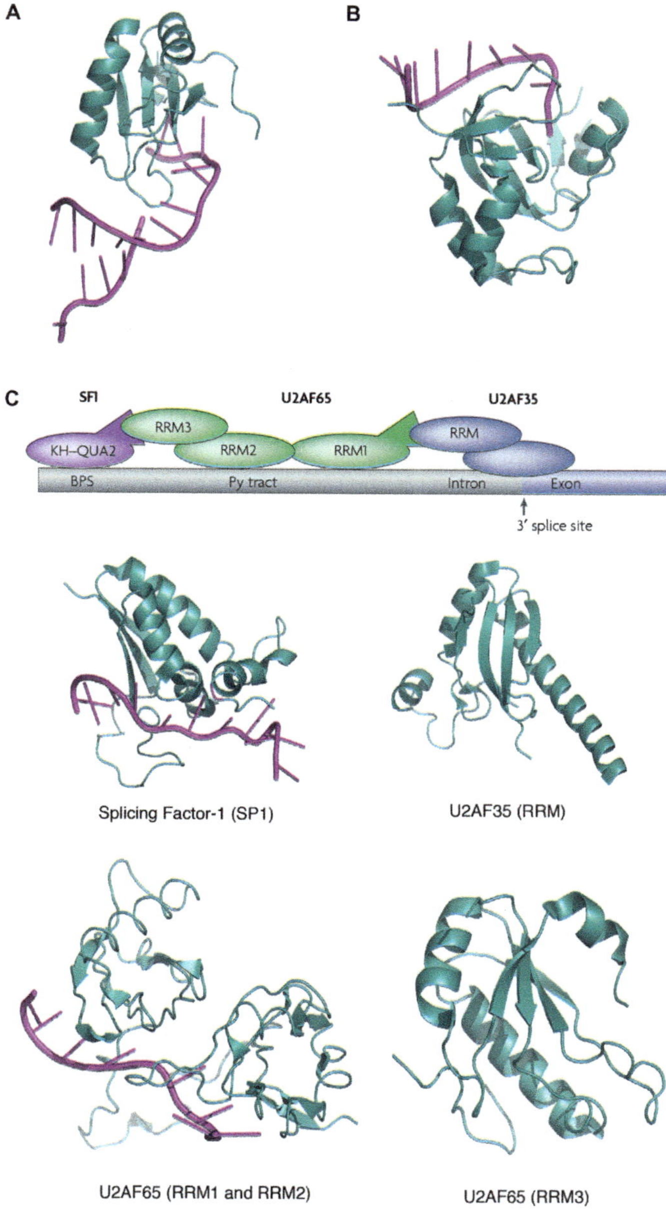

Figure 7.16 RNA binding domains. (A) RNA recognition motif (RRM) in the U1A spliceosomal protein bound to RNA (Protein Data Bank (PDB): 1URN). (B) PAZ domain of eIF2c1 protein bound to siRNA-like duplex (PDB:1SI3). (C) Various spliceosome components assemble through RNA–protein and protein–protein interactions: Splicing Factor (SF)-1 recognizes intron branch site RNA (PDB: 1K1G), U2AF35 RRM (PDB: 1JMT), U2AF65 RRM1 and RRM2 (PDB: 2YH1), and U2AF65 RRM3 (PDB: 1O0P).
Reprinted by permission from Macmillan Publishers Ltd.: *Nat. Rev. Mol. Cell Biol.*, 2007, **8**, 479, © 2007.[32]

contain aromatic residues, instead relying on hydrogen bonding, electrostatic interactions and shape complementarity. In another distinct example, dsRNA binding domains (dsRBD) interact almost entirely with the 2′-hydroxyl and phosphate backbone, rather than the nucleobases.

The oligonucleotide/oligosaccharide (OB)-fold is also fairly common in RNA binding proteins and is composed of a 5-stranded antiparallel β-barrel capped with a short α-helix.[34] For example, in the S1 domain, named for the ribosomal S1 protein, nucleotide recognition is again performed by two β-strands and surrounding loops. Similar structures are involved in the PAZ and PIWI protein domains crucial to RNA interference and miRNA processing.[35] In these structures, the β-barrel recognizes the two-nucleotide overhand of the dsRNA complexes described earlier and positions the end for cleavage by Argonaute or Dicer proteins (Figure 7.16B).

Recognition of a specific RNA sequence is generally achieved by linking multiple RNA-binding domains or even multiple RNA-binding proteins. For example, in the Pumilio (Puf) family of proteins, each domain contains three amino acids that code for a specific nucleotide and can be combined with up to seven other domains to recognize large sequences.[36] Furthermore, a combination of RNA–protein and protein–protein interactions is critical for spliceosome recognition and function. During assembly at the 3′ splice site, two U2AF proteins recognize the polypyrimidine tract through two RRM domains. At the same time, a third RRM domain interacts with the protein splicing factor (SF)1, which is bound through an expanded KH domain at the branch-point sequence (Figure 7.16C).[37] In this latter example, specificity is achieved through protein–protein interactions. A complete picture of domain linkages and protein interactions will continue to form as more RNA–protein structures become available.

7.2.4 Oligonucleotide Interactions with Small Biological Molecules

Naturally occurring small molecule–oligonucleotide interactions have most commonly been observed in antibiotic natural products, which are synthesized by microbes to target critical oligonucleotides in neighboring organisms. Antibiotics that bind DNA include both intercalating agents and groove binders, while RNA-targeted antibiotics most often target bulge structures in the ribosome. These classes of small molecules have served as an inspiration source for many synthetic receptors and are discussed in Section 7.3.

The attention paid to toxic antibiotics masks the fact that specific small molecule–RNA interactions can be an integral part of normal cell activity. In particular, the recognition of metabolites by short RNA molecules termed riboswitches and ribozymes can regulate a number of cellular processes.[38,39] Riboswitches are regulatory elements of mRNA containing a highly structured aptamer domain that recognizes and binds a particular metabolite.

The resulting conformation change in the aptamer domain effects a larger alteration in the mRNA structure to either promote or inhibit translation. The metabolite targets can range from large nucleotides and nucleotide analogues such as flavin mononucleotide (FMN) to very small ligands such as lysine and glycine to metal cations such as Mg^{2+}. These activities are often part of an inhibitory feedback loop, where a threshold concentration of a metabolite down-regulates translation of proteins involved in its synthesis. Not surprisingly, the recognition of nucleobases and their analogs commonly involves hydrogen bonding and base-stacking interactions, while anionic groups such as phosphates are often bound through metal cation bridging.

Ribozymes, as the name implies, are catalytic RNA molecules and best known for their function in the ribosome and in self-cleavage or splicing processes. The *glmS* ribozyme in Gram-positive bacteria is an excellent example of how these functions can combine.[40] The *glmS* ribozyme is located in the 5′-untranslated region of the *glmS* gene mRNA sequence, which codes for the glucosamine-6-phosphate synthetase (GlmS) protein. The *glmS* ribozyme is sensitive to the concentration of glucosamine-6-phosphate (GlcN6P), with physiological concentrations activating a latent self-cleavage mechanism in which a single phosphodiester bond is broken (Figure 7.17A). Recognition of GlcN6P is achieved through direct and water-mediated hydrogen bonds between the sugar hydroxyl and amine groups and the active site nucleobases and backbone oxygens (Figure 7.17B). In addition, the G1 guanine base contributes stacking interactions with the sugar ring and a hydrogen bond to the phosphate group of GlcN6P (Figure 7.17C). GlcN6P is critical for bacterial cell wall biosynthesis and its tight regulation is necessary for appropriate sporulation.

7.3 Synthetic Receptors for Oligonucleotides and Nucleic Acids

7.3.1 Overview

Nucleic acids offer a unique set of recognition challenges and opportunities for synthetic receptors relative to other biological targets. To begin with, the chemical space is rather limited, consisting of four monomers with similar hydrogen bonding and aromatic surfaces, a relatively consistent sugar moiety, and a highly negatively charged backbone. Furthermore, at the level of secondary structure, DNA is found almost entirely in a single (B-form) conformation, though this regularity also allows for the programmable recognition of DNA by minor-groove binding oligomers (Section 7.3.5.3). RNA adopts a much more diverse range of secondary and tertiary structures; however, these structures are highly dynamic and often difficult to predict. Despite these challenges, several research groups have been able to identify both small molecule and bio-oligomer-derived synthetic

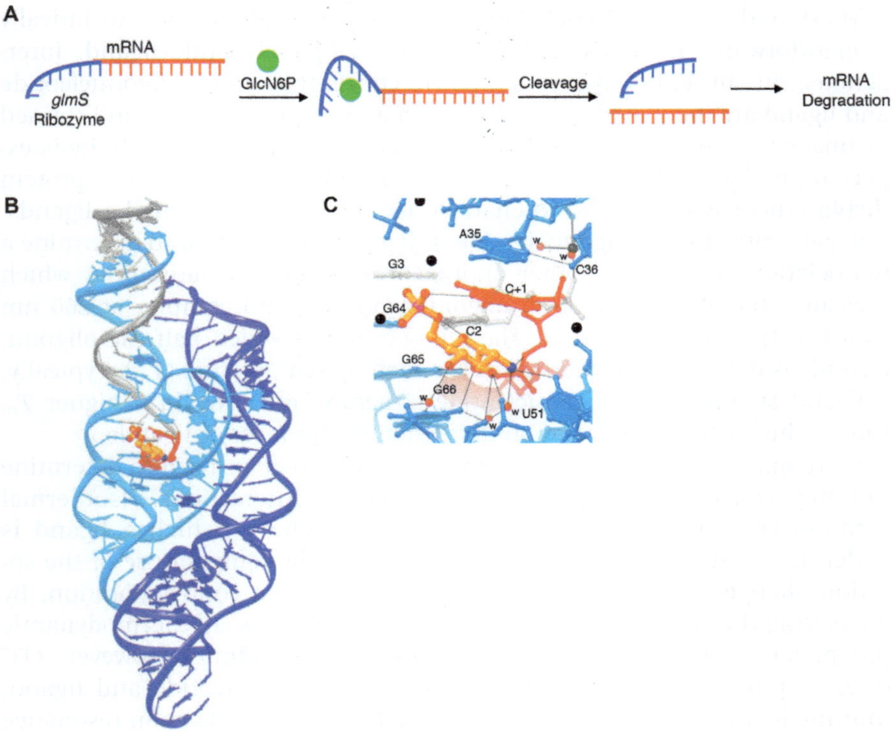

Figure 7.17 (A) Binding of GlcN6P to *glmS* ribozyme catalyzes cleavage and subsequent mRNA degradation. (B) Structure of GlcN6P (yellow) bound to the *glm*S ribozyme. (C) Recognition of GlcN6P through sugar and nucleobase hydrogen bonds, water-mediated hydrogen bonds (red W), and stacking with the G1 guanine nucleobase. The RNA structure is heavily stabilized by Mg^{2+} ions (black).
From A. R. Ferre-D'Amare, *Q. Rev. Biophys.*, 2010, **43**, 423, reproduced with permission.[40]

ligands. While specificity remains a common obstacle, several powerful recognition systems have emerged that can provide excellent tools for researchers interested in nucleic acid recognition. This section focuses on select examples with the intention of serving as a resource for readers new to the field, as well as identifying trends to inform future work in the field.

7.3.2 Methods

7.3.2.1 Binding Characterization

See Blakeley *et al.*[41] for a review of binding characterization methods. Some biophysical techniques, including electrophoresis mobility shift assay

(EMSA) and thermal denaturation are low-throughput but technically straightforward methods to characterize oligonucleotide–ligand interactions. In an EMSA assay, a known concentration of oligonucleotide and ligand are incubated together, run on a nondenaturing gel, and stained or imaged to quantify each electrophoresis band. Decreased mobility is expected for oligonucleotides bound to large ligands or proteins, while protein displacement is often used to characterize the binding of smaller ligands. The concentration of oligonucleotide or ligand can be varied to determine a dissociation constant. Another approach is thermal denaturation, which measures the absorption of an oligonucleotide–ligand complex at 260 nm as a function of temperature. The temperature at which half the oligonucleotide is denatured is recorded as the melting temperature (T_m). Typically, a ligand stabilizes an oligonucleotide structure, resulting in a higher T_m value, which can give relative binding energies for a set of ligands.

More quantitative label-free techniques can also be utilized to determine binding constants of oligonucleotide–ligand interactions. Isothermal titration calorimetry (ITC) is a titration technique in which a ligand is added to a solution of receptor. Upon binding, the temperature of the solution changes and is measured in comparison to a control solution. By measuring the change in heat at known concentrations, all thermodynamic parameters can be calculated, including binding affinity; however, ITC often requires a high concentration of both oligonucleotide and ligand, limiting its throughput. An alternative method, surface plasmon resonance (SPR) requires the immobilization of the ligand onto a gold chip, while a solution of the receptor is passed over the surface (Figure 7.18). Upon binding, the refraction angle changes, which is measured at equilibrium to calculate binding affinity or to measure the rates of complex association

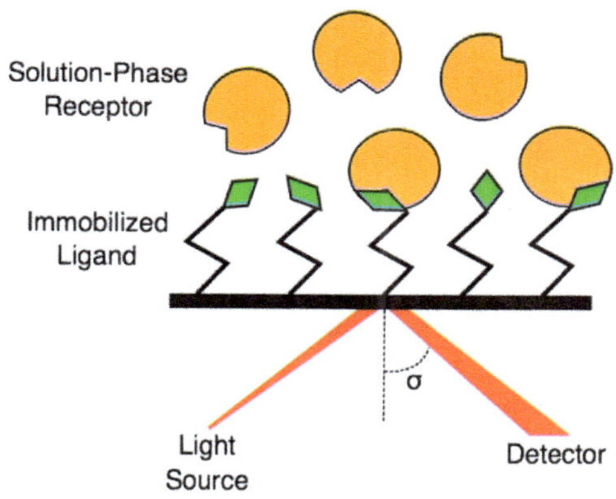

Figure 7.18 Surface plasmon resonance. Ligands are immobilized onto a gold surface, and a solution of receptor is passed over the surface. Polarized light is reflected off the chip and measured by a detector. By plotting the angle of refraction (σ) as a function of time, dissociation constants as well as on and off rates can be calculated.

and dissociation. Although one component must be immobilized on the chip, SPR requires a small amount of ligand and oligonucleotide. Another common method to calculate binding affinity is nuclear magnetic resonance (NMR) titration. Typically, the oligonucleotide is in solution and the change in chemical shift is monitored at varying ligand concentrations. For larger oligonucleotides, isotopic labeling may be required to increase sensitivity and overcome spectral overlap. While NMR requires relatively large amounts of the oligonucleotide and ligand, the technique can be used to measure weak binding constants.

Optical techniques offer increased sensitivity and quantitation, although the potential impact of introducing a covalent label into the structure of a binding species must always be considered. Fluorescent oligonucleotides can be generated through the incorporation of environmentally sensitive fluorescent nucleobase derivatives, such as 2-aminopurine, which allow the oligonucleotide to be titrated with a binding candidate that alters stacking interactions. Alternatively, fluorescent indicator displacement assays monitor the change in emission intensity caused by displacement of a fluorescent indicator upon titration with a binding candidate. Another popular optical method monitors the change in fluorescence polarization or anisotropy value due to binding of a fluorescently labeled ligand by an oligonucleotide (Figure 7.19A). Lastly, Förster resonance energy transfer (FRET) monitors the efficiency of energy transfer from a donor fluorophore to a suitable acceptor on an appropriately labeled oligonucleotide. The transfer efficiency is high over distances of ~10–100 Å, and the technique can be used to extract binding constants as well as gain information about structural changes caused by the binding interaction (Figure 7.19B).

7.3.2.2 Structural Characterization

See Weeks[42] for a review of chemical mapping techniques. Chemical mapping techniques obtain secondary structural information by modifying oligonucleotides with various chemical reagents. These techniques are particularly useful for identifying residues that are important for oligonucleotide recognition by comparing the chemical map of the oligonucleotide in the presence and absence of ligand. Some modifying regents, such as dimethyl sulfate and bisulfate, are base-specific and convey specific recognition information such as base stacking and hydrogen bonding interactions. Another technique, hydroxyl radical probing, utilizes radical promoters to abstract protons from the ribose sugar, affording information on the solvent accessibility of the backbone.[43] Other techniques, including in-line probing, 2′-hydroxyl acylation analyzed by primer extension (SHAPE), or RNase cleavage, provide information on the local flexibility of an oligonucleotide. In-line probing is typically performed at alkaline pH, where flexible regions of the oligonucleotide can undergo nucleophilic attack by the 2′-hydroxyl, cleaving the sugar–phosphate backbone. SHAPE utilizes various reagents with differential reactivity toward the 2′-hydroxyl to provide secondary structure

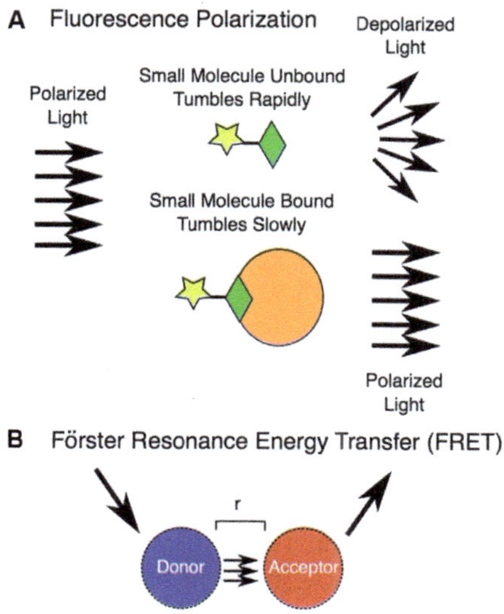

Figure 7.19 Fluorescence methods. (A) Fluoresence polarization. Polarized light is applied to a solution of small molecule and receptor. If the small molecule is bound, the tumbling motion is slow and the light remains polarized. If the small molecule is not bound, the tumbling motion is fast and the light is depolarized. (B) Förster resonance energy transfer (FRET). Within a certain radius (r), energy from an excited donor fluorophore is transferred to an acceptor fluorophore, and the transfer efficiency is measured.

models of RNA.[44] For instance, 1-methyl-7-nitroisatoic anhydride measures general local nucleotide flexibility, while *N*-methylisatoic anhydride reacts with nucleotides that experience slow dynamics. Similarly, RNase enzymes cleave the sugar–phosphate backbone at single-stranded or unstructured regions.

7.3.3 Nucleic Acid Targeting using Synthetic Oligonucleotides

Some of the simplest methods for nucleic acid detection harness the recognition power of nucleobases themselves through the use of synthetic nucleotide derivatives. For example, recognition of dsDNA was first achieved using a triplex-forming oligonucleotide (TFO).[45] TFOs were shown to bind in the major groove of double-stranded oligopurine–oligopyrimidine sequences of DNA in either a parallel fashion through Hoogsteen interactions or antiparallel through reverse-Hoogsteen interactions (Figure 7.20). The utility of TFOs is limited, however, due to the necessity of cytosine becoming protonated to form two hydrogen bonds in the Hoogsteen arrangement and the propensity of G-rich sequences to form alternate structures in the presence of metals. Binding affinities can be improved through synthetic

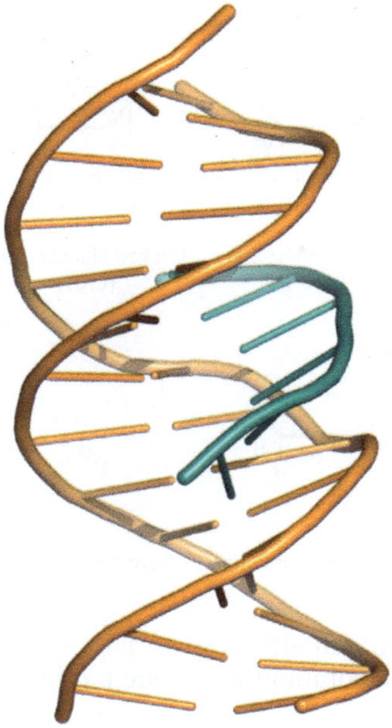

Figure 7.20 A nuclear magnetic resonance structure of duplex polypyrimidine:-polypurine sequences bound by triplex-forming oligonucleotide. Upon binding, there are significant changes in the conformation of purine residues, the 5′ and 3′ junctions, and groove widths associated with protonated cytosine residues (Protein Data Bank: 1BWG).

modifications, most commonly of the sugars [*e.g.* 2′-methoxy (2′-OMe) and locked nucleic acids (LNA)] or the backbone [*e.g.* phosphorothioate (PS) and peptide nucleic acids (PNA)] (Figure 7.21). These modifications also increase biological stability and have allowed TFOs to selectively target a variety of DNA-dependent processes, including transcription and mutagenesis, both in cell culture and in animal models.[46] Sequence-specific labeling can also be achieved through the conjugation of biotin or fluorescent tags to the TFOs.

Modified oligonucleotide derivatives have been extensively exploited to target single-stranded RNA, with applications for *in situ* labeling, steric blocking, and degradation.[47] In the latter case, it is important to note that the traditional mRNA degradation pathways do not accept complexes containing a modified strand. For example, duplexes of fully modified oligomers, particularly with modifications at the 2′-hydroxyl, are not adequate substrates of the RNA-induced silencing complex central to the siRNA pathway (see Section 7.2.1). Hybrid oligonucleotides that contain 2′-modified ends and a phosphorothioate core, however, can be substrates for RNaseH-catalyzed degradation and similarly decrease protein expression levels. At the same time, fully modified oligonucleotides can participate in sequence-specific targeting and steric blocking of both splicing and translation machinery, respectively. Anti-miRNA oligonucleotides

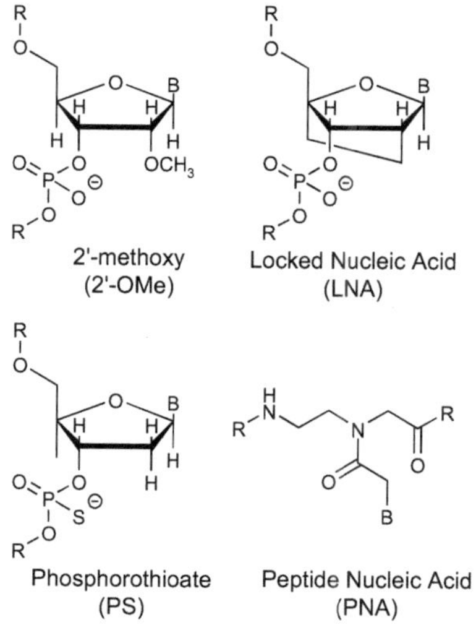

Figure 7.21 Oligonucleotide modifications. Top: common 2′-hydroxyl modifications; Bottom: common backbone modifications.

have also been shown to be highly effective at either preventing pri-/pre-miRNa processing or hybridizing to the mature miRNa to block its binding to the cognate RNA target.[48,49] A notable example is miravirsen, a combination LNA and phosphorthioate derivative that blocks the function of human miR-122 in hepatitis C virus infection[50] and recently completed phase 2 clinical trials.[51]

Despite this success, delivery and stability remain daunting challenges in the development of therapeutically relevant oligonucleotides. Nanoparticle–oligonucleotide conjugates, pioneered by Mirkin and co-workers, offer a promising alternative.[52,53] In this work, the derivatization of gold nanoparticles leads to a dense array of unmodified oligonucleotides with a number of remarkable properties, including high binding affinity, co-operative binding, and sensitive detection. Highly specific optical detection *in vitro* is possible due to the SPR signals available with the nanoparticles, and, in the simplest configuration can detect single nucleotide polymorphisms through target-induced aggregation. In cell culture, oligonucleotide–nanoparticle conjugates are not only taken up into cells but can also be tuned for gene knockdown. Live cell gene detection is possible through the incorporation of fluorescently labeled complementary "flare" sequences that are quenched when bound to the nanoparticle but released upon target binding (Figure 7.22). Alternatively, the narrow peak fingerprint of surface-enhanced Raman scattering (SERS) spectroscopy has allowed for sensitive and multiplexed detection of oligonucleotides in a variety of systems, including the plasmonic nanoprobes used as "molecular sentinels" by Vo-Dinh and co-workers.[54] Such a system was recently shown to detect as few as 6 attomol of ssDNA from the dengue virus (Figure 7.23).[55]

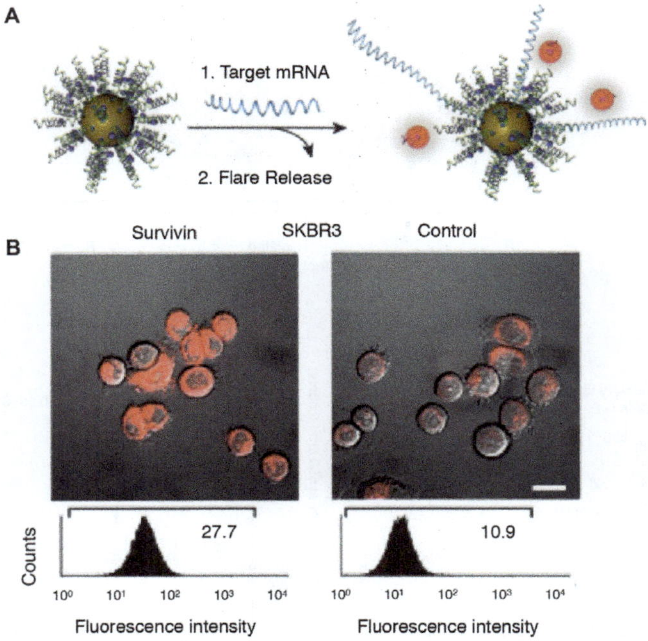

Figure 7.22 (A) Nanoflares are quenched when hybridized to an oligonucleotide-functionalized gold nanoparticle. Binding of target mRNA leads to release of the nanoflare and subsequent signal increase. (B) SKBR3 breast cancer cells stained with survivin-nanoflares (left) *versus* control (right).
Reprinted with permission from *J. Am. Chem. Soc.*, 2012, **134**, 1376, © 2012 American Chemical Society.[52]

7.3.4 Nucleic Acid Targeting using Synthetic Peptides

Several groups have demonstrated that synthetic peptides and peptide conjugates can selectively bind nucleic acids, particularly RNA targets where branched, three-dimensional structures are common. Select examples of these different approaches are discussed below.

As expected, many oligonucleotide-targeted synthetic peptides are enriched in positively charged and aromatic residues. Work by Cline and Waters has demonstrated that folded β-hairpin peptides containing lysine and tryptophan residues create a nucleotide binding cleft and that dimers or conjugates of these hairpin peptides interact with ssDNA, dsDNA and RNA.[56] For example, a β-hairpin peptide dimer binds an 11-mer ssDNA with a dissociation constant of ~3 μM, an affinity that is comparable to DNA-binding proteins containing the β-sheet-based OB fold (see Section 7.2) and consistent with a combination of aromatic and electrostatic interactions. More recently, Waters and co-workers described a β-hairpin peptide–intercalator conjugate with similar affinity toward RNA and

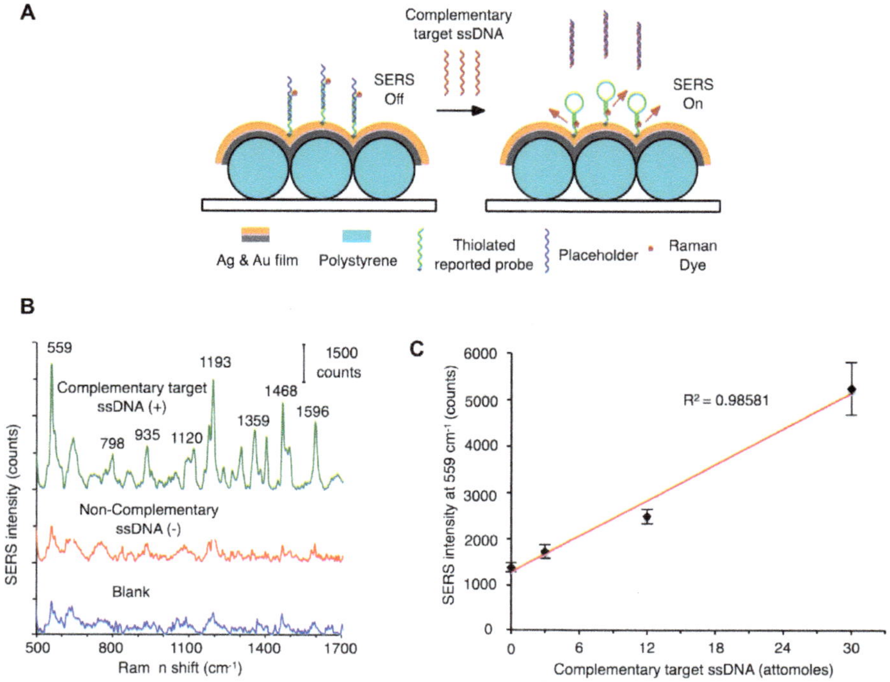

Figure 7.23 (A) DNA bioassay using surface-enhanced Raman scattering (SERS). To begin, the reporter probe is complexed to a placeholder strand, which maintains distance between the SERS dye and the film. Upon target ssDNA binding to the placeholder, the reporter probe is able to form a hairpin structure that brings the SERS dye near the film and generates an "on" signal. (B) Example SERS spectra demonstrating strong, discrete signals in the presence of the target. (C) Attomole detection of ssDNA.
Adapted from ref. 55 with permission from the Royal Society of Chemistry.

achieved selective binding of a tetraloop-stem-bulge RNA sequence in which all three local secondary structures were bound simultaneously (Figure 7.24).

Indeed, conjugates of aromatic intercalators and natural peptides have often led to increased affinity and specificity. For example, Beal and co-workers have developed helix-threading-peptides in which an aromatic heterocyclic intercalator is derivatized on either side with specific peptide sequences (Figure 7.25).[57] Upon intercalation of the aromatic residue, the two peptide arms interact with both the major and minor groove of the helix, creating a binding preference for proximity to loop or bulge regions. Miller and co-workers developed peptide-based ligands for AT-rich DNA by taking advantage of the ability of cysteine to participate in reversible disulfide bonds.[58] A library of cysteine- and quinolone-containing peptides was immobilized on beads and then combined with a second, solution-based

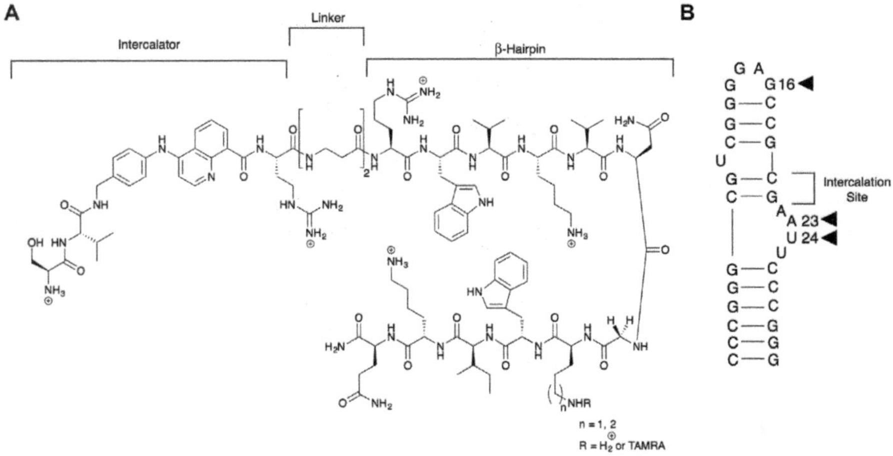

Figure 7.24 (A) β-Hairpin peptide-intercalator conjugate enriched with lysine and tryptophan residues. (B) Binding sites on the tetraloop-stem-bulge structure were determined by RNase digestion and are denoted with triangles. The ligand bound to the target RNA with a low micromolar dissociation constant.
Adapted from ref. 56 with permission from the Royal Society of Chemistry.

peptide library to generate >10 000 possible peptide dimers. Dynamic combinatorial chemistry was employed to select the disulfide-linked dipeptides that bound mostly strongly to a DNA target. The resin-bound and solution-free cysteine-containing peptides were allowed to interact reversibly in the presence of a fluorescently labeled DNA sequence. Over time, the most stable dipeptide binders were enriched and beads that were fluorescently labeled with the DNA target were isolated (Figure 7.26). The identified peptides were synthesized in free solution and found to bind the sequence with low micromolar dissociation constants. This technique has also been used in a number of RNA-based screens, including successful targeting of the HIV-1 frameshift-inducing RNA.[59]

Peptides with discrete side chain and backbone modifications have also been successfully employed. Santos and co-workers, for example, developed large libraries of branched peptides containing a boronic acid side chain (Figure 7.27A).[60] It was hypothesized that the boronic acid would form a reversible covalent bond with the 2′-hydroxyl group of RNA, producing increased binding affinity and selectivity over DNA. On-bead high-throughput screening of >40 000 unique sequences identified **BPBA1** (Figure 7.27B) as a selective ligand for HIV-1 RRE RNA ($K_D = 1.4$ μM) compared to DNA and other HIV RNA variants. RNase digestion revealed a broad footprint for the interaction (Figure 7.27C), and the ligand exhibited cellular uptake and low toxicity to mammalian cells. Alternative peptide backbone modifications have been investigated by Luebke and co-workers

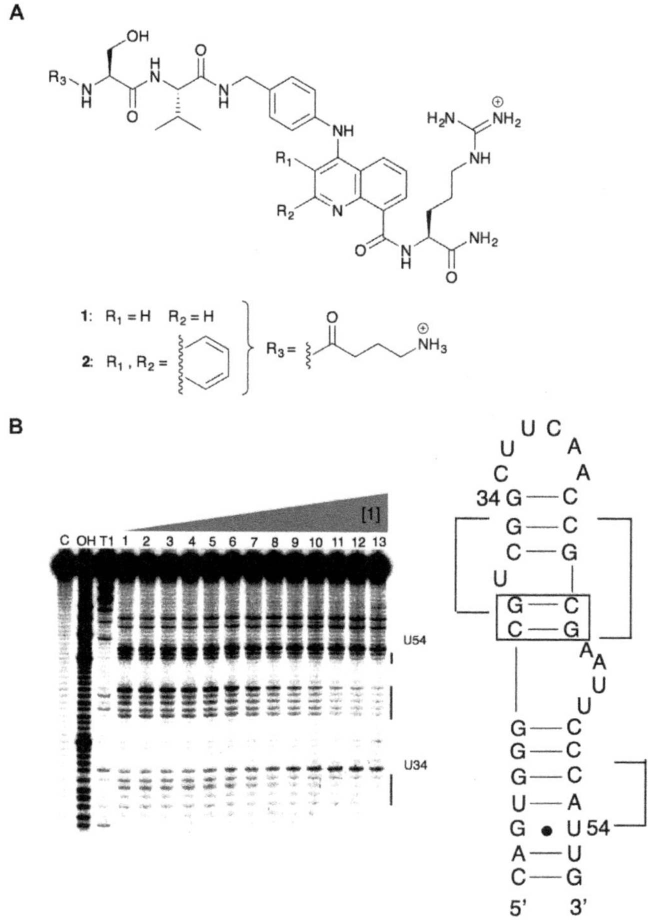

Figure 7.25 (A) Library of helix-threading peptides (HTPs) that includes aromatic heterocyclic intercalators. (B) RNase digestion in the presence of HTP **1** is consistent with binding to a stem region, presumably by the quinolone, as well as longer range interactions with the peptide arms. Adapted from ref. 57 with permission from the Royal Society of Chemistry.

who targeted hairpin miR-21 RNA by screening a library of protease-resistant peptoid derivatives immobilized on a microarray (Figure 7.28).[61] After immobilization of the peptoids, 5′-fluorescently labeled miR-21 RNA was washed over the microarray to identify high-affinity ligands. While 891 library members (12%) were found to bind miR-21, only two library members did so selectively in the presence of a closely related hairpin derivative. Additional structure-activity studies led to the peptoid in Figure 7.28A, which was found to have a dissociation constant of 12 μM for miR-21 and to specifically inhibit processing of a miR-21 primary transcript by Drosha.[62]

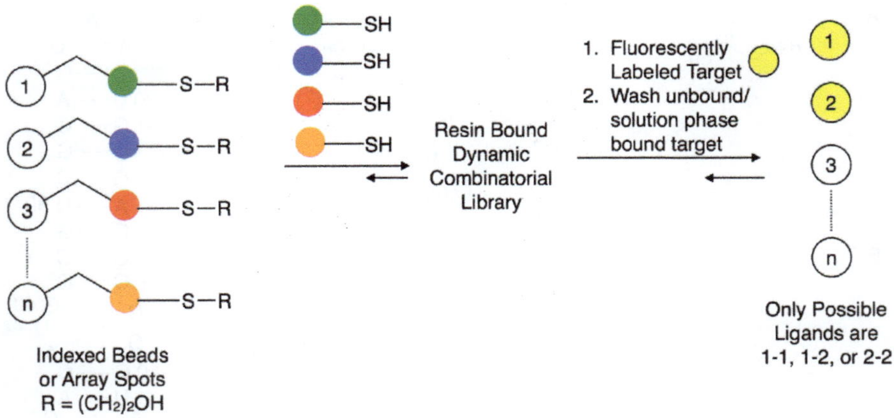

Figure 7.26 Selection of peptide dimers. Indexed beads or spots are functionalized with one cysteine-containing peptide. A pool of cysteine-containing peptides is then allowed to react reversibly with each bead in the presence of a fluorescently labeled target. Target-bound fluorescent beads are isolated.
Reprinted with permission from *Org. Lett.*, 2006, **8**, 1803, © 2006 American Chemical Society.[58]

7.3.5 DNA Targeting using Small Organic Molecules

7.3.5.1 General Principles

From the beauty of its structural simplicity to its role as the starting point of the central dogma, DNA has long been an attractive target for molecular recognition. Given the regular structure and limited surface availability of B-form DNA, it is perhaps not surprising that the majority of DNA-targeted ligands can be classified as either intercalators or minor groove binders (Figure 7.29). In the first class, aromatic molecules bind non-specifically between base pairs using mainly aromatic stacking, van der Waals, and hydrophobic interactions. Applications have included DNA detection and staining, along with a wide range of chemotherapeutics.[63] Minor groove binding ligands rely on shape and hydrogen bonding complementarity to "read" the pattern of groove hydrogen bond donors and acceptors programmatically and have demonstrated both gene regulation and *in vivo* xenograft activity. Nonetheless, selectivity at the genomic level remains a significant challenge. As mentioned previously, specific sequence targeting requires binding of approximately 18 bases.[64] Both specificity and affinity have been improved through the combination of intercalating and groove-binding moieties to generate threading intercalators (Figure 7.29). The targeting of more unique DNA structural motifs such as G-quadruplex or i-motif structures offers additional routes to specificity, and such ligands have also demonstrated biological activity. It is worth noting that some of the most biologically active DNA-targeting small molecules interact irreversibly with DNA through covalent modifications, often inducing DNA

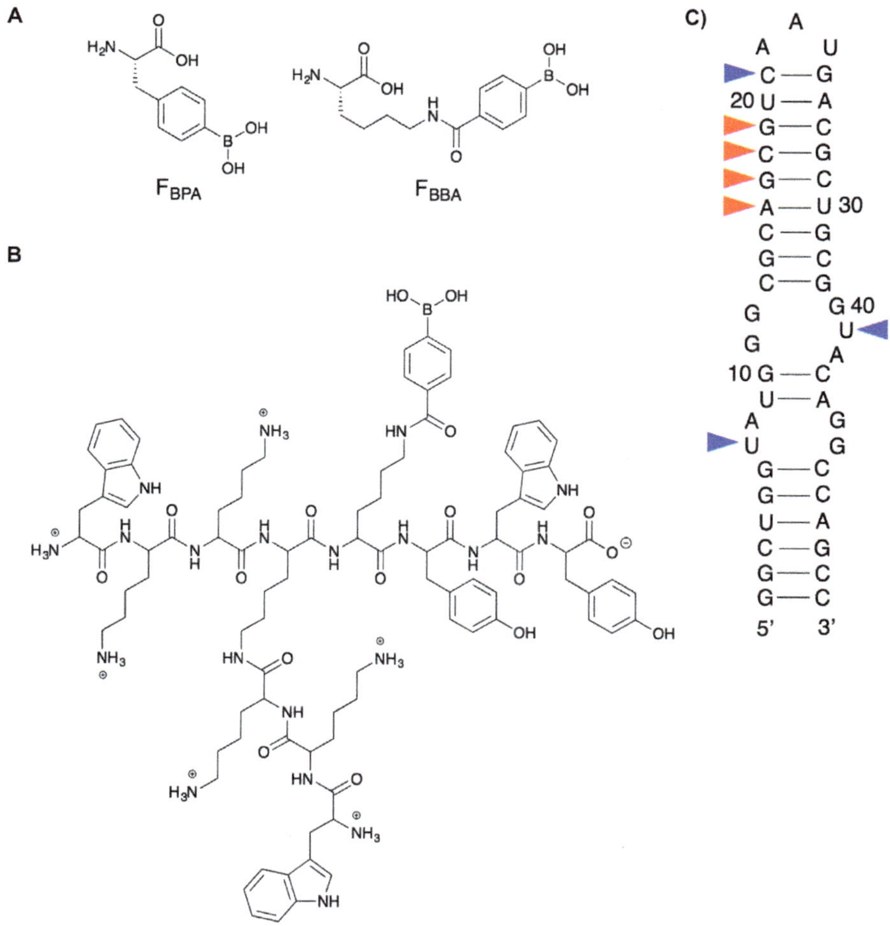

Figure 7.27 (A) Unnatural amino acids (F_{BPA} and F_{BBA}) containing boronic acids can form reversible, covalent bonds with the 2′-hydroxyl of RNA. (B) Structure of the lead peptide, **BPBA1**, which binds to the HIV-1 Rev-response element (RRE) RNA with a dissociation constant of 1.4 μM. (C) The binding site of **BPBA1** was identified by RNase digestion of a 52-mer of HIV-1 RRE RNA. Blue triangles represent areas of decreased cleavage and red triangles represent areas of increased cleavage. Adapted from ref. 60 with permission from the Royal Society of Chemistry.

damage and leading to cell death. The focus here is on noncovalent recognition of DNA, the discussion organized primarily by binding mode.

7.3.5.2 Intercalators

Perhaps the earliest identification of DNA intercalation was the conclusion by Lerman in 1961 that acridine and proflavine impacted the hydrodynamics of DNA by binding "between adjacent nucleotide-pair layers", thus

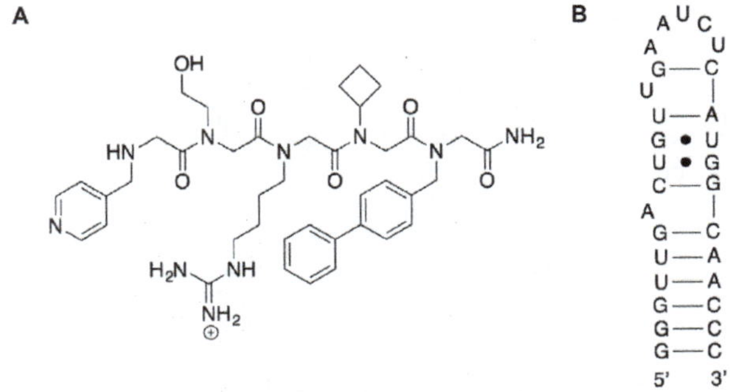

Figure 7.28 (A) Peptoid identified in a screen that bound to the miRNA-21 hairpin with a dissociation constant of 12 μM. The peptoid exhibited selectivity over a closely related hairpin. (B) The secondary structure of miRNA-21 used in the screen. The structure consists of a hairpin loop and two one-nucleotide bulge structures.

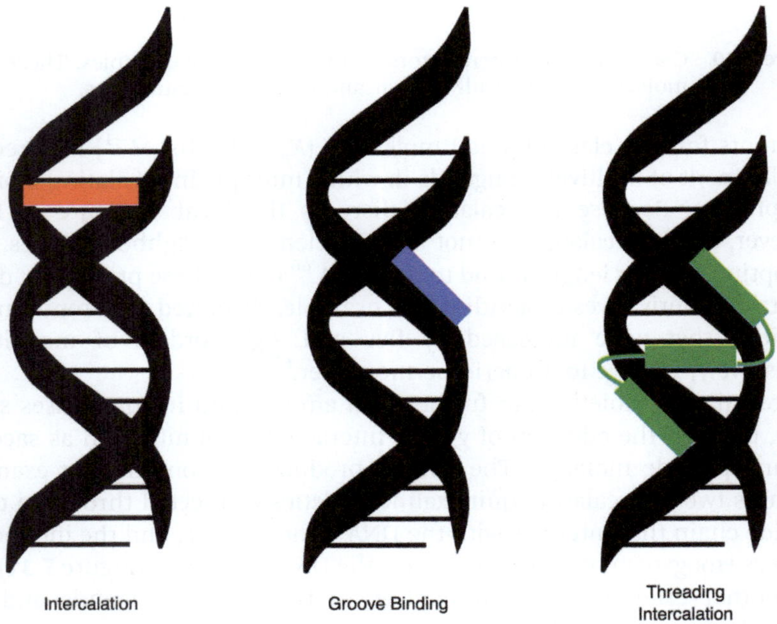

Figure 7.29 Modes of DNA–small molecule interactions.

extending or unwinding the helix.[65] Another important characteristic of many DNA intercalators is optical activity, particularly a large increase in fluorescence intensity upon DNA binding. A multitude of planar, polycyclic aromatic small molecules have been identified as DNA intercalators (Figure 7.30), with the binding affinity largely attributed to van der Waals or hydrophobic interactions.[66] In general, smaller aromatic systems require positively charged moieties to achieve adequate DNA binding affinity while systems with larger π-rich and hydrophobic surfaces do not. Modest binding

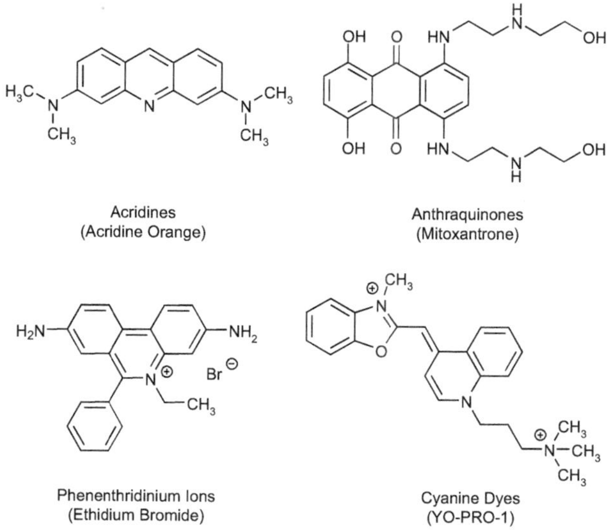

Figure 7.30 Classes of DNA intercalators and representative examples. These small molecules are typically planar and rich in aromatic rings.

constants for this class of small molecules ($K_a = 10^4$–10^6 M^{-1}) inspired the development of multivalent ligands in which multiple intercalating moieties are present. Because intercalation disrupts the local structure of DNA, however, two intercalators cannot bind efficiently at neighboring sites, with the optimal spacer length found to be >10 Å.[66] Using these principles, dimer and trimer derivatives of acridine, for example, displayed apparent binding affinities that were increased by five and eight orders of magnitude, respectively, relative to an acridine monomer.[67]

Intercalating moieties can further gain affinity, and in some cases specificity, through the addition of groove-interacting subunits such as saccharide and peptide moieties. The natural product echinomycin, for example, contains two intercalating quinoxaline moieties connected through a cyclic peptide chain that interacts with the DNA minor groove, and the interaction induces Hoogsteen base pairing in specific DNA sequences (Figure 7.31).[68,69] In another example, the combination of two acridine ligands and two synthetic minor groove binders by Dervan and co-workers led to a sequence-specific bis-intercalator.[70] Intercalators flanked by groove binders are generally referred to as "threading interacalators" and often offer impressively slow dissociation constants. Iverson and co-workers have developed a series of naphthalene diimide intercalator conjugates linked by peptide and alkyl linkers (Figure 7.32).[71,72] Specifically, the peptide linkers (Gly_3Lys or β-Ala_3Lys) were found to bind in the minor groove of DNA with sequence specificity while alternating alkyl linkers bound the major groove. An optimized tetraintercalator was found to achieve both 14 base-pair sequence specificity and an exceptionally long dissociation half-life of 57 days.[71] While applications of organic intercalators often include their use as DNA

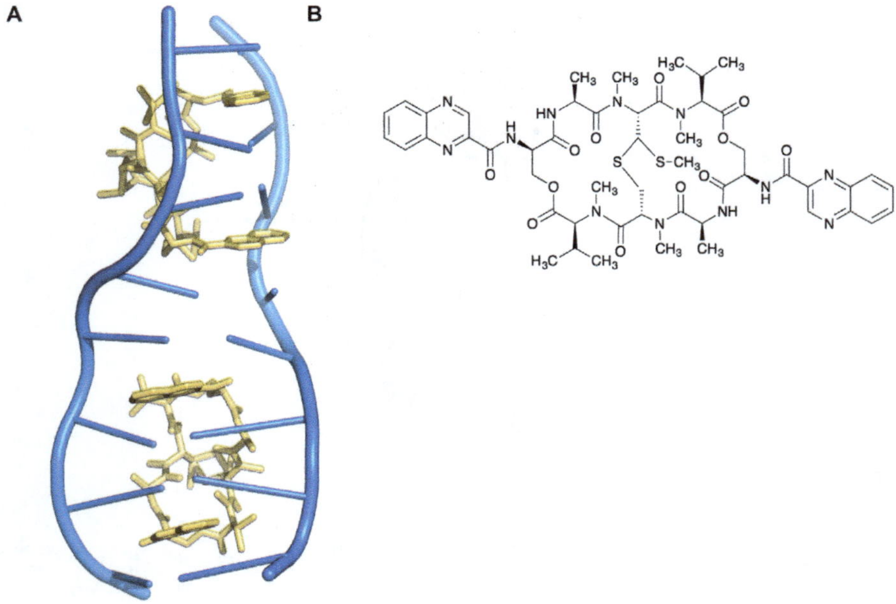

Figure 7.31 Echinomycin is a bis-intercalator. (A) Crystal structure of echinomycin bound to DNA in the presence of manganese. The terminal quinoxalines intercalate into the DNA and the peptide backbone acts as a linker (Protein Data Bank: 3GO3). (B) Structure of the cyclic peptide echinomycin, a natural product.

indicators or stains, it is important to note that the conjugation of inorganic complexes to classic intercalators has produced several potent anticancer agents.[73] In these conjugates the aromatic ligands, and, in some cases the metal centers themselves, interact with the DNA bases. In addition, ruthenium and rhodium complexes with aromatic intercalating ligands have allowed in-depth studies of DNA-mediated charge transport.[74]

7.3.5.3 Groove Binders

Inspired by a variety of natural products, most groove-binding small molecules bind to the narrower minor groove of B-form DNA and prefer AT-rich tracts.[66] These ligands are generally polyaromatic small molecules linked to give a curved recognition surface, as seen in Hoescht 33258 and furamidine derivatives (Figure 7.33). Importantly, sequence-specific recognition with small molecules has been observed almost exclusively through binding the minor groove, largely inspired by the natural products distamycin and netropsin (Figure 7.34).[64,75] Both small molecules contain amide-linked oligomers of *N*-methylpyrrole and specifically interact with the minor groove of A/T tracts, possibly due to the narrowing of the minor groove in A/T-rich *versus* G/C-rich sequences or the negative electrostatic potential of A/T *versus* G/C floors in the minor groove. Hydrogen-bonding interactions are observed between the amide hydrogen donors and N/O hydrogen bond acceptors on

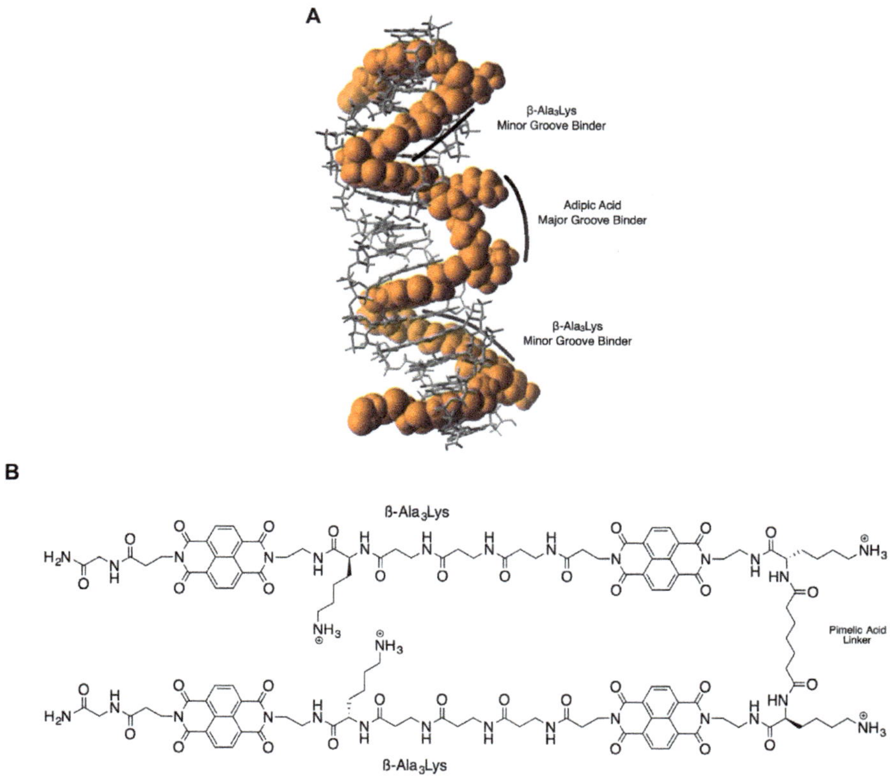

Figure 7.32 Napthalene diimide ligands are tetraintercalators. The β-Ala_3Lys peptides bind in the minor groove and the various-sized linkers bind in the major groove. (A) Nuclear magnetic resonance structure of the tetraintercalator binding to a 14 nucleotide DNA sequence. (B) Structure of the naphthalene diimide ligand with a 57 day dissociation half-life. Reprinted with permission from *J. Am. Chem. Soc.*, 2013, **135**, 12783, © 2013 American Chemical Society.[71]

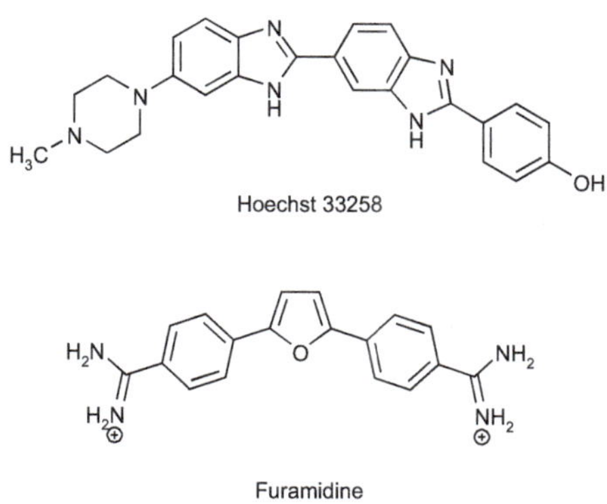

Figure 7.33 Representative small molecules demonstrating the aromatic residues and generally curved shape characteristic of minor groove-binding ligands.

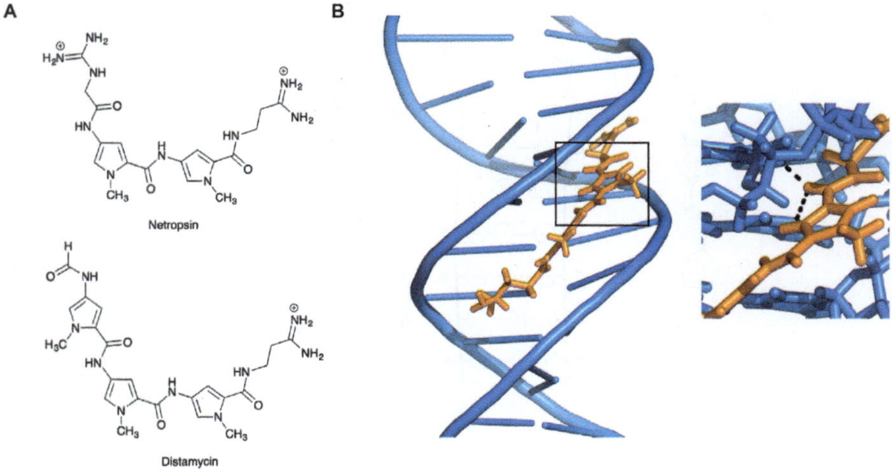

Figure 7.34 (A) Netropsin and distamycin are natural products that bind to the minor groove of DNA. (B) Left: crystal structure of netropsin bound to the minor groove of DNA. Right: an N–H in netropsin forms a bifurcated hydrogen bond with thymine and adenine for minor groove recognition (Protein Data Bank: 6BNA).

the edge of the nucleobases. The contribution of the cationic group(s) is expected to be minimal in both structures, while the influence of van der Waals interactions is predicted to be significant. Importantly, both netropsin and distamycin can form 1 : 1 and 2 : 1 complexes in the minor groove. X-ray crystal structures of the latter case support a model in which each small molecule binds one strand of the DNA double helix. Another significant advance arose with the development of lexitropsin derivatives in which replacement of *N*-methylpyrrole by *N*-methylimidazole allowed for recognition of G–C base pairings through hydrogen bonding interactions between the imidazole acceptor and the guanine exocyclic N2 atom donor, though decreased affinity was also observed. An important step in assessing the binding mode and preferences of these early recognition systems was the development of affinity cleavage techniques in which appended EDTA–iron complexes induce site-specific oxidative degradation of DNA.[76]

The combination of *N*-methylpyrrole and *N*-methylimidazole recognition along with a two-strand, antiparallel-binding model yielded a class of high-affinity, high-specificity small molecules termed pyrrole–imidazole (Py–Im) polyamides. Dervan and co-workers developed a set of "pairing rules" in which opposing heterocycles recognized specific base pairings (Figure 7.35A).[77,78] Specifically, Im/Py pairings recognized G–C base pairs while Py/Im pairings recognized C–G base pairs. Py/Py pairs were found to be degenerate for A–T or T–A base pairings. Importantly, the amide bond linkages allowed for rapid and high-yielding solid-phase synthesis of large libraries. While a number of alternative heterocycle building blocks have been investigated, decreases in affinity have led the most robust systems to

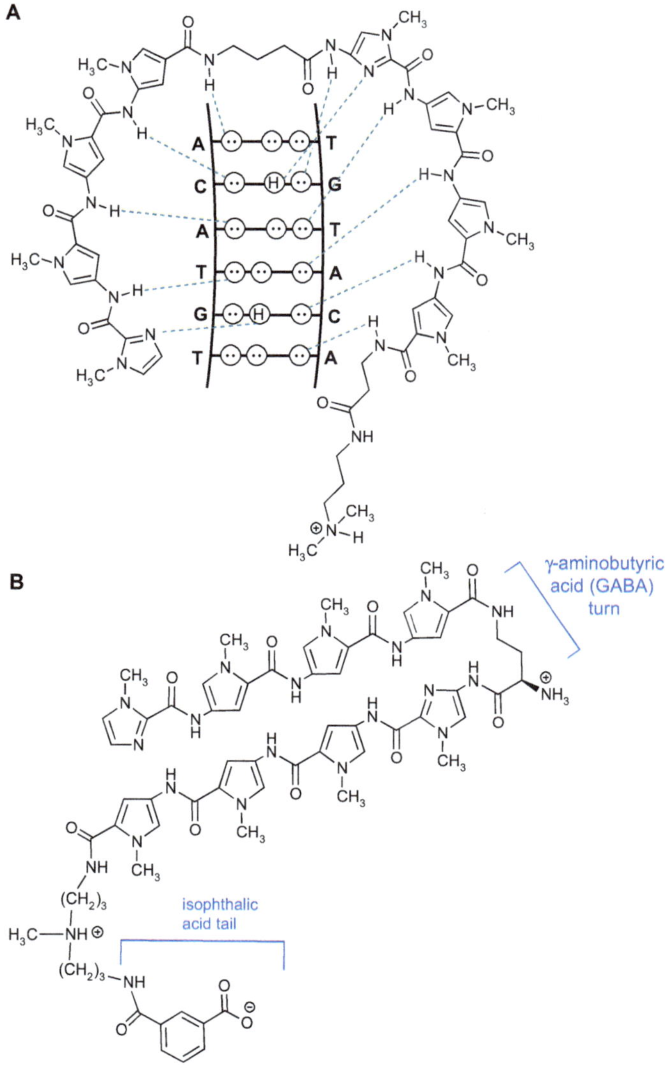

Figure 7.35 (A) *N*-methylpyrrole (Py) and *N*-methylimidazole (Im) pairs can specifically recognize nucleobases in the minor groove of DNA. Im/Py pairs recognize G–C base pairs, Py/Im pairs recognize C–G base pairs, and Py/Py pairs recognize A–T or T–A base pairs. (B) A representative biologically active pyrrole–imidazole polyamide, which recognizes the androgen receptor response element. The two tetramers are connected by a γ-aminobutyric acid (GABA) turn with an α-amine subunit for low animal toxicity, and an isophthalic acid tail is added for increased cellular uptake.

contain only the imidazole and pyrrole subunits. This common design features two heterocyclic tetramers linked by a γ-aminobutyric acid (GABA) turn region to form a hairpin structure, which recognizes up to six base pairs based on the four heterocycles and the A/T or T/A preference of both the turn and tail units (Figure 7.35B). Longer sequences can also be targeted but require the incorporation of β-alanine monomers, which provide the required spacing but impart flexibility to relax the over-curvature of Py–Im polyamides relative to the minor groove of DNA. The advent of facile DNA sequencing allows for nearly global evaluation of specificity using a modified Bind-N-Seq method in which specific sequences are "pulled down" from a large pool by biotin-tagged Py–Im polyamides and then identified by sequencing.[79] Recent high-resolution X-ray crystal structures have revealed a conformational change in DNA upon binding of Py–Im polyamides wherein the minor groove is expanded and the major groove compressed by about four ångströms, producing a kink in the helical axis (Figure 7.36).[80] The high affinity ($K_D = 10^{-12}$–10^{-8} M) and unique specificity of Py–Im polyamides represent an unquestionably powerful aqueous molecular recognition system.

These characteristics inspired biological studies of Py–Im polyamides, including investigations into their ability to disrupt DNA–protein

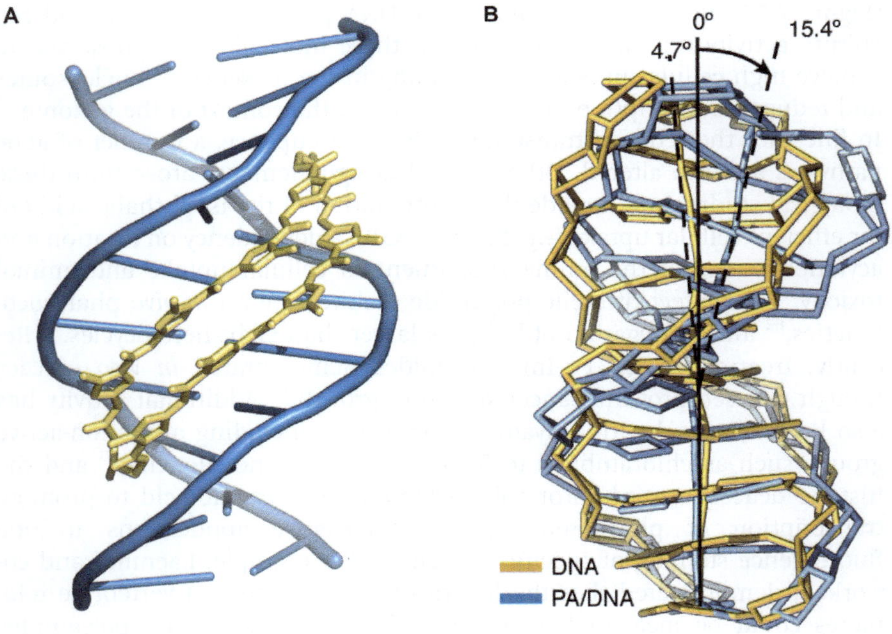

Figure 7.36 (A) Crystal structure of pyrrole–imidazole polyamide bound to the minor groove of DNA (Protein Data Bank: 3OMJ). (B) The yellow structure represents unbound DNA, and the blue structure represents polyamide-bound DNA, where the polyamide has been removed. Upon binding to DNA, the pyrrole–imidazole polyamide bends the DNA, compressing the major grove by 15.4° and producing a kink in the helical axis.

Reprinted with permission from *J. Am. Chem. Soc.*, 2010, **132**, 14521. © 2010 American Chemical Society.[80]

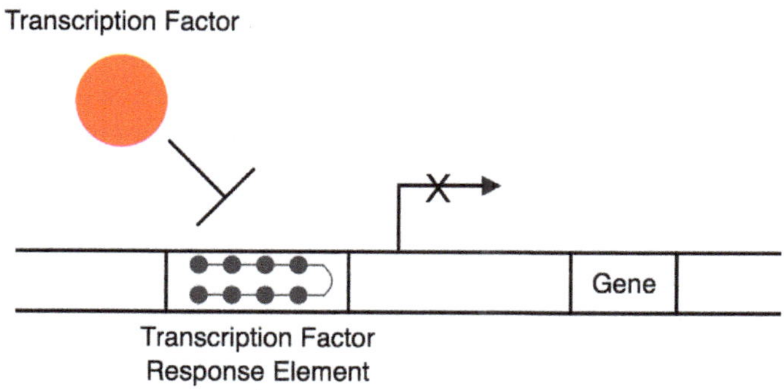

Figure 7.37 Pyrrole–imidazole polyamides are designed to recognize transcription factor response elements. Upon binding, the polyamide prevents the transcription factor from recognizing the DNA element and alters the expression of genes.

interactions. It was hypothesized that Py–Im polyamides could inhibit the binding of specific transcription factors and thus regulate gene expression (Figure 7.37). Both *in vitro* inhibition of DNA–protein interactions and cell culture activity was observed. In short, these molecules were observed to achieve high cellular uptake, DNA binding in the presence of nucleosomes and reduced transcription factor occupancy in the context of the genome.[78] In line with the reduced transcription factor occupancy, a number of gene pathways are also altered. Other critical design elements arose from these biological studies and include the incorporation of the isophthalic acid tail for efficient cellular uptake (*e.g.* Figure 7.35B), a dependency on location and acylation status of the amine substituent in cellular uptake and animal toxicity,[81] the effect of cyclic polyamide architecture on *in vivo* pharmacokinetics,[82] and aggregation of hairpins larger than eight heterocycles.[83] Recently, treatment with Py–Im polyamides demonstrated *in vivo* efficacy though reduced prostate cancer xenograft growth.[84] Additional activity has also been observed with polyamide conjugates, including alkylation-active groups such as chlorambucil to further promote gene silencing[85] and the histone deacetylase inhibitor suberoylanilide hydroxamic acid to promote transcription of pluripotency genes. Alternative applications include fluorescence staining of specific sequences. For example, Laemmli and coworkers demonstrated that the length of invertebrate and vertebrate telomeres could be measured using fluorescently labeled Py–Im polyamides targeted to telomere repeat sequences (Figure 7.38).[86] Despite this progress in biological activity, specificity may remain a problem, as potential off-target effects on gene expression and alternative mechanisms such as replication stalling have also been proposed.[87]

At the same time, there is a paucity of research into *major* groove binding small molecules, despite its prevalence as a protein-binding site.[88] To date, most reports of major groove interactions have involved natural products that intercalate with DNA bases and covalently modify residues in the major

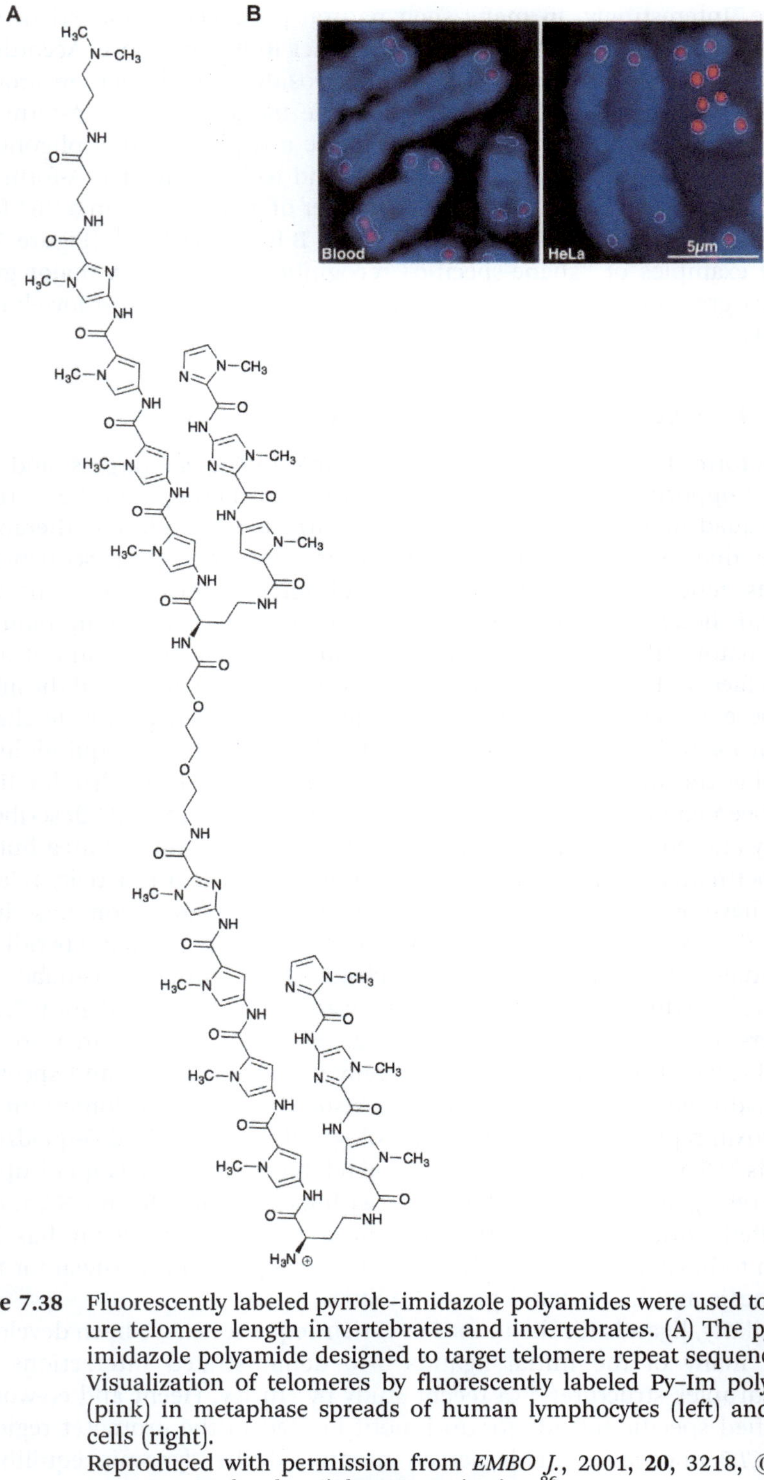

Figure 7.38 Fluorescently labeled pyrrole–imidazole polyamides were used to measure telomere length in vertebrates and invertebrates. (A) The pyrrole-imidazole polyamide designed to target telomere repeat sequences. (B) Visualization of telomeres by fluorescently labeled Py–Im polyamide (pink) in metaphase spreads of human lymphocytes (left) and HeLa cells (right).
Reproduced with permission from *EMBO J.*, 2001, **20**, 3218, © 2001 European Molecular Biology Organization.[86]

groove. Interestingly, in many such natural products, cyclohexyl or sugar residues were found to interact with the bases in major groove. Accordingly, Arya and co-workers found that aminoglycosides, which lack an aromatic residue for intercalation, can recognize the major groove of A-form DNA sequences and in some cases even displace a triplex-forming oligonucleotide.[89] Indeed, a neomycin dimer was found to bind GC-rich A-form DNA as well as contiguous AT-rich DNA, the latter of which exists in a "B* form" with a narrower major groove relative to B-form DNA[90,91] (Figure 7.39). These examples of "shape-specific" recognition of the DNA major groove suggest great promise for the development of future major groove-binding ligands.

7.3.5.4 Ligands for Alternative DNA Structures

Non-B-form DNA structures are interesting biological targets and offer unique opportunities for selective recognition.[92] Perhaps the best studied are G-quadruplex structures, which have attracted attention as therapeutic targets due to their predicted formation in telomeric repeat sequences as well as gene promoter regions.[93] G-quadruplex ligands tend to be large, aromatic heterocyclic structures that offer more surface area than traditional intercalators (Figure 7.40). Generally, G-quadruplex ligands bind at an external face of the quadruplex through π–surface interactions, and the affinity can be enhanced by electrostatic interactions between positively charged side arms and the grooves of the quadruplex structure (*e.g.* quindoline).[94] Arguably the most detailed recognition studies of a G-quadruplex ligand have been conducted with the tetracationic porphyrin TMPyP4 described by Hurley and co-workers, which boasts not only a large surface area but also four permanent positive charges.[95] A number of structure–activity relationships have been investigated for binding, selectivity, and telomerase inhibition. For example, the position of the nitrogen in the *meso*-pyridinium substituent was found to impact selectivity for different G-quadruplex structures, while increased bulkiness was not tolerated (Figure 7.41).[95] Sanders and co-workers increased the aromatic surface area to yield TMPyPz, which resulted in a dramatically increased affinity and specificity for quadruplex over duplex DNA, as measured by SPR.[96] Telomestatin and quarfloxin represent two of the best biologically characterized G-quadruplex ligands.[93] The large, neutral macrocycle telomestatin binds G-quadruplexes in the oncogenic *MYC* promoter region without binding B-form DNA and has exhibited antitumor properties in mouse models. Quarfloxin has been shown to block the binding of nucleolin to a G-quadruplex crucial for rRNA biogenesis and has entered phase II clinical trials.

Similarly, ligands for the i-motif in C-rich sequences have been developed. While many of the initial ligands also demonstrated interactions with G-quadruplex structures,[97] a recent study by Hurley, Hecht and co-workers identified specific ligands for an i-motif located in the promoter region of the *BCL2* oncogene.[98] This screen monitored the dynamic equilibrium

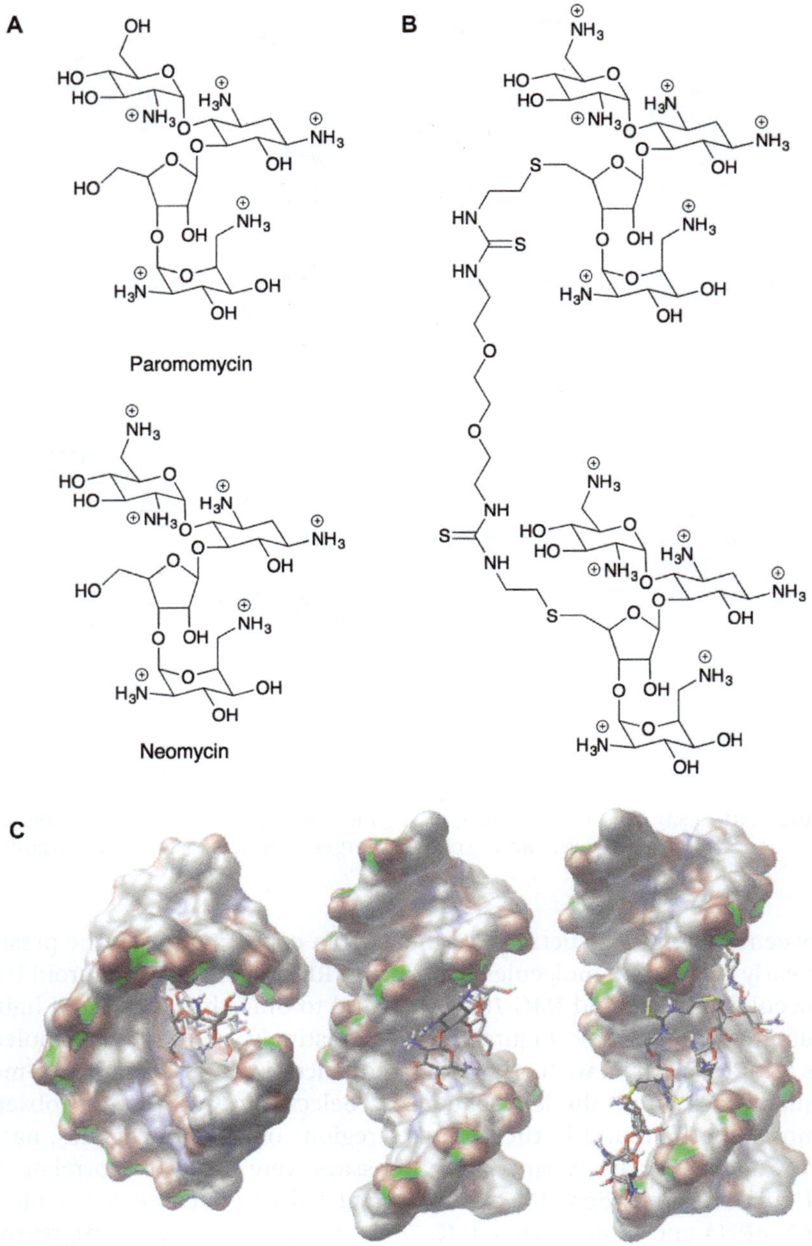

Figure 7.39 Aminoglycosides can recognize the major groove of DNA. (A) Structures of selected aminoglycosides, paromomycin and neomycin, that bind to the major groove of A-form DNA. (B) Structure of neomycin dimer that binds to GC-rich A-form DNA. (C) Computer-generated models of neomycin bound to different forms of DNA. Left: neomycin bound to A-form DNA. Middle: neomycin bound to B-form DNA. Right: neomycin dimer bound to major groove in B-form DNA.
Adapted from ref. 88 with permission from the Royal Society of Chemistry.

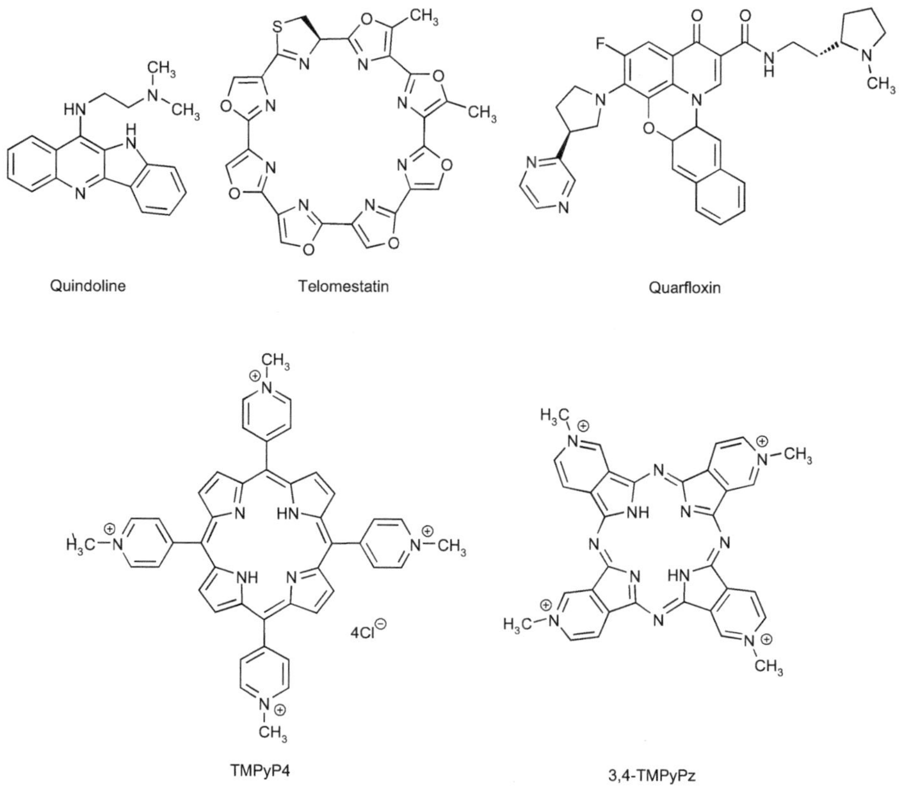

Figure 7.40 Examples of G-quadruplex ligands, which are characteristically large, aromatic-rich, and span a larger surface area than traditional intercalators.

between the i-motif structure and the hairpin conformation in the presence of nearly 2000 small molecules by using a FRET reporter. Two steroid-based molecules, IMC-48 and IMC-76, were found to bind the i-motif and hairpin structures, respectively (Figure 7.42). Interestingly, neither small molecule was found to interact with other *BCL2* sequences including a mutant i-motif, G-quadruplex, and duplex sequences. Selectivity was further observed against i-motifs found in the promoter regions of other oncogenes, namely *MYC* and *VEGF*. The *in vitro* binding assays were found to correlate with cell culture data, where IMC-48 treatment led to increased levels of both *BCL2* mRNA and protein while IMC-76 led to decreased levels. Impressively, neither small molecule affected mRNA levels of *MYC* or *VEGF*. The combined *in vitro* and cell culture specificity render this paradigm particularly promising for biologically relevant targeting of non-canonical DNA structures.

In addition to G-quadruplex and i-motifs, a number of other non-canonical DNA structures have been targeted. These include three-way junctions targeted with iron(II)-centered helicates and Holliday junctions targeted with

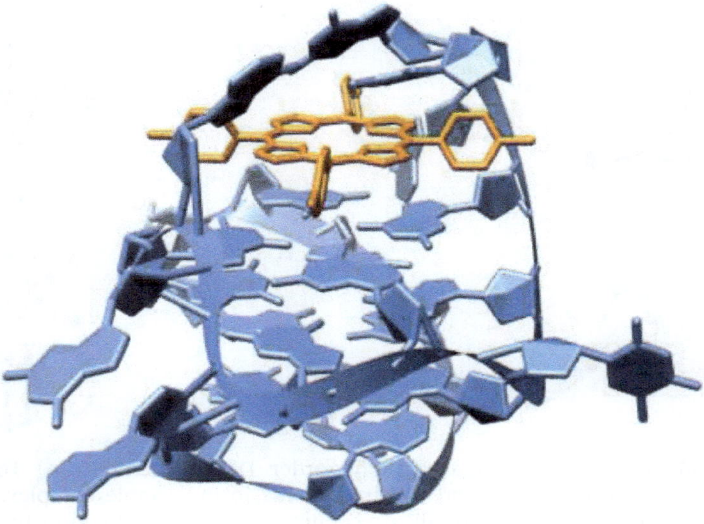

Figure 7.41 Nuclear magnetic resonance structure of the tetracationic porphyrin TMPyP4 with c-*MYC* derived G-quadruplex. TMPyP4 participates in π- and electrostatic interactions to stabilize the G-quadruplex and subsequently decreases the transcription level of *MYC* (Protein Data Bank: 2A5R).
Adapted from ref. 92 with permission from the Royal Society of Chemistry.

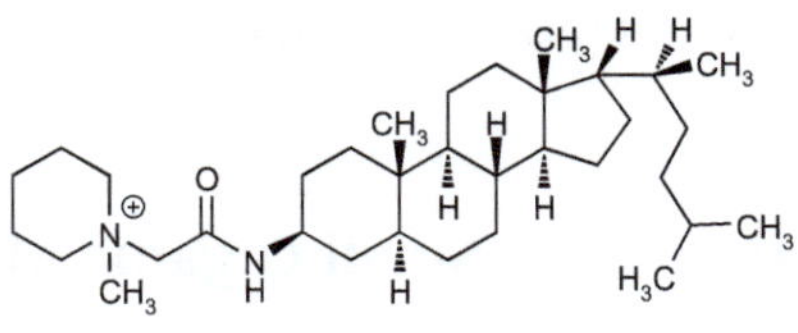

IMC-48

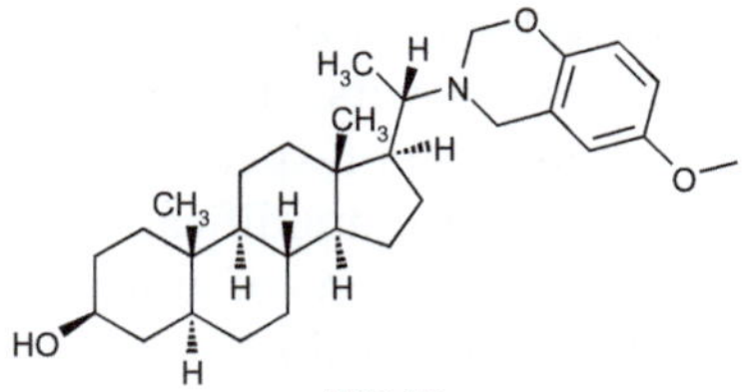

IMC-76

Figure 7.42 Förster resonance energy transfer-based dynamic screening against the C-rich motif located in the *BCL2* promoter identifies two leads, i-motif-selective ligand IMC-48 and hairpin-selective ligand IMC-76, which act as transcriptional activators and repressors, respectively.

multivalent acridine derivatives (Figure 7.43),[92] as well as extended aromatic ligands to target triple-helix structures.[99] Continued progress toward targeting unique motifs in DNA structure offers increased opportunities for

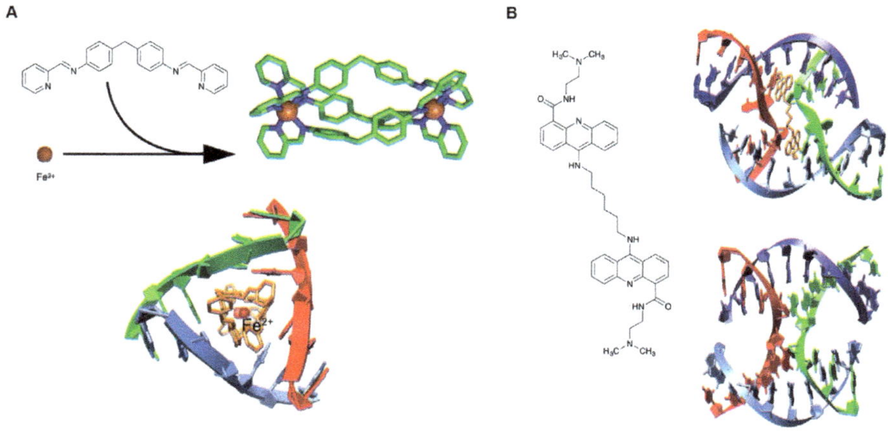

Figure 7.43 Examples of targeted higher-order DNA structures. (A) Top: three organic ligands and two iron atoms form a stable complex. Bottom: the multivalent complex recognizes a three-way junction in DNA. (B) Left: structure of acridine derivative. Top right: the acridine derivative recognizes Holliday junctions in DNA. Bottom right: unbound structure of a Holliday junction in DNA.
Adapted from ref. 92 with permission from the Royal Society of Chemistry.

selectivity in a genomic context. These strategies, along with combined intercalating and groove interactions are likely to drive the next generation of studies.

7.3.6 RNA Targeting using Small Organic Molecules

7.3.6.1 General Principles

Selective targeting of RNA with small organic molecules is a challenging task that requires synthetic systems that exhibit both high affinity and specificity. The polyanionic nature of RNA favors strong electrostatic interactions, which are high affinity yet inherently non-specific. However, RNA is enriched with naturally occurring nucleobase modifications and numerous non-canonical base pairs, providing a range of unique recognition surfaces. Furthermore, the 2′-hydroxyl group and single-stranded nature of RNA allow folding of the oligonucleotide into distinct secondary and tertiary structures. The higher-order structure creates distinctive binding pockets for small molecules. At the same time, the range of electrostatic interactions within RNA pockets is restricted compared to protein pockets due to the negatively charged backbone and lack of charge on the nucleobase. Specificity thus relies on weaker hydrogen bonding, hydrophobic effects, and shape complementarity. An additional barrier to achieving high specificity for a single RNA species is the high concentration of competing ribosomal RNA, which comprises 80–90% of all cellular RNA.

Most of these general principles are exemplified by studies of aminoglycoside-RNA recognition.[100] Aminoglycosides are protonated at physiological pH and form ionic interactions with the negatively charged sugar–phosphate backbone. Typically, the dissociation constants (K_d) range from low μM to high nM, with affinity correlating to the number of amine substituents; however, aminoglycosides exhibit low specificity and can bind to multiple sites within a single RNA. While aminoglycosides are useful tools to study general RNA interactions *in vitro*, the low specificity hinders their use *in vivo*.

The flexibility of RNA combined with a lack of RNA structural information creates another formidable challenge. RNA is a dynamic macromolecule with the potential for multiple active conformations. Upon binding to proteins, RNAs undergo structural rearrangement to adopt a less flexible, stabilized conformation.[101] Larger RNAs, such as the ribosome, utilize proteins as chaperones to aid in the sequential folding of the biologically active conformation. However, the inherent flexibility of unbound RNA makes structural determination difficult. Indeed, many of the RNA X-ray crystal structures are found in RNA–protein complexes or truncated RNA structures where the flexible regions are removed. Furthermore, X-ray crystallography usually only captures a single conformation, which may not accurately represent the solution-state structure of a highly flexible molecule.[102] As an alternative method, NMR spectroscopy has the potential to elucidate the solution-state structure of RNA ensembles, but the method is currently limited to oligonucleotide structures of <100 nucleotides without isotopic labeling.[103,104] The scarcity of high-resolution RNA structures hinders rational design and docking efforts, often making screening a necessity to identify small molecule RNA ligands.

7.3.6.2 Computational Methods

See Disney *et al.*[105] for a review of computational methods. First developed using protein models, computational docking predicts small molecule–macromolecule interactions and calculates the relative complexation energy. The scoring read-out can predict biological affinity or activity and prioritize small molecules for experimental validation. Typical docking parameters for proteins have proven unsuccessful for RNA due to its altered flexibility, limited non-covalent binding forces, and higher solvation. However, the recent increase in NMR and X-ray diffraction crystal structures for RNA in the Protein Data Bank has led to refined docking parameters that are more suitable for RNA recognition. Most structure-based drug design also employs molecular dynamics calculations to predict flexibility within a macromolecule. An applied force field allows the macromolecule to sample a finite space within a limited time frame. The low-energy conformers can be grouped into dynamic ensembles and used for docking studies of small molecule-RNA recognition.

Al-Hashimi and co-workers, for example, generated a dynamic ensemble of HIV-1 RNA transactivation response element (TAR) conformers by

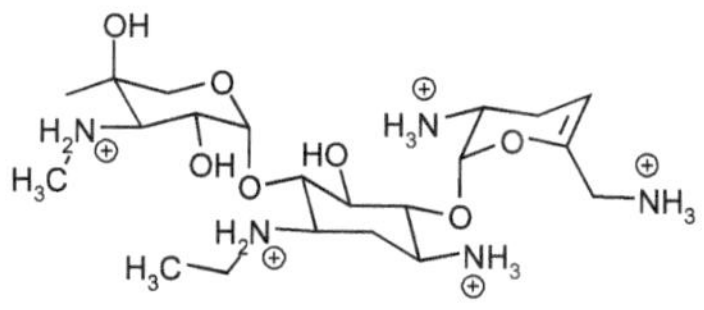

Figure 7.44 A library of 51 000 compounds was virtually docked against an ensemble of transactivation response element (*TAR*) RNA structures and a lead compound, netilmicin, was identified. The small molecule bound *TAR* with a dissociation constant of 1.35 μM and led to a reduction of HIV-1 replication in cell culture.

evaluating a large pool of potential structures using molecular dynamics.[106] The ensemble was narrowed to 20 structures by selecting conformers that were consistent with solution-based, time-averaged NMR residual dipolar coupling data. A library of ~51 000 small molecules was docked against the TAR ensemble with the top 57 commercially available hits tested for *in vitro* binding activity. Six small molecules were experimentally validated with dissociation constants ranging from 55 nM to 122 μM. One lead small molecule, netilmicin, showed antiviral activity in cell culture, significantly inhibiting HIV-1 replication (Figure 7.44).

Using an alternative computational method, James and co-workers developed a virtual screen for telomerase RNA (hTR), where flexibility was allowed in both the RNA and ligand.[107] To identify a small molecule that binds to the hTR, a virtual screening program, Molecular Recognition with a Driven Dynamics Optimizer (MORDOR), was used. A library of ~3000 US Food and Drug Administration (FDA)-approved drugs was docked against hTR with 13 small molecules confirmed to bind to hTR by standard NMR experimental methods. After a chemical similarity search of the lead small molecules, another round of computational docking was completed. The two rounds of docking and chemical similarity provided a total of 48 ligands that were verified to bind to the hTR with 10 exhibiting selectivity over the ribosomal A-site with low to high micromolar affinity.

As a final example, Zimmerman, Baranger, and co-workers rationally designed a small molecule to disrupt the r(CUG) repeat-MBNL1 complex in mytonic dystrophy 1.[108] The small molecule combined acridine, a known DNA intercalator, and triaminotriazine, a molecule known to recognize T–T or U–U base pairs. Molecular modeling was used to identify binding sites in the RNA and suggested a "stacked intercalator" binding mode. The triaminotriazine–acridine unit exhibited a high nanomolar dissociation constant and destabilized the r(CUG) repeat-MBNL1 complex (Figure 7.45).

The combination of molecular dynamics and docking has identified several small molecule leads for various RNA targets. Although the methods require the RNA structure to be known, computational programs provide a valuable predictive tool to guide *in vitro* small molecule–RNA studies.

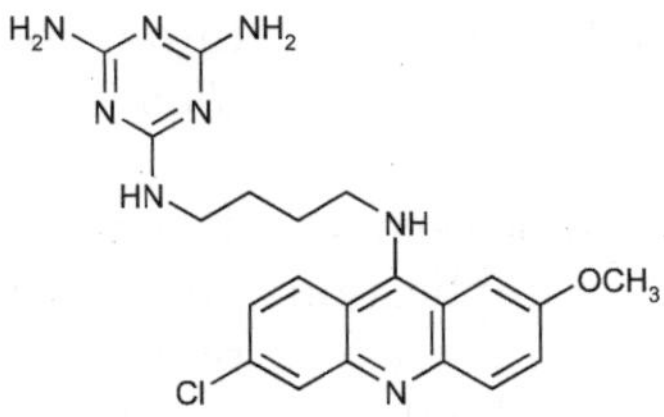

Figure 7.45 A computational approach was used to rationally design a triaminotriazine–acridine unit to disrupt the r(CUG) repeat-MBNL1 complex in mytonic dystrophy 1. The small molecule exhibited a stacked intercalator binding mode and a high nanomolar dissociation constant.

7.3.6.3 Hybrid Experimental-computation Prediction Methods

A series of hybrid experimental-computational methods has been developed to identify small molecules that specifically bind to secondary regions in a sequence-dependent manner. One outcome of this technology is modular assembly of multivalent conjugates of small molecules that can recognize combinations of secondary structures, potentially increasing RNA-ligand specificity and affinity. Toward this end, Disney and co-workers designed a two-dimensional combinatorial screening (2DSC) microarray to simultaneously screen a small molecule library against an RNA library and select for specific small molecule recognition.[109] Initial efforts screened four aminoglycosides against a library of >4000 internal loop and stem structures. Many trends emerged, including recognition of internal loops with C–A pairs by kanamycin derivatives. The work was extended to include peptoids and small molecule RNA-targeted libraries to screen against a variety of motifs and biological targets.

As a result, Disney and co-workers have developed several computational prediction programs to complement their experimental results. One such program, RNA Privileged Space Predictor (RNA-PSP), analyzes small molecule trends for statistically significant sequence-dependent recognition.[110] Initially, the program was developed by screening an aminoglycoside library against a library of RNA hairpins. The program predicted sequence-specific recognition by several aminoglycosides with low nanomolar dissociation and up to 200-fold specificity. Another program, Structure–Activity Relationships through Sequencing (StARTS), utilizes information from RNA-PSP and allows the prediction of small molecule binding affinity to an RNA motif.[111] The program was applied to four benzimidazole-containing small molecules to understand heterocyclic recognition of secondary RNA structures.[112] It was discovered that benzimidazole small molecules with greater steric bulk preferred intercalation into RNA stems over similar DNA structures. Lastly, Privileged Chemical Space Predictor program (PCSP) predicts small molecule features that are significant for RNA target recognition.[113]

The technology was applied to a high-throughput microarray screen of peptoids against the group 1 intron of *Candida albicans* to develop structure–activity relationships.

The technology developed by Disney and co-workers has also been applied to RNA targets with disease implications, including myotonic dystrophies 1 (DM1). A kanamycin A derivative, which prefers to bind to pyrimidine-rich internal loops, was identified by the group's high-throughput 2DSC microarray.[114] The kanamycin A derivative was modularly assembled to target the multiple internal loops of the r(CUG) repeats characteristic of DM1 (Figure 7.46). The multivalent small molecule exhibited *in vitro* and cell culture activity and was further tested in animal models of DM1. Statistically significant reduction in mis-splicing was observed when compared to the control mice, consistent with the predicted *in vivo* activity.

An alternative technique, quantitative structure-activity relationships utilizes experimental data to computationally predict the biological activity of related small molecules. This computational method has been used often to identify lead protein–ligand interactions by predicting variables such as

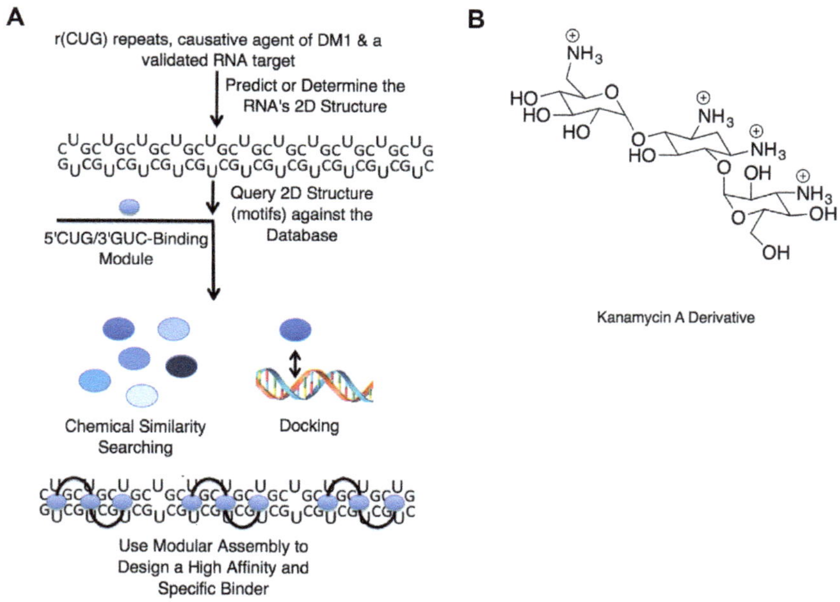

Figure 7.46 (A) A prediction-based screening approach was used to identify an inhibitor of the r(CUG) repeat-MBNL1 complex in mytonic dystrophy 1. The secondary structure of the r(CUG) repeats was compared to an RNA motif-small molecule database derived from a two-dimensional combinatorial screening microarray. (B) A Kanamycin A derivative was identified and modularly assembled to create a multivalent inhibitor of the r(CUG) repeat-MBNL1 complex with cell culture and *in vivo* activity.
Adapted from ref. 105 with permission from the Royal Society of Chemistry.

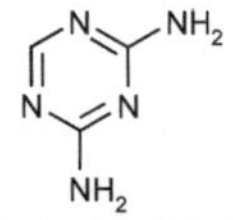
1,3,5-triazine-2,4-diamine

Figure 7.47 Quantitative structure–activity relationships were derived for a group of 1,3,5-triazine-2,4-diamines that bind to the A-site of the ribosome. The half maximal inhibitory concentration (IC_{50}) values were predicted with an r^2 of ~0.6, and the significance of electrostatic and hydrogen-bonding interactions for A-site recognition was emphasized.

half maximal inhibitory concentration (IC_{50}) and partition coefficient (log P) before experimental verification. Sekhar *et al.* applied the method to a class of selected aminoglycoside mimetics from the literature, 1,3,5-triazine-2,4-diamines, that bind to the A-site of the ribosome with known IC_{50} values (Figure 7.47).[115] One subset of the small molecules was used to develop structure-activity relationships. The resulting comparative molecular similarity index analysis and comparative molecular field analysis models were validated with a separate subset, predicting the IC_{50} values with an r^2 of 0.653 and 0.632, respectively. Several structure–activity relationships were derived, including the dependence of binding affinity on electrostatic interactions and an essential hydrogen-bonding network for recognition. Similar studies were performed with the T-box riboswitch and oxazolidinones, identifying hydrogen bonds and hydrophobic interactions important in recognition.[116] Importantly, this method depends on knowing the structure of the RNA target. As more RNA structures are solved, the method could prove to be a valuable tool in small molecule optimization.

7.3.6.4 *Experimental Screening for RNA Binding*

When the structure of the RNA is not known, various screening methods can be applied to identify small molecule ligands. Screening has three components: target validation, library selection, and assay design. Most RNA targets are validated in cell culture or *in vivo* by RNA interference, a well-known method that knocks-down an RNA of interest. Equally important is the selection and design of a small molecule library. There is a significant need to improve the quality of RNA-targeted libraries by understanding how small molecules recognize RNA. Both commercially available high throughput screening libraries and fragment-based screening libraries are enriched with protein-binding small molecules, leading to poor hit rates for RNA targets. Focused libraries have recently been utilized, but these sets are typically much smaller. The last component of screening, assay development, consists of several techniques with varying throughput, specificity, and sensitivity to identify small molecule leads.

(a) Microarrays. Microarray screening utilizes a printer to spot thousands of functionalized small molecules onto a glass slide for simultaneous

Figure 7.48 Microarray technology allows high-throughput screening of thousands of compounds simultaneously with small amounts of receptor. Typically, the small molecules are immobilized onto a microarray slide and a fluorescent- or radio-labeled macromolecule is allowed to hybridize to the array.

evaluation (Figure 7.48).[117] The technique has the potential to be high-throughput and requires a small amount of ligand and receptor. An early proof-of-concept study screened a focused-library of aminoglycosides against an A-site ribosome mimic.[118] The small molecules were immobilized with a polyethylene glycol linker, and the slide was incubated with a fluorescently tagged A-site mimic. The fluorescence intensities at each immobilization site were compared and ranked in order with the greater intensity correlating to the highest binding affinity. The microarray results did not correlate to a similar ranking determined by a mass spectroscopy assay, suggesting that immobilization affected the binding mode and subsequent read-out. This pioneering work highlighted the need to immobilize the ligands using functional groups that are not involved in RNA binding interactions.

Later work by Schneekloth and co-workers used a similar microarray screen to identify small molecules that bind to HIV-1 RNA TAR.[119] A commercially available library of 20 000 diverse small molecules was immobilized on the slide using terminal amines or alcohols. The slide was washed with a cyanine-5-labeled TAR RNA sequence as well as a fluorescently labeled microRNA hairpin and DNA sequences as controls. Three small molecules were identified from the screen with a hit rate of 0.02%. The lead small molecule (Figure 7.49) was validated *in vitro* with a K_d of 2.4 μM and half maximal effective concentration (EC_{50}) of 28 μM in cell culture. Computer docking of the small molecule identified van der Waals forces, hydrogen bonding, and hydrophobic interactions to be important for recognition and activity. Despite the low hit rate, this screening identified an RNA-targeted small molecule with cell culture activity. With improved libraries and further

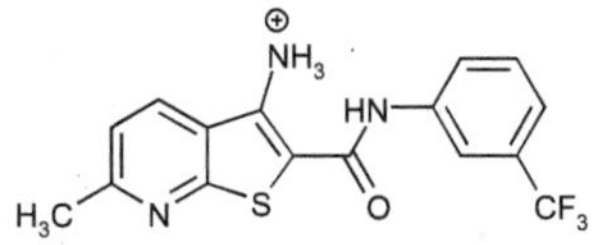

Figure 7.49 Microarray technology was used to screen 20 000 small molecules for binding to a fluorescently labeled- transactivation response element (TAR) RNA. The thienopyridine lead compound was identified and found to have a dissociation constant of 2.4 μM and a half maximal effective concentration (EC_{50}) of 28 μM in cell culture.

knowledge of immobilization techniques, microarrays could provide a high-throughput screening method for small molecule-RNA discovery.

(b) NMR. NMR spectroscopy is typically used to screen fragment-based libraries comprised of drug-like small molecules <300 Da. The fragments bind to multiple sites of the biological target with dissociation constants in the milimolar to micromolar range, which are then covalently combined into one scaffold to comprise the lead small molecule. Using various NMR techniques, multiple small molecules can be screened in a single assay, and, if hits are identified, each small molecule is screened separately. Despite its utility, NMR assays are resource-demanding, since these methods require larger quantities of ligand, receptor, and instrument time.

Campos-Olivas, Gonzalez, and co-workers used NMR spectroscopy to screen a fragment-based small molecule library of 355 small molecules against telomeric repeat-containing RNA (TERRA).[120] All small molecules contained at least one fluorine atom, which enabled ^{19}F-NMR to be used for screening. Fluorine-NMR lacks background noise from solvent and buffers and has narrow signals with very large chemical shift dispersion, making it an attractive method. Upon addition of TERRA, the spectra of 20 small molecules were altered, with a total hit rate of 5.6%. The lead small molecules (Figure 7.50) had binding affinities in the 120–1900 μM range with varied selectivity for TERRA over transfer RNA, telomeric DNA, and duplex DNA. Some small molecules identified in the screen displayed structure-specific recognition by favoring the propeller-like parallel G-quadruplex conformation in contrast to the antiparallel conformation.

In a similar NMR assay, Fesik and co-workers screened ~10 000 small molecules against a 29-mer mimic of the ribosome A-site to identify aminoglycoside mimetics.[121] In contrast to the TERRA screen, standard ^{1}H-NMR was utilized to monitor changes in the imino protons of the RNA, which is feasible with a small RNA due to limited spectral interference in the imino region. A hit rate of 3% was obtained with dissociation constants ranging from 70 μM to 3 mM. Two classes of small molecules emerged: benzimidazoles and 2-aminoquinolines with an optimized 2-aminoquinoline competing with neamine and paromomycin for binding to the A-site mimic (Figure 7.51). Additionally, data from NMR nuclear Overhauser effect (NOE)

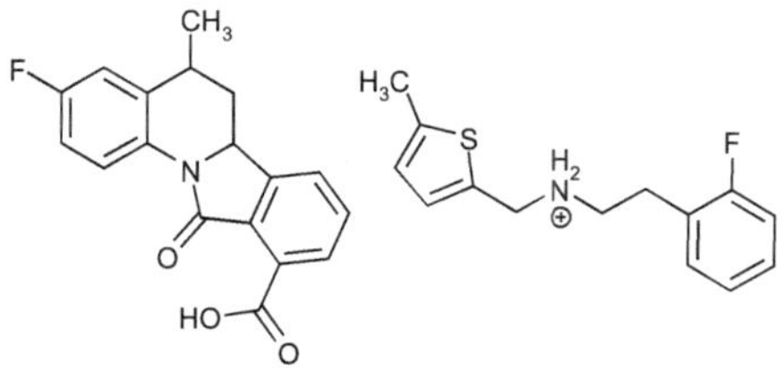

Figure 7.50 An ^{19}F-nuclear magentic resonance assay was developed to identify small molecules that bound to the telomeric repeat-containing RNA (TERRA). Each small molecule in the library contained at least one fluorine atom, which was used for detection. In the screen, 20 compounds were identified to bind to TERRA with dissociation constants in the micromolar range. Representative small molecules are shown.

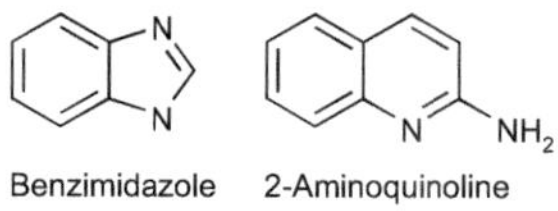

Figure 7.51 A ^{1}H-nuclear magnetic resonance screen was developed to identify compounds that bind to a 29-mer A-site ribosomal mimic. Two classes of small molecules were enriched, benzimidazole and 2-aminoquinoline, with dissociation constants in the low micromolar to millimolar range.

experiments were used to describe docking of the lead small molecule into the A-site, providing recognition information. Despite the resource demand, hit rates for NMR screens have been greater than other screening techniques.

(c) Equilibrium Dialysis. A dialysis membrane with a predetermined molecular weight cut-off separates two chambers, such that only the smaller of the two binding components, the ligand or receptor, can equilibrate across the barrier. After equilibrium is reached, the concentrations in the chambers can be measured to determine binding affinity. Smith and co-workers used equilibrium dialysis to identify small molecules that compete for the thiamine pyrophosphate riboswitch.[122] An alternative riboswitch metabolite, thiamine, was radiolabeled and allowed to diffuse across the dialysis membrane and reach thermodynamic equilibrium (Figure 7.52A). A cocktail of five ligands was added to the system, and if the thermodynamic equilibrium of thiamine was disturbed, it was inferred that a small molecule in the cocktail displaced thiamine from its binding site. A total of 1300 fragments were tested, with 20 hits giving a modest hit rate of 2%. Many of the hits were aromatic and/or mimics of nucleobases that had binding affinities in the micromolar range (Figure 7.52B). Equilibrium dialysis is a quantitative screen that requires little RNA; however, the RNA must have a native ligand for competition.

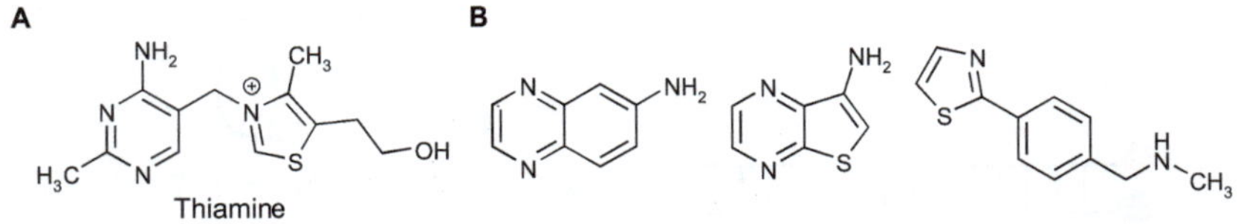

Figure 7.52 An equilibrium dialysis competition assay was developed for the thiamine pyrophosphate riboswitch. (A) The native metabolite, thiamine, was radiolabeled for detection and allowed to diffuse across the membrane, reaching a thermodynamic equilibrium. (B) 1300 ligands in cocktails of five were screened for disruption of the thiamine equilibrium. 20 hits were identified and three representative lead compounds are shown.

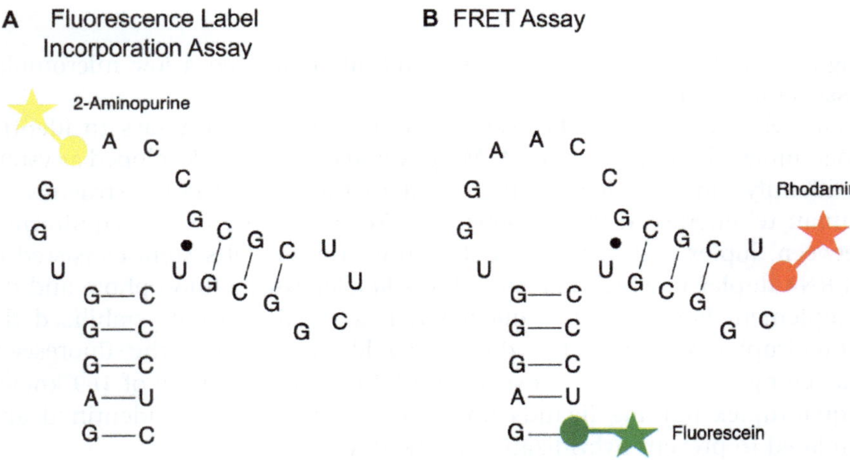

Figure 7.53 Fluorescence is a powerful tool to detect small molecule–RNA interactions. (A) 2-Aminopurine was incorporated into an RNA of interest for screening. Upon binding of a small molecule, changes in fluorescence can be observed and utilized to calculate dissociation constants. (B) Similarly, a Förster resonance energy transfer (FRET) pair can be incorporated into an RNA of interest for screening. The increase or decrease in FRET can be used to calculate dissociation constants.

(d) Optical Methods. Due to its high-sensitivity, ease of detection, and quantitative read-out, fluorescence is used often in chemical biology to detect small molecule–macromolecule interactions. Bergmeier and coworkers used a 2-aminopurine-labeled T-box riboswitch model to screen a 27-membered, focused library of oxazolidinone small molecules.[123] The fluorophore was judiciously incorporated into a bulge region of the RNA, such that its emission signal changed upon binding by a small molecule (Figure 7.53A). Three small molecule leads were identified. Subsequently, a FRET-based assay, in which a fluorescein donor was incorporated onto the 3′ end of the RNA and a rhodamine acceptor in the nearby loop, was used to determine the binding affinities of optimized ligands

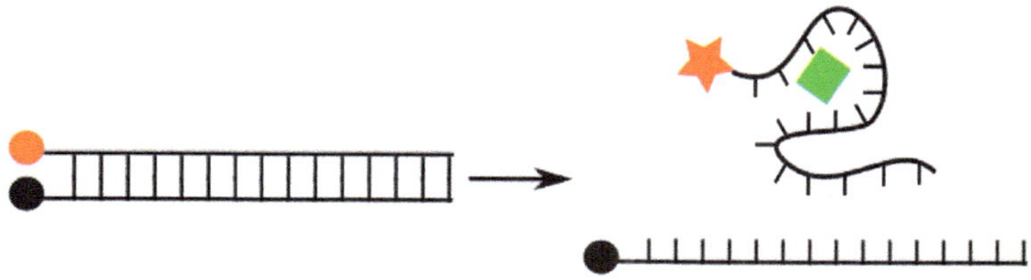

Figure 7.54 A fluorescence indicator displacement assay was utilized to identify compounds that bound to the G-quadruplex of the human telomerase RNA component. The ends of the RNA stem were labeled with a fluorophore and quencher. In an unbound state, the fluorescence is quenched. However, upon binding of a small molecule, the stem unwinds and the fluorescence is switched on. The screen identified 11 small molecules that bind to the G-quadruplex.

(Figure 7.53B). The best lead small molecule displayed a low micromolar dissociation constant.

Fluorescence can also be used in competition-based assays to identify small molecules that target RNA. Mergny and co-workers developed a system to identify small molecules that bind to the G-quadruplex structure of human telomerase RNA component.[124] To mimic the *in vivo* equilibrium between duplex and quadruplex structures, the model system consisted of an RNA duplex in which one strand was labeled with a fluorophore and the complement labeled with a quencher. If a small molecule stabilized the G-quadruplex structure, the duplex would unwind and the fluorescent quenching effect would be lost (Figure 7.54). A small library of 169 known G-quadruplex binding ligands was screened, with 11 hits identified and validated to prevent hybridization of the duplex.

Several other fluorescence-based assays have been developed for screening small molecule libraries. Hickey and Hammond developed a fluorescence polarization assay to screen small molecules that displace the metabolite *S*-adenosyl methionine (SAM) in SAM-1 riboswitches.[125] In their assay, SAM was labeled with a cyanine-5 fluorophore through a C8 linker and displacement assays were optimized for a 384-well plate assay. In another method, Lafontaine and co-workers developed a dual molecular beacon assay to screen the transcriptional effects of small molecule binding to the adenine riboswitch.[126] A hairpin structure was developed that was complementary to the mRNA expressed upstream or downstream of the riboswitch. The two ends of the hairpin were labeled with a fluorophore and quencher and, depending upon the transcriptional state, created a fluorescence readout, allowing monitoring of small molecule effects. Additionally, Hammond, Herr, and co-workers developed a microfluidic mobility shift assay against a fluorescently labeled SAM-I riboswitch.[127] The assay utilizes laser-induced fluorescence, which is highly sensitive, requires small amounts of material, and allows calculation of binding affinity. The recent development and validation of many fluorescence-based RNA assays demonstrates the utility of fluorescence for small molecule screens.

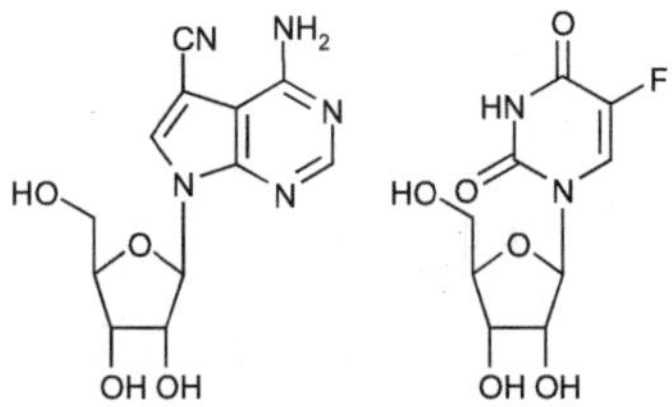

Figure 7.55 A library of ~60 000 compounds was screened against the catalytic hammerhead ribozyme by utilizing a luciferase reporter in cell culture. Two compounds, toyocamycin and 5-fluoro-uridine, were identified to alter gene expression by 365-fold and 110-fold, respectively.

(e) Catalytic Activity. Ribozymes are RNAs that have catalytic activity, which is advantageous for screening by providing a biochemical read-out. Czarnik and co-workers screened a small molecule library for inhibition of self-splicing group 1 introns in the micro-organism *Pneumocystis carinii.*[128] Upon addition of radiolabeled guanine triphosphate (GTP), the self-splicing activity is initiated and the ^{32}P label is incorporated into the RNA. Precipitation and filtration separates the GTP and RNA, allowing quantification of the activity. A library of 1400 small molecules was screened to disrupt the catalytic activity of the ribozyme with a hit rate of ~5%. Mulligan and co-workers developed a similar assay to screen a small molecule library against a catalytic hammerhead ribozyme.[129] A modified version of the ribozyme was efficiently expressed and self-cleaved in mammalian cells, and utilizing a luciferase reporter, the cell culture activity could be monitored. A library of 58 076 small molecules was screened, providing 15 hits, most of which were nucleoside analogs (Figure 7.55). The most obvious limitation to this screening approach is that only a small subset of RNA structures have catalytic activity.

7.3.6.5 Small Molecules with Selectivity for RNA Secondary Structures

Small molecules that recognize RNA secondary structures, with varying specificity, have been identified through various computational, prediction-based, and screening methods. In most instances, the motif is present in a biological target, where small molecule recognition could have therapeutic potential. However, many of the studies are limited to recognition *in vitro*, and rarely has cell culture or *in vivo* activity been observed. Examples for each secondary motif are given below. For more detailed descriptions, readers are directed to reviews by Thomas and Hergenrother[100] and Guan and Disney.[130]

(a) Stem. Base-paired stem regions of RNA structure adopt an A-form double helix with a deep major groove, shallow minor groove, and 3′-*endo*

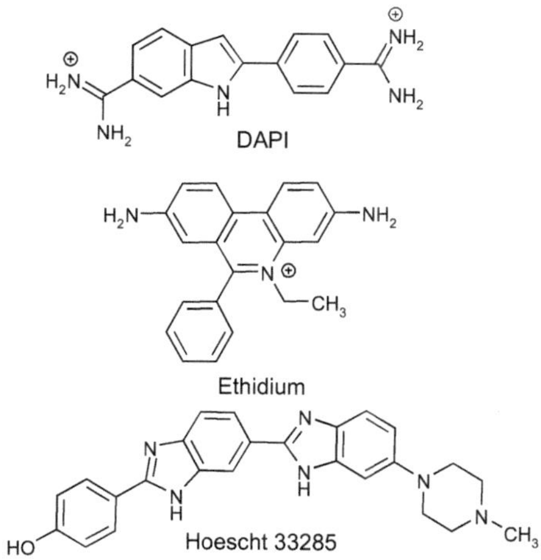

Figure 7.56 Examples of common DNA intercalators that also bind to RNA. The small molecules are typically aromatic-rich and cationic.

sugar pucker.[100] Typically, small aromatic-rich molecules recognize stems by intercalation. For instance, DNA intercalators, including Hoescht small molecules, ethidium, and 4′,6-diamidino-2-phenylindole (DAPI) bind to RNA, but, in most cases, more weakly than they bind to DNA (Figure 7.56). Similarly, small molecules developed to recognize the minor groove of DNA generally exhibit minimal binding to RNA. Therefore, the principles of DNA stem recognition do not fully extend to RNA, and new small molecules that prefer RNA over DNA are required.

An initial study by Wilson and co-workers focused on a small library of three diversified scaffolds: planar aromatic xylene, non-planar nitrogen-rich piperazine, and flexible propyl chain.[131] Thermal denaturation studies were performed with a poly(A)-poly(U) RNA and a poly(dA)-poly(dT) DNA with each of the small molecules at pH 6.3, 4.9, and 3.9. Five of the 19 small molecules that were screened exhibited selectivity for RNA over DNA with several apparent trends. The incorporation of basic sites for protonation was found to be important for preferential RNA recognition. Furthermore, non-aromatic rings exhibited decreased binding to DNA, potentially due to increased steric sensitivity. Thus, small molecules with positive charge and steric bulk could potentially recognize RNA over DNA.

Exploiting these trends, Wilson and co-workers developed a subsequent library of six diphenylfuran small molecules, which were functionalized with dicationic or tetracationic substituents.[132] An initial thermal denaturation screen analyzed the change in T_m of each small molecule in the presence of a poly(A)-poly(U) RNA duplex. One small molecule, furimidazoline, increased the T_m of the duplex by 14.4 °C, almost twice that of the other small molecules (Figure 7.57). A battery of spectroscopic and physical assays indicated intercalation into the RNA duplex and a strong sensitivity to salt concentration. This binding mode was supported by molecular modeling, which

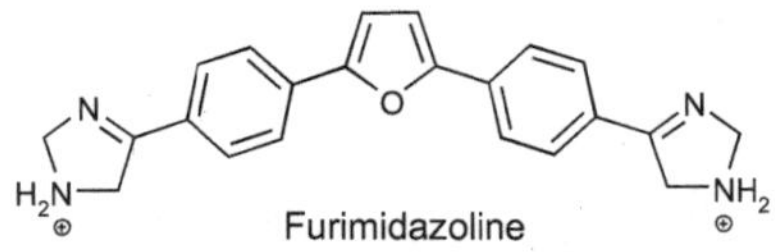

Figure 7.57 A small library of diphenylfuran derivatives was screened by thermal denaturation to identify RNA ligands. The lead compound, furimidazoline, is a planar dicationic small molecule that is proposed to bind in the major groove and hydrogen bond with the RNA backbone.

predicted both cationic substituents of the furimidazoline to be in the major groove. The significant increase in thermal stability was attributed to the planarity of the imidazoline groups, which were able to slide deeper into the major groove, with hydrogen bonds between the anionic phosphate oxygen of the backbone and –NH groups of the imidazoline. Thus, the recognition of an RNA stem was achieved through intercalation and backbone recognition, although the precise reason for the specificity over DNA is not presently known.

(b) Internal Loops. Internal loops occur in stem regions when nucleotides are excluded from base pairing on both strands.[100] Three factors affect the stability of internal loops: symmetry, size, and sequence. Generally, asymmetric loops are unstructured and flexible in solution, which also serves to destabilize surrounding stem structures. In contrast, symmetrical loops can be stabilizing or destabilizing depending upon their size and sequence. For instance, there is a significant entropic penalty for formation of a 2×2 or 3×3 loop; however, enthalpic gains from non-canonical base pairs can compensate. Furthermore, a 2×2 loop with a GA opposite an AG stabilizes nearby stem structures, while a 2×2 loop with CU opposite UC is destabilizing. Due to the sequence dependence and unique structure, there is potential to specifically recognize internal loops with small molecules.

One of the best explored RNA targets is the ribosome—an essential ribozyme in prokaryotes and eukaryotes that is responsible for the translation of messenger RNA (mRNA) into protein. The prokaryotic ribosome is $\sim$65% RNA and consists of a large subunit (50S), which catalyzes peptide bond formation, and a small subunit (30S), which recognizes mRNA and proofreads incoming tRNAs. Several FDA-approved antibiotics target the active site of the prokaryotic ribosome with varying selectivity over the eukaryotic ribosomes. The best known are the aminoglycosides, which bind to the A-site of the prokaryotic large subunit. Generally, aminoglycosides alter the recognition of cognate tRNAs, leading to mis-incorporation of amino acids and the formation of mutated proteins. Despite the potency of aminoglycosides, these ligands exhibit off-target effects that lead to oto- and nephrotoxicity, and recently, several resistance mechanisms have emerged. Thus, there is a need to improve aminoglycoside binding affinity and selectivity.

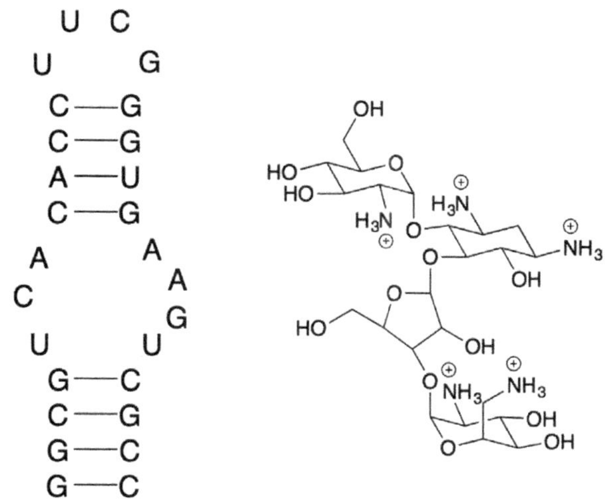

Figure 7.58 Paromomycin is an aminoglycoside antibiotic that targets ribosomal RNA. A nuclear magnetic resonance structure of paromomycin bound to a 27-mer A-site ribosomal mimic identified a 3×4 internal loop and the two closing base pairs as being critical for recognition.

Puglisi and co-workers solved an NMR structure of the aminoglycoside paromomycin binding to a 27-mer A-site ribosomal mimic, identifying structural motifs involved in aminoglycoside recognition (Figure 7.58).[133] The RNA has a 3×4 internal loop flanked by an upper and lower stem as well as a terminal tetraloop. Utilizing NOE data, paromomycin was found to bind to the major groove of the A-site mimic within the internal loop structure. All seven nucleotides of the internal loop are required for aminoglycoside recognition, including two additional closing base pairs in the lower stem. Upon binding, the three adenine residues undergo rearrangement to form a non-canonical A–A base pair and a single bulged adenine, potentially contributing to the observed antibiotic activity.

Disney and co-workers have extensively studied the recognition of internal loops by aminoglycosides and benzimidazole small molecules through various systematic approaches.[134] Recently, the group designed a computational program, INFORNA, which incorporates the recognition information from their 2DCS microarray and StARTs program to identify small molecules that specifically recognize motifs in RNA targets. The database included 792 RNA motif–small molecule interactions and 1048 precursor miRNAs from the miRBase database. Small molecule binders were identified for 22 miRNA precursors that had precedence in disease, including three small molecule–RNA complexes with fitness scores >75. The structural motif recognized by each lead small molecule was an internal loop along with its closing base pairs, which were involved in pre-miRNA recognition by Drosha. A small

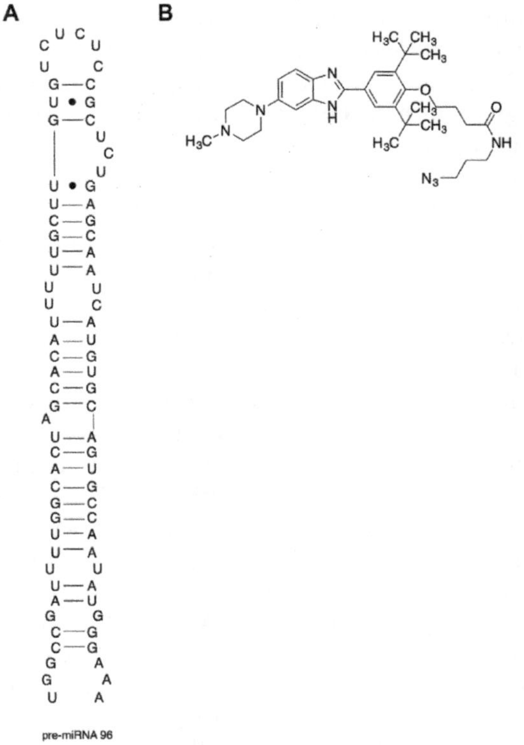

Figure 7.59 (A) The predicted secondary structure of pri-miRNA96. The 1×1 internal loop is recognized by Drosha, leading to processing of the miRNA. (B) A computational program, INFORNA, identified a lead compound that specifically recognizes the 1×1 internal loop. The benzimidazole compound had *in vitro* and cell culture activity with specificity against other miRNAs.

molecule with a fitness score of 100 (Figure 7.59) was observed to target the pre-miRNA-96, which is upregulated in cancer and represses the translation of Forkhead box protein O1 (FOXO1). Enzymatic mapping studies using RNase III and T1 confirmed the secondary structure of pri-miRNA-96 and indicated binding to a 1×1 internal loop. In the presence of the lead small molecule, Drosha cleavage of pri-miRNA was inhibited, ultimately leading to increased FOXO1 protein expression. To investigate selectivity, the expression of 149 disease-associated miRNAs was measured and a selective 19-fold decrease in miRNA-96 levels was observed. The small molecule recognition of a 1×1 internal loop within the pri-miRNA exhibited promising *in vitro* and cell culture activity.

(c) Bulge. Bulge motifs occur when nucleotides are excluded from base pairing on a single strand of a duplex.[100,135] The size of bulge regions can vary from one to several nucleotides with increasing size leading to further destabilization of the surrounding stem structure. Bulge structures are thought to be inherently flexible due to the lack of base-pairing; however, the nucleobases can stack with the stem or with themselves to form structured regions. Bulge nucleobases can be exposed to solvent, interact with hydrophobic regions of adjacent helices, or act as a "flap" residue, covering a ligand-binding site.

A potential antibiotic target, the antiterminator T-box riboswitch, is found in gram-positive bacteria and contains a seven-nucleotide bulge that is recognized by tRNA. The antiterminator form of the riboswitch allows the expression of essential genes downfield, including aminoacyl-tRNA synthetases and other genes involved in amino acid biosynthesis and transport. In the absence of tRNA, the structure folds into an alternate terminator hairpin structure, halting the transcription of the downfield genes. An NMR structure of a 29-mer anti-terminator model RNA (AM1A) revealed both flexible and structured regions within the bulge.[100,136]

Bergmeier, Hines, and co-workers synthesized a library of 27 3,4,5-trisubsituted oxazolidinones to screen against the T-Box riboswitch antiterminator.[123] The AM1A model and a C11U mutant (AM1A-C11U) were used in the initial fluorescence-based screen to identify small molecules that specifically recognized the AM1A bulge (Figure 7.60). Titration experiments were performed where the ninth residue in AM1A and AM1A-C11U was labeled with 2-aminopurine and the change in fluorescence was monitored upon addition of ligand. Although caution must be taken when comparing the fluorescence of two different model systems, five small molecules were identified to have differential binding. A FRET assay, with a fluorescein donor and rhodamine acceptor incorporated into the RNA structures, was developed to validate lead small molecules and determine dissociation constants.

Further studies uncovered two additional ligands (Figure 7.61).[137,138] **rac-1** demonstrated eight-fold selectivity for AM1A over AM1A-C11U with a K_d of

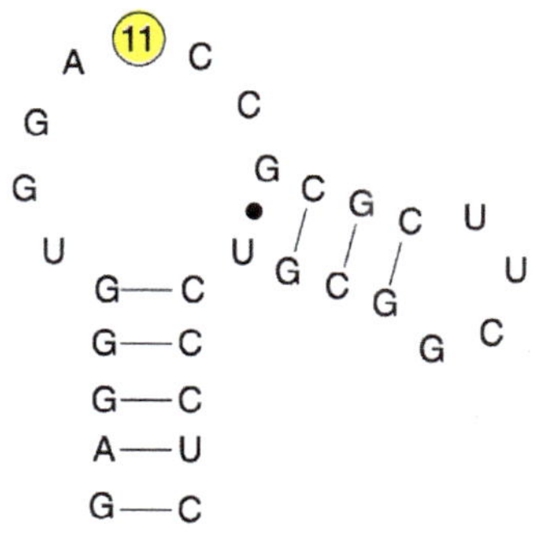

Figure 7.60 The T-box riboswitch is found in the 5′ untranslated region of bacteria and codes for essential genes downfield. Screens against the T-box riboswitch focused on the anti-terminator structure with a one nucleotide mutant serving as a control. Nucleotide 11 (yellow circle) is cytosine (C) in the native structure and uracil (U) in the mutant.

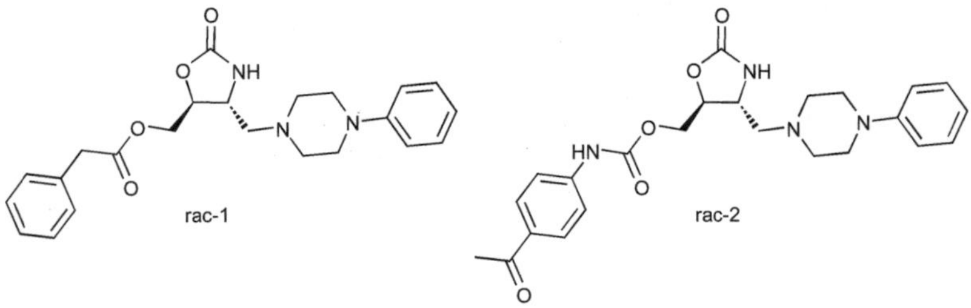

Figure 7.61 Oxazolidinones, **rac-1** and **rac-2**, bind to the bulge region of the antiterminator T-box riboswitch with dissociation constants of 13 μM and 0.9 μM, respectively. The oxazolidinones were proposed to have two binding modes as **rac-1** leads to inhibition of translation and **rac-2** leads to promotion of translation *in vitro*.

13 μM and 100 μM, respectively. The other small molecule, **rac-2**, had a very similar structure to **rac-1**, but did not display any selectivity for the AM1A model, with a K_d of 0.9 μM. Enzymatic cleavage assays, utilizing RNase A and RNase T1 showed different cleavage patterns for **rac-1** and **rac-2**, corresponding to altered binding modes. Interestingly, an *in vitro* antiterminator transcription system showed that **rac-1** inhibited the tRNA dependent antitermination, while **rac-2** promoted antitermination in the presence and absence of tRNA. The structures of **rac-1** and **rac-2** were docked into the AM1A model using the Glide module of First Discovery 2.7. The most stable conformation for **rac-1** formed interactions with the major groove of the lower stem, as well as the 5′ end of the bulge. In contrast, **rac-2** bound exclusively to the bulge region, supporting their differential binding modes and altered *in vitro* activity.

(d) Hairpin Loop. Hairpin or apical loops are regions of unpaired nucleotides that are formed when RNA folds upon itself.[100] Despite being the second most prevalent RNA motif, it is unknown how many different hairpin loops exist with varied size and sequence. The most stable apical loops are comprised of six or seven nucleotides. However, it is well known that the stability of apical loops is sequence-dependent, with the GNRA tetraloop being an exceptionally stable small hairpin.

The HIV-1 genome is comprised of several RNA elements, including the trans-activating response element (TAR) (Figure 7.62A). *In vivo*, Tat, an HIV-1 protein, binds to the bulged region of TAR, while cyclin T1 binds to the hairpin hexaloop. The Tat-TAR-cyclin T1 complex dramatically increases the transcription of the HIV-1 genome; therefore, disruption of the complex is a potential anti-HIV-1 strategy. Recognition of TAR by Tat is mediated through an arginine-rich arm, which binds to the bulge region of the structure. Upon binding, TAR undergoes a conformation change and further interactions are mediated in the major groove. Several small molecules have shown to disrupt the TAR-tat interaction *in vitro,* including neomycin (Figure 7.62B).[100] NMR structures determined that neomycin binds to the

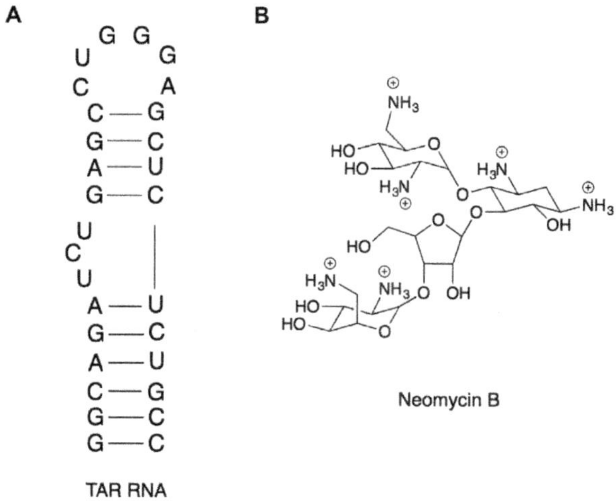

Figure 7.62 (A) Secondary structure of a transactivation response element (TAR) RNA construct used for *in vitro* studies. TAR is a promising target in the HIV-1 RNA genome and has a characteristic stem–loop structure with a three nucleotide bulge. (B) Neomycin B is an aminoglycoside that disrupts the interaction between TAR and its protein binding partner, Tat, with a half maximal inhibitory concentration (IC_{50}) of 0.92 μM *in vitro*. However, the small molecule lacks specificity due to its polycationic structure.

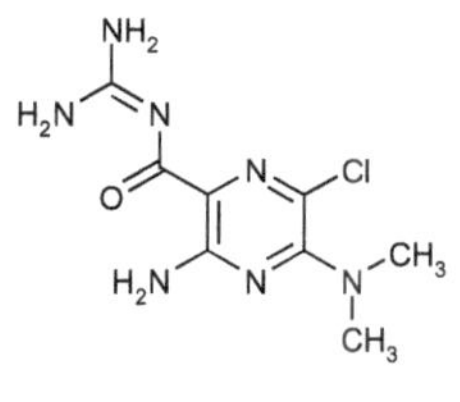

Figure 7.63 A virtual screen against an ensemble of transactivation response element (TAR) structures identified 5-(*N*,*N*′)-dimethylamiloride as a lead compound for binding the apical loop of TAR. The small molecule had moderate affinity with a dissociation constant of 122 μM.

minor groove of TAR, and, through a non-competitive mechanism, decreases the lifetime of the complex. Thus, small molecules that bind and alter the conformation of TAR can modify structural elements essential for Tat recognition.

Two methods for discovering small molecules binders of TAR have already been described above. As discussed in Section 7.3.6.2, Al-Hashimi and co-workers used a computational screen followed by fluorescence assays to identify commercially-available ligands.[106] One of the hits, 5-(*N*,*N*′)-dimethylamiloride (Figure 7.63), was found to recognize the apical hexaloop. An alternative microarray screening approach of Schneekloth and co-workers uncovered a chemically distinct thienopyridine ligand (Figure 7.49) with the ability to selectively bind the TAR hairpin loop within the highly structured 5′ untranslated region of HIV-1.[119]

7.3.7 Nucleic Acid Targeting with Inorganic Coordination Complexes

The rich field of nucleic acid recognition by inorganic coordination complexes is beyond the scope of this chapter, and the interested reader is directed to recent review articles.[73,139,140] Many types of transition metal complexes, especially those containing platinum centers, are known to form coordination bonds with cellular DNA and exhibit potent anticancer activity. A key mechanistic step appears to be irreversible reaction of the metal center with the N7 atoms of neighboring guanine or adenine bases.[141] Ongoing research is exploring a multitude of new molecular designs, especially multinuclear complexes, and further clinical success is anticipated.[142]

Supramolecular chemists have also actively pursued small inorganic complexes as mimics of nuclease enzymes. A wide range of zinc, copper, and lanthanide complexes have been developed and studied for ability to cleave the phosphate backbone of nucleic acids.[143] Some of these biomimetic catalysts are described in Chapters 1 and 6 of this book.

7.4 Future Directions and Challenges

While much progress has been made in the field of oligonucleotide recognition, the design of high affinity and in particular high specificity ligands remains a daunting challenge. As with all biological recognition systems, the ultimate goal is selective target engagement *in vivo*. A major complication is the sheer volume of competing genomic DNA and ribosomal RNA in a cell and the recognition problem becomes even harder if cell penetration is a required ligand performance criterion. Enhanced binding selectivity is gained by identifying and optimizing multivalent interactions, using both structure-guided and dynamic combinatorial methods. In addition, recent structural and physical studies of DNA and RNA have uncovered novel elements of nucleic acid structure, including unique backbone conformations, non-canonical base pairs, and low-population conformations that allow for more specific targeting. A current technical hindrance is the protein-centric nature of high-throughput screening libraries and binding assays. Recent advances in structural prediction and analysis tools, as well as novel screening techniques, will soon lead to new classes of nucleic acid binding small molecules. Additional opportunities are anticipated as new biological roles of nucleic acids are elucidated, particularly the importance of non-coding RNA in transcriptional regulation, development, and disease. The recent and continued strides in the field of small molecule–nucleic acid recognition have the potential to significantly impact our knowledge of fundamental biology, diagnostics, and therapy.

References

1. W. Saenger, *Principles of Nucleic Acid Structure*, Springer-Verlag, New York, 1984.
2. E. N. Nikolova, E. Kim, A. A. Wise, P. J. O'Brien, I. Andricioaei and H. M. Al-Hashimi, *Nature*, 2011, **470**, 498.
3. T. Hermann and E. Westhof, *Chem. Biol.*, 1999, **6**, R335.
4. J. L. Battiste, H. Mao, N. S. Rao, R. Tan, D. R. Muhandiram, L. E. Kay, A. P. Frankel and J. Williamson, *Science*, 1996, **273**, 1547.
5. E. T. Kool, *Annu. Rev. Biophys. Biomol. Struct.*, 2001, **30**, 1.
6. J. Kuriyan, B. Konforti and D. Wemmer, *The Molecules of Life: Physical and Chemical Principles, Garland Science*, Taylor & Francis Group, New York, 2012.
7. N. V. Hud, *Nucleic Acid-Metal Ion Interactions*, RSC Publishing, Cambridge, UK, 2009.
8. S. E. Butcher and A. M. Pyle, *Acc. Chem. Res.*, 2011, **44**, 1302.
9. A. L. Feig and O. C. Uhlenbeck, *The Role of Metal Ions in RNA Biochemistry*, Cold Spring Harbor Laboratory Press, Cold Spring Harbor, New York, 1999.
10. X. Shui, L. McFail-Isom, G. G. Hu and L. D. Williams, *Biochemistry*, 1998, **37**, 8341.
11. J. M. Berg, J. L. Tymoczko and L. Stryer, *Biochemistry*, 6th edn, Freeman, New York, 2006.
12. R. Rohs, X. Jin, S. M. West, R. Joshi, B. Honig and R. S. Mann, *Annu. Rev. Biochem.*, 2010, **79**, 233.
13. A. V. Vologodskii and N. R. Cozzarelli, *Annu. Rev. Biophys. Biomol. Struct.*, 1994, **23**, 609.
14. I. V. Kovtun and C. T. McMurray, *Cell Res.*, 2008, **18**, 198.
15. G. Varani and W. H. McClain, *EMBO Rep.*, 2000, **1**, 18.
16. H. Grosjean, *DNA and RNA Modification Enzymes: Structure, Mechanism, Function and Evolution*, Landes Bioscience, Austin, TX, 2009.
17. M. Helm, *Nucleic Acids Res.*, 2006, **34**, 721.
18. R. Thapar, A. P. Denmon and E. P. Nikonowicz, *Wiley Interdiscip. Rev.: RNA*, 2014, **5**, 49.
19. B. P. Belotserkovskii, S. M. Mirkin and P. C. Hanawalt, *Chem. Rev.*, 2013, **113**, 8620.
20. R. M. Mitton-Fry, S. J. DeGregorio, J. Wang, T. A. Steitz and J. A. Steitz, *Science*, 2010, **330**, 1244.
21. P. H. von Hippel and O. G. Berg, *J. Biol. Chem.*, 1989, **264**, 675.
22. A. C. Cheng, W. W. Chen, C. N. Fuhrmann and A. D. Frankel, *J. Mol. Biol.*, 2003, **327**, 781.
23. R. Wintjens, C. Biot, M. Rooman and J. Liévin, *J. Phys. Chem. A*, 2003, **107**, 6249.
24. M. Rooman, J. Lievin, E. Buisine and R. Wintjens, *J. Mol. Biol.*, 2002, **319**, 67.
25. P. J. Mitchell and R. Tjian, *Science*, 1989, **245**, 371.

26. A. Jolma, J. Yan, T. Whitington, J. Toivonen, K. R. Nitta, P. Rastas, E. Morgunova, M. Enge, M. Taipale, G. Wei, K. Palin, J. M. Vaquerizas, R. Vincentelli, N. M. Luscombe, T. R. Hughes, P. Lemaire, E. Ukkonen, T. Kivioja and J. Taipale, *Cell*, 2013, **152**, 327.
27. C. Helsen, S. Kerkhofs, L. Clinckemalie, L. Spans, M. Laurent, S. Boonen, D. Vanderschueren and F. Claessens, *Mol. Cell. Endocrinol.*, 2012, **348**, 411.
28. C. D. Cooper, J. A. Newman and O. Gileadi, *Biochem. Soc. Trans.*, 2014, **42**, 130.
29. P. C. Hollenhorst, L. P. McIntosh and B. J. Graves, *Annu. Rev. Biochem.*, 2011, **80**, 437.
30. D. Panne, T. Maniatis and S. C. Harrison, *Cell*, 2007, **129**, 1111.
31. T. H. Dickey, S. E. Altschuler and D. S. Wuttke, *Structure*, 2013, **21**, 1074.
32. B. M. Lunde, C. Moore and G. Varani, *Nat. Rev. Mol. Cell Biol.*, 2007, **8**, 479.
33. C. Maris, C. Dominguez and F. H. Allain, *FEBS J.*, 2005, **272**, 2118.
34. V. Arcus, *Curr. Opin. Struct. Biol.*, 2002, **12**, 794.
35. A. Lingel and M. Sattler, *Curr. Opin. Struct. Biol.*, 2005, **15**, 107.
36. X. Wang, J. McLachlan, P. D. Zamore and T. M. Hall, *Cell*, 2002, **110**, 501.
37. P. Selenko, G. Gregorovic, R. Sprangers, G. Stier, Z. Rhani, A. Krämer and M. Sattler, *Mol. Cell*, 2003, **11**, 965.
38. J. Zhang, M. W. Lau and A. R. Ferre-D'Amare, *Biochemistry*, 2010, **49**, 9123.
39. B. J. Tucker and R. R. Breaker, *Curr. Opin. Struct. Biol.*, 2005, **15**, 342.
40. A. R. Ferre-D'Amare, *Q. Rev. Biophys.*, 2010, **43**, 423.
41. B. D. Blakeley, S. M. DePorter, U. Mohan, R. Burai, B. S. Tolbert and B. R. McNaughton, *Tetrahedron*, 2012, **68**, 8837.
42. K. M. Weeks, *Curr. Opin. Struct. Biol.*, 2010, **20**, 295.
43. F. Ding, C. A. Lavender, K. M. Weeks and N. V. Dokholyan, *Nat. Methods*, 2012, **9**, 603.
44. G. M. Rice, C. W. Leonard and K. M. Weeks, *RNA*, 2014, **20**, 846.
45. M. Duca, P. Vekhoff, K. Oussedik, L. Halby and P. B. Arimondo, *Nucleic Acids Res.*, 2008, **36**, 5123.
46. K. M. Vasquez, L. Narayanan and P. M. Glazer, *Science*, 2000, **290**, 530.
47. R. Kole, A. R. Krainer and S. Altman, *Nat. Rev. Drug. Discovery*, 2012, **11**, 125.
48. J. Li, S. Tan, R. Kooger, C. Zhang and Y. Zhang, *Chem. Soc. Rev.*, 2014, **43**, 506.
49. J. Weiler, J. Hunziker and J. Hall, *Gene Ther.*, 2006, **13**, 496.
50. L. F. R. Gebert, M. A. E. Rebhan, S. E. M. Crivelli, R. Denzler, M. Stoffel and J. Hall, *Nucleic Acids Res.*, 2014, **42**, 609.
51. H. L. A. Janssen, H. W. Reesink, E. J. Lawitz, S. Zeuzem, M. Rodriguez-Torres, K. Patel, A. J. van der Meer, A. K. Patick, A. Chen, Y. Zhou, R. Persson, B. D. King, S. Kauppinen, A. A. Levin and M. R. Hodges, *N. Engl. J. Med.*, 2013, **368**, 1685.

52. J. I. Cutler, E. Auyeung and C. A. Mirkin, *J. Am. Chem. Soc.*, 2012, **134**, 1376.
53. N. L. Rosi and C. A. Mirkin, *Chem. Rev.*, 2005, **105**, 1547.
54. T. Vo-Dinh, A. M. Fales, G. D. Griffin, C. G. Khoury, Y. Liu, H. Ngo, S. J. Norton, J. K. Register, H. N. Wang and H. Yuan, *Nanoscale*, 2013, **5**, 10127.
55. H. T. Ngo, H. N. Wang, A. M. Fales, B. P. Nicholson, C. W. Woods and T. Vo-Dinh, *Analyst*, 2014, **139**, 5655.
56. L. L. Cline and M. L. Waters, *Org. Biomol. Chem.*, 2009, 7, 4622.
57. M. Krishnamurthy, B. D. Gooch and P. A. Beal, *Org. Biomol. Chem.*, 2006, **4**, 639.
58. B. R. McNaughton and B. L. Miller, *Org. Lett.*, 2006, **8**, 1803.
59. L. O. Ofori, T. A. Hilimire, R. P. Bennett, N. W. Brown, H. C. Smith and B. L. Miller, *J. Med. Chem.*, 2014, **57**, 723.
60. W. Zhang, D. I. Bryson, J. B. Crumpton, J. Wynn and W. L. Santos, *Org. Biomol. Chem.*, 2013, **11**, 6263.
61. S. Chirayil, R. Chirayil and K. J. Luebke, *Nucleic Acids Res.*, 2009, **37**, 5486.
62. J. P. Diaz, R. Chirayil, S. Chirayil, M. Tom, K. J. Head and K. J. Luebke, *RNA*, 2014, **20**, 528.
63. R. Palchaudhuri and P. J. Hergenrother, *Curr. Opin. Biotechnol.*, 2007, **18**, 497.
64. S. Neidle, *Nat. Prod. Rep.*, 2001, **18**, 291.
65. L. S. Lerman, *J. Mol. Biol.*, 1961, **3**, 18.
66. H. Ihmels and D. Otto, in *Topics in Current Chemistry, Supramolecular Dye Chemistry*, ed. F. Würthner, Springer, Berlin, 2005.
67. P. Laugaa, J. Markovits, A. Delbarre, J. B. Le Pecq and B. P. Roques, *Biochemistry*, 1985, **24**, 5567.
68. G. Ughetto, A. H. Wang, G. J. Quigley, G. A. van der Marel, J. H. van Boom and A. Rich, *Nucleic Acids Res.*, 1985, **13**, 2305.
69. D. Mendel and P. B. Dervan, *Proc. Natl. Acad. Sci. U. S. A.*, 1987, **84**, 910.
70. E. J. Fechter, B. Olenyuk and P. B. Dervan, *Angew. Chem., Int. Ed.*, 2004, **43**, 3591.
71. A. Rhoden Smith and B. L. Iverson, *J. Am. Chem. Soc.*, 2013, **135**, 12783.
72. A. Rhoden Smith, B. A. Ikkanda, G. G. Holman and B. L. Iverson, *Biochemistry*, 2012, **51**, 4445.
73. H.-K. Liu and P. J. Sadler, *Acc. Chem. Res.*, 2011, **44**, 349.
74. J. K. Barton, E. D. Olmon and P. A. Sontz, *Coord. Chem. Rev.*, 2011, **255**, 619.
75. P. B. Dervan, *Science*, 1986, **232**, 464.
76. M. W. Van Dyke, R. P. Hertzberg and P. B. Dervan, *Proc. Natl. Acad. Sci. U. S. A.*, 1982, **79**, 5470.
77. P. B. Dervan, R. M. Doss and M. A. Marques, *Curr. Med. Chem.: Anti-Cancer Agents*, 2005, **5**, 373.
78. M. S. Blackledge and C. Melander, *Bioorg. Med. Chem.*, 2013, **21**, 6101.

79. J. L. Meier, A. S. Yu, I. Korf, D. J. Segal and P. B. Dervan, *J. Am. Chem. Soc.*, 2012, **134**, 17814.
80. D. M. Chenoweth and P. B. Dervan, *J. Am. Chem. Soc.*, 2010, **132**, 14521.
81. F. Yang, N. G. Nickols, B. C. Li, J. O. Szablowski, S. R. Hamilton, J. L. Meier, C.-M. Wang and P. B. Dervan, *J. Med. Chem.*, 2013, **56**, 7449.
82. J. A. Raskatov, A. E. Hargrove, A. Y. So and P. B. Dervan, *J. Am. Chem. Soc.*, 2012, **134**, 7995.
83. A. E. Hargrove, J. A. Raskatov, J. L. Meier, D. C. Montgomery and P. B. Dervan, *J. Med. Chem.*, 2012, **55**, 5425.
84. F. Yang, N. G. Nickols, B. C. Li, G. K. Marinov, J. W. Said and P. B. Dervan, *Proc. Natl. Acad. Sci. U. S. A.*, 2013, **110**, 1863.
85. T. Bando and H. Sugiyama, *Acc. Chem. Res.*, 2006, **39**, 935.
86. K. Maeshima, S. Janssen and U. K. Laemmli, *EMBO J.*, 2001, **20**, 3218.
87. T. F. Martínez, J. W. Phillips, K. K. Karanja, P. Polaczek, C.-M. Wang, B. C. Li, J. L. Campbell and P. B. Dervan, *Nucleic Acids Res.*, 2014, **42**, 11546.
88. P. L. Hamilton and D. P. Arya, *Nat. Prod. Rep.*, 2012, **29**, 134.
89. D. P. Arya, R. L. Coffee Jr. and L. Xue, *Bioorg. Med. Chem. Lett.*, 2004, **14**, 4643.
90. S. Kumar, L. Xue and D. P. Arya, *J. Am. Chem. Soc.*, 2011, **133**, 7361.
91. S. Kumar, M. Newby Spano and D. P. Arya, *Biopolymers*, 2014, **101**, 720.
92. D. R. Boer, A. Canals and M. Coll, *Dalton Trans.*, 2009, 399.
93. S. Balasubramanian, L. H. Hurley and S. Neidle, *Nat. Rev. Drug Discovery*, 2011, **10**, 261.
94. D. Monchaud and M. P. Teulade-Fichou, *Org. Biomol. Chem.*, 2008, **6**, 627.
95. D.-F. Shi, R. T. Wheelhouse, D. Sun and L. H. Hurley, *J. Med. Chem.*, 2001, **44**, 4509.
96. D. P. Goncalves, R. Rodriguez, S. Balasubramanian and J. K. Sanders, *Chem Commun.*, 2006, 4685.
97. H. A. Day, P. Pavlou and Z. A. Waller, *Bioorg. Med. Chem.*, 2014, **22**, 4407.
98. S. Kendrick, H.-J. Kang, M. P. Alam, M. M. Madathil, P. Agrawal, V. Gokhale, D. Yang, S. M. Hecht and L. H. Hurley, *J. Am. Chem. Soc.*, 2014, **136**, 4161.
99. Y. Xie, V. K. Tam and Y. Tor, *The Chemical Biology of Nucleic Acids*, John Wiley & Sons, Ltd, 2010.
100. J. R. Thomas and P. J. Hergenrother, *Chem. Rev.*, 2008, **108**, 1171.
101. R. Schroeder, A. Barta and K. Semrad, *Nat. Rev. Mol. Cell Biol.*, 2004, **5**, 908.
102. B. L. Golden, *Methods Mol. Biol.*, 2007, **363**, 239.
103. B. Furtig, C. Richter, J. Wohnert and H. Schwalbe, *ChemBioChem*, 2003, **4**, 936.
104. H. M. Al-Hashimi, *J. Magn. Reson.*, 2013, **237**, 191.
105. M. D. Disney, I. Yildirim and J. L. Childs-Disney, *Org. Biomol. Chem.*, 2014, **12**, 1029.

106. A. C. Stelzer, A. T. Frank, J. D. Kratz, M. D. Swanson, M. J. Gonzalez-Hernandez, J. Marta, J. Lee, I. Andricioaei, D. M. Markovitz and H. M. Al-Hashimi, *Nat. Chem. Biol.*, 2011, **7**, 553.
107. I. G. Pinto, C. Guilbert, N. B. Ulyanov, J. Stearns and T. L. James, *J. Med. Chem.*, 2008, **51**, 7205.
108. J. F. Arambula, S. R. Ramisetty, A. M. Baranger and S. C. Zimmerman, *Proc. Natl. Acad. Sci. U. S. A.*, 2009, **106**, 16068.
109. J. L. Childs-Disney, M. Wu, A. Pushechnikov, O. Aminova and M. D. Disney, *ACS Chem. Biol.*, 2007, **2**, 745.
110. D. J. Paul, S. J. Seedhouse and M. D. Disney, *Nucleic Acids Res.*, 2009, **37**, 5894.
111. S. P. Velagapudi, S. J. Seedhouse and M. D. Disney, *Angew. Chem. Int. Ed.*, 2010, **49**, 3816.
112. S. P. Velagapudi, S. J. Seedhouse, J. French and M. D. Disney, *J. Am. Chem. Soc.*, 2011, **133**, 10111.
113. S. J. Seedhouse, L. P. Labuda and M. D. Disney, *Bioorg. Med. Chem. Lett.*, 2010, **20**, 1338.
114. J. L. Childs-Disney, R. Parkesh, M. Nakamori, C. A. Thornton and M. D. Disney, *ACS Chem. Biol.*, 2012, **7**, 1984.
115. Y. N. Sekhar, M. R. S. Nayana, N. Sivakumari, M. Ravikumar and S. K. Mahmood, *J. Mol. Graphics Modell.*, 2008, **26**, 1338.
116. I. Maciagiewicz, S. Zhou, S. C. Bergmeier and J. V. Hines, *Bioorg. Med. Chem. Lett.*, 2011, **21**, 4524.
117. J. E. Bradner, O. M. McPherson and A. N. Koehler, *Nat Protoc.*, 2006, **1**, 2344.
118. M. D. Disney and P. H. Seeberger, *Chem. – Eur. J.*, 2004, **10**, 3308.
119. J. Sztuba-Solinska, S. R. Shenoy, P. Gareiss, L. R. Krumpe, S. F. Le Grice, B. R. O'Keefe and J. S. Schneekloth Jr., *J. Am. Chem. Soc.*, 2014, **136**, 8402.
120. M. Garavis, B. Lopez-Mendez, A. Somoza, J. Oyarzabal, C. Dalvit, A. Villasante, R. Campos-Olivas and C. Gonzalez, *ACS Chem. Biol.*, 2014, **9**, 1559.
121. L. Yu, T. K. Oost, J. M. Schkeryantz, J. Yang, D. Janowick and S. W. Fesik, *J. Am. Chem. Soc.*, 2003, **125**, 4444.
122. E. Cressina, L. Chen, C. Abell, F. J. Leeper and A. G. Smith, *Chem. Sci.*, 2011, **2**, 157.
123. J. Means, S. Katz, A. Nayek, R. Anupam, J. V. Hines and S. C. Bergmeier, *Bioorg. Med. Chem. Lett.*, 2006, **16**, 3600.
124. L. Lacroix, A. Seosse and J. L. Mergny, *Nucleic Acids Res.*, 2011, **39**, e21.
125. S. F. Hickey and M. C. Hammond, *Chem. Biol.*, 2014, **21**, 345.
126. R. Chinnappan, A. Dube, J. F. Lemay and D. A. Lafontaine, *Nucleic Acids Res.*, 2013, **41**, e106.
127. K. Karns, J. M. Vogan, Q. Qin, S. F. Hickey, S. C. Wilson, M. C. Hammond and A. E. Herr, *J. Am. Chem. Soc.*, 2013, **135**, 3136.
128. H. Y. Mei, M. Cui, S. T. Sutton, H. N. Truong, F. Z. Chung and A. H. Czarnik, *Nucleic Acids Res.*, 1996, **24**, 5051.

129. L. Yen, M. Magnier, R. Weissleder, B. R. Stockwell and R. C. Mulligan, *RNA*, 2006, **12**, 797.
130. L. Guan and M. D. Disney, *ACS Chem. Biol.*, 2012, 7, 73.
131. A. W. McConnaughie, J. Spychala, M. Zhao, D. Boykin and W. D. Wilson, *J. Med. Chem.*, 1994, **37**, 1063.
132. M. Zhao, L. Ratmeyer, R. G. Peloquin, S. Yao, A. Kumar, J. Spychala, D. W. Boykin and W. D. Wilson, *Bioorg. Med. Chem.*, 1995, **3**, 785.
133. D. Fourmy, M. I. Recht, S. L. Blanchard and J. D. Puglisi, *Science*, 1996, **274**, 1367.
134. S. P. Velagapudi, S. M. Gallo and M. D. Disney, *Nat. Chem. Biol.*, 2014, **10**, 291.
135. T. Hermann and D. J. Patel, *Structure*, 2000, **8**, R47.
136. M. S. Gerdeman, T. M. Henkin and J. V. Hines, *J. Mol. Biol.*, 2003, **326**, 189.
137. R. Anupam, A. Nayek, N. J. Green, F. J. Grundy, T. M. Henkin, J. A. Means, S. C. Bergmeier and J. V. Hines, *Bioorg. Med. Chem. Lett.*, 2008, **18**, 3541.
138. C. M. Orac, S. Zhou, J. A. Means, D. Boehm, S. C. Bergmeier and J. V. Hines, *J. Med. Chem.*, 2011, **54**, 6786.
139. N. Cutillas, G. S. Yellol, C. de Haro, C. Vicente, V. Rodríguez and J. Ruiz, *Coord. Chem. Rev.*, 2013, **257**, 2784.
140. I. Omae, *Coord. Chem. Rev.*, 2014, **280**, 84.
141. J. Reedijk, *Eur. J. Inorg. Chem.*, 2009, **10**, 1303.
142. B. W. Harper, F. Li, R. Beard, K. B. Garbutcheon-Singh, N. S. Ng and J. R. Aldrich-Wright, in *Supramolecular Systems in Biomedical Fields*, ed. H.-J. Schneider, RSC, Cambridge, 2013.
143. E. García-España, I. Piantanida and H.-J. Schneider, in *Supramolecular Systems in Biomedical Fields*, ed. H.-J. Schneider, RSC, Cambridge, 2013.

CHAPTER 8

Synthetic Receptors for Amino Acids and Peptides

DEBRABATA MAITY AND CARSTEN SCHMUCK*

University of Duisburg-Essen, Faculty of Chemistry, Universitätsstrasse 7, 45141, Essen, Germany
*Email: carsten.schmuck@uni-due.de

8.1 Properties of Amino Acids and Peptides

Amino acids and peptides have neutral, acidic, or basic side chains, thus they have different iso-electric points and different electrostatic charges at physiological pH. The side chains also control the molecular polarity and associated properties such as hydrophobicity. In natural systems, amino acids and proteins are recognized by protein receptors or enzymes that employ a mixture of non-covalent interactions and hydrophobic effects. Chemists have tried to mimic nature and develop synthetic receptors, sometimes called hosts, that recognize their complementary guest molecules in a highly selective fashion. Chiral recognition is also an important goal, since most amino acids and peptides can exist as different stereoisomers. In water, amino acids and many small peptides are well solvated and it is difficult to produce stable supramolecular complexes simply using polar non-covalent interactions. In biological systems, guest binding often occurs within the cleft of an enzyme active site or receptor binding site. Hydrophobic effects help drive guests into these shape-selective clefts and non-covalent interactions such as hydrogen bonds (H-bonds) are significantly stronger within these shielded microenvironments.

Monographs in Supramolecular Chemistry No. 14
Synthetic Receptors for Biomolecules: Design Principles and Applications
Edited by Bradley D. Smith

Published by the Royal Society of Chemistry, www.rsc.org

Small peptides typically have flexible structures with several low energy conformations. Peptide conformational energies are dominated by the folding inclination of the polyamide backbone, and typically the amide group has a strong preference to adopt a planar s-*trans* conformation with the NH group opposite to the carbonyl oxygen. The one exception is the tertiary amide group of proline, which has nearly equal s-*cis* and s-*trans* conformational energies. In certain cases the conformation of a peptide can be affected by stabilizing intramolecular interactions such as salt bridges. Sometimes the rate of peptide conformational exchange may be slow enough to control the kinetics of peptide association with a receptor.

8.2 Biological Receptors for Amino Acids and Peptides

In living systems, molecular recognition interactions involving amino acids or peptides are vital factors in many biochemical and medicinal processes.[1,2] Amino acids have biological roles as biosynthetic building blocks or as signaling molecules. For example, glutamate and γ-aminobutyric acid (GABA) are well known as neurotransmitters. In these cases, the amino acids associate with membrane-bound protein receptors and trigger changes in receptor shape and activity with subsequent signaling across the membrane. Shown in Figure 8.1 is an example of the non-covalent interactions (H-bonding and van der Waals contacts) between a G-coupled protein

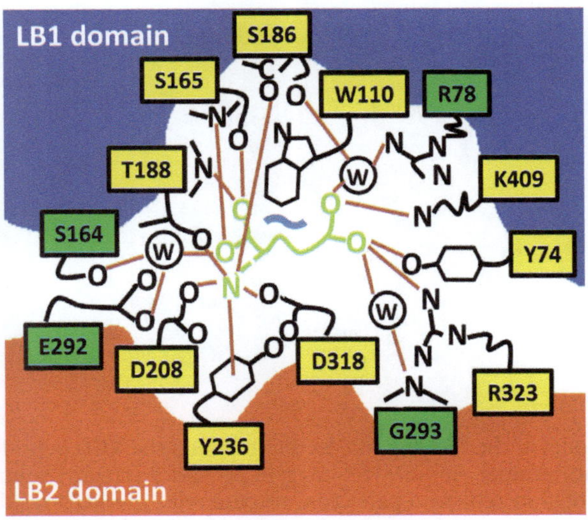

Figure 8.1 Schematic of the binding of glutamate (green) in a G-protein coupled glutamate receptor with red lines showing H-bonding and blue lines showing van der Waals contacts.
Adapted from *Br. J. Clin. Pharmacol.*, 2009, **156**, 869, © 2009 The Authors. Journal compilation © 2009 The British Pharmacological Society.[3]

receptor and a bound glutamate molecule.[3] These interactions are the basis for the selective molecular recognition.

Similarly, peptides often act as signaling molecules or hormones. For example, endorphins are small neuropetides produced by the central nervous system to relieve stress or enhance pleasure. They associate with opiate receptors in the brain to produce signaling cascades. Membrane-bound receptors also associate strongly with short peptidyl segments that are part of a larger protein chain. A well-known example is recognition of the peptidyl sequence RGD (arginine-glycine-aspartic acid; Arg-Gly-Asp) by integrin receptors. These transmembrane receptors are composed of α- and β-subunits and integrin association with the RGD sequence produces connective contacts between cells and the extracellular matrix, often leading to signal transduction inside the cell. Shown in Figure 8.2 is a model highlighting the non-covalent RGD/integrin interactions including salt bridges with the cationic Arg residue and coordination of the anionic Asp residue to a Ca^{2+} cation.[4]

Vancomycin is the best-known member of a large family of glycopeptide antibiotics that are used for the treatment of resistant bacterial infections. It is the drug of last resort for the treatment of methicillin-resistant *Staphylococcus aureus*.[5,6] Its interaction with a small peptidyl component of the bacteria cell wall is a classic example of substrate–receptor complexation.[7,8] Vancomycin binds strongly to a muropeptidic precursor terminating in L-Lys-D-Ala-D-Ala and prevents transpeptidase-mediated cross-linking of the bacterial cell wall, leading to a cessation of growth and eventual destruction of

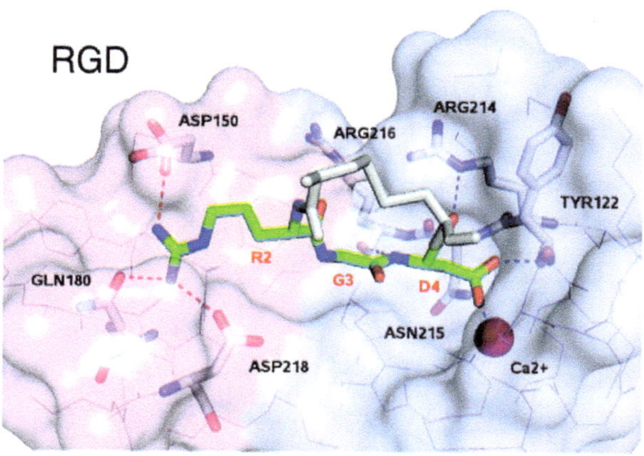

Figure 8.2 Model of the binding interaction between the RGD peptide (Arg-Gly-Asp) and binding site of αvβ3-integrin. α- and β-integrin subunits are represented in pink and pale cyan, respectively. The RGD residues are shown in green, and nitrogen and oxygen atoms in blue and red, respectively. Ca^{2+} is represented by a red sphere. Integrin and ligand residues involved in binding are labeled with three- and one-letter codes, respectively. Dotted lines denote H-bonds between ligands and integrin. Reproduced with permission from *J. Cell Sci.*, 2011, **124**, 515, © 2011 The Company of Biologists Ltd.[4]

the bacterial cell by lysis.[9,10] Vancomycin is a tricyclic molecule consisting of vancosamine, glucose, and seven unusual amino acid residues, which are constrained by several phenolic oxidative cross-links.[11] Nuclear magnetic resonance (NMR) studies of vancomycin/Ac-D-Ala-D-Ala and vancomycin/Ac_2-L-Lys-D-Ala-D-Ala complexes have revealed the structural basis of the complexation. A series of H-bonds is formed between the vancomycin amide groups and the peptide (Figure 8.3, top). These intermolecular hydrogen bonds are further stabilized by hydrophobic interactions between the hydrocarbon

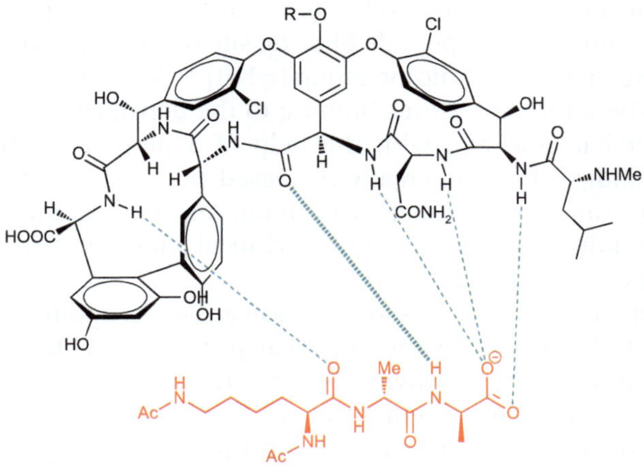

Ac_2-L-Lys-D-Ala-D-Ala

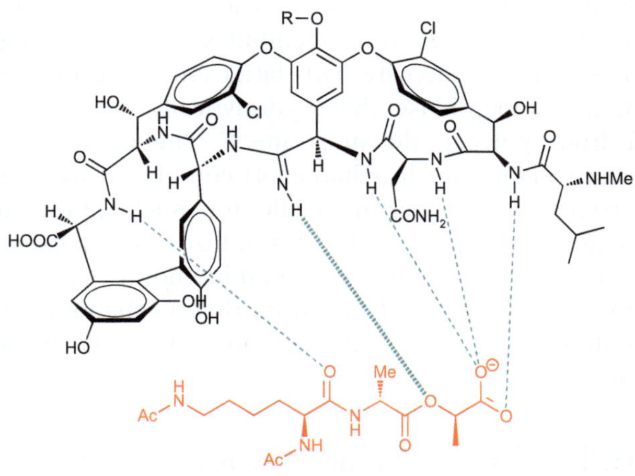

Ac_2-L-Lys-D-Ala-D-Lac

Figure 8.3 Complex structures showing (top) vancomycin and mimic of the normal bacteria cell wall peptidyl fragment Ac_2-L-Lys-D-Ala-D-Ala, (bottom) modified vancomycin analogue and mimic of the drug-resistant bacteria cell wall peptidyl fragment Ac_2-L-Lys-d-Ala-D-Lac.

portion of vancomycin and the alanine methyl groups of the dipeptide.[12,13] In recent years, resistant bacteria have emerged with an altered cell wall precursor that avoids vancomycin recognition. Instead of a terminal D-Ala-D-Ala moiety, the cell wall precursor contains the depsipeptide moiety D-Ala-D-Lac. Hence, an amide group is replaced by an ester group and accordingly a H-bond donor (–NH) is replaced by an H-bond acceptor (–O–). This replacement of an attractive H-bond by a repulsive dipole–dipole interaction weakens the vancomycin association constant by a factor of 1000. However, the latest round in this never-ending battle with the microbes has seen the development of new vancomycin analogues. The group of Boger has synthetically modified the peptide binding site of vancomycin by replacing a carbonyl oxygen with an amidine group (=NH).[14] As shown at the bottom of Figure 8.3, the amidine restores binding to the depsipeptide component in the resistant bacteria, and with the help of a peripheral chlorobiphenyl group (not shown) there is greatly increased potency against vancomycin-sensitive and vancomycin-resistant bacteria. Antibiotic studies are ongoing but this breakthrough may extend the clinical lifetime of the vancomycin family of drugs.

Peptides are often substrates for protease enzymes, and for many decades enzymologists have studied the chemical principles behind the substrate recognition and catalytic reactivity. X-ray crystal structures show that enzyme active sites contain convergent catalytic functionality (acids, bases, and nucleophiles) that accelerate amide-bond hydrolysis and promote enzyme turnover. The catalytic functional groups operate in concerted fashion as the reaction progress and often the motion of the protein's ternary structure is coupled with the reaction coordinate to achieve selective stabilization of the reaction transition state, which enables catalytic turnover. Enzyme active sites often contain a metal cation to facilitate substrate binding and provide Lewis acid activation of reactivity. This latter effect is exhibited by carboxypeptidase A, an exopeptidase that hydrolytically cleaves the C-terminal amino acid from polypeptide substrates.[15] The active site of carboxypeptidase A (see Figure 4.9 in Chapter 4) contains a zinc cation that (a) activates a coordinated water molecule for nucleophilic attack at the substrate's reactive carbonyl and (b) stabilizes the subsequent anionic tetrahedral intermediate which leads to amide bond cleavage and product release. The research effort that produced this mechanistic understanding included catalysis studies of synthetic molecules with metal centers as models of the enzyme active site.

8.3 Introduction to Synthetic Receptors

Artificial receptors that are capable of selectively binding a given amino acid or small peptide under physiological conditions, are interesting biomimetic systems for studying the underlying supramolecular chemistry. Early molecular designs were primarily based on H-bonding interactions and the work was, therefore, limited to organic solvents. However, the field has progressed significantly in recent years and it is now possible to design small synthetic receptors which function in aqueous solution.[16–19] This has opened the way for prospective applications in proteomics, medical diagnostics, and drug delivery.[20–26]

8.4 Synthetic Receptors for Amino Acids

Amino acids play central roles, both as building blocks of proteins and as intermediates in metabolism.[27] The correct amount of lysine (Lys) in the diet is essential for proper growth and metabolic functions.[28] Histidine (His) is essential in proteins as metal binding sites (*e.g.*, zinc), often acting together with cysteine (Cys) or other amino acids.[29] Tryptophan (Trp) is required not only for the production of proteins, but also the B vitamin niacin and the neurotransmitters serotonin and melatonin.[30] A deficiency in some amino acids causes various medical conditions, for example, a lack of Cys results in slow growth, hair depigmentation, edema, lethargy, liver damage, and muscle and fat loss.[31] As a consequence, effective methods for amino acid analysis are expected to help improve human health as well as the diagnosis and treatment of disease. A logical research plan is to develop analytical methods based on supramolecular chemistry.[32–34] Surprisingly, there are few synthetic receptors that can recognize amino acids in their natural form. Recent advances in amino acid recognition have exploited electrostatic interactions, hydrogen bonding, metal coordination, nanoparticle recognition, and specific reactions with chemical probes. Selected examples of these strategies are discussed in the following sections.

8.4.1 Amino Acid Recognition Mediated Mainly by Electrostatic Interaction

The strength of H-bonds, often used for molecular recognition in organic solvents, decreases rapidly with increasing polarity of the solvent. To achieve amino acid recognition in aqueous media, researchers have exploited ionic interactions to enhance the binding affinity of simple hydrogen bonding motifs. Positively charged receptors have been developed based on molecules containing guanidinium, imidazolium, and viologen groups.

8.4.1.1 Receptors Based on Guanidinium Groups

In 1999, Schmuck reported a class of guanidinium receptors for binding *N*-acetyl-α-amino acids in aqueous media.[35] The 2-(guanidiniocarbonyl)pyrrole receptor **8.1** (Figure 8.4) was designed in such a way that the guanidinium cation binds to oxo anions and the pyrrole unit also forms a N–H···O hydrogen bond to the bound carboxylate. Receptor **8.1** binds *N*-acetyl-α-amino acid carboxylates in 40% H_2O–dimethyl sulfoxide (DMSO) with binding constants ranging from $K_a = 360–1700\ M^{-1}$, depending on the

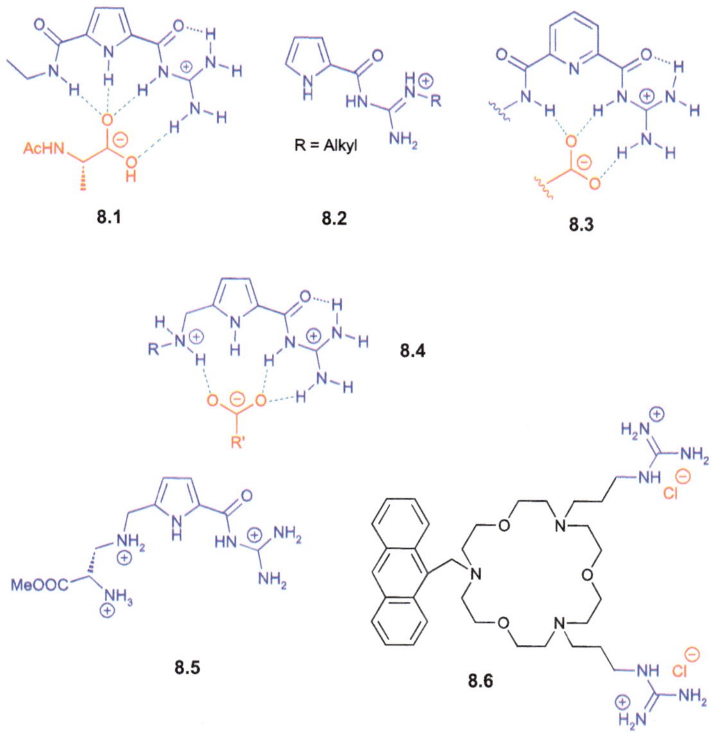

Figure 8.4 Receptors based on guanidinium groups.

structure of the amino acid side chain. For example, *N*-acetyl-phenylalanine is bound much stronger than the corresponding alanine or lysine derivatives. Molecular modeling calculations indicated that the differences in complex stability were due to secondary interactions of the side chains with the receptor. Later, Schmuck *et al.* reported *N′*-alkylated guanidiniocarbonyl pyrroles **8.2**, which efficiently bind amino acid carboxylates in water (pH 6.1).[36] A series of guanidiniocarbonylpyridine receptors (**8.3**) were also evaluated for binding *N*-acetyl-amino acids in aqueous DMSO.[37] However, the electrostatic repulsion between the pyridine nitrogen lone pair and the bound carboxylate makes anion binding less efficient than the analogous pyrrole receptors. A later prototype was bis-cationic guanidiniocarbonylpyrrole receptor, **8.4**, which efficiently binds *N*-acetyl alanine carboxylate with log $K_a = 3.9$ in water (pH 6.0).[38] Due to the increased charge density in this dication, the binding constant is five times larger than for the parent guanidiniocarbonyl pyrrole monocation **8.1** (log $K_a = 3.2$). A tris-cationic receptor **8.5** binds amino acid carboxylates efficiently in water.[39] Notably, it binds *N*-acetyl-L-alanine carboxylate very strongly with an association constant of $K_a \geq 10^5$ M^{-1} in 40% water/DMSO. Furthermore, **8.5** shows an unexpected cooperative 2:1 complex formation with *N*-acetyl glutamate but not aspartate in 90% water/DMSO (v/v). Hence the tris-cation **8.5** is capable of differentiating between structurally similar glutamate and aspartate. The smaller distance between the two carboxylates in aspartate compared to glutamate might prevent the formation of 2:1 complexes due to steric/electrostatic repulsions.

Suzuki and co-workers reported a fluorescent receptor (**8.6**) based on triaza-18-crown-6 ether combined with two guanidinium groups, which could bind several amino acids in aqueous methanol (pH 9.5) and showed fluorescence enhancement response by a photoinduced electron transfer (PET) mechanism.[40] The binding site for the ammonium cation, the triaza-18-crown-6 ether unit, was connected to two guanidinium groups which function as a carboxylate binding site. The anthracene unit was used as a fluorescent reporter for amino acid sensing. The fluorescence response of receptor **8.6** is GABA > lysine > n-$BuNH_3^+$, as only GABA and lysine have both a carboxylate and an ammonium group and thus interact with both binding sites, whereas the n-$BuNH_3^+$ cation can only bind to the crown ether moiety.

8.4.1.2 Receptors Based on Imidazolium Groups

You and Lan and co-workers reported 1,1′-binaphthyl-based imidazolium receptors for highly selective recognition of tryptophan in aqueous solutions (pH 7.4) *via* synergistic effects of multiple hydrogen bonding and electrostatic interactions.[41] The binding affinity and selectivity of the structurally open cleft-like receptor (*R*)-**8.7** with L-Trp ($K_a = 1.7 \times 10^4\ M^{-1}$) was larger than homologues (*R*)-**8.8**–**8.11** (Figure 8.5). These results demonstrated that the C-2 hydrogen atom of the imidazolium ring plays a key role. The macrocyclic (*R*)-**8.9** exhibited remarkably high enantioselective recognition of tryptophan with a K_D/K_L of 6.2, but the selectivity was not retained with other aromatic amino acids.

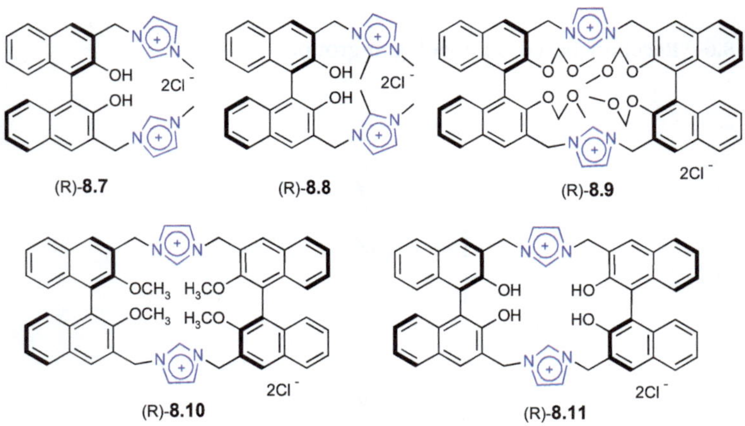

Figure 8.5 Receptors based on imidazolium groups.

8.4.1.3 Receptors Based on Viologen Groups

Ramaiah and co-workers reported a novel donor–acceptor conjugate **8.12** for the selective recognition of Trp in aqueous medium (Figure 8.6).[42] The viologen-linked pyrene **8.12** forms a novel fluorescent intramolecular charge-transfer (ICT) complex and was tested for recognition of different amino acids. Upon addition of Trp, a decrease in the ICT fluorescence intensity at 475 nm was observed, along with a concomitant increase in the pyrene fluorescence at 400 nm. This fluorescence change was due to 1:1 complexation between the conjugate **8.12** and Trp with an association constant of $K_a = 1300\ M^{-1}$. Significantly weak association was observed for Phe ($K_a = 73\ M^{-1}$) and Tyr ($K_a = 52\ M^{-1}$) and no binding of other amino acids was seen.

8.4.2 Amino Acid Recognition Mediated Mainly by Hydrogen Bonding

Chan and co-workers investigated the binding affinities of a cholic-acid-based fluorescent neutral receptor **8.13** toward amino acids in a buffered CH_3OH/H_2O system (pH = 7.4) (Figure 8.7).[43] The emission spectrum of **8.13** was quenched ~20% upon the addition of L-aspartate, glutamate, *N*-acetyl-L-aspartate, and *N*-acetyl-L-glutamate. The preorganization of **8.13** permits three-point binding with the guests, leading to stronger complexation of acidic amino acids.

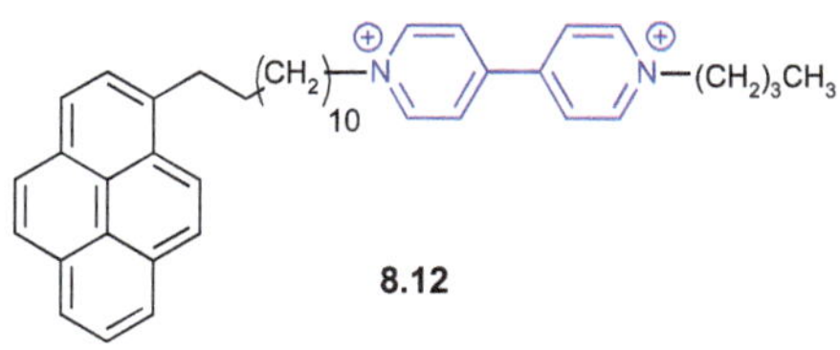

8.12

Figure 8.6 Receptor based on a viologen group.

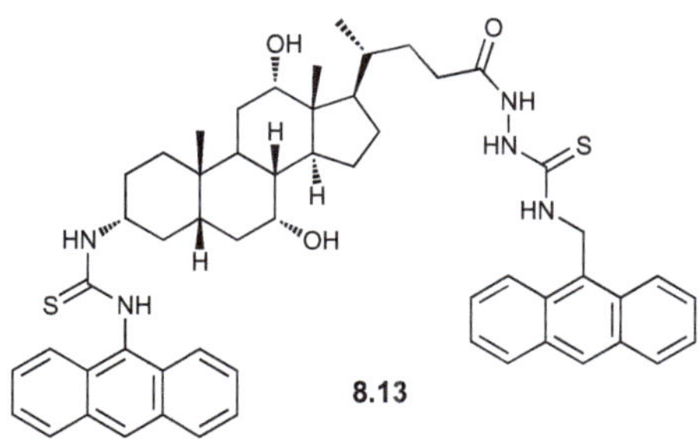

8.13

Figure 8.7 Receptor mainly based on hydrogen bonding.

8.4.3 Amino Acid Recognition Mediated by Metal Coordination

The most common applications of synthetic receptors for amino acids are either membrane transport or chemical sensing. The most straightforward way to associate reversibly with an underivatized α-amino acid in water is to use a synthetic receptor that incorporates a metal center that can chelate to the amino nitrogen and the carboxyl oxygen.

8.4.3.1 Receptors Based on Cu-complexes

In 2003, Fabbrizzi and co-workers reported an off–on "chemosensing ensemble" for selective recognition of histidine over other natural amino acids using an indicator-displacement assay (IDA).[44] Coumarin 343, fluorescein, and eosin Y were examined as the fluorescent indicator. The indicator was coordinated by the two Cu^{2+} centers in receptor **8.14** (Figure 8.8) at pH 7, which quenched its emission. The added analyte displaced the indicator and restored its full fluorescence. The **8.14**/coumarin ensemble did not discriminate His from glycine (Gly). However, the fluorescein-containing ensemble satisfactorily discriminated His with full recovery of fluorescence. The highest sensing selectivity was observed for **8.14**/eosin Y ensemble.

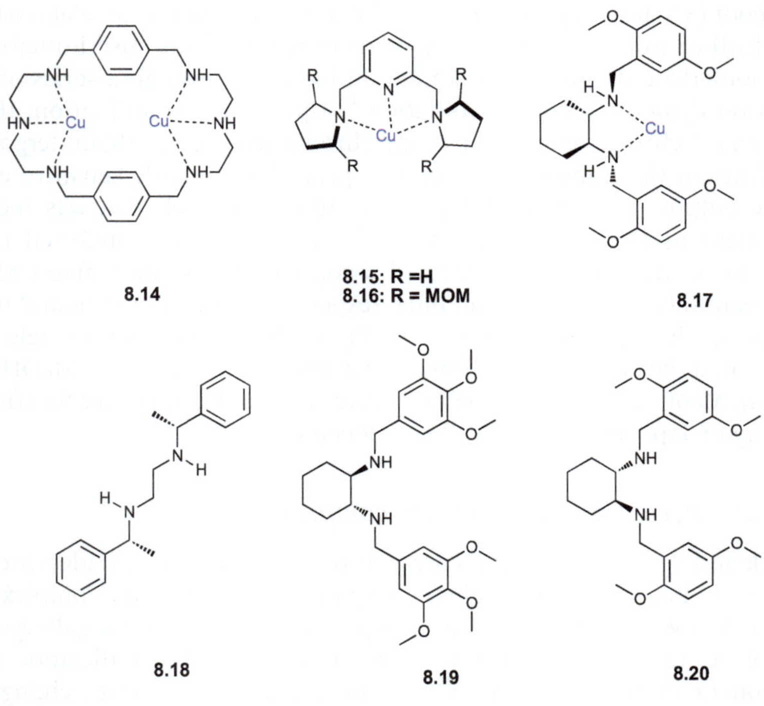

Figure 8.8 Copper-containing receptors for amino acid recognition based on indicator-displacement assays.

The observed stability trend of Gly > alanine (Ala) > phenylalanine (Phe) > valine (Val) > leucine (Leu) > proline (Pro) seems to be related to the increasing steric repulsive effects exerted by the amino acid side chain. The Anslyn group reported using coordinatively unsaturated metal complexes to discriminate His from other zwitterionic α-amino acids by means of an IDA using 5(6)-carboxyfluorescein as an indicator.[45] The titrations of **8.15** or **8.16** complexes into solutions of the indicator caused a visual color change in buffered MeOH/H_2O (3 : 1) solution. Colorimetric recognition of His was achieved using this IDA method, and the selectivity was attributed to formation of a 2 : 1 His/Cu^{2+} complex rather than simple indicator displacement. Anslyn and co-workers also reported an operationally simple colorimetric recognition method based on competitive dynamic metal coordination in H_2O/MeOH (1 : 1) solutions.[46] The dye, pyrocatechol violet (PV), was used in this indicator displacement assay to effectively compete with the amino acid guest for open coordination sites on (S,S)-**8.17**. Titration of (S,S)-**8.17** with PV gave a 1 : 1 complex with a shift in λ_{max} from 445 to 645 nm, resulting in a change in color from pale yellow to intense blue. The reversal of the spectral change was observed upon addition of amino acids to the (S,S)-**8.17**/PV complex, signaling PV displacement. The relationship between enantiomeric excess and absorbance in buffered 1 : 1 MeOH/H_2O (pH 7.0) was remarkably linear, demonstrating uniform sensitivity over the entire enantiomeric excess range. Enantiomeric excess curves for Leu generated with both (S,S)-**8.17** and (R,R)-**8.17** exhibit a near mirror-image relationship.

In further work, Anslyn and co-workers reported the discrimination of enantiomeric and structurally similar amino acids through a series of IDAs based on dynamic metal coordination.[47] Upon Cu^{2+} coordination, the indicators PV, chromoxane cyanin R and chrome azurole S (CAS) undergo large red shifts in their absorbance spectra, providing a highly sensitive colorimetric output. The enantioselectivity exhibited in each IDA was receptor dependent as the complexes **8.18**–Cu^{2+} and **8.19**–Cu^{2+} preferred L configurations, while **8.20**–Cu^{2+} favored D configurations. Since free CAS and Cu^{2+}-bound CAS are yellow and blue, respectively, the ratio of bound to free indicator, which depends on the stability of the receptor–amino acid complex, could be assessed by visual inspection (buffered 1 : 1 MeOH:H_2O, pH 7.8). All of the IDAs exhibited a relative chemoselective order for complex stability of Trp > Phe > Leu ∼ Val > *tert*-leucine (*t*Leu).

8.4.3.2 Receptors Based on Rh-complexes

Buryak and Severin reported an array of IDAs for colorimetric identification of the 20 natural amino acids.[48] Organometallic Cp*Rh complex **8.21** (Figure 8.9) was employed as the receptor unit and the dyes gallocyanine, xylenol orange, and calcein blue were employed as the indicators. A key component of the sensor array was the utilization of pH to change the selectivity of the sensing ensemble. The absorption at 750 nm of an IDA composed of **8.21** and gallocyanine was used to classify the amino acids as

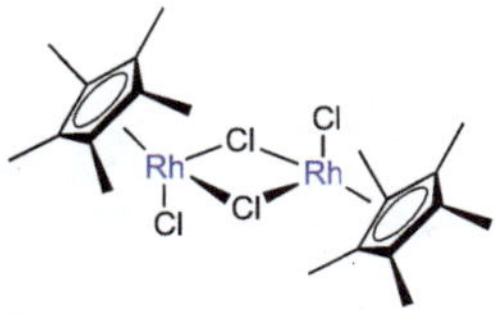

8.21

Figure 8.9 Rhodium-containing receptor for amino acid recognition based on indicator-displacement assays.

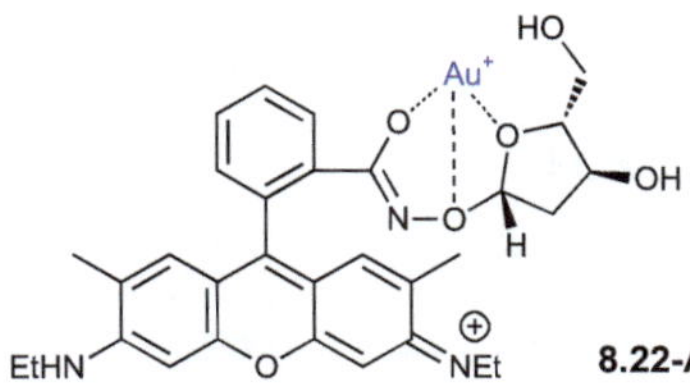

8.22-Au⁺

Figure 8.10 Au^+-containing receptor for amino acid recognition.

either high-affinity [consisting of His, Cys, Met, Asp, Asn, ΔA (750 nm) < 0.06] or low affinity {the remaining 15 amino acids [0.55 > ΔA(750 nm) > 0.06]}. Each member of the first group was analyzed by four IDAs (using two different dyes, xylenol orange, and calcein blue, and two different pH values), and each member of the second group was analyzed by five IDAs (gallocyanine as the indicator, at five different pH values). Linear discriminant analysis (LDA) and principal component analysis (PCA) permitted identification of the amino acids.

8.4.3.3 Receptors Based on Au-complexes

Yang *et al.* reported a "chemosensing ensemble" method for the detection of Cys and homocysteine (Hcy) in water (MeOH 1%).[49] A complex of Au^+ and a 2-deoxyribose functionalized rhodamine hydroxylamine was used for this purpose (Figure 8.10). In the presence of Au^+ only Cys and Hcy induced significant enhancements in fluorescence intensities of **8.22**. The recognition of Cys was attributed to binding of the Cys thiol to the **8.22**–Au^+ probe. The corresponding binding constant was $K_a = 6.7 \times 10^3$ M^{-1} and detection of Cys was possible even at the 100 nM level. The binding of Cys to **8.22**–Au^+ also caused a red coloration that could be viewed with the naked eye. No other amino acids exhibited similar color changes with **8.22**–Au^+ under the same conditions.

8.4.4 Amino Acid Recognition Based on Nanoparticles

Patel and Menon investigated the recognition of Lys, Arg or His by novel *p*-sulfonatocalix[4]arene thiol functionalized gold nanoparticles in aqueous

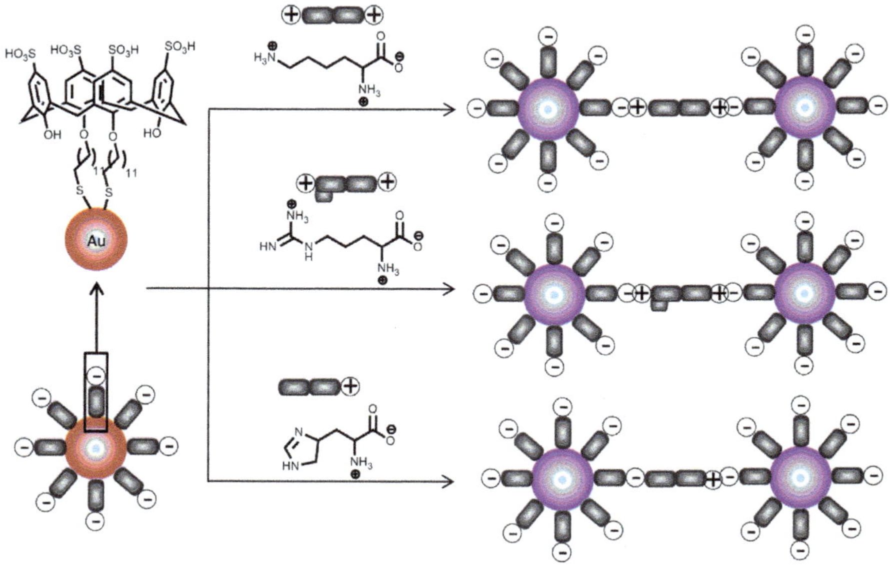

Figure 8.11 Schematic representation of the amino acid (Lys, Arg, or His)-induced aggregation of calix-capped gold nanoparticles.

solution.[50] Eleven different aqueous solutions of amino acids were added to calix-capped gold nanoparticles in phosphate buffered saline. Thirty minutes after addition of Lys, Arg or His, the nanoparticle solutions changed color from red to purple due to aggregation of the nanoparticles (Figure 8.11). However, after 24 hours, the color had disappeared due to aggregate sedimentation.

8.4.5 Reaction-mediated Amino Acid Recognition

In recent years, there has been increased attention towards the development of reaction-based molecular probes. This approach exploits highly selective and usually irreversible chemical reactions, induced by the target amino acids. The chemical reactions are often promoted by the non-covalent host–guest interactions between receptor and amino acids, but the final result is a chemical reaction. Typically, these reaction-based molecular probes are used as dosimeters in batch-wise chemical sensing protocols.

8.4.5.1 Amino Acid Recognition Based on Imine Formation

Feuster and Glass developed a method for the detection of unprotected amino acids in aqueous solutions by the formation of highly fluorescent iminium ions.[51] The coumarin aldehyde **8.23** and its 4-butyl derivative **8.24** were studied for their interactions with amino acids using absorption spectroscopy under high-salt aqueous conditions (Figure 8.12). The aldehydes reacted with amino acids to form iminium ions and an internal

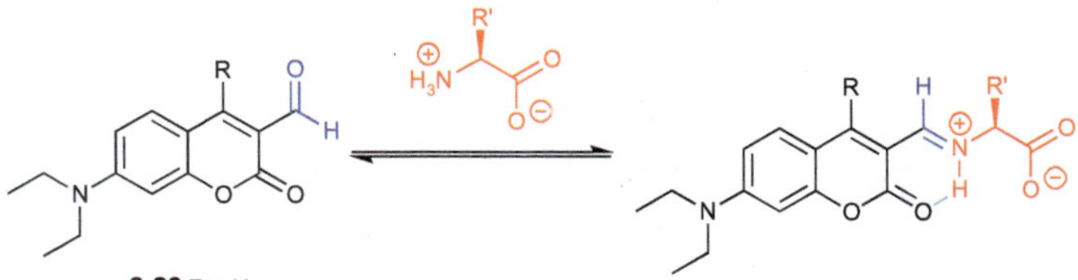

Figure 8.12 Reaction of coumarin receptors with unprotected amino acids.

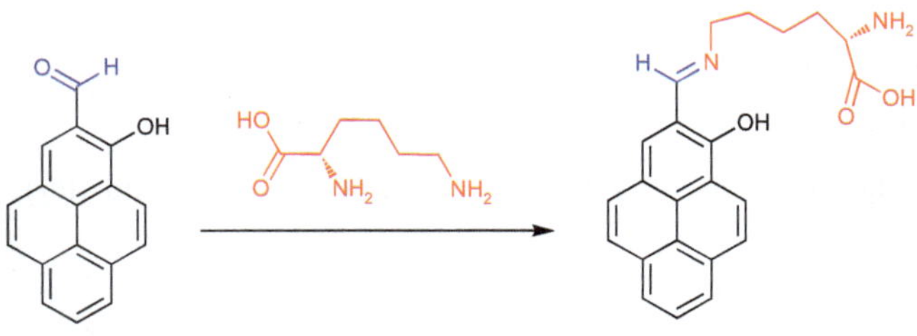

Figure 8.13 Recognition of lysine by imine bond formation.

hydrogen bond to the coumarin carbonyl group modulated the fluorescence. The equilibrium constant for iminium formation with glycine was very low for **8.23** ($K_{eq} < 1$) but substantially more favorable for **8.24** ($K_{eq} = 4.0$), with a much larger (34 nm) red shift in absorption. The fluorescence intensity of **8.24** increased by 26-fold at 513 nm upon titration with glycine.

Zhou *et al.* used a pyrene-based probe **8.25** for recognition of Lys in aqueous media (buffered CH_3CN/water, 1 : 9, pH 7.4) (Figure 8.13).[52] A distinct colorimetric change of **8.25** from light yellow to pink occurred only in the presence of Lys. The dissociation constant for the 1 : 1 complex formed between **8.25** and Lys was calculated to be $K_d = 2.8 \times 10^{-3}$ M. In addition, the fluorescence enhancement was selective for Lys and was not observed with the other tested amino acids. Computations suggested that the observed colorimetric change was caused by a stacked dimeric form of the products in which an intramolecular H-bond interaction exists between the C=N and OH group mediated by an H-bond water relay.

8.4.5.2 Amino Acid Recognition Based on Thiazinane or Thiazolidine Formation

Zhang *et al.* demonstrated that a ruthenium(II) complex **8.26** can act as a luminescence probe for the recognition and detection of Cys and Hcy in buffered DMSO/water, 9 : 1, pH 7.0 (Figure 8.14).[53] The weakly luminescent probe rapidly reacted with Cys and Hcy to yield the corresponding thiazolidine and thiazinane derivatives, accompanied by a remarkable luminescence enhancement and a large blue-shift of the maximum emission

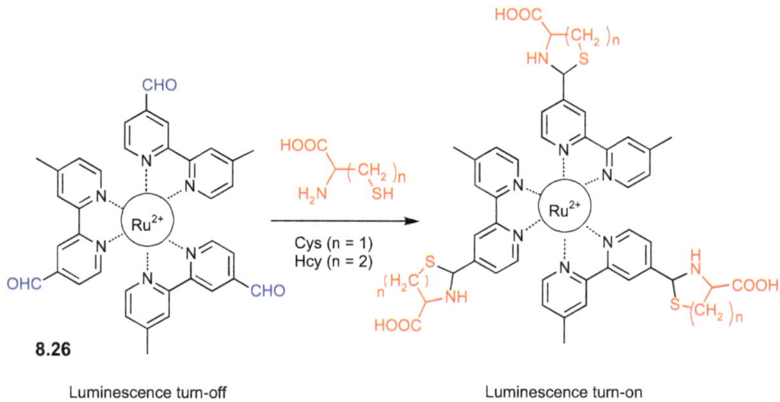

Figure 8.14 Reaction-based recognition of amino acids (Cys and Hcy).

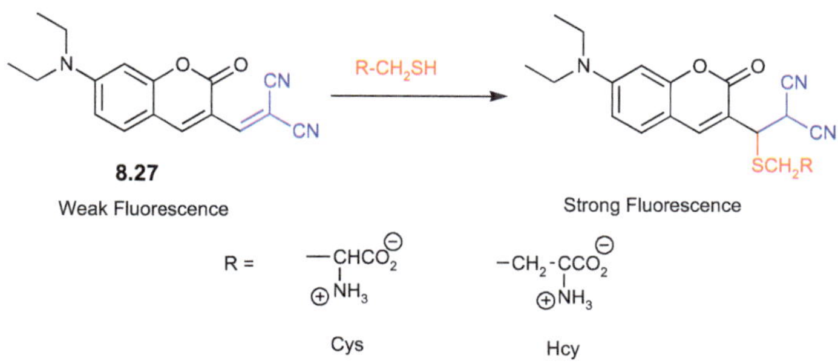

Figure 8.15 Reaction of **8.27** with sulfur-containing amino acids (Cys, Hcy, and glutathione).

wavelength from 720 to 635 nm. The luminescence response of the probe was highly specific to Cys/Hcy, even in the presence of various other amino acids, proteins, and DNA.

8.4.5.3 Amino Acid Recognition Based on Conjugate Addition

Kim and co-workers synthesized a weakly fluorescent coumarin–malonitrile probe **8.27** for selective recognition of Cys, Hcy, and glutathione (GSH) (Figure 8.15).[54] A fluorescence enhancement was caused by conjugate addition of the thiol group to the α,β-unsaturated malonitrile unit of **8.27** in buffered DMSO/water (1 : 2, pH 7.4). The order of fluorescence increase was Cys (19) > Hcy (12) > GSH (5.6), with micromolar sensitivity both *in vitro* and *in vivo*.

8.4.5.4 Amino Acid Recognition Based on Sulfonate Ester Cleavage

Lu *et al.* developed a reactive probe **8.28** for the recognition of Cys by incorporating the 2,4-dinitrobenzenesulfonyl (DBS) group as a quencher into

8.28 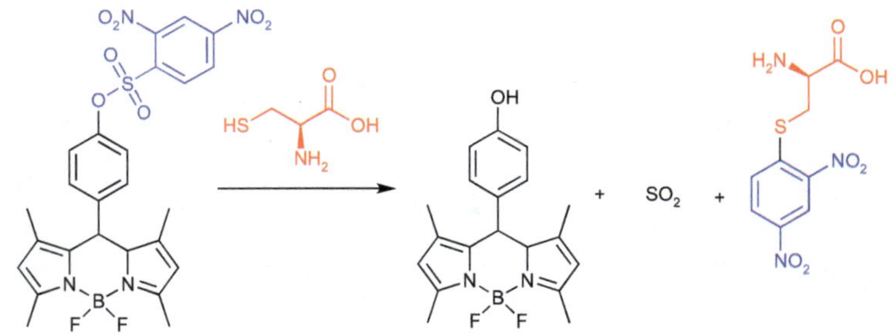

Figure 8.16 Reaction-based recognition of cysteine with **8.28.**

the boron-dipyrromethane (BODIPY) skeleton.[55] This BODIPY-based probe itself is non-fluorescent due to a strong intramolecular charge transfer process. Upon reaction with Cys, the probe produced a rapid and large fluorescence enhancement through the removal of the DBS unit by nucleophilic aromatic substitution (Figure 8.16).

8.4.5.5 Amino Acid Recognition Based on Disulfide Cleavage

Zhu *et al.* exploited the disulfide group contained in a naphthalimide-based colorimetric fluorescent probe **8.29** for recognition of sulfur-containing amino acid (Cys, Hcy, and GSH) in buffered ethanol/water (1 : 9 solution, pH 7.4).[56] Upon addition of GSH to the solution of **8.29**, the maximum absorption peak showed an 85 nm red shift and the color of the solution turned from colourless to jade green. The maximum emission peak underwent a red-shift of 48 nm, and the ratio of fluorescence intensities (F_{533}/F_{485}) ($\lambda_{ex} = 400$ nm) changed from 0.5 to 5.7. These changes are due to a disulfide cleavage process that produces a highly fluorescent 4-aminonaphthalimide moiety (Figure 8.17). Live cell imaging experiments demonstrated the potential utility of **8.29** for determination of biomolecular thiols in living systems.

8.5 Synthetic Receptors for Peptides

Peptides have a variety of different functional groups (amide backbone and side-chain functional groups) that can be used as binding sites. Thus, various non-covalent interactions such as H-bonding, metal coordination, salt bridges, hydrophobic stacking, or π–π interactions can be exploited for peptide recognition. Interactions with the functional groups in the amino acid side chains are of special interest as this not only increases binding affinity, but also recognition selectivity. An ultimate goal for the field is to develop molecular design algorithms that lead to a synthetic receptor for any specific peptide sequence. The following sections systematically discuss synthetic receptors for peptides based on hydrophobic interactions, electrostatic interactions, and metal coordination.

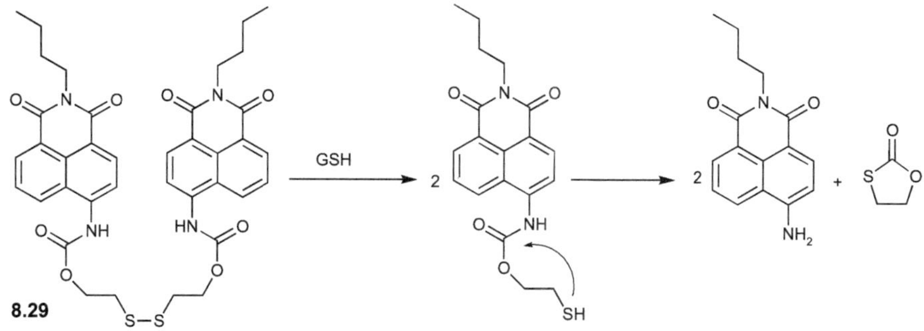

8.29

Figure 8.17 Reaction of **8.29** with thiol-containing amino acids. GSH: glutathione.

8.5.1 Peptide Recognition Mediated by Hydrophobic Interactions

The driving force for the molecular interaction of two nonpolar molecules in water is often an increase in the entropy upon complex formation. Nonpolar molecules in water disturb the H-bonded network of bulk water molecules. The water molecules at the surface of a nonpolar organic molecule are restricted in their orientation, as they find neighboring water molecules only in one direction, away from the organic molecule. Hence they have a higher degree of order compared to water molecules in the bulk phase. The number of such unfavorable water molecules directly depends on the surface area of the organic molecule. The larger the solvent-exposed surface the more water molecules are affected. When two organic molecules aggregate the overall solvent exposed surface area is reduced and hence a number of formerly ordered solvent molecules are released into the bulk phase. This provides an entropic and sometimes also an enthalpic driving force for association.

Molecular association in water, due to the hydrophobic effect, is comparatively insensitive to molecular shape. However, shape selectivity can be imparted if the receptor molecule possesses a relatively rigid hydrophobic cavity that forces a size filter on the binding partner. Alternatively, substrate selectivity can be achieved if the hydrophobic binding site is augmented by salt bridges or a network of highly directional H-bonds.

8.5.1.1 Cyclodextrin-based Receptors

Cyclodextrins are a class of cyclic oligosaccharides consisting of six to eight D-glucose units linked by α-1,4-glycosidic bonds. Cyclodextrins form stable inclusion complexes with many organic molecules in water,[57–60] and they are well known to encapsulate the hydrophobic side chains of the amino acids Phe, tyrosine (Tyr), and Trp, respectively.[61] However, binding constants in water are often modest ($\sim 10^2$ M^{-1}). Bis-cyclodextrin hosts exhibit improved binding ability and molecular recognition properties through the cooperative binding of two adjacent hydrophobic cavities. Furthermore, the linker between the two cyclodextrins not only sets the distance and orientation of the two cavities, but it can provide further functional groups for additional binding interactions to the substrate. There have been several reports of cyclodextrin-based receptors for the sequence-selective recognition of peptides containing hydrophobic amino acids in water.

In 1996, Still and Breslow used a combinatorial approach to investigate the binding properties of β-cyclodextrin hosts to di- and tripeptides.[62] They screened two orange-colored nickel salophen-based hosts **8.30** and **8.31** with two or one β-cyclodextrins, respectively, against a tripeptide library synthesized on a solid support (a hydrophilic poly-(ethyleneglycol)polystyrene (TentaGel)) (Figure 8.18). The general structure of the library was AA^3-AA^2-AA^1-$NH(CH_2)_2$-TentaGel, prepared by split-mix synthesis using 29 different amino acids at each position, providing a library size of $29^3 = 24\,389$ tripeptides. The screening results in water at pH 7 showed that the two receptors favor peptides containing the L-Phe-D-Pro dipeptide unit. Subsequent ^{1}H NMR titration studies of H_2N-L-Phe-D-Pro-$CONHCH_3$ with unsubstitued β-cyclodextrin in D_2O revealed a binding constant of 180 M^{-1} for the s-*trans* conformer of L-Phe-D-Pro. With the help of molecular modeling the authors concluded that hydrophobic encapsulation of the phenyl ring by the cavity, together with interactions of the proline ring with the rim of the cyclodextrin, accounted for the high stability of the complex.

Extending this work, Breslow *et al.* later reported the β-cyclodextrin dimers **8.32–8.37** (Figure 8.19) and studied the peptide sequence selectivity in aqueous buffer (pH 9.0).[63] Host **8.37**, having a short disulfide linker between the two primary cyclodextrin faces, bound the simple dipeptide Trp-Trp ($K_a = 1200$ M^{-1}) better than the other dimers. For example, dimer **8.36** contains a longer linker and binds significantly weaker to Trp-Trp (96 M^{-1}). Overall, the affinity and sequence selectivity of these hosts depended on both the length of the linker and the mode of attachment to the cyclodextrin (primary or secondary face).

Liu *et al.* examined the binding of bridged bis(β-cyclodextrin) **8.38**, and the corresponding Cu^{2+} (**8.39**) and Ni^{2+} (**8.40**) complexes, with some oligopeptides in buffered aqueous solution (pH 7.4) (Figure 8.20).[64] Host **8.38** strongly binds glutamic acid (Glu-Cys-Gly with a binding constant of $K_a = 6850$ M^{-1}). This strong binding was interpreted as a cooperative interaction with the two β-cyclodextrin cavities. In contrast, Glu-Glu contains three carboxylate

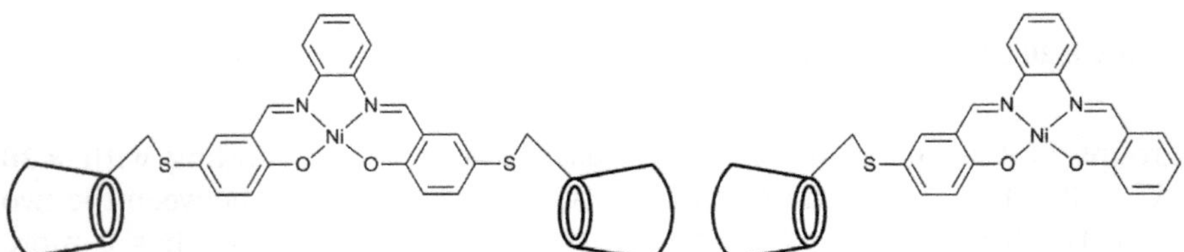

8.30

8.31

Figure 8.18 Cyclodextrin–nickel salophen complexes for recognition of L-Phe-D-Pro containing peptides in water.

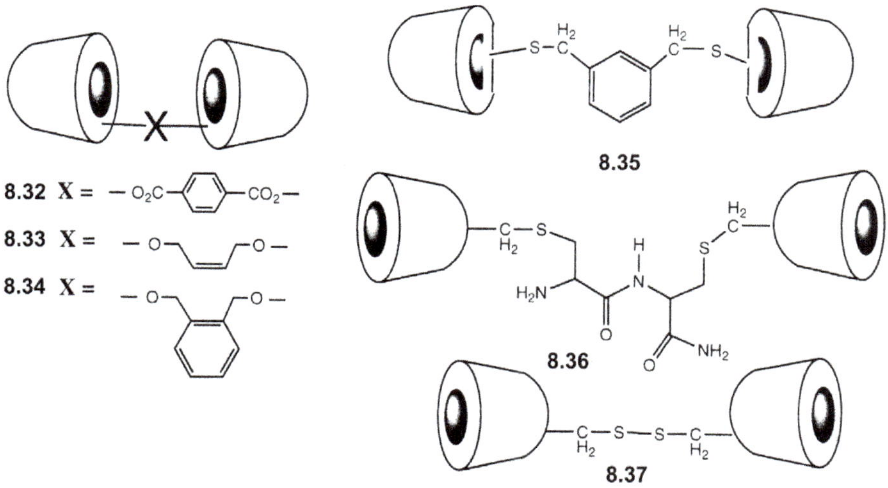

Figure 8.19 Cyclodextrin-based receptors for peptides.

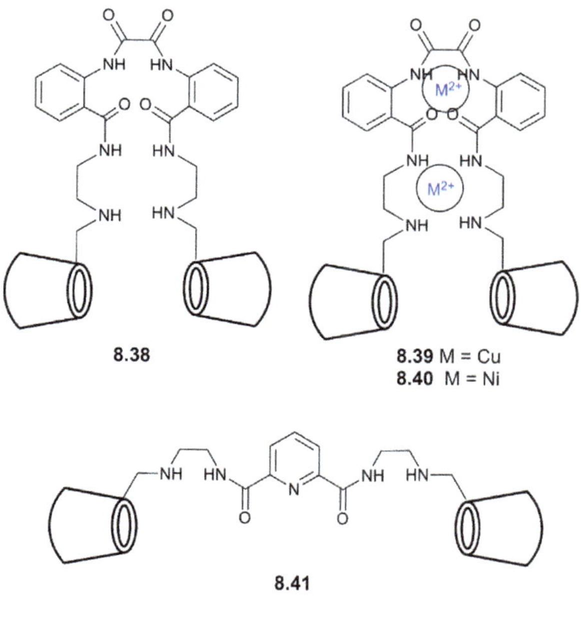

Figure 8.20 Bis-cyclodextrin receptors used for binding dipeptides.

groups and is more hydrophilic, thus showing poorer binding with **8.38** ($K_a = 135\ M^{-1}$). Furthermore, these hosts could differentiate between the two isomeric dipeptides Leu-Gly and Gly-Leu. The former peptide has a more linear conformation which fits better into the two cyclodextrin cavities than the Leu-Gly dipeptide that possesses a large isobutyl branch. The longer tripeptide Gly-Gly-Gly fits even better as it perfectly spans the distance between the two cyclodextrin cavities within the host. Later, the same authors incorporated a fluorescent pyridine-2,6-dicarboxamide linker between two β-cyclodextrins (**8.41**) and studied its complexation behavior with some aliphatic oligopeptides in aqueous buffer solution at two different pH values, namely 2.0 and 7.2. The 2,6-disubstituted pyridine group in the linker acted as an effective binding motif for the peptide carboxylate fragment.[65] Binding was strongest for Gly-Leu ($K_a = 1208\ M^{-1}$) at neutral pH, and affinity decreased in acid. In neutral media, the host adopts an extended conformation and the peptide guests bind into the two cavities of the host. In acidic media, the protonated pyridinium ring can partially bind into one cyclodextrin cavity, so that guest binding occurs only with the remaining free cavity.

8.5.1.2 Cucurbituril-based Receptors

Cucurbiturils are another interesting class of container molecules with a hydrophobic cavity. They are made of glycouril monomers linked together by methylene bridges and they bind a wide range of neutral and cationic guests due to hydrophobic and cation–dipole interactions (Chapter 2) (Figure 8.21).[66,67] Cucurbiturils have been used for recognition of peptides having aromatic side chains, preferably at the N-terminal position. Sequence specificity is mediated by the simultaneous inclusion of the aromatic side chain within the nonpolar cavity of the cucurbituril and electrostatic attraction of the cationic N-terminal ammonium group of the peptide with the negative dipoles of the carbonyl groups of the host. For example, Urbach and colleagues reported charge-mediated recognition of N-terminal tryptophan within peptides using synthetic cucurbit[8]uril host **Q8**.[68] Due to its large cavity size **Q8** can bind two guest molecules at the same time. The binding of a charge transfer complex formed by an electron-rich and an electron-poor aromatic ring in the hydrophobic cavity is especially favorable. Host **Q8** binds the electron poor methyl viologen (MV) to form a 1 : 1 complex. This **Q8**:MV complex can then further bind to the electron-rich indole side chain of tryptophan ($K_a = 4.3 \times 10^4\ M^{-1}$, pH 7.4). This termolecular complex is eight-fold and 19-fold more stable than the termolecular complex with the other aromatic amino acids Phe and Tyr, respectively.

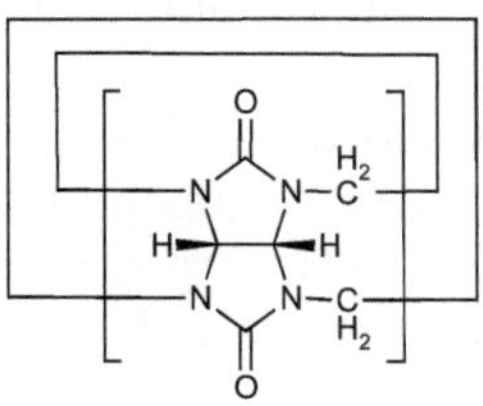

Q6 n = 6

Q7 n = 7

Q8 n = 8

Figure 8.21 Cucurbit[*n*]uril (**Qn**) host.

In later work, Urbach and co-workers demonstrated selective recognition and cooperative dimerization of N-terminal aromatic peptides by **Q8** alone.[69] They found that **Q8** selectively binds Trp-Gly-Gly and Phe-Gly-Gly as a 1 : 2 complex ($K_a = 10^9 - 10^{11}$ M^{-2}). Peptide recognition was driven by the simultaneous inclusion of two hydrophobic aromatic side chains in the cavity of **Q8** and the electrostatic interaction of the proximal N-terminal ammonium group with the carbonyl groups of the host. More recent work by Dang *et al.* has used **Q8** to promote dimerization of two inactive caspase oliopeptide stands, each with N-terminal Phe-Gly-Gly sequences, to create a supramolecular caspase dimer as an active enzyme.[70]

Kim and co-workers studied cucurbit[6]uril (**Q6**) and cucurbit[7]uril (**Q7**) for sequence selective recognition and self-sorting of dipeptides in aqueous media.[71] The cavity in **Q6** is relatively small and does not accommodate aromatic side chains, but forms strong complexes with dipeptides having aliphatic residues at the N-terminus. The larger **Q7** forms very strong complexes with dipetides having aromatic residues at the N-terminus. Only 1 : 1 complexes are formed and the smaller phenyl ring is preferred over the larger indole ring. The preference for N-terminal Phe is very strong, as demonstrated by a recent crystal structure of a **Q7** insulin complex showing that the N-terminus of the protein unfolds to enable **Q7** encapsulation of the N-terminal Phe residue.[72]

8.5.1.3 Hydrophobic Macrocycle Receptors

In early work Tomeiro and Still developed a chromophore-tagged polycyclic receptor and studied its binding properties to different peptides in water, again using a combinatorial approach.[73] Host **8.42** was synthesized from trimesic acid and an azacylooctane-1,2-diamine (Figure 8.22). To facilitate screening, a water soluble rhodamine dye was attached to one of the diamines to create **8.43.** Screening revealed that the hydrophobic cavity of the receptor accepts peptides with hydrophobic side chains of similar size (Leu, Val), and a preference for the adjacent amino acid to have opposite absolute configuration and an amide side chain [glutamine (Gln), asparagine (Asn)]. Furthermore, the receptor strongly binds the anionic tripeptide Asp-Asp-Asp due to electrostatic interactions with the cationic ammonium groups of the linking amide units.

Fujita and co-workers developed a self-assembled coordination cage, **8.44,** which can accommodate specific oligopeptides in its cavity (Figure 8.23).[74] Host **8.44** selectively binds Ac-Trp-Trp-Ala-NH_2 with a high association constant of $K_a > 10^6$ M^{-1}. Furthermore, it could recognize the Trp-Trp-Ala sequence in oligopeptides having more than three amino acid residues. NMR and X-ray analysis explained the specific sequence selective recognition, which is due to multiple cooperative interactions, *i.e.*, π–π stacking of the indole moiety to the electron-deficient triazine ligand and the interaction of the methyl group of alanine by CH–π interaction with another part of the cage wall.

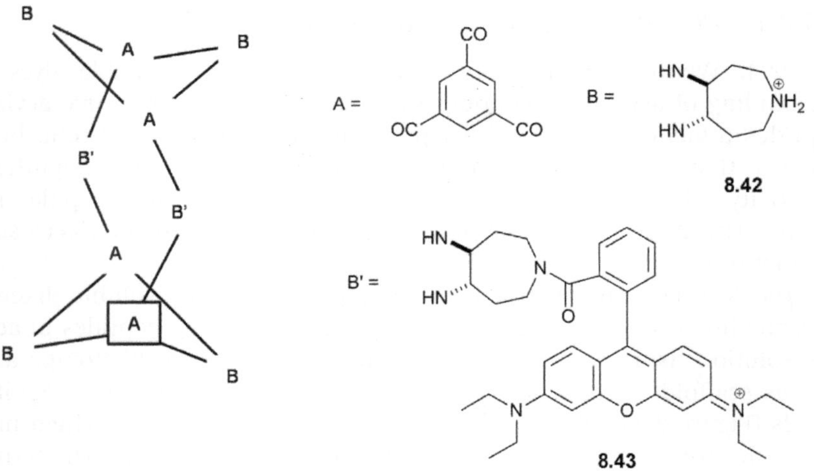

Figure 8.22 Macrocyclic hosts **8.42** and **8.43**, which preferentially bind hydrophobic peptides in water.

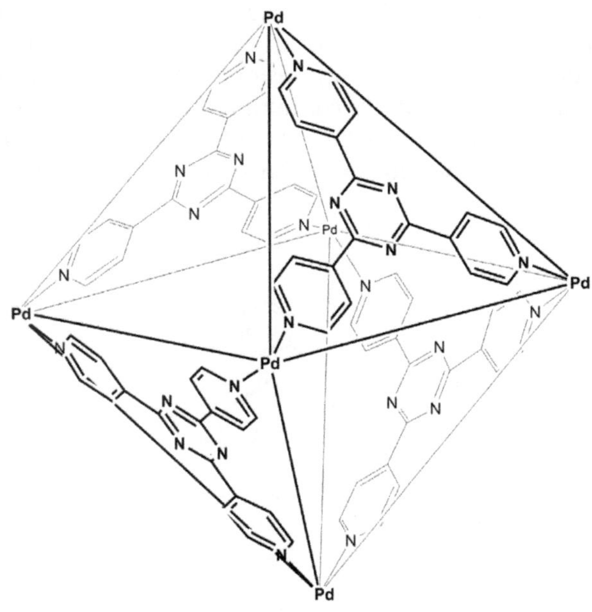

Figure 8.23 Self-assembled coordination cage **8.44** for peptide binding in water.

8.5.2 Peptide Recognition Mediated Mainly by Electrostatic Interactions

In nature, non-covalent binding interactions between charged peptides and proteins often involve the formation of an ion pair between the cationic groups of arginine or lysine and the anionic carboxylate groups of either aspartate or glutamate.[75]

8.5.2.1 Receptors Based on Natural Amino Acids

In 1994, Wennemers and Still used commercially available dyes for the binding of acylated tripeptides in water.[76] Screening of 45 000 acylated peptides attached to hydrophilic poly(ethylene glycol)-polystyrene beads showed that some dyes bound to certain peptides with significant selectivity. There was a clear preference for peptides having polar side chains. The driving force for binding was mainly based on electrostatic interactions.

Later, Krattiger and Wennemers developed rigid water-soluble diketopiperazine hosts for the recognition of arginine-containing peptides in aqueous solution using combinatorial chemistry.[77] The rigid diketopiperazine acts as scaffold for appending two peptidic receptor arms as recognition motifs (Figure 8.24). The peptidic arms contain aspartic acid, which has a preference for arginine and also increases the water solubility. The termini of the arms were acetylated for receptor **8.45** and functionalized with succinic acid for receptor **8.46.** Binding affinities of receptor **8.45** and **8.46** were studied in Tris buffer at pH 7.2 using a solid-phase bound tripeptide library Ac-AA^3-AA^2-AA^1-TentaGel which was prepared using split-and-mix synthesis with 31 different D- and L-amino acids in each position. While receptor **8.45** binds tripeptides containing two Arg residues at any position, receptor **8.46** preferentially binds to tripeptides containing the sequence Arg-Arg-X [where X = Arg, serine (Ser), threonine (Thr) or Cys]. The binding constant between receptor **8.46** and the peptide Ac-Arg-Arg-Arg-NHPr was $K_a = 13\,900\ M^{-1}$ at pH 8.

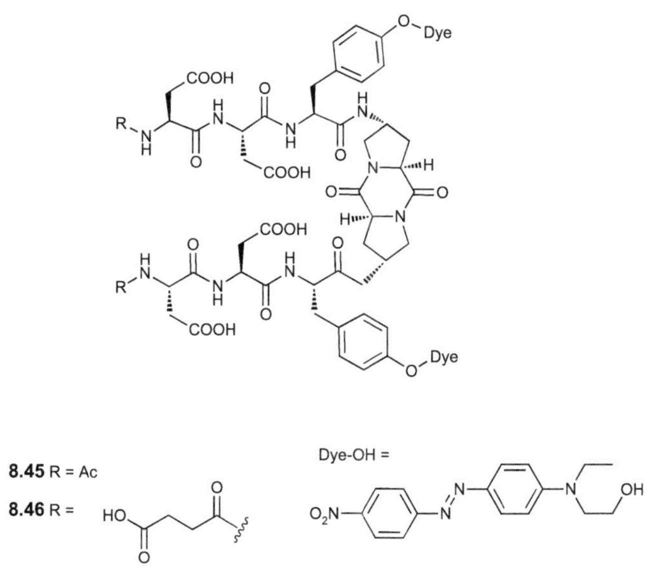

Figure 8.24 Diketopiperazine-based receptors.

8.5.2.2 Receptors Based on Modified Guanidinium Groups

The guanidine group is strongly basic. The high pK_a of 13.5 for the guanidinium cation is due to the delocalization of positive charge over all three nitrogen atoms.[78] Furthermore, the guanidinium cation has six potential hydrogen bond donors, which makes it highly soluble in water and allows for the formation of H-bond-assisted ion pairs. Thus, arginine is often involved in biological recognition events with proteins or nucleic acids.[79–84] With the advent of supramolecular chemistry, modifications of simple alkylguanidinium cations were introduced to improve their anion-binding properties. For example, acylguanidinium cations have higher affinity for anions compared to the parent guanidinium cation due to increased acidity of the NH groups (pK_a 6–7 *vs.* 13, respectively).[85] Furthermore, the guanidiniocarbonyl pyrrole cation provides an additional H-bond donor for increased affinity, and allows efficient oxoanion-binding in aqueous media.[86,87]

Jensen *et al.* developed a library of tweezer receptors **8.47** with a guanidinium head group and two variable peptidic arms (Figure 8.25).[88] The guanidinium group was introduced to bind the C-terminus of a peptide in aqueous media. Upon screening of the library with several tripeptides, selective binding was observed. The binding constant with peptide **f** was $K_a = 8.2 \times 10^4\ M^{-1}$ in aqueous media (pH 8.8). A neutral sulfonylated tweezer failed to bind the peptide, clearly demonstrating the importance of ion pair formation between the guanidinium cation and the carboxylate terminus of the receptor. However, host **8.47** showed poor selectivity towards different stereoisomers of the tripeptide and no binding of the side chain deprotected tripeptide was observed. This suggests that hydrophobic interaction of the protecting groups with the host was the reason for efficient binding.

Later, the same group developed a small library of "unsymmetrical" tweezer receptors like **8.48**, again with a guanidinium head group as a carboxylate binding site, but now with two different peptidic arms.[89] They screened this library using the bacterial cell wall precursor tripeptide mimic *N*-Ac-Lys-D-Ala-D-Ala as substrate. Binding studies on solid support showed that the association constant for resinbound tweezer **8.48** and tripeptide *N*-Ac-Lys-D-Ala-D-Ala complex was 1350 M^{-1}.

Schmuck and co-workers developed 2-(guanidiniocarbonyl)pyrrole cations as highly efficient binding motifs for carboxylates in aqueous media.[90,91] For example, the artificial receptor **8.49** was developed for the tetrapeptide Val-Val-Ile-Ala (Ile: isoleucine) using a combinatorial approach (Figure 8.26).[92] The tetrapeptide substrate represents a model for the self-aggregation of the amyloid-β-peptide within the brains of Alzheimer's disease patients. This sequence constitutes the C-terminal part of the amyloid-β-peptide and is one of two parts responsible for the strong self-assembly of this 42-amino-acid peptide. For this study, the N-terminus of a tripeptide arm was modified with a cationic guanidiniocarbonyl pyrrole group, which was designed to form an ion pair with the negatively charged C-terminal carboxylate of the

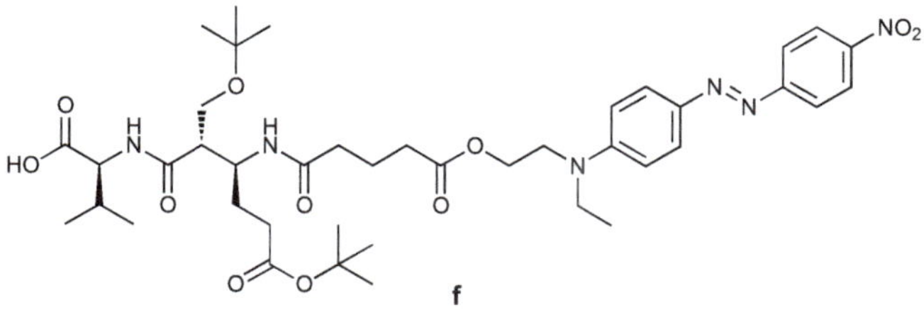

f

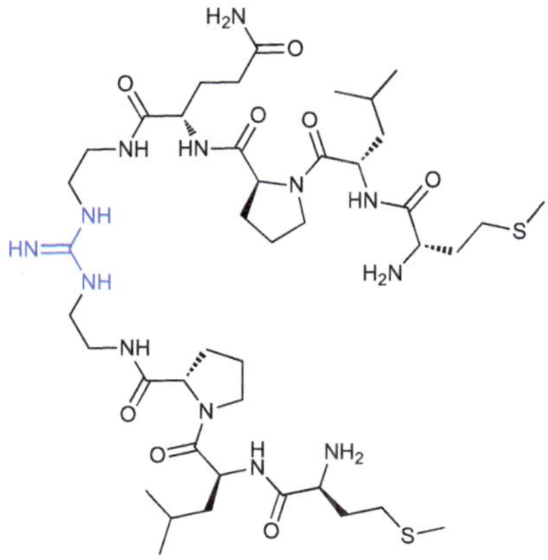

8.47

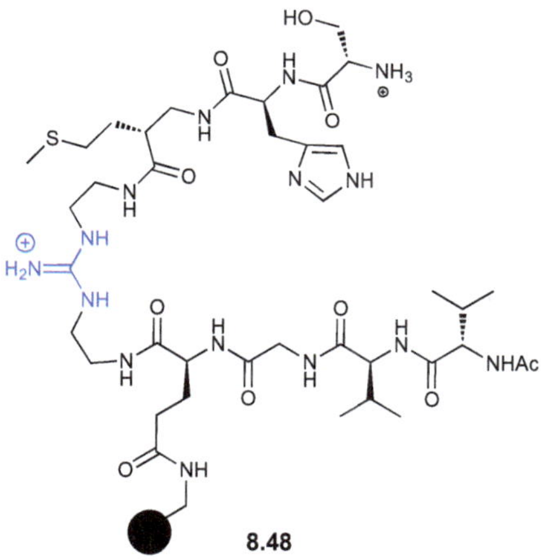

8.48

Figure 8.25 Cationic guanidinium-based receptors.

tetrapeptide guest. Additional H-bonds and electrostatic or hydrophobic interactions between the host and the bound tetrapeptide stabilize the complex and impart substrate selectivity. On-bead screening of a library of 512 receptors with the general structure of **8.49** showed that the best receptor contained Lys(Boc) next to the guanidinium cation. Most likely, additional hydrophobic interactions with the nonpolar amino acid residues

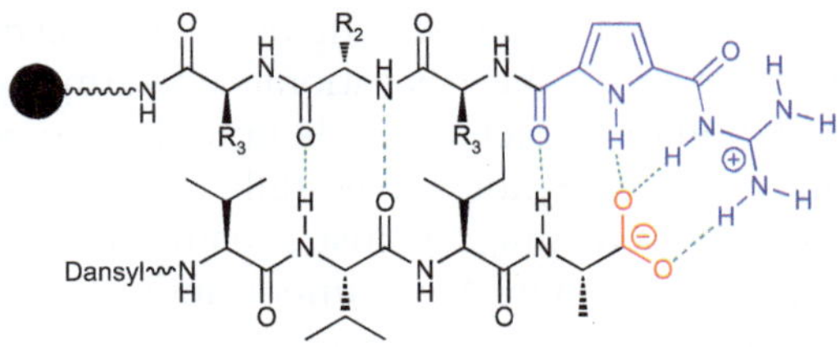

8.49

Figure 8.26 General structure of 2-(guanidiniocarbonyl)pyrrole functionalized receptor **8.49** and its interaction with a tetrapeptide.

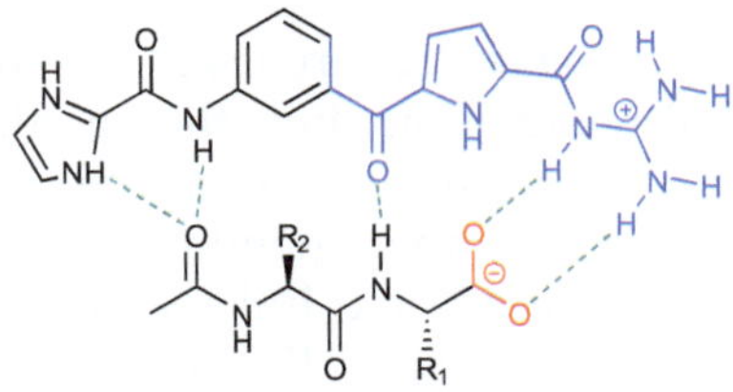

8.50

Figure 8.27 2-(Guanidiniocarbonyl)pyrrole modified receptor **8.50**.

of the substrate improved binding. Subsequent studies in solution showed that a tris-cationic receptor GCP-Lys-Leu-Lys had high affinity for the tetrapeptide with negative side chains, suggesting that electrostatic interactions are the main driving force for complexation.

In 2004, Schmuck and Geiger designed and synthesized a very simple guanidiniocarbonyl pyrrole modified receptor **8.50** for dipeptides (Figure 8.27).[93] Spectroscopic titration studies showed that **8.50** forms a 1 : 1 complex with Ac-Ala-Ala-OH in aqueous DMSO. The stability of the complex is due to ion pairing and additional hydrogen bonding between the imidazole group and the peptide backbone. Accordingly, dipeptides were bound significantly better than single amino acids. For different dipeptides the stability of the complexes increased with increasing hydrophobicity of the side chains in the order Gly < Ala < Val.

In other work, a combinatorial library of hosts **8.49** was tested for binding to the hydrophilic tetrapeptide (EKAA) in water.[94] Screening of the library revealed that receptors containing two lysine residues next to the guanidinium cation formed the strongest complexes with binding affinities $K_a > 10^4\ M^{-1}$, whereas neutral or anionic receptors showed only weak or no binding under these conditions. Later, the substrate selectivity of the same receptor library **8.49** was further explored by screening a second tetrapeptide substrate with the inverse sequence AAKE.[95] A quantitative on-bead binding assay confirmed that the "inverse" substrate is also efficiently bound by these receptors (K_a ~6000 M^{-1}) due to long-range electrostatic interactions. The work demonstrated that the tris-cationic one-armed flexible peptide receptor (GCP-Lys-Lys-Phe) efficiently binds anionic tetrapeptides, even in water.[96] Furthermore, the receptor successfully discriminated very similar substrates as shown by screening against a combinatorial library of 320 different tetrapeptide substrates. Most importantly, the receptor showed stereoselective discrimination against D-Ala and L-Ala within certain positions of the tetrapeptide substrate. The receptor was also capable of distinguishing a peptide from a depsipeptide because an attractive H-bond to a peptide amide NH was replaced by dipole–dipole repulsion with the corresponding depsipeptide ester oxygen. This change in affinity mimics the molecular basis of bacterial resistance development against the antibiotic vancomyin.

In other work, Schmuck *et al.* reported several ditopic RGD receptors combining two artificial recognition sites, the guanidiniocarbonyl pyrrole cation for the binding of the anionic aspartate and a bisphosphonate moiety for the binding of the guanidinium cation of arginine (Figure 8.28).[97] Either flexible or rigid linkers were used to connect both binding sites. Receptors with flexible linker **8.51** and **8.52** showed only moderate affinity for the H-RGD-OH tripeptide as the receptors undergo intramolecular self-association. NMR titration also showed that the receptor **8.53** was unable to bind the RGD tripeptide as it prefers to form stable homodimers. Most likely the linker in **8.53** was too short to bind the tripeptide. Only receptor **8.54** with a rigid aromatic linker binds the RGD tripeptide efficiently in aqueous buffer as this receptor does not self-associate (neither intra- nor intermolecularly). Weak binding was observed for the free H-RGD-OH tripeptide (K_a < 1000 M^{-1}), whereas significantly higher binding was observed for Ac-RGD-NH_2 (K_a = 2700 M^{-1}), which is the tripeptide sequence found within proteins. Molecular modeling studies confirmed that the most likely binding mode involves an ion pair between the cationic guanidiniocarbonyl pyrrole cation and the

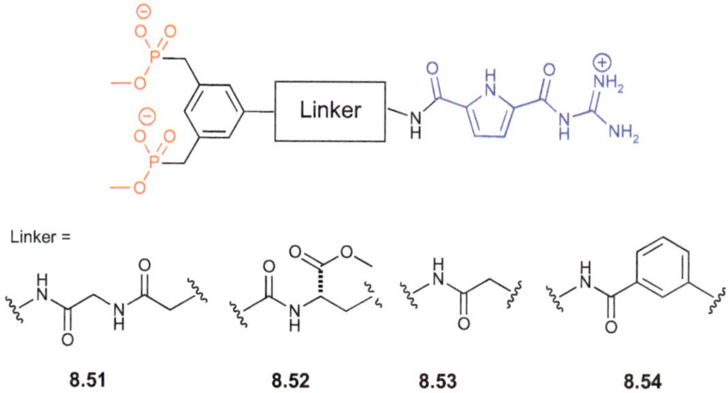

Figure 8.28 Ditopic receptors for RGD tripeptides.

carboxylate group of the aspartate side chain and an ion pair of the bisphosphonate with the arginine side chain. An additional hydrogen bond between the glycine amide carbonyl of the substrate and the aniline amide NH group in the receptor provides extra stability.

8.5.2.3 Receptors Based on Crown Ethers

Crown ethers are commonly used to form complexes with ammonium cations, although the binding affinity is modest in aqueous solvents. Hossain and Schneider designed a ditopic receptor **8.55** containing an 18-crown-6 connected to a peralkylammonium group (Figure 8.29).[98] NMR experiments showed moderate association with peptides in water ($K_a > 100\ M^{-1}$). The ammonium group of the free N-terminus of the peptide binds to the crown ether, whereas the carboxylate of the free C-terminus forms an ion pair with the peralkyl ammonium cation of the host. The receptor had selectivity for Gly-Gly-Phe over Phe-Gly-Gly, most likely due to additional π-stacking interactions between the phenylalanine and the benzocrown ether. The receptor also showed length selectivity for Gly-Gly-Gly ($K_a = 200\ M^{-1}$) over

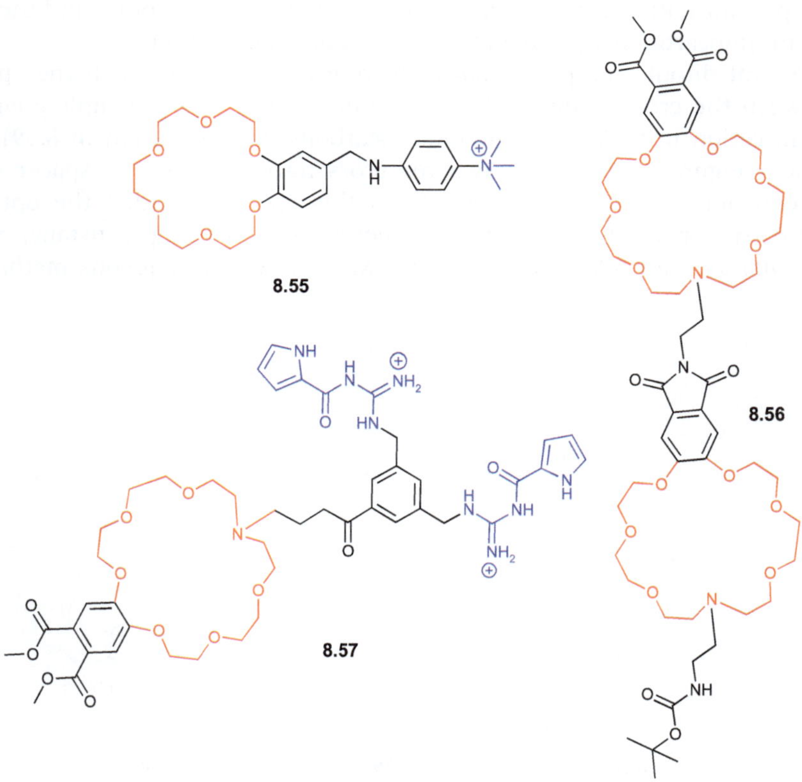

8.55

8.56

8.57

Figure 8.29 Peptide receptors containing crown ethers.

Gly-Gly-Gly-Gly ($K_a = 45\ M^{-1}$). The distance between the two charged termini in the tetrapeptide is too long to match the complementary binding sites in the host.

Mandl and König reported a double crown ether-based host **8.56** for detecting lysine and distinguishing lysine residues in different positions in small isomeric peptides.[99] The host consists of two crown ether units connected to a luminescent phthalic ester and phthalimide moiety, respectively. Receptor **8.56** was used for recognition of lysine side chains in different positions of isomeric tripeptides: H-Lys-Gly-Gly-OMe, H-Gly-Lys-Gly-OMe, and H-Gly-Gly-Lys-OMe. Interaction of the ammonium group of the lysine with the crown ether suppresses the PET process within the phthalimide moiety, leading to a significant fluorescence enhancement. A comparison of the three peptides showed weaker binding when the Lys moiety was further from the N-terminus. Later, the same group developed a synthetic receptor **8.57** for the tetrapeptide Ac-Ser-Asp-Lys-Pro, having two carboxylate groups and a lysine ammonium side chain as potential binding sites.[100] The receptor contained a luminescent crown ether for binding the lysine side chain and two pyrrole-guanidinium units for binding the carboxylates. The receptor exhibited high 1 : 1 peptide affinity ($K_a = 7 \times 10^3\ M^{-1}$) as determined by fluorescence titrations experiments. Control experiments showed that the ion pair formation between the two carboxylates of the peptide and the two guanidiniocarbnonyl pyrrole cations was essential for binding.

Several ditopic receptors **8.58–8.60** were evaluated in which the spacer between the crown ether and the cationic binding site (a simple guanidinium cation in **8.58** and a guanidiniocarbonyl pyrrole cation in **8.59**) was varied (Figure 8.30).[101] The two receptors having a xylylene spacer were specific for the tetrapeptide EGGG as this spacer provided the optimal geometry for binding as 1 : 1 complexes. The binding constants were $K_a = 400\ M^{-1}$ for **8.58** and $K_a = 13\,000\ M^{-1}$ for **8.59** in aqueous methanol.

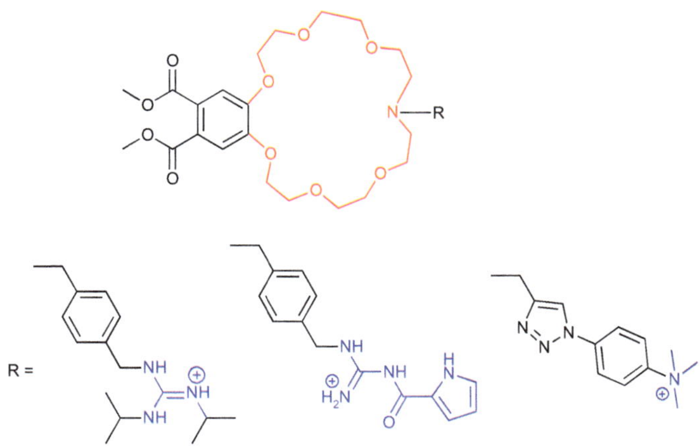

8.58 8.59 8.60

Figure 8.30 Peptide receptors containing crown ethers.

These data underline again the superior binding properties of the guanidiniocarbonyl pyrrole cation compared to a simple guanidinium cation. Peptides containing glutamine instead of glutamate did not show any fluorescence response, confirming the requirement for ion pairing. Other peptides, GEGG and GGEG, which have a smaller distance between the two binding sites, also showed weaker binding. Receptor **8.60**, having a triazolyl spacer was found to be selective for the tripeptide Gly-Gly-Gly ($K_a = 600\ M^{-1}$) in aqueous methanol.

8.5.3 Peptide Recognition Mediated by Metal Coordination

Incorporation of a metal cation into a synthetic receptor as a peptide binding site is one of the most effective ways to achieve high complex stability in aqueous solution. Oxygen- and nitrogen-containing functional groups such as phosphate, carboxylate, or imidazole form strong coordination bonds to transition metal cations in water. Thus, the receptor design strategy is to create a suitable organic scaffold that will chelate the metal cation and also provide additional functionality for secondary non-covalent interactions.

8.5.3.1 Receptors Based on Zn-complexes

Ait-Haddou *et al.* reported a Zn^{2+}-containing tweezer-like receptor **8.61** that exhibited high affinity for aspartic acid ($K_a = 1.5 \times 10^5\ M^{-1}$).[102] The binding affinity was dominated by coordination of the carboxylate with the metal center with the convergent guanidinium groups enhancing the amino acid selectivity (Figure 8.31).

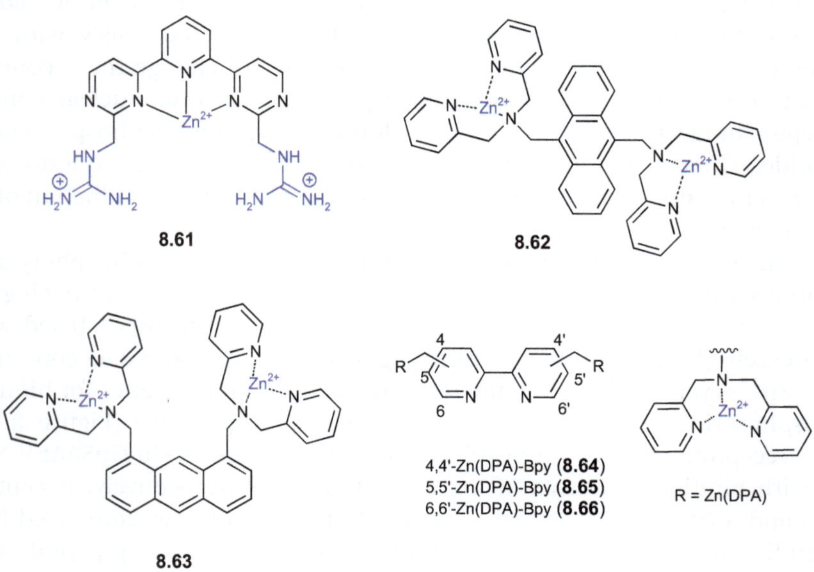

Figure 8.31 Zn^{2+} complexes for recognition of phosphorylated peptides.

In the last two decades, Zn^{2+} complexes with a bis(2-pyridylmethyl)amine [Zn(DPA)] unit have attracted considerable attention due to their phosphate-recognition properties. For example, Hamachi and co-workers demonstrated fluorescent detection of phosphorylated peptides under aqueous conditions.[103,104] They investigated anthracene based Zn(DPA) complexes **8.62** and **8.63** as peptide hosts. Addition of phosphorylated tyrosine (p-Tyr) or inorganic phosphate produced a fluorescence increase due to a PET mechanism. Addition of tyrosine and several types of small anions did not show any significant change in emission intensity. The receptors were tested for peptide sensing and four to five-fold fluorescence enhancements were observed upon addition of phosphorylated peptides with high binding affinity ($K_a \sim 10^7$ M^{-1}). No fluorescence change was observed upon addition of non-phosphorylated peptides. Additional control studies showed that electrostatic interactions between the positively charged Zn(DPA) host and the negatively charged peptide was a major factor for both binding and the concomitant fluorescence change. Furthermore, cooperative coordination of both Zn(DPA) sites in **8.62** and **8.63** was required for optimal phosphate binding. The practical potential of the fluorescent receptors was demonstrated by monitoring an enzyme-catalyzed dephosphorylation reaction.

Later, the same group used Zn(DPA) complexes to identify multi-phosphorylated peptides using a ditopic recognition strategy.[105] They connected two Zn(DPA) units to bipyridine at either the 4,4′-, 5,5′-, or 6,6′-positions, respectively (**8.64, 8.65, 8.66**), for targeting two phosphorylated amino acids separated by a certain distance within a single peptide sequence. Upon interaction with an oligopeptide containing two phosphorylated serine residues, the dinuclear receptors having 4,4′- or 5,5′-substitution patterns induced significant amounts of α-helicity in the peptide. The peptide folding allowed the two phosphate groups to interact simultaneously with the ditopic receptors. Accordingly, the receptors did not change the secondary structure of a mono-phosphorylated peptide, and a mononuclear control receptor induced very little helical conformation in the bis-phosphorylated peptide. Thus, a ditopic interaction between the Zn(DPA) complexes with two p-Ser residues was required for high binding affinity and induction of an α-helix conformation.

König *et al.* reported several receptors for identifying phosphorylated peptides with different i + 3 amino acid side chains under physiological conditions.[106] The N-terminus of the peptides (**g** and **h**) was labeled with fluorescein (Flu) to monitor the binding. Receptors **8.67** and **8.68** contained two (Zn^{2+}–cyclen) triazine complexes and guanidinium moieties for binding phosphoserine and carboxylate functionalities, respectively (Figure 8.32). Both receptors exhibited high affinity for peptides such as Flu-GpSAAEV-NH_2 (**g**) with binding constants of log $K_a = 8.0$ and 7.8, respectively. Receptors **8.69** and **8.70** have two bis(Zn^{2+}–cyclen) triazine complexes connected by a peptidic linker. These receptors bind the histidine-containing peptide Flu-GpSAAHV-NH_2 (**h**) stronger (log $K_a = 7.5$ and 6.5, respectively) than glutamate-containing peptide **g** (log $K_a = 5.0$ and 4.9, respectively). Apparently, the

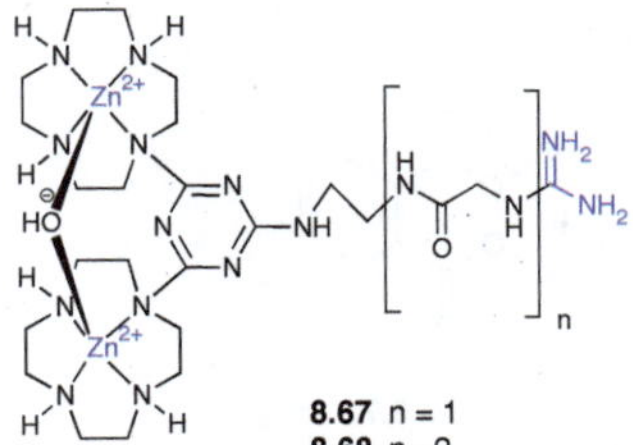

8.67 n = 1
8.68 n = 2

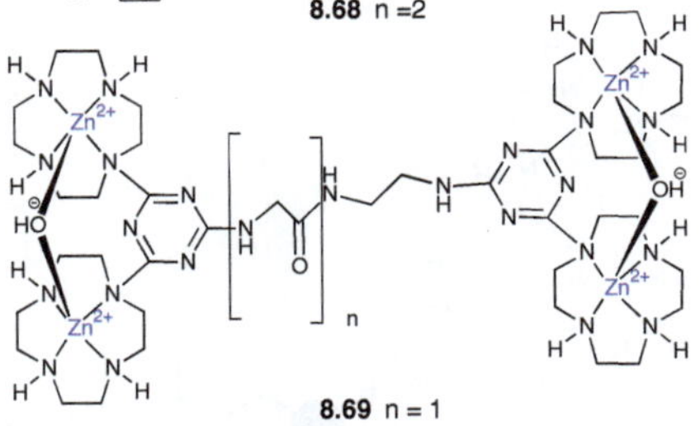

8.69 n = 1
8.70 n = 2

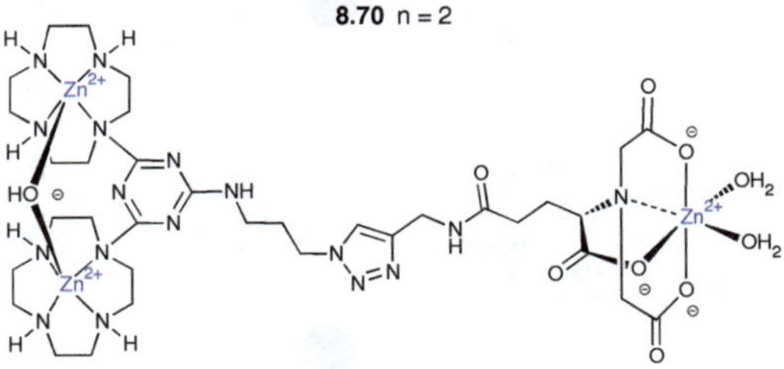

8.71

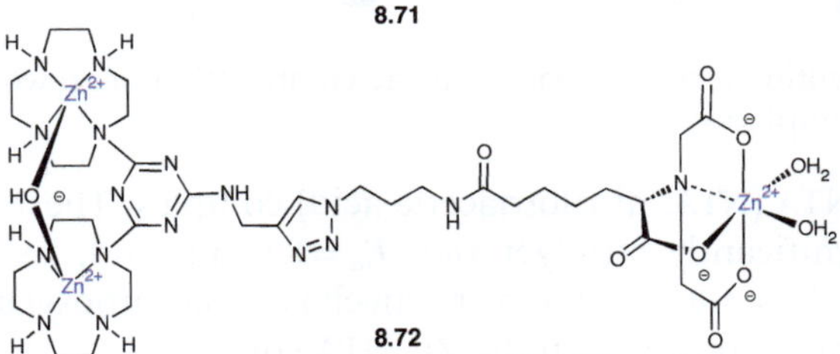

8.72

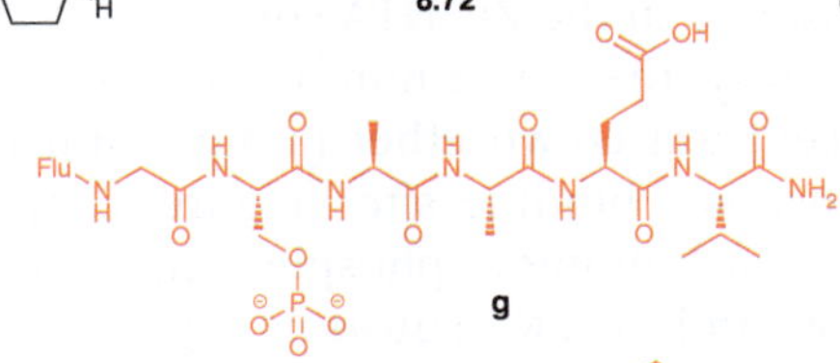

g

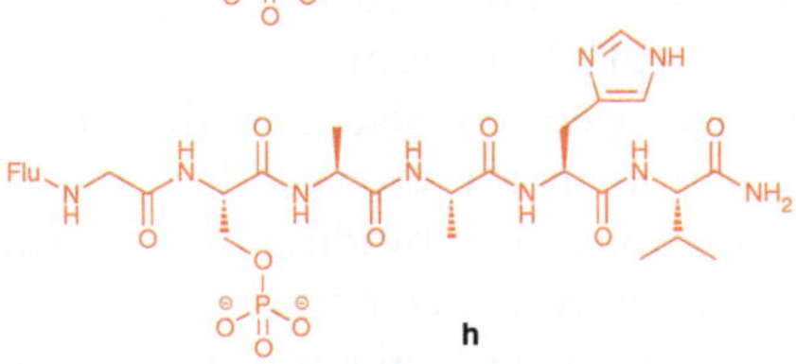

h

Figure 8.32 Zn^{2+} complexes for recognition of phosphorylated peptides.

imidazole side chain of the peptide **h** coordinates to the second bis(Zn^{2+}–cyclen) triazine present in **8.69** and **8.70**, providing extra stability. The third set of receptors **8.71** and **8.72** combine a bis(Zn^{2+}–cyclen) triazine complex

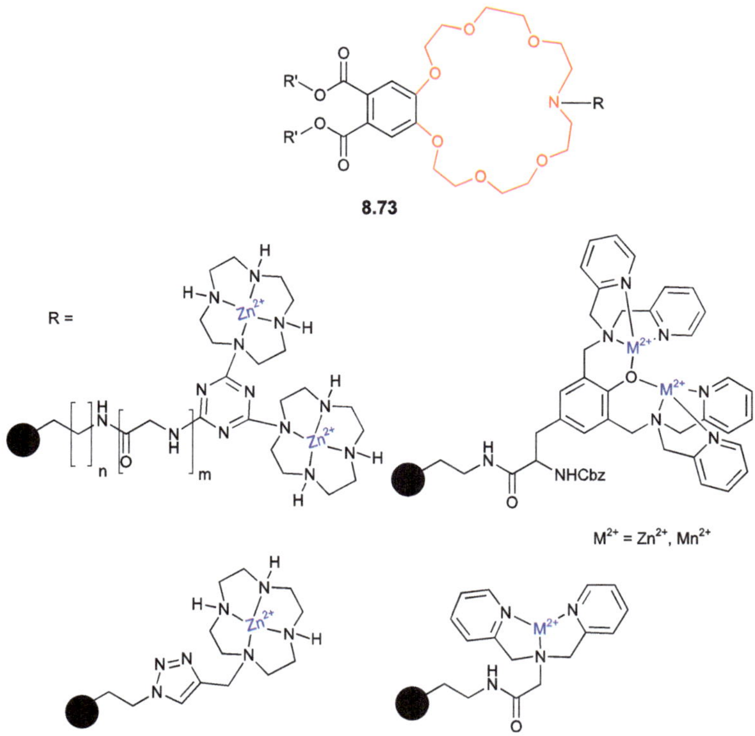

Figure 8.33 Peptide receptors based on the combination of crown ether and metal complexes.

with a Zn^{2+}–NTA (NTA: nitrilotriacetic acid) complex. These receptors bind peptide **g** significantly stronger (log $K_a = 8.0$ and 7.9, respectively) than peptide **h** (log $K_a = 4.8$ and 4.6, respectively) as the carboxylate functionality in peptide **g** can coordinate to the Zn–NTA complex.

Later, the same group synthesized a number of synthetic receptors (**8.73**) connecting a fluorescent aza crown ether (ammonium binding site) with metal complexes (phosphate binding site) (Figure 8.33).[107] They screened this series of receptors for binding to phosphoserine, the oligopeptide Tyr-pSer-Pro-Thr-Ser-Pro-Ser and the PKG substrate Arg-Lys-Arg-pSer-Arg-Ala-Glu in aqueous buffer solution (pH 7.4). Significant fluorescence enhancements were observed and the data were consistent with simultaneous binding of the lysine ammonium cation in the crown ether and coordination of the phosphate to the metal complex. The binding of the lysine side chain to the crown was responsible for fluorescence enhancement.

The same group also functionalized methoxypyrrole amino acids (MOPAS) with a histidine-coordinating Zn–NTA complex to produce a peptide β-sheet binder (**8.74**) in DMSO (Figure 8.34).[108] Intramolecular hydrogen bonds between the MOPAS amide NH to the methoxy groups are responsible for the planarity of the host and binding to the pentapeptide H_2N-His-Leu-Leu-Val-Phe-OMe induced a β-sheet conformation in the peptide. In other

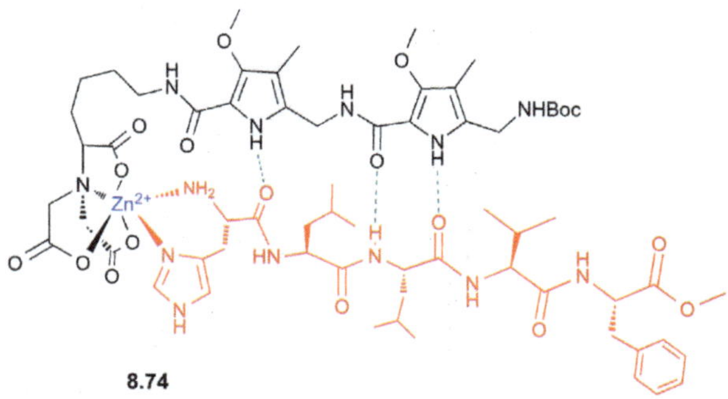

8.74

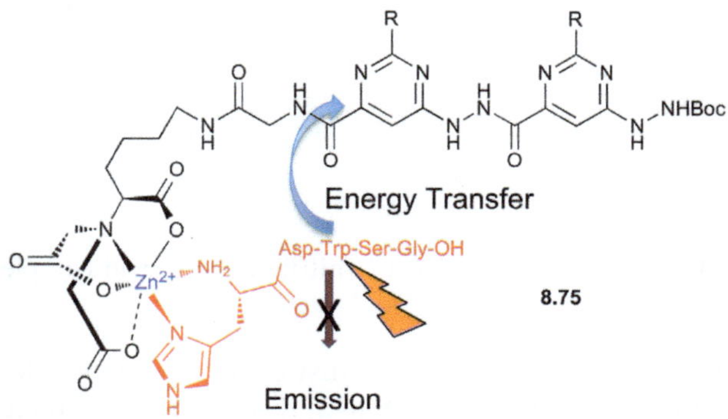

8.75

Figure 8.34 Histidine-coordinating Zn^{2+}–nitrilotriacetic acid complex receptors.

work, a histidine-coordinating Zn–NTA complex was combined with a fluorescent pyrimidine hydrazine acid (PHA).[109] The receptor **8.75** binds to the pentapeptides H-His-Asp-Trp-Ser-Gly-OH and H-His-Thr-Trp-Asp-Asp-OH with binding constants of log $K_a = 4.6$ and log $K_a = 4.0$, respectively. The binding is primarily due to coordination of the Zn^{2+}–NTA by the N-terminal histidine with additional weak secondary interactions. Receptor **8.75** quenched the fluorescence of the tryptophan group in both peptides due to an efficient fluorescence resonance energy transfer from Trp to the PHA chromophore.

8.5.3.2 Receptors Based on Cu-complexes

Mallik and co-workers used a Cu^{2+}–iminodiacetic acid complex for binding histidine-containing flexible peptides in an aqueous buffer (pH 7.0).[110] The Cu^{2+} complex **8.76** strongly binds substrate **I** which contains three Gly-His dipeptides in the correct arrangement and distance ($K_a = 1.2 \times 10^6$ M^{-1}) (Figure 8.35). Related substrates with shorter or longer histidine-containing peptide arms or a simple histidine-containing peptide showed only weak binding affinity.

König and co-workers combined a benzo crown ether with a pendant copper imidodiacetic acid complex to develop luminescent receptor **8.77** for

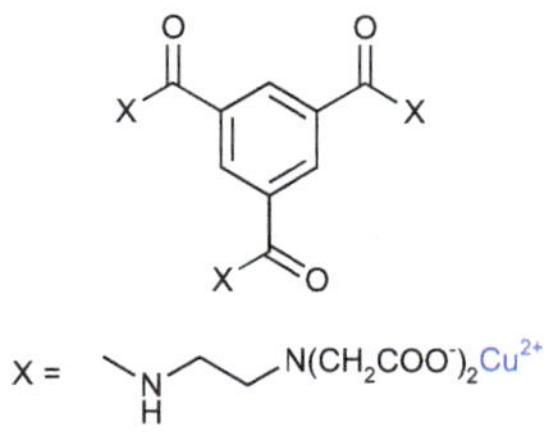

8.76

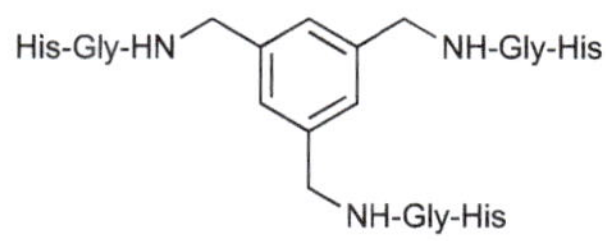

I

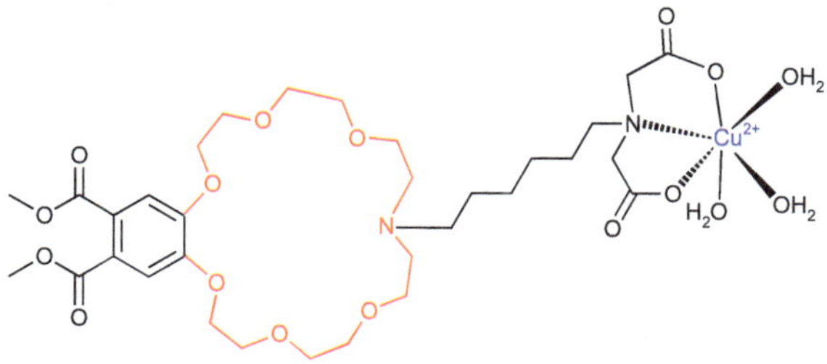

8.77

Figure 8.35 Histidine-coordinating Cu^{2+}–nitrilotriacetic acid complex receptors.

small peptides containing N-terminal histidine.[111] Receptor **8.77** selectively forms a 1 : 1 complex with His-Lys-OMe and the binding constant was found to be log $K_a = 4.22$. The crown ether in **8.77** is responsible for binding of the lysine side chain and the pendant copper imidodiacetic acid complex coordinates to the histidine. Binding was accompanied by a significant enhancement of emission intensity. The receptor however did not bind the acylated dipeptide Ac-His-Lys-OMe, showing the importance of the free N-terminus.

Wright and Anslyn developed a tweezer-like receptor **8.78** in which a Cu^{2+}–metal complex preorganizes two variable peptide arms for the recognition of tripeptides in buffered water/methanol solution (pH 7.4) (Figure 8.36).[112] The Cu^{2+} also presents an additional binding site for His, Cys or Met in the peptide N-terminus. The peptidic arms were enriched with acidic amino acid residues for efficient ion pairing to complementary charged groups in the guest. Absorbance studies showed that the receptor was indeed selective for tripeptides: L-Xxx-L-Lys-L-Lys, with Xxx = His, Cys, and Met, with association constants $\sim 10^6$ M^{-1}.

Based on this study, Wright *et al.* also developed a library, **8.79**, and tested it for the discrimination of some simple tripeptides and tripeptide mixtures.[113] The structure of the first arm was kept constant as Asp-Gly-Lys, and the second arm was combinatorially varied using all natural amino acids except Cys, resulting in a library size of $19^3 = 6859$ members. Thirty library members were randomly chosen for a recognition study with four tripeptides (His-Lys-Thr, His-Glu-Thr, His-Gly-Thr, and Gly-His-Thr) as well as three mixtures containing two of the tripeptides. The receptors were placed in a

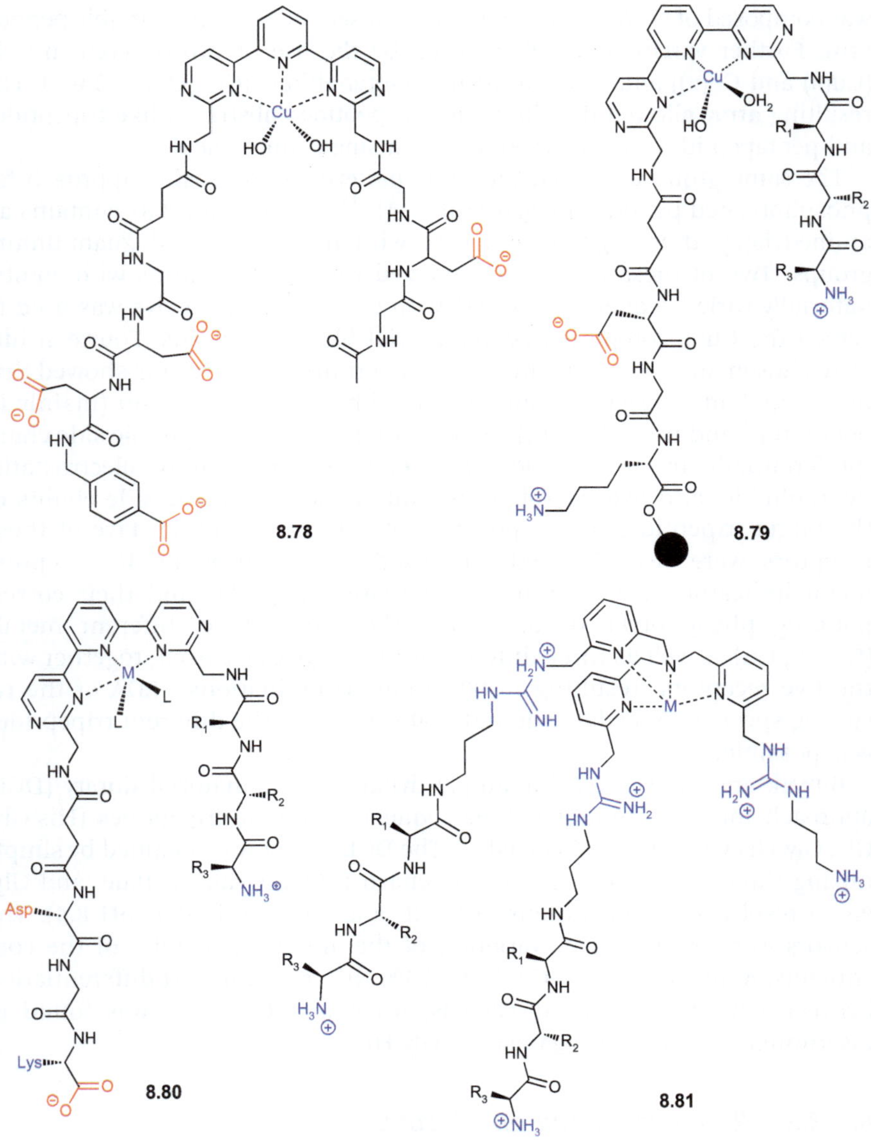

Figure 8.36 Metal complex receptors for peptides.

microarray chip and used in an indicator uptake assay using Orange G as dye for visualization. PCA of the responses revealed that this receptor array was capable of successfully discriminating the tripeptides. The same research group used a similar approach successfully to study tachykinins and similar peptides as analytes in buffer solution (pH 7.4).[114] Six library members **8.80** were identified that bound to neurokinin A, a decapeptide that is an essential neurotransmitter in the central nervous system. In this case the sensing array was composed of six receptors with different sequences in the variable peptide arm. Further variation was introduced by the use of two different metals (Cu(II) and Cd(II)) and three different counter-anions (Cl^-, OTf^-, OAc^-). The resulting array allowed discrimination of peptide substrates, like tripeptides and pentapeptides which varied only by a single amino acid.

The same group also used the same pattern-based sensing approach for phosphorylated peptides in water (pH 7.4).[115] The receptor **8.81** contains an amine-tris(pyridine) ligand as a core with three appended guanidinium groups. Two of these were functionalized with peptidic arms with combinatorially varied sequences. An IDA with Celestine Blue as dye was used to screen the Cu(II)-complexed combinatorial library on beads. Fourteen hits were chosen and their sequences were determined. The results showed that most hits had at least one amino acid with a polar side chain (mainly in position R^1 and R^2 of the arm) and one nonpolar or hydrophobic side chain (preferentially in position R^3). Hence, a combination of electrostatic, hydrophobic, and hydrogen-bonding interactions with the side chains of the target tripeptides are responsible of substrate binding. Five of these receptors were resynthesized and used for an array of 45 receptor–metal–indicator ensembles to differentiate tripeptides and their corresponding phosphorylated forms. For the array, three different metals (Co, Ni, Cu) as well as three different indicator dyes we used, together with the five receptors (resulting in 90 unique combinations). LDA of the resulting spectroscopic data showed that sensing of the different tripeptides was possible.

Buryak and Severin used a simple dynamic combinatorial library (DCL) approach for discriminating three sequence-isomeric tripeptides (His-Gly-Gly, Gly-His-Gly, and Gly-Gly-His).[116] The DCL sensor was obtained by simply mixing three metal-binding dyes, Arsenazo I, Methylcalcein Blue, and Glycine Cresol Red with solutions of $CuCl_2$ and $NiCl_2$ in buffer (pH 8.4). The sensors were optimized by variation of the amount and ratio of the constituents. A mixture of 25% Cu^{2+} and 75% Ni^{2+} gave the best differentiation between Gly-His-Gly and Gly-Gly-His, whereas 100% Cu^{2+} was found to discriminate His-Gly-Gly against Gly Gly-His.

8.5.3.3 Receptors Based on Rh-complexes

Buryak and Severin have also developed the rhodium complex **8.21** (Cp*Rh) for sequence-specific detection of peptides containing His and Met in aqueous phosphate buffer solution (pH 7.0) (Figure 8.9).[117] An IDA was created using an ensemble of the rhodium complex and the indicator azophloxine. There was high selectivity for peptides with His/Met residues near the N-terminus with binding constants $>10^3$ M^{-1}. This was due to simultaneous coordination of the free N-terminus of the amino acid and the respective side chain with the rhodium.

Later, a sensor array was developed combining three transition-metal complexes [$PdCl_2$(en)], [{$RhCl_2$Cp*}$_2$], and [{$RuCl_2$(p-cymene)}$_2$] with six different fluorescent dyes.[118] A sensor array of only six metal–dye combinations could differentiate 10 different dipeptides at low-micromolar concentrations in aqueous buffer. This sensor array was also able to distinguish a mixture of the nonapeptide, bradykinin, and the decapeptide, kallidin. Moreover, the cross-reactive sensor could also identify and quantify the pharmacologically interesting dipeptides, carnosine, and homocarnosine, in human serum.

8.5.4 Peptide Recognition by Molecularly Imprinted Polymers

Molecularly imprinted polymers have been pursued for several decades as a way to create insoluble polymers with high-affinity binding sites for small molecules. In recent years, the field has developed imprinted nanoparticles that can be dispersed into solution, and this work has led to systems that act like "plastic antibodies" with an ability to selectively recognize relatively large biomolecules. Recent work by the Shea group has produced methods of preparing imprinted nanoparticles that recognize peptides. One of their prototype systems was designed to sequester melittin, a 26-amino-acid peptide that lyses red blood cells. The imprinted nanoparticle was produced by polymerizing a collection of acrylamide monomers with different functional groups, in the presence of the peptide and a small concentration of surfactant.[119] The peptide template was removed by extensive dialysis and a range of binding studies demonstrated high and selective melittin affinity.[120] Subsequent biological testing showed that the imprinted nanoparticles were able to sequester free melittin from blood and prevent blood cell lysis.[121] Furthermore, the imprinted nanoparticles were able to clear melittin from the blood stream of living animals and dramatically decrease animal mortality.[122] This effect was not observed with non-imprinted control nanoparticles. The generality of the imprinting process is very attractive and it should be applicable to other peptides. The work opens the door for new classes of plastic antidotes that are able to bind and neutralize toxic peptides.

8.6 Future Directions and Challenges

The goal of obtaining strong and selective binding of amino acids or peptides by artificial receptors has fascinated many research groups over the years. Each group developed a different approach, which has led to an extensive set of artificial receptors, but in many ways the field is still in its infancy. Only in recent years has it become possible to produce synthetic receptors that operate efficiently in water and now the challenge is to augment the basic recognition process with performance features that enable practical outcomes. The recognition of amino acids utilizing indicator-displacement assays, metal cation coordination, or specific reactions (*e.g.*, imine formation) has been achieved. Due to the similarity in structure and reactivity, it is still challenging to develop receptors that can discriminate amino acids with high selectively and sensitivity. For example, it is still difficult to distinguish the three biological important thiols Cys, Hcy and GSH in certain biomedical samples. However, progress to date has been encouraging and indicates that improved amino acid receptors will soon be developed.

During the past decade, high affinity and sequence-selective recognition of small peptides in water has been achieved using synthetic receptors. Moreover, there is now a good understanding of the major supramolecular factors that enhance peptide recognition by synthetic receptors. Conversely, there is clearly room for improvement and the field has to progress considerably to achieve practically useful peptide recognition performance in complex matrices such as biological fluids. The recent *in vivo* success using peptide-imprinted nanoparticles is encouraging. Eventually, synthetic peptide receptors are likely to be employed as pharmaceutical agents, antidotes for toxic peptides, or as tools in cell biology research and chemical genetics.

Another important future direction is the development of chemical indicators that report the levels of amino acid or peptides in various types of biomedical or environmental settings. In some cases the indicators will be incorporated into analytical and diagnostic devices for batch sampling. In other cases, the indicators will be part of sensing devices for real-time monitoring. Not only is the recognition and detection of peptides an important topic but there is also a need to create enzyme mimics that catalyze chemical reactions with peptide substrates. Artificial proteases are expected to have many applications, ranging from environmental remediation to pharmaceutical activity. Catalyzed amide bond hydrolysis is very challenging[123] and to date most work has focused on the easier problem of ester hydrolysis.[124] An ambitious but potentially very useful example of the latter is a synthetic agent that could cleave the D-Ala-D-Lac depsipetide ester bond that is present in the cell wall of vancomycin-resistant bacteria.[125] Success in achieving this antibiotic application would have substantial global impact.

The field has made tremendous progress over the past 20 years, and it is exciting to contemplate what will become possible in the next decade.

References

1. H.-J. Böhm and G. Klebe, *Angew. Chem., Int. Ed. Engl.*, 1996, **35**, 2588.
2. E. A. Meyer, R. K. Castellano and F. Diederich, *Angew. Chem., Int. Ed.*, 2003, **42**, 1210.
3. P. Wellendorph and H. Bräuner-Osborne, *Br. J. Clin. Pharmacol.*, 2009, **156**, 869.
4. A. Corti and F. Curnis, *J. Cell Sci.*, 2011, **124**, 515.
5. J. E. Geraci and P. E. Herman, *Mayo Clin. Proc.*, 1983, **58**, 88.
6. M. Foldes, R. Munro, T. C. Sorrell, S. Shanker and M. Toohey, *J. Antimicrob. Chemother.*, 1983, **11**, 21.

7. J. C. J. Barna and D. H. Williams, *Annu. Rev. Microbiol.*, 1984, **38**, 339.
8. J. H. Rao, J. Lahiri, L. Isaacs, R. M. Weis and G. M. Whitesides, *Science*, 1988, **280**, 708.
9. H. R. Perkins, *Biochem. J.*, 1969, **111**, 195.
10. G. M. Sheldrick, P. G. Jones, O. Kennard, D. H. Williams and G. A. Smith, *Nature*, 1978, **271**, 223.
11. J. S. Rajagopalan, C. M. Harris and T. M. Harris, *Bioorg. Chem.*, 1995, **23**, 54.
12. D. H. Williams, *Acc. Chem. Res.*, 1984, **17**, 364.
13. D. H. Williams and B. Dardsley, *Angew. Chem., Int. Ed.*, 1999, **38**, 1172.
14. A. Okano, A. Nakayama, A. W. Schammel and D. L. Boger, *J. Am. Chem. Soc.*, 2014, **136**, 13522.
15. D. W. Christianson and W. N. Lipscomb, *Acc. Chem. Res.*, 1989, **22**, 62.
16. M. W. Peczuh and A. D. Hamilton, *Chem. Rev.*, 2000, **100**, 2479.
17. H. J. Schneider, *Adv. Supramol. Chem.*, 2000, **6**, 185.
18. H. J. Schneider, *Angew. Chem.*, 1993, **105**, 890; *Angew. Chem., Int. Ed. Engl.*, 1993, **32**, 848.
19. T. H. Webb and C. S. Wilcox, *Chem. Soc. Rev.*, 1993, **22**, 383.
20. M. Maue and T. Schrader, *Angew. Chem.*, 2005, **117**, 2305; *Angew. Chem., Int. Ed.*, 2005, **44**, 2265.
21. W. L. Wong, K.-H. Huang, P.-F. Tei, C.-S. Lee and H.-L. Kwong, *Chem. Commun.*, 2005, 384.
22. Z. Zhong and E. V. Anslyn, *J. Am. Chem. Soc.*, 2002, **124**, 9014.
23. S. E. Schneider, S. N. O'Neil and E. V. Anslyn, *J. Am. Chem. Soc.*, 2000, **122**, 542.
24. K. Niikura, A. Metzger and E. V. Anslyn, *J. Am. Chem. Soc.*, 1998, **120**, 8533.
25. H. Yin and A. D. Hamilton, *Angew. Chem.*, 2005, **117**, 4200; *Angew. Chem., Int. Ed.* 2005, **44**, 4130.
26. D. L. Dong, R. Liu, R. Sherlock, M. H. Wigler and H. P. Nestler, *Chem. Biol.*, 1999, **6**, 133.
27. S. Shahrokhian, *Anal. Chem.*, 2001, **73**, 5972.
28. H. Yoshida, Y. Nakano, K. Koiso, H. Nohta, J. Ishida and M. Yamaguchi, *Anal. Sci.*, 2001, **17**, 107.
29. G. N. Chen, X. P. Wu, J. P. Duan and H. Q. Chen, *Talanta*, 1999, **49**, 319.
30. G. M. Mackay, C. M. Forrest, N. Stoy, J. Christofides, M. Egerton, T. W. Stone and L. G. Darlington, *Eur. J. Neurol.*, 2006, **13**, 30.
31. H. Refsum, P. M. Ueland, O. Nygård and S. E. Vollset, *Annu. Rev. Med.*, 1989, **49**, 31.
32. C. S. Lee, P. F. Teng, W. L. Wong, H. L. Kwong and A. S. C. Chan, *Tetrahedron*, 2005, **61**, 7924.
33. J. Wang, M. Chtrathi and B. Tian, *Anal. Chem.*, 2000, **72**, 5774.
34. D. Vardanega and C. Giradet, *Chem. Phys. Lett.*, 2009, **469**, 172.
35. C. Schmuck, *Chem. Commun.*, 1999, 843.
36. C. Schmuck and V. Bickert, *Org. Lett.*, 2003, **5**, 4579.
37. C. Schmuck and U. Machon, *Chem. – Eur. J.*, 2005, **11**, 1109.

38. C. Schmuck and S. Graupner, *Tetrahedron Lett.*, 2005, **46**, 1295.
39. C. Schmuck and L. Geiger, *J. Am. Chem. Soc.*, 2005, **127**, 10486.
40. S. Sasaki, A. Hashizume, D. Citterio, E. Fujiia and K. Suzuki, *Tetrahedron Lett.*, 2002, **43**, 7243.
41. L. Yang, S. Qin, X. Su, F. Yang, J. You, C. Hu, R. Xie and J. Lan, *Org. Biomol. Chem.*, 2010, **8**, 339.
42. M. Hariharan, S. C. Karunakaran and D. Ramaiah, *Org. Lett.*, 2007, **9**, 417.
43. S. Liu, L. Fang, Y. He, W. Chan, K. Yeung, Y. Cheng and R. Yang, *Org. Lett.*, 2005, **7**, 5825.
44. M. A. Hortalá, L. Fabbrizzi, N. Marcotte, F. Stomeo and A. Taglietti, *J. Am. Chem. Soc.*, 2003, **125**, 20.
45. J. F. Folmer-Andersen, V. M. Lynch and E. V. Anslyn, *Chem. - Eur. J.*, 2005, **11**, 5319.
46. J. F. Folmer-Andersen, V. M. Lynch and E. V. Anslyn, *J. Am. Chem. Soc.*, 2005, **127**, 7986.
47. J. F. Folmer-Andersen, M. Kitamura and E. V. Anslyn, *J. Am. Chem. Soc.*, 2006, **128**, 5652.
48. A. Buryak and K. Severin, *J. Am. Chem. Soc.*, 2005, **127**, 3700.
49. Y. Yang, S. Shim and J. Tae, *Chem. Commun.*, 2010, **46**, 7766.
50. G. Patel and S. Menon, *Chem. Commun.*, 2009, 3563.
51. E. K. Feuster and T. E. Glass, *J. Am. Chem. Soc.*, 2003, **125**, 16174.
52. Y. Zhou, J. Won, J. Y. Lee and J. Yoon, *Chem. Commun.*, 2011, **47**, 1997.
53. R. Zhang, X. Yu, Z. Ye, G. Wang, W. Zhang and J. Yuan, *Inorg. Chem.*, 2010, **49**, 7898.
54. H. Kwon, K. Lee and H.-J. Kim, *Chem. Commun.*, 2011, **47**, 1773.
55. J. Lu, C. Sun, W. Chen, H. Ma, W. Shi and X. Li, *Talanta*, 2011, **83**, 1050.
56. B. Zhu, X. Zhang, Y. Li, P. Wang, H. Zhang and X. Zhuang, *Chem. Commun.*, 2010, **46**, 5710.
57. R. Breslow, *Acc. Chem. Res.*, 1991, **24**, 159–164.
58. J. Szejtli, *Chem. Rev.*, 1998, **98**, 1743.
59. W. Saenger, *Angew. Chem., Int. Ed. Engl.*, 1980, **19**, 344.
60. G. Wenz, *Angew. Chem., Int. Ed. Engl.*, 1994, **33**, 803.
61. R. Breslow and B. Zhang, *J. Am. Chem. Soc.*, 1992, **114**, 5882.
62. M. Maletic, H. Wennemers, D. Q. McDonald, R. Breslow and W. C. Still, *Angew. Chem., Int. Ed. Engl.*, 1996, **35**, 1490.
63. R. Breslow, Z. Yang, R. Ching, G. Trojandt and F. Odobel, *J. Am. Chem. Soc.*, 1998, **120**, 3536.
64. Y. Liu, Y.-L. Zhao, Y. Chen, F. Ding and G.-S. Chen, *Bioconjugate Chem.*, 2004, **15**, 1236.
65. Y. Liu, G.-S. Chen, Y. Chen, F. Ding, T. Liu and Y.-L. Zhao, *Bioconjugate Chem.*, 2004, **15**, 300.
66. L. Isaacs, *Acc. Chem. Res.*, 2014, **47**, 2052.
67. J. W. Lee, S. Samal, N. Selvapalam, H.-J. Kim and K. Kim, *Acc. Chem. Res.*, 2003, **36**, 621.

68. M. E. Bush, N. D. Bouley and A. R. Urbach, *J. Am. Chem. Soc.*, 2005, **127**, 14511.
69. L. M. Heitmann, A. B. Taylor, P. J. Hart and A. R. Urbach, *J. Am. Chem. Soc.*, 2006, **128**, 12574.
70. D. T. Dang, H. D. Nguyen, M. Merkx and L. Brunsveld, *Angew. Chem., Int. Ed.*, 2013, **52**, 2915.
71. M. V. Rekharsky, H. Yamamura, Y. H. Ko, N. Selvapalam, K. Kim and Y. Inoue, *Chem. Commun.*, 2008, 2236.
72. J. M. Chinai, J. M. Chinai, A. B. Taylor, L. M. Ryno, N. D. Hargreaves, C. A. Morris, P. J. Hart and A. R. Urbach, *J. Am. Chem. Soc.*, 2011, **133**, 8810.
73. M. Tomeiro and W. C. Still, *J. Am. Chem. Soc.*, 1995, **117**, 5887.
74. S. Tashiro, M. Tominaga, M. Kawano, B. Therrien, T. Ozeki and M. Fujita, *J. Am. Chem. Soc.*, 2005, **127**, 4546.
75. P. Gund, *J. Chem. Educ.*, 1972, **49**, 100.
76. H. Wennemers and W. C. Still, *Tetrahedron Lett.*, 1994, **35**, 6413.
77. P. Krattiger and H. Wennemers, *Synlett*, 2005, **4**, 0706.
78. J. Klein and A. Medlik, *J. Chem. Soc., Chem. Commun.*, 1973, 275.
79. M. D. Best, S. L. Tobey and E. V. Anslyn, *Coord. Chem. Rev.*, 2003, **240**, 3.
80. P. A. Gale, *Coord. Chem. Rev.*, 2003, **240**, 191.
81. R. J. Fitzmaurice, G. M. Kyne, D. Douheret and J. D. Kilburn, *J. Chem. Soc., Perkin Trans. 1*, 2002, 841.
82. F. P. Schmidtchen and M. Berger, *Chem. Rev.*, 1997, **97**, 1609.
83. P. Blondeau, M. Segura, R. Perez-Fernandez and J. de Mendoza, *Chem. Soc. Rev.*, 2007, **36**, 198.
84. K. A. Schug and W. Lindner, *Chem. Rev.*, 2005, **105**, 67.
85. R. D. Dixon, J. S. Geib and A. D. Hamilton, *J. Am. Chem. Soc.*, 1992, **114**, 365.
86. C. Schmuck, *Chem. – Eur. J.*, 2000, **6**, 709.
87. C. Schmuck, *Coord. Chem. Rev.*, 2006, **205**, 3053.
88. K. B. Jensen, T. M. Braxmeier, M. Demarcus, J. G. Frey and J. D. Kilburn, *Chem. – Eur. J.*, 2002, **8**, 1300.
89. J. Shepherd, T. Gale, K. B. Jensen and J. D. Kilburn, *Chem.–Eur. J.*, 2006, **12**, 713.
90. C. Schmuck and J. Lex, *Org. Lett.*, 1999, **1**, 1779.
91. C. Schmuck and V. Bickert, *Org. Lett.*, 2003, **5**, 4579.
92. C. Schmuck and M. Heil, *ChemBioChem*, 2003, **4**, 1232.
93. C. Schmuck and L. Geiger, *J. Am. Chem. Soc.*, 2004, **126**, 8898.
94. C. Schmuck, M. Heil, J. Scheiber and K. Baumann, *Angew. Chem., Int. Ed.*, 2005, **44**, 7208.
95. C. Schmuck and M. Heil, *Chem. – Eur. J.*, 2006, **12**, 1339.
96. C. Schmuck and P. Wich, *Angew. Chem., Int. Ed.*, 2006, **45**, 4277.
97. C. Schmuck, D. Rupprecht, M. Junkers and T. Schrader, *Chem. – Eur. J.*, 2007, **13**, 6864.
98. M. A. Hossain and H.-J. Schneider, *J. Am. Chem. Soc.*, 1998, **120**, 11208.
99. C. P. Mandl and B. König, *J. Org. Chem.*, 2005, **70**, 670.

100. A. Späth and B. König, *Tetrahedron*, 2010, **66**, 6019.
101. A. Späth and B. König, *Tetrahedron*, 2010, **66**, 1859.
102. H. Ait-Haddou, S. L. Wiskur, V. M. Lynch and E. V. Anslyn, *J. Am. Chem. Soc.*, 2001, **123**, 11296.
103. A. Ojida, Y. Mito-oka, M. Inoue and I. Hamachi, *J. Am. Chem. Soc.*, 2002, **124**, 6256.
104. A. Ojida, Y. Mito-oka, K. Sada and I. Hamachi, *J. Am. Chem. Soc.*, 2004, **126**, 2454.
105. A. Ojida, M. Inoue, Y. Mito-oka and I. Hamachi, *J. Am. Chem. Soc.*, 2003, **125**, 10184.
106. A. Grauer, A. Riechers, S. Ritter and B. König, *Chem. - Eur. J.*, 2008, **14**, 8922.
107. A. Späth, C. Koch and B. König, *Eur. J. Inorg. Chem.*, 2010, 2926.
108. M. Kruppa, C. Bonauer, V. Michlová and B. König, *J. Org. Chem.*, 2005, **70**, 5305.
109. X. Li, S. Miltschitzky, A. Grauer, V. Michlová and B. König, *Tetrahedron*, 2006, **62**, 12191.
110. S. Sun, M. A. Fazal, B. C. Roy and S. Mallik, *Org. Lett.*, 2000, **2**, 911.
111. M. Kruppa, C. Mandl, S. Miltschitzky and B. König, *J. Am. Chem. Soc.*, 2005, **127**, 3362.
112. A. T. Wright and E. V. Anslyn, *Org. Lett.*, 2004, **6**, 1341.
113. A. T. Wright, E. V. Anslyn and J. T. McDevitt, *J. Am. Chem. Soc.*, 2005, **127**, 17405.
114. A. T. Wright, N. Y. Edwards, E. V. Anslyn and J. T. McDevitt, *Angew. Chem., Int. Ed.*, 2007, **46**, 8212.
115. T. Zhang, N. Y. Edwards, M. Bonizzoni and E. V. Anslyn, *J. Am. Chem. Soc.*, 2009, **131**, 11976.
116. A. Buryak and K. Severin, *J. Comb. Chem.*, 2006, **8**, 540.
117. A. Buryak and K. Severin, *Angew. Chem., Int. Ed.*, 2004, **43**, 4771.
118. S. Rochat, J. Gao, X. Qian, F. Zaubitzer and K. Severin, *Chem. - Eur. J.*, 2010, **16**, 104.
119. Y. Hoshino, T. Kodama, Y. Okahata and K. J. Shea, *J. Am. Chem. Soc.*, 2008, **130**, 15242.
120. Y. Hoshino, T. Urakami, T. Kodama, H. Koide, N. Oku, Y. Okahata and K. J. Shea, *Small*, 2009, **5**, 1562.
121. Y. Hoshino, H. Koide, T. Urakami, H. Kanazawa, T. Kodama, N. Oku and K. J. Shea, *J. Am. Chem. Soc.*, 2010, **132**, 6644.
122. Y. Hoshino, H. Koide, K. Furuya, W. W. Haberaecker, S. H. Lee, T. Kodama, H. Kanazawae, N. Okuc and K. J. Shea, *Proc. Nat. Acad. Sci. U. S. A.*, 2012, **109**, 33.
123. D. Monti and S. Riva, *Biocatal. Biotransform.*, 2001, **19**, 251.
124. J. Zhang, X. Meng, X. Zeng and X. Yu, *Coord. Chem. Rev.*, 2009, **253**, 2166.
125. G. Chiosis and I. G. Boneca, *Science*, 2001, **293**, 1484.

CHAPTER 9

Synthetic Receptors for Protein Surfaces

MOUMITA RAY, AKASH GUPTA AND VINCENT M. ROTELLO*

Department of Chemistry, University of Massachusetts, Amherst, 710 North Pleasant Street, Amherst, MA, USA
*Email: rotello@chem.umass.edu

9.1 Introduction

Protein–protein assemblies are central to many cellular processes, including cell proliferation, differentiation, signal transduction, and programmed cell death.[1] Protein–protein interactions (PPIs) are crucial components underpinning the structure and function of many enzymatic and signal transduction systems. Knowledge of the supramolecular forces involved in PPIs provides a starting point for efforts to engineer synthetic receptors for protein surfaces.[2] Crystallographic structures of proteins co-crystallized with their counterparts help to determine the structural epitopes involved in PPIs, as do mutagenesis studies of protein–protein interfaces. The following sections discuss the chemical and physical properties of protein surfaces that contribute to PPIs, providing context for later discussions of synthetic systems.

9.2 Structure and Properties of Protein Surfaces

There are multiple types of PPIs, including homo- and heterodimeric/oligomeric assemblies, as well as systems where supramolecular assembly

Monographs in Supramolecular Chemistry No. 14
Synthetic Receptors for Biomolecules: Design Principles and Applications
Edited by Bradley D. Smith

Published by the Royal Society of Chemistry, www.rsc.org

imparts ternary protein structure. These various protein–protein complexes all have common characteristics, which are listed below.

9.2.1 Protein–Protein Interface Area

The normal interface area for PPIs varies from 1200 to 2000 Å^2, with 600 Å^2 being considered the cut-off for effective interaction.[3] PPIs with relatively small interfaces of 1150–1200 Å^2 are usually less stable,[4] while larger protein–protein interfaces in the range of 2000–4660 Å^2 are mostly seen within enzyme-inhibitor complexes.[5]

9.2.2 The Chemical Nature of the Interface

PPIs are governed by a cooperative network of non-covalent interactions. Often the hydrophobicity of nonpolar amino acid residues is the dominant component. Indeed, surface hydrophobicity can be used as a structural search parameter to identify regions on a protein's surface most likely to interact with a binding partner.[6] The surface hydrophobicity often occurs as a patch of neighboring nonpolar residues. The number of patches in a PPI can vary from one to 15 and a patch size is usually in the range of 200–400 Å.[7] Hydrophobic association leads to the release of water molecules at the interface, resulting in an increase in entropy that promotes the free energy of binding. Usually, hydrophobic interactions are the major stabilizers of a protein–protein complex; however, they have very little directionality.[8] The recognition selectivity is often provided by additional electrostatic and dipolar interactions.[9] These interactions are angle- and distance-dependent, although they are weakened by the desolvation penalty associated with their burial. There is evidence that charged amino acids can significantly enhance the rates of diffusion-controlled association of protein–protein complexes.

9.2.3 Hot Spots

Within a protein–protein interface, a small subset of specific residues or "hot spots" is often crucial for the complex stability and selectivity.[10] Alanine-scanning mutagenesis has been widely used to identify these hot spots and analyze the associated PPIs. This experimental method removes the side-chain atoms after the β-carbon, allowing direct probing of side-chain interactions.[11,12] Using this method, it has been determined that only a few amino acids contribute significantly to PPIs. Hot spots can often be multipurpose, with the same hot spot binding to different interacting partners to carry out distinct functions. Elucidating the structural features of a hot spot is not only necessary for understanding a single PPI, but also helpful for identifying the complementary binding site within the structure of a large binding partner.

Systematic analyses of hot spots show trends in amino acid composition, with tryptophan (21%, Trp), arginine (13.3%, Arg), and tyrosine (12.3%,

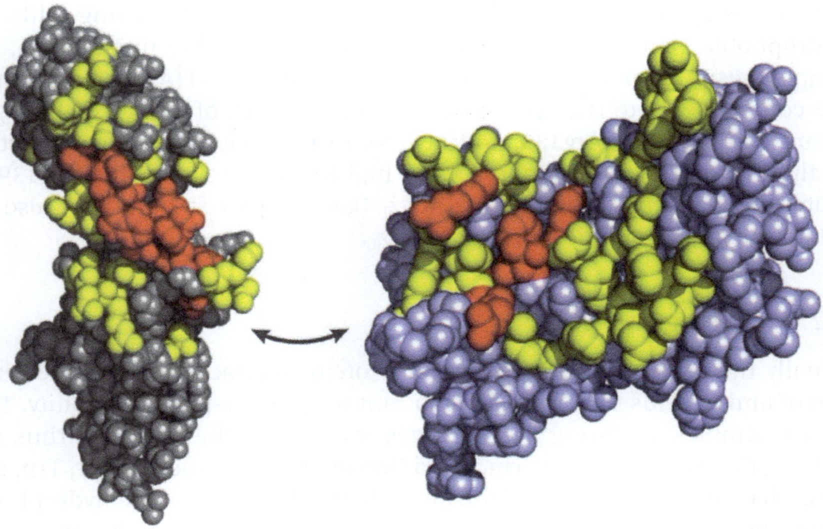

Figure 9.1 The protein–protein complex of human growth hormone binding protein (hGHbp) and growth hormone (hGH) (Protein Data Bank (PDB) ID: 3HHR) is separated to expose the interacting surface. The proteins are colored either purple (hGH) or grey (hGHbp) except for the "hot spot" (red) and "rim" (yellow) residues.
Reprinted by permission from Macmillan Publishers Ltd: *Nat. Rev. Cancer*, ref. 15, © 2014.

Tyr)[13] regularly identified as hot spots. For example, at the interface of human growth hormone and the growth hormone binding protein two of the hot spot residues are Trp, as shown in Figure 9.1.[14,15] The abundance of Trp is due to its large size and aromatic nature.[16] This amino acid can contribute to aromatic–π interactions and is a hydrogen bond (H-bond) donor. In addition, Trp has a large hydrophobic surface and protects weak H-bonds from water.[17] Moreover, mutation of Trp to alanine (Ala) creates a large cavity that destabilizes the protein–protein complex.[18] Arg is another amino acid that is commonly identified as a hot spot. This amino acid can form H-bonded salt bridges because of the positive charge on the guanidinium group. Tyr can be an effective hot spot because of its hydrophobic surface. This amino acid can also form aromatic–π interactions along with H-bonds at the protein–protein interface. In addition to the above three amino acids, isoleucine (Ile) can act as a hot spot (9.62%). However, leucine (Leu), phenylalanine (Phe), serine (Ser), and threonine (Thr) are not considered good candidates for hot spots. The heterodimeric interface of the Bcl-x_L-Bak protein complex is a classic example of hot spot interactions. The crystal structure[19] of the Bcl-x_L-Bak complex indicates that the interacting region of Bak is a 16-amino-acid peptide stretch present at the BH3 region of Bak.[20] Although the Bak peptide is a random coil in solution, it adopts an amphipathic α-helical structure on binding to the hydrophobic cleft of Bcl-x_L. The hydrophobic residues on the Bak peptide, such as valine (Val)74,

Leu78, Ile81, and Ile85 are mostly responsible for interacting with the hydrophobic cleft of Bcl-x_L. Along with the hydrophobic interactions, the charged residues on the Bak, such as Arg76, aspartic acid (Asp)83, and Asp84 are complements to the oppositely charged residues of Bcl-x_L, such as glutamic acid (Glu)129, Arg139, and Arg100, respectively. A structure function analysis shows that mutation of the highly conserved residue Arg139 to glutamine (Gln) not only reduces Bcl-x_L-Bak complex stability, but also inhibits the anti-apoptotic activity of Bcl-x_L.

9.2.4 O-ring Structure

Usually the hot spot within a protein–protein interface is surrounded by a rim of amino acids that are less important for protein–protein stability. This area resembles an O-ring and excludes water from the interface, thus stabilizing the complex. The O-ring motif favors residues, such as Tyr, Trp, and Asn, that are capable of establishing both H-bonding and hydrophobic interactions. Recently, a comprehensive study of the complex formation between lysozyme HEL and antibody FVD1.3 was conducted with a focus on the O-ring dynamics.[21] Simulation experiments showed that complex formation depended on five crucial water molecules buried deep within the HEL–FVD1.3 interface. The hot spot Tyr101 and the "warm spots" Asp100 and Gln121 interact with these water molecules (Figure 9.2) through H-bonds and stabilize the protein–protein complex. These interfacial water molecules are sheltered from the bulk solvent surrounding the protein complex, thus supporting the concept of an O-ring.

The following sections review how the concepts of hot spots and O-rings have been employed to create synthetic receptors that can control the biological functions of target proteins.

9.3 Synthetic Receptors for Protein Surface Recognition

The unique distribution of functionality on the surface of proteins provides potential targets for protein surface recognition. Synthetic receptors that can bind selectively to a specific region of a protein surface are expected to be useful tools for modulating protein behavior.[22] However, there are certain challenges that need to be considered while designing synthetic receptors for protein surfaces. On a molecular scale, the normal interfacial surface area for effective PPI is large (1200–2000 Å) and often not continuous. Therefore, expansive and multivalent synthetic scaffolds are required to engage with the exposed protein surface and compete with the natural binding partner. Pharmaceutically, this can be achieved by using antibodies that have molecular weights well over 150 kDa. The same high affinity for target proteins can be obtained using nucleic acid aptamers, which are discussed in Chapters 1 and 2. However, the relatively high production costs

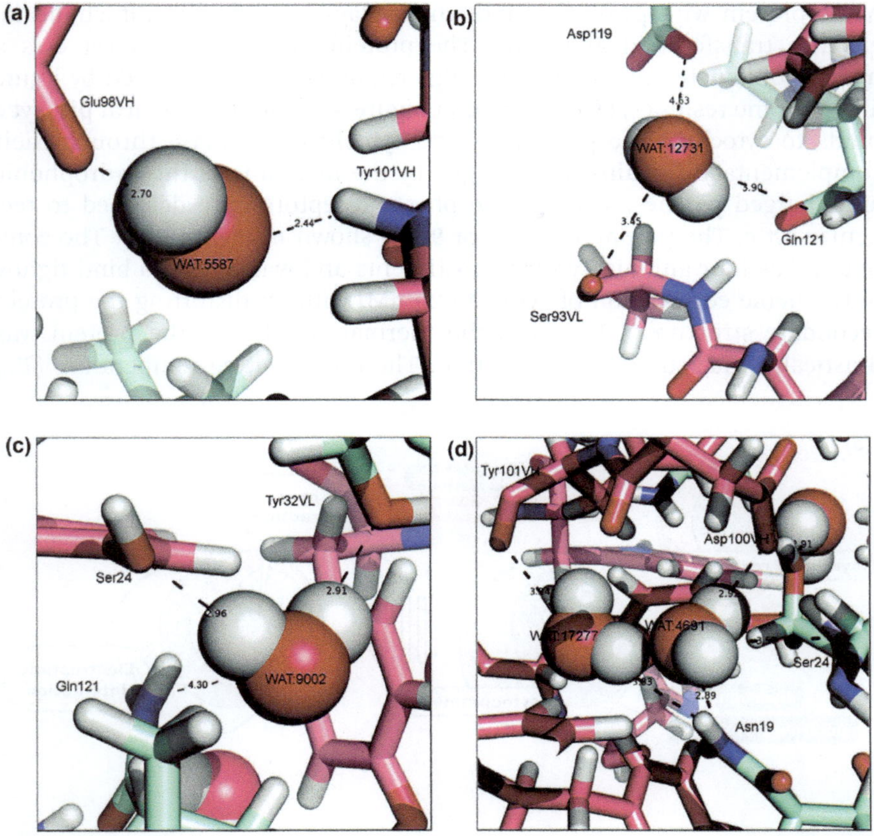

Figure 9.2 Molecular representation of the five important water residues and the interfacial amino acid residues buried within the binding interface of antibody FvD1.3 and HEL lysozyme.
Reprinted with permission from ref. 21. © 2007 American Chemical Society.

and biochemical fragility of these large oligomeric molecules are limitations for many types of applications, and in these cases there is a need for smaller and more robust chemical agents with protein affinity. Synthetic receptors such as porphyrins, molecular tweezers and clips, calixarenes, and nanoparticles have been shown to bind to protein surfaces both specifically and selectively. A common structural feature is a central nanoscale scaffold with multiple polar functionalities that can span across a highly irregular protein surface.

9.3.1 Protein Surface Recognition by Porphyrin Scaffolds

Pioneering studies by Hamilton and co-workers examined synthetic receptors for the surface of cytochrome c (cyt c).[23–27] Cyt c (12 kDa) is a cationic heme protein with pI of 10.5 and which plays key roles in mitochondrial electron transfer and apoptosis. The protein binding site of cyt c is a hydrophobic patch near the heme edge region that is surrounded by lysine and arginine residues, as illustrated in Figure 9.3a. At physiological pH, cyt c binds to cytochrome c peroxidase and cytochrome oxidase through their complementary domains. Encouraged by the matching of the hydrophobic and charged patches, synthetic porphyrin receptors were designed to recognize cyt c. The porphyrin receptor **9.1** is shown in Figure 9.3b. The compound has four anionic Tyr-Asp substituents and was found to bind tightly to the heme edge region of cyt c (K_d 20 nM) without distorting the protein secondary structure.[23] However, the thermal stability of the protein was drastically affected by the interaction. The cyt c melting temperature (T_m)

(a)

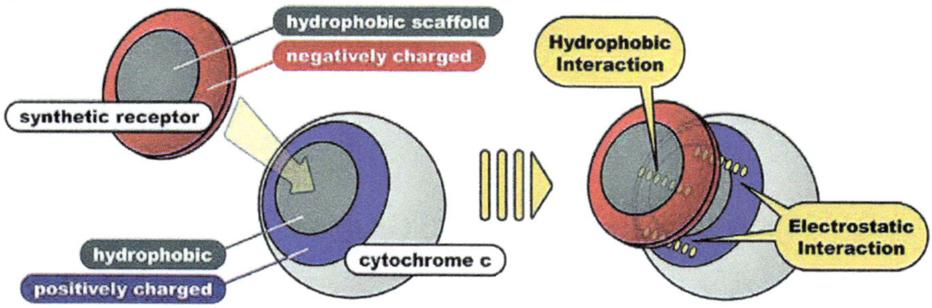

(b)

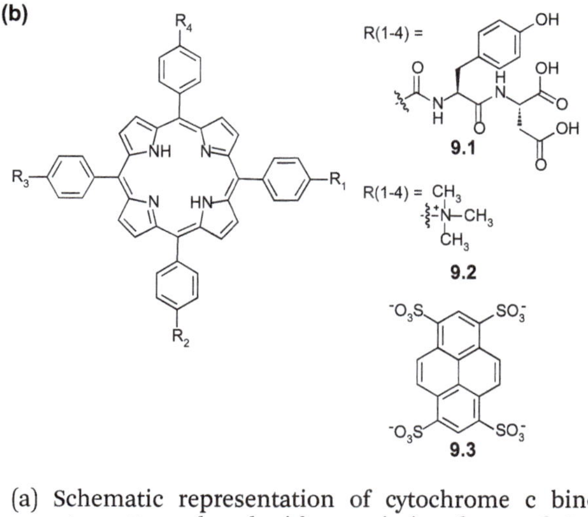

Figure 9.3 (a) Schematic representation of cytochrome c binding by synthetic receptors. Reproduced with permission from ref. 25. © 2003 Elsevier Science Ltd. (b) Structures of compounds employed in the denaturation studies. Reproduced with permission from ref. 23. © 2002 WILEY-VCH Verlag GmbH, Weinheim, Fed. Rep. of Germany.

decreased from the usual T_m of 85 °C to 70 °C on titrating with one molar equivalent of **9.1**. In addition, **9.1** was able to refold denatured cyt c with ~90% reversibility. The ability of **9.1** to destabilize the folded protein and also stabilize the unfolded state is due to complementary hydrophobic and electrostatic interactions with the cationic cyt c. Receptor **9.1** is octaanionic and disrupts the Coulombic forces between the surface lysine residues and other regions of the folded protein. Another porphyrin molecule, **9.2** (Figure 9.3b), with four peripheral cationic functional groups, did not interact with cyt c due to charge mismatch, and **9.3**, an anionic control compound interacted weakly with no effect on the native protein structure.

In a similar study, Hamilton and co-workers used another set of tetraphenylporphyrin derivatives to selectively denature cyt c at room temperature.[24] Initial studies showed that **9.4a**, a close analogue of **9.1** (Figure 9.4) bound strongly to cyt c and also decreased the T_m by 20 °C (from 85 °C to 65 °C). However, these two compounds had no effect on the native structure of the protein at room temperature. Conversion of **9.1** to the zinc porphyrin derivative, **9.1a**, led to minimal receptor aggregation in aqueous buffer due to the axial coordination of water to the Zn(II). Receptor **9.1a** interacted strongly with cyt c and decreased T_m by ~20 °C, but it did not cause any change in the secondary structure of the protein at room temperature. In contrast, the Cu(II)-porphyrin derivatives (**9.1b** and **9.4b**) were found to undergo extensive dimerization in water and the porphyrin aggregation greatly enhanced the unfolding of cyt c. Receptor **9.4b** thermally denatured the protein, decreasing the T_m by 50 °C, while **9.1b** denatured cyt c even at room temperature. The enhanced denaturation property of aggregated Cu(II)-porphyrin dimers is due to the larger size and greater charge producing stronger interactions with the hydrophobic surfaces of the unfolded protein. Furthermore, these large denaturation effects are very selective for cyt c when compared to other proteins with different T_m and pI values.

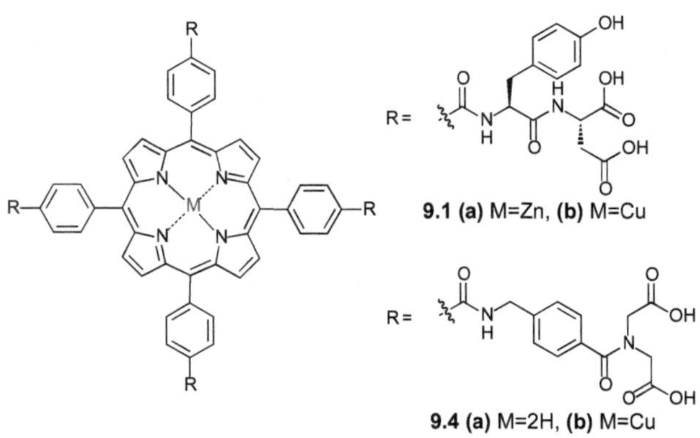

Figure 9.4 Chemical structures employed in denaturation studies. Reprinted with permission from ref. 24. © 2003 American Chemical Society.

To further improve the protein surface recognition by these porphyrin receptors, the hydrophobic surface area along with the negative charge density was increased. The porphyrin core was changed from tetraphenylporphyrin, with a diameter of 15.5 Å, to tetrabiphenylporphyrin, which measures 24.0 Å[25] (Figure 9.5a). Figure 9.5b shows the anionic substituents that were appended to the periphery of the scaffolds. The binding affinities for cyt c were investigated by fluorescence quenching measurements. The dissociation constants (K_d) for each of the receptors are listed in Table 9.1. Receptor **9.10** has exceptionally strong binding affinity towards cyt c, due to the presence of 16 carboxylic groups on its surface. In addition, receptor **9.1** shows higher binding affinity for cyt c compared to receptors **9.7–9.9**. This is due to the flexibility of the functional groups within receptor

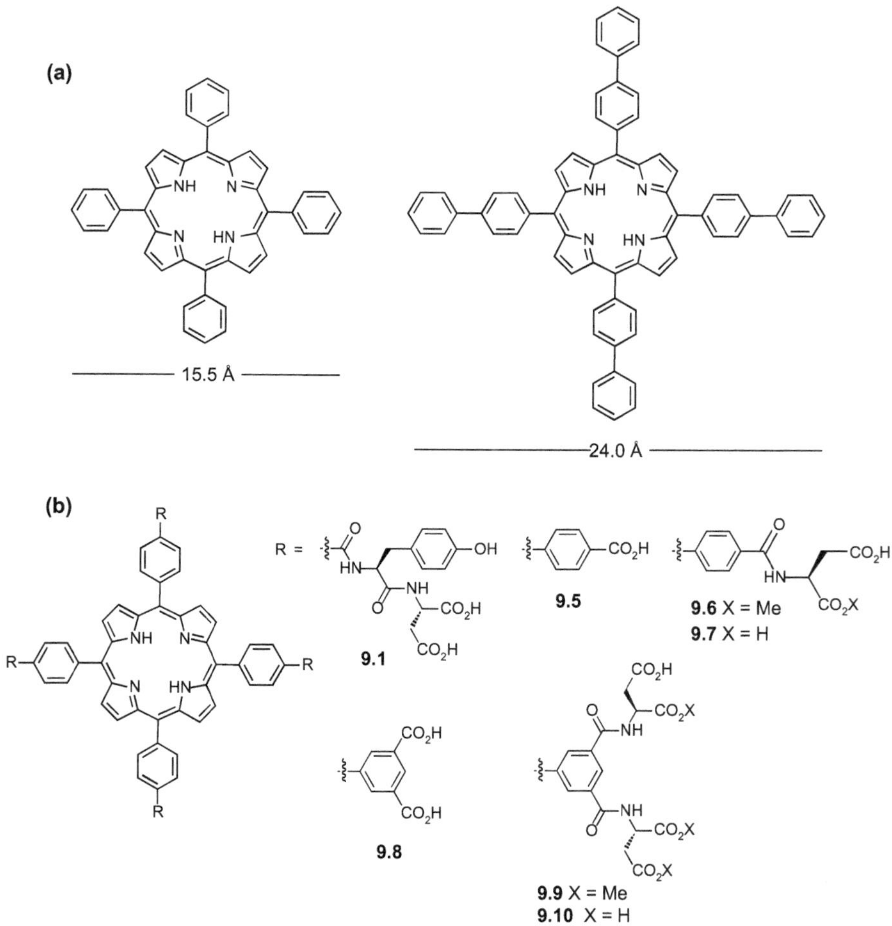

Figure 9.5 (a) Comparison of tetraphenylporphyrin and tetrabiphenylphorphyrin scaffolds. (b) Chemical structures employed in the unfolding studies of cytochrome c.
Reprinted with permission from ref. 25. © 2003 Elsevier Science Ltd.

Table 9.1 Dissociation constants and charge of synthetic receptors. Reprinted from ref. 25, © 2003, with permission from Elsevier Science Ltd.

Receptor	K_d (nM)	Receptor charge
9.1	220 ± 17	−8
9.5	17 000 ± 840	−4
9.6	12 000 ± 950	−4
9.7	1300 ± 130	−8
9.8	1500 ± 170	−8
9.9	1700 ± 97	−8
9.10	0.67 ± 0.34	−16

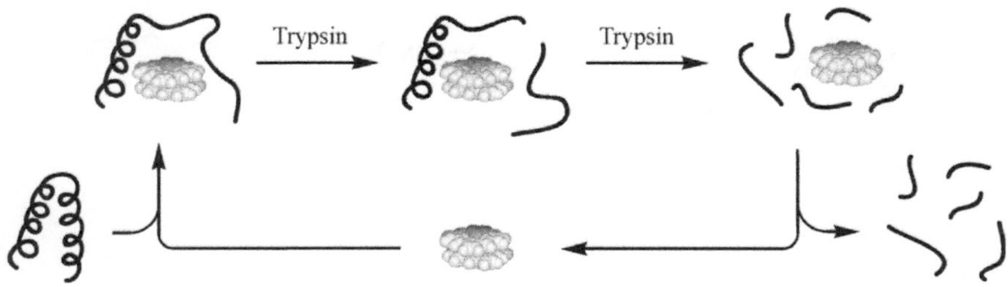

Figure 9.6 Proposed catalytic cycle for the accelerated digestion of cytochrome c by porphyrin dimer.
Reprinted with permission from ref. 26. © 2004 American Chemical Society.

9.1 which permit an induced fit with the protein surface. In terms of cyt c denaturation, receptor **9.10** decreased the T_m by ~50 °C, with the enhanced electrostatic interaction between receptor and protein facilitating the denaturation.

Not only do these metallo-porphyrin receptors promote unfolding of cyt c, they also facilitate catalytic digestion by proteolytic enzymes such as trypsin.[26] Cyt c is a stable protein and resistant to proteolysis. However, in the presence of receptor **9.4b**, cyt c can be digested by trypsin in 15 minutes. Furthermore, receptor **9.4b** and trypsin remains active even after each digestion cycle is complete. The proteolytic stability of cyt c is due to Met80 coordination with the heme, which stabilizes the protein secondary structure. Receptor **9.4b** is able to disrupt the heme ligation to Met80, thus accelerating proteolysis by trypsin. Furthermore, the receptor promotes unfolding of the α-helices and other parts of the cyt c, allowing better access to the trypsin active site. The catalytic cycle in Figure 9.6 is entropically driven by the proteolytic digestion which releases the porphyrin dimer back into the solution. Further structure–activity relationship studies of these anionic copper porphyrins looked at larger heme proteins such as myoglobin (17.2 kDa, pI 7.2) and hemoglobin (64.5 kDa, pI 6.9)[27] (Figure 9.7a), and a new copper porphyrin (Figure 9.7b) was identified for the follow-up

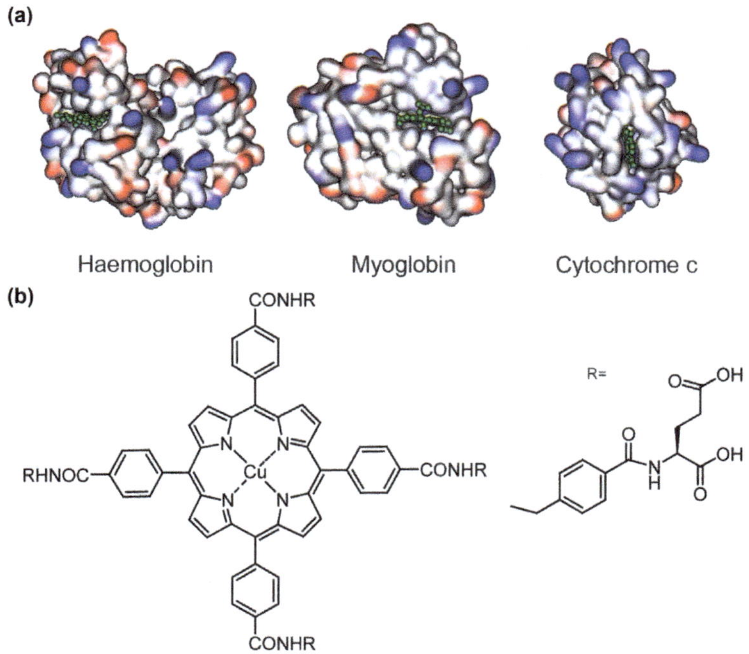

Figure 9.7 (a) The relative sizes of cytochrome c (12 kDa, Protein Data Bank (PDB) ID: 3CYT), myoglobin (17.2 kDa, PDB ID: 1MBN) and hemoglobin (64.5 kDa, PDB ID: 2HCO); green: prosthetic heme; blue: basic residue; grey: neutral residue; red: acidic residue. (b) Receptor structure. Reproduced with permission from ref. 27 with permission from the Royal Society of Chemistry.

experiments. The studies showed that a catalytic amount of the copper porphyrin receptor accelerates the digestion of myoglobin and hemoglobin. The copper porphyrin acts as a transient unfolding agent that selectively promotes the cleavage of cationic proteins by proteases. This type of selective protein degradation is exhibited by the protease family of ClpA/ClpP.[28] In this natural system, tagged proteins are first unfolded by ClpA through an adenosine triphosphate-dependent denaturation procedure before they are digested in the proteolytic chamber of ClpP.

In other work, Hamilton and co-workers have reported pattern-based detection of proteins using a series of synthetic tetraphenylporphyrin (TPP) derivatives[29] containing one or more of the differently charged carboxamide substituents in Figure 9.8. These TPP derivatives are excellent candidates for protein surface recognition due to their large hydrophobic surface area and highly charged peripheral residues. Four iron-containing proteins were chosen for detection: ferredoxin (pI 2.75), cytochrome c551 (pI 6.72), myoglobin (pI 7.2), and cyt c (pI 10.6). It is known that the Fe(II) centers can readily quench the red porphyrin fluorescence on binding. Addition of the proteins to solutions of TPP receptor led to complex formation and receptor

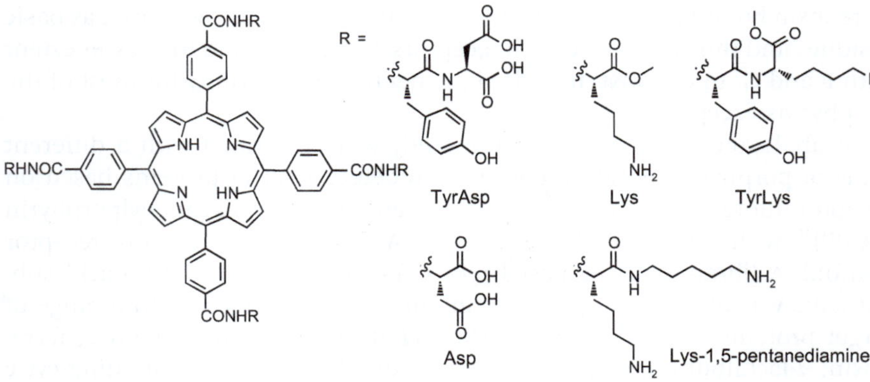

Figure 9.8 Scaffold and peptidic components used to make porphyrin receptors employed in the pattern-based detection of different proteins.
Reprinted with permission from ref. 29. © 2004 American Chemical Society.

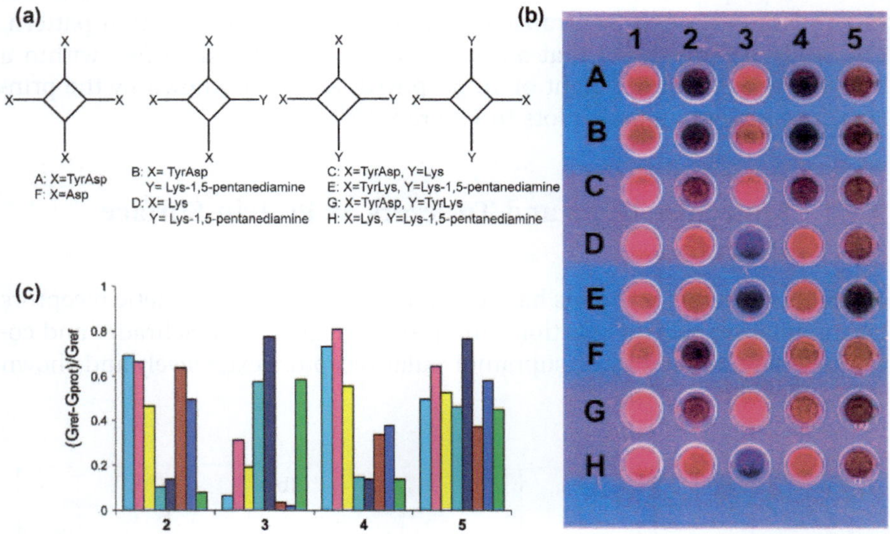

Figure 9.9 (a) Schematic structure of porphyrin receptors A–H. (b) Picture of part of the 96-well quartz plate irradiated with ultraviolet light (302 nm) and producing red fluorescence. Each well contains a solution of porphyrins A–H; in the wells of columns 2–5 were added three equivalents of the following. 2: cytochrome c (pI 10.6, MW≈12 500);3: ferredoxin (pI 2.75, MW≈5500); 4: cytochrome c551 (pI 4.7, MW≈9000); 5: myoglobin (pI 6.8, MW≈18 000). (c) Fingerprints of cytochrome c (2), ferredoxin (3), cytochrome c551 (4), and myoglobin (5) based on the eight-porphyrin array.
Reprinted with permission from ref. 29. © 2004 American Chemical Society.

fluorescence quenching. The fluorescence attenuation results shown in Figure 9.9 reflect the affinities of the porphyrin receptors for the target proteins. Cyt c interacted with anionic receptors A, B, C, F, and G, while ferredoxin bound to D, E, and H. Cytochrome c551 has acidic as well as basic residues and thus interacted with receptors A, B, and C, and to a lesser extent with F and G. In contrast, myoglobin is neutral, with affinity for most of the porphyrins except F.

An analogous protein recognition study was performed with a different array of porphyrin receptors designed to detect a set of proteins based on receptor fluorescence changes. Synthetic tetra-*meso*-carboxyphenylporphyrin (TCPP)[30] with a hydrophobic area >300 Å^2 was used as the core receptor scaffold. A library of TCPP-based receptors with different amino acid substituents was observed to produce unique binding patterns with a range of target proteins having variable surface characteristics such as cyt c, ferredoxin, α-lactalbumin, and lysozyme. As expected, the iron-containing cyt c and ferredoxin produced different degrees of receptor fluorescence quenching. But altered receptor fluorescence was also observed with the metal-free proteins α-lactalbumin and lysozyme. In these cases, association of the porphyrin receptor to the protein surface permitted energy transfer from the porphyrin to amino acid residues such as Trp and Tyr and the varying amount of energy transfer produced a unique recognition pattern. The study demonstrated that all the proteins could be identified within a mixture by using either eight or 16 porphyrin arrays, as shown by the principal component analysis plots in Figure 9.10.

9.3.2 Molecular Clips and Tweezers as Protein Surface Receptors

Molecular tweezers and clips have emerged as promising synthetic receptors for modulating enzyme reactions and protein aggregation. Schrader and coworkers have studied these supramolecular receptors extensively and shown

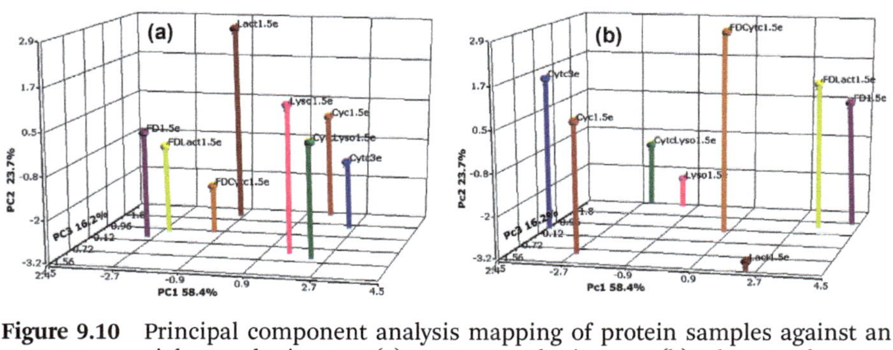

Figure 9.10 Principal component analysis mapping of protein samples against an eight-porphyrin array (a) or a 16-porphyrin array (b). The samples were tested on a 16-porphyrin array, while (a) is extracted from (b) by taking the data matrix of the first eight porphyrin receptors.
Reprinted with permission from ref. 30. © 2006 American Chemical Society.

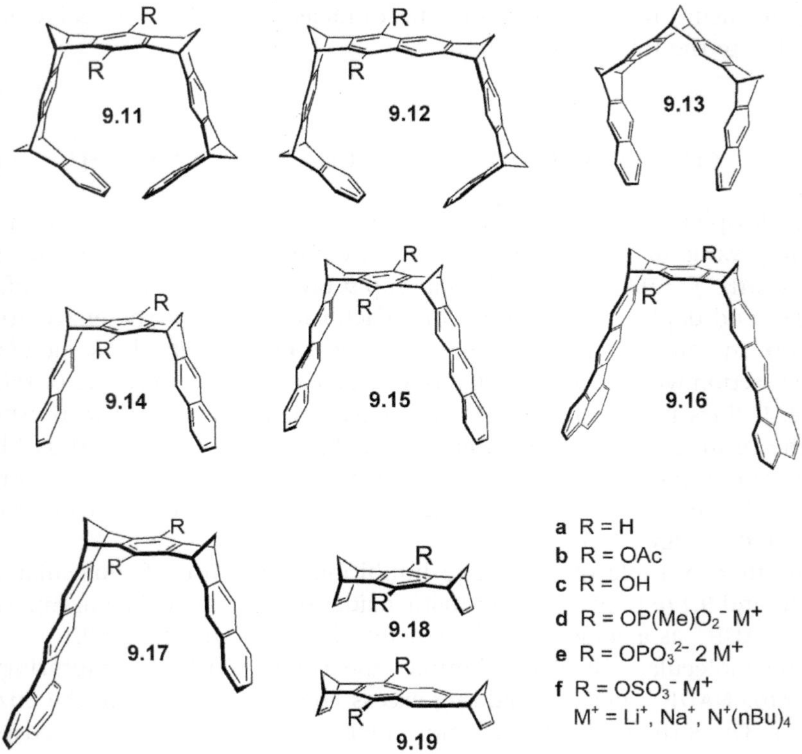

Figure 9.11 Structures of molecular tweezers, molecular clips, and bridges. Reprinted with permission from ref. 31. © 2012 American Chemical Society.

that they can bind to a range of cationic and neutral guests, including several types of biomolecules.[31] The tweezers and clips possess a central parallel or torus-shaped cavity with a surrounding belt of convergent aromatic rings. The clips with sidewalls can bind to aromatic cations such as pyridinium salts, while tweezers with torus-shaped cavities can bind to amino acids, such as lysine and arginine (Figure 9.11).[31] In general, tweezers **9.11** and **9.12** form more stable complexes than the molecular clips **9.13**, **9.14**, and **9.15**, because of the higher amount of aromatic stacking. The tweezers **9.11** can bind the alkane chains of aliphatic guest molecules inside their cavity *via* CH–π interactions, while tweezers **9.12** can bind aromatic guests *via* both CH–π and π–π interactions. Molecular clips **9.13**, **9.14**, and **9.15** can also accommodate aromatic guests that stack against the aromatic sidewalls, whereas the largest clips, **9.16** and **9.17**, can bind polyaromatic guest molecules.

The guest binding ability of molecular clips and tweezers has been used to inhibit the enzymatic function of alcohol dehydrogenase (ADH),[32] and glucose-6-phosphate-dehydrogenase (G6PD).[33] ADH catalyzes the oxidation of ethanol using nicotinamide adenine dinucleotide $(NAD)^+$ as cofactor, according to the following reaction:

$$CH_3{-}CH_2{-}OH + NAD^+ \xrightarrow[pH=9.0]{ADH,\ 37\,^\circ C} CH_3{-}CHO + NADH + H^+ \quad (9.1)$$

The phosphate substituted clip **9.14e** inhibits this enzymatic reaction with an IC_{50} value of 1.5 mM and almost an equimolar inhibitor/cofactor ratio $(IC_{50}/NAD^+) = 0.8$.[32] The molecular clip **9.14e** acts as a trap for the cofactor NAD^+ and depletes the amount of cofactor available for the enzymatic reaction. In contrast, the molecular tweezers **9.11e** inhibits the same enzymatic reaction with an $IC_{50} = 180\ \mu M$ and a smaller inhibitor/cofactor ratio of 0.09. It does not bind to NAD^+ but is assumed to associate with exposed lysine residues that are proximal to the ADH active site, and thus blocks substrate entry. Enzymatic activity can be switched back on by adding an external lysine derivative to complex with the tweezers, with further cycling depicted in Figure 9.12.

Another example of enzymatic inhibition is the use of molecular clips **9.14d** and **9.14e** to prevent enzymatic oxidation of G6P by the enzyme G6PD with $NADP^+$ as a cofactor.[33] The reaction is shown in Scheme 9.1.

The molecular clip **9.14d** inhibits the G6PD activity by capturing the cofactor $NADP^+$; however, adding excess cofactor can activate the enzyme again. The structurally related molecular clip **9.14e**, with appended phosphate groups, proved to be a much better inhibitor. Kinetic studies suggest that clip **9.14e** uses two paths to prevent substrate binding. It forms either a ternary complex with the enzyme and the cofactor, or a binary complex with the enzyme. The two modes of G6PD inhibition are illustrated by the molecular dynamics simulations in Figure 9.13.

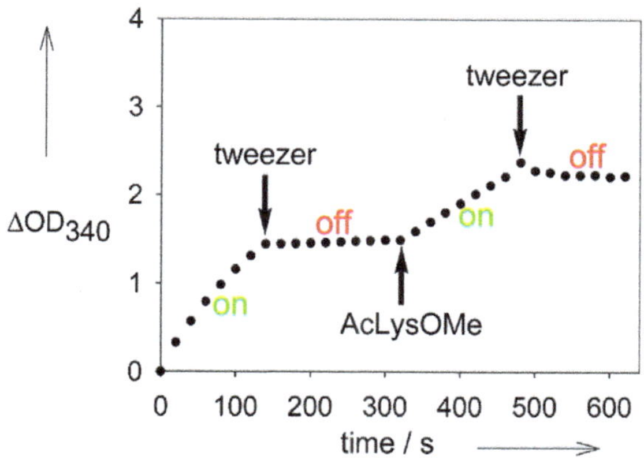

Figure 9.12 Enzyme switching: alternating addition of tweezer and lysine derivative turns alcohol dehydrogenase enzyme activity on and off.
Reprinted with permission from ref. 32. © 2008 American Chemical Society.

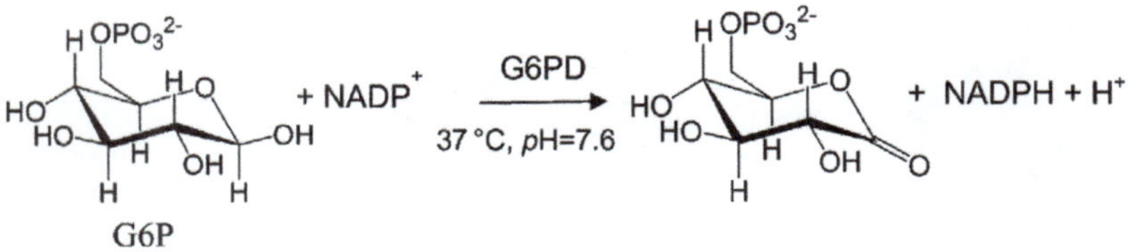

Scheme 9.1

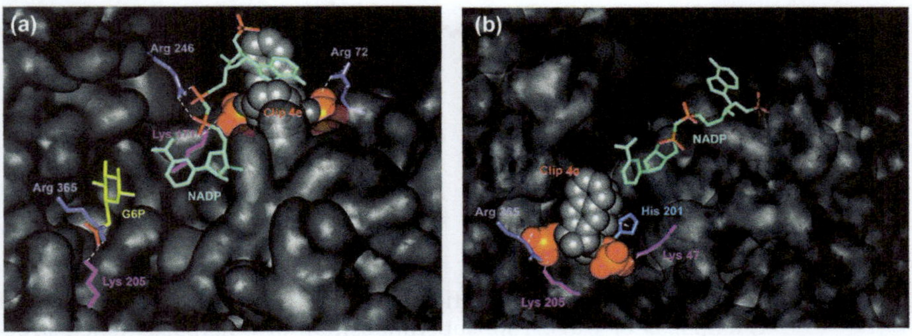

Figure 9.13 Molecular dynamics simulations illustrating the two modes of inhibition of G6PD by clip. (a) Ternary complex of enzyme, clip, and cofactor; (b) binary enzyme complex of clip occupying the substrate-binding site. Reproduced with permission from ref. 33. © 2009 WILEY-VCH Verlag Gm bH & Co. KGaA, Weinheim and ref. 31 © 2012 American Chemical Society.

Most recently, molecular tweezer **9.11e** has been used to modulate the PPIs of 14-3-3 proteins, which are eukaryotic adaptor proteins that regulate the activity of kinase C-Raf and regulate the tumor suppressor protein p53.[34] In humans, seven different types of 14-3-3 proteins are present. Among them, 14-3-3σ contains 17 lysine residues. Molecular tweezers like **9.11e** with negatively charged hydrogen phosphate groups can target these lysine residues and modulate the interaction of 14-3-3σ with its binding partners, a phosphorylated (C-Raf) protein and an unphosphorylated protein (ExoS) (Figure 9.14a). The isothermal titration calorimetry (ITC) results in Figure 9.14(b) reveal an exothermic process that comprises two independent binding events with different affinities. Protein crystallography shows that tweezer **9.11e** primarily targets the surface-exposed Lys214, which is a hot spot for 14-3-3σ binding with its partner proteins (Figure 9.14c).

Finally, tweezers **9.11e** has been shown to recognize aggregated protein surfaces such as Aβ, which are the misfolded and aggregated proteins associated with Alzheimer's disease.[35] Nuclear magnetic resonance (NMR) and mass spectrometry studies proved that tweezers **9.11e** complexes with accessible Lys16 and Lys28 residues on Aβ and, unlike control compound **9.18e**, reduces the propensity for β-sheet formation (Figure 9.15). In fact, **9.11e** is able to dissolve preformed Aβ fibrils. Overall, supramolecular

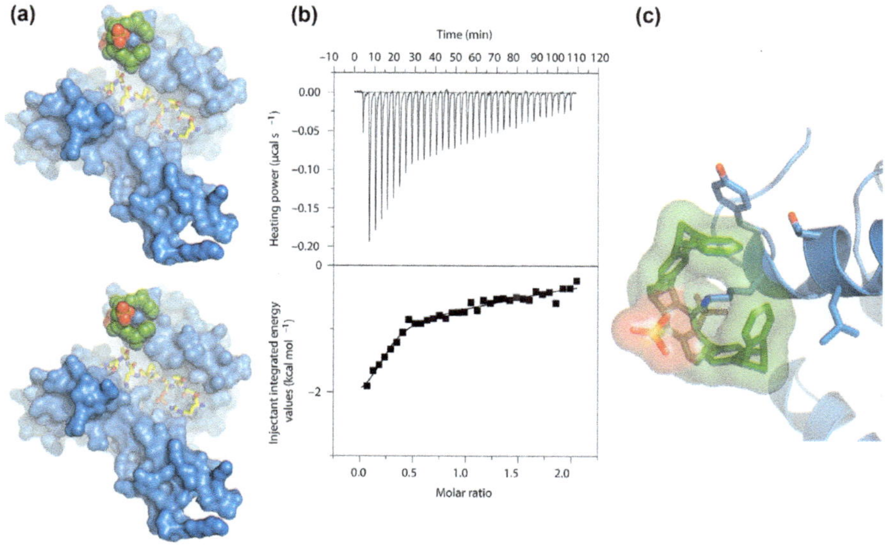

Figure 9.14 (a) Binding site competition between molecular tweezers and C-RafpS259 or ExoS. Top: superimposition of C-RafpS259 (stick structure) binding to the complex of tweezers (green and red) and 14-3-3σ (blue surface). Bottom:superimposition of ExoS interaction motif (stick structure) binding to the complex of tweezers (green and red) and 14-3-3σ (blue surface). (b) Isothermal titration calorimetry of 14-3-3σ titrated with tweezers shows exothermic binding. (c) Detailed view of tweezers binding with Lys214 from 14-3-3σ.
Reprinted by permission from Macmillan Publishers Ltd: *Nat. Chem.*, ref. 34, © 2013.

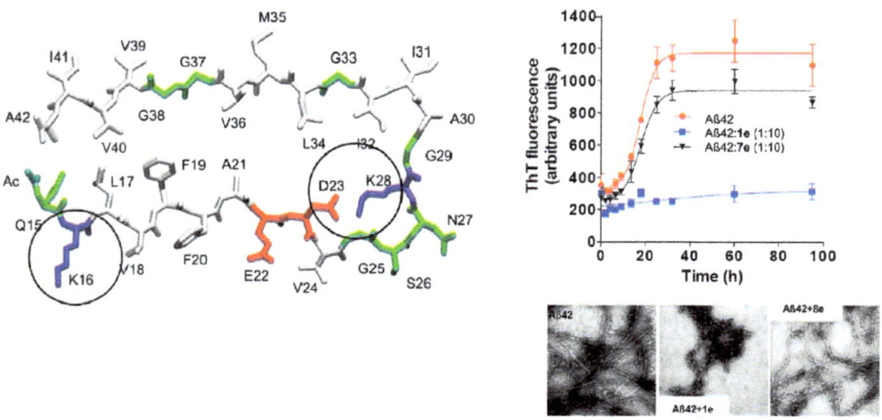

Figure 9.15 (a) Misfolded part of Aβ42 in fibrils with critical lysine residues K16 and K28 that are complexed by phosphate tweezers. (b) Aβ42 aggregation is inhibited by one of the tweezers. (c) Corresponding transmission electron microscopy pictures demonstrating exclusive suppression of fibril formation by tweezers.
Reprinted with permission from ref. 31. © 2012 American Chemical Society.

complexation of critical lysine residues by synthetic receptors appears to be a promising strategy to redirect the aggregation of misfolded proteins into nontoxic species.

9.3.3 Protein Surface Recognition by Calixarene Scaffolds

Calixarenes are another class of synthetic receptors that are useful for recognizing protein surfaces. These cone-shaped structures have an upper rim that can be functionalized by attaching multiple recognition domains.[36] Hamilton and co-workers have used the calix[4]arene scaffold to create receptors for the protein surface of cyt c.[37] The multivalent calixarene design mimics the structures of antibodies, which have six hyper-variable loops to contact with interacting antigens. Three polyanionic calix[4]arenes (**9.20a–c**) were prepared each with four copies of the constrained cyclic peptide (A) (Figure 9.16). Affinity chromatography and gel filtration indicated that **9.20a** bound strongly to cyt c, and that binding depended strongly on the nature of the substituents on the lower rim of the calixarene. The binding of **9.20a** inhibited the reduction of cyt c from its Fe(III) form to Fe(II), suggesting that the receptor interacted with the four lysines near the heme edge of the protein.

In another study, the Hamilton group used calix[4]arenes as surface binding inhibitors of the serine protease enzyme, α-chymotrypsin (ChT).[38] It was proposed that negatively charged calix[4]arenes would interact with a cationic patch on the surface of the protein. Initially, three different types of calix[4]arenes derivatives were synthesized, each projecting four peptide loops composed of either: (1) Gly-Asp-Gly-Asp, (2) Gly-Asp-Gly-Tyr, or (3) Gly-Lys-Gly-Lys. The peptide loops totaled ~450 Å^2 in area and offered different amounts of hydrophobicity (Tyr) and negative charge (Asp) for interaction with the ChT surface. Submicromolar inhibitors were uncovered, some

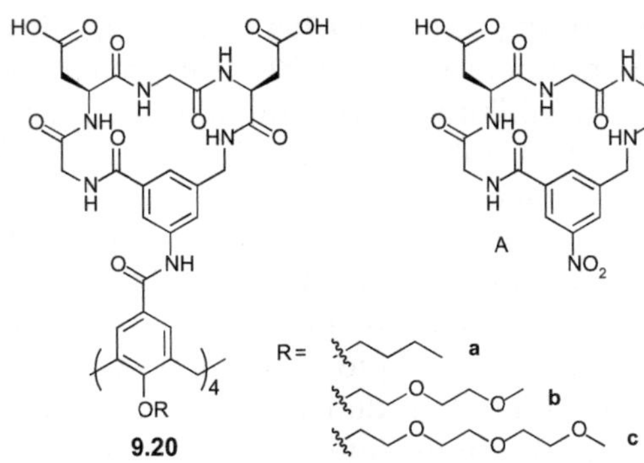

Figure 9.16 Left: chemical structures of calixarene receptors employed in protein surface recognition. Right: chemical structure of the cyclic peptide A. Reproduced with permission from ref. 37. © 1998 John Wiley & Sons, Inc.

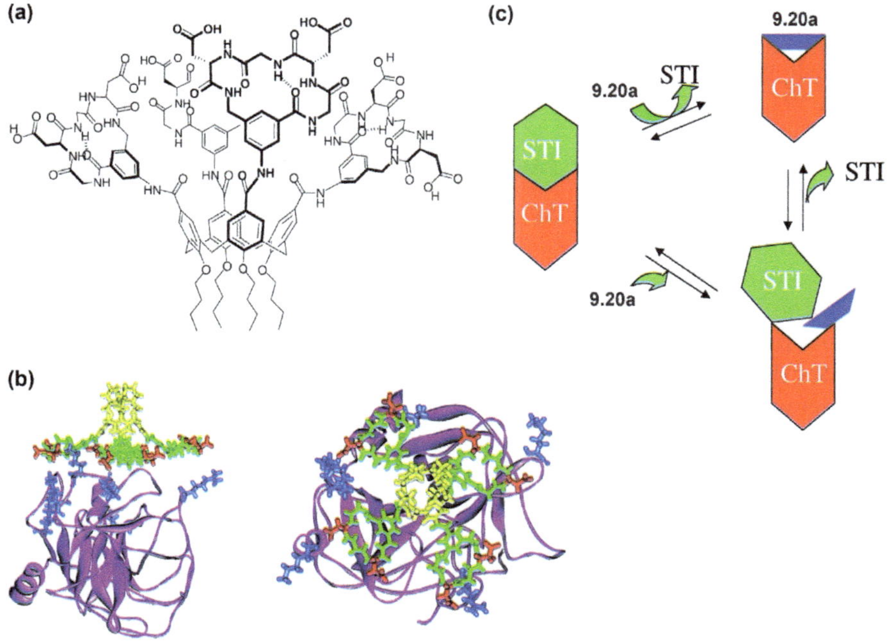

Figure 9.17 (a) Chemical structure of the calixarene receptor **9.20a**. (b) Calculated structure for the interaction of receptor with α-chymotrypsin (ChT). (c) Possible mechanisms for the interaction of receptor **9.20a** with ChT–STI complexes. (STI: soybean trypsin inhibitor).
Reproduced with permission from ref. 39. © 2002 The National Academy of Sciences.

exhibiting slow binding kinetics in an analogous manner to several of the natural protein's proteinase inhibitors. Shown in Figure 9.17a is the structure of receptor **9.20a** with eight carboxylic acids, and in Figure 9.17b is a calculated structure showing receptor **9.20a** interacting with ChT. The synthetic receptor disrupts interaction of ChT with proteinaceous inhibitors such as soybean trypsin inhibitor (STI).[39] Initially, a ternary complex of receptor **9.20a** bound to ChT–STI is formed, followed by a time-dependent displacement of the STI (Figure 9.17c).

Another application of synthetic receptors is restoration of protein ability to self-oligomerize. An example is the use of calix[4]arene to induce tetramerization of mutant p53.[40] The protein p53 is a transcription factor and tumor suppressor that induces apoptosis when DNA is damaged. Human p53 is structurally divided into four domains. Two identical protein chains, composed of α-helix and β-strands, intertwine to form a dimer that further stacks as a tetramer. The dimer is stabilized by a crucial ion-pair interaction between Arg337 and Asp352.[41] This ion-pair stabilization is missing with the inherited mutation of p53-R337H, where Arg337 is replaced by histidine. As a result, the p53 tetramer is destabilized, inducing a proclivity to tumorigenesis. A synthetic calix[4]arene receptor has been designed to bridge the four monomers of p53-R337H and retain the tetrameric integrity. The receptor consists of a conical calix[4]arene platform with four upper-rim guanidiniomethyl residues that interact with negatively charged residues on the p53 surface.

The broad supramolecular rules for predicting how calixarenes associate with protein surfaces are just beginning to emerge. Recently, McGovern *et al.* used NMR and X-ray crystallography to characterize the binding of *p*-sulfonatocalix[4]arene ($sclx_4$, structure shown in Figure 9.18) to cyt c.[42,43] As discussed above, cyt c is a lysine-rich protein, and X-ray crystallography shows binding of three separate $sclx_4$ molecules to three sites on cyt c: Lys89, Lys4, and Lys22. Each $sclx_4$ complex exhibits a different mixture of polar and nonpolar interactions. Interestingly, the $sclx_4$ molecules are also closely packed between different protein chains, thus mediating PPI. It is likely that calixarenes will be very useful in the future as synthetic receptors for modulating multicomponent protein assemblies.

9.3.4 Nanoparticles as Protein Surface Receptors

A synthetic receptor for proteins must have a large surface area for receptor–protein interaction and exhibit multiple electrostatic, hydrophobic and H-bonding sites for specific protein surface recognition.[44] Nanoparticles (NPs)[45] can serve as highly efficient synthetic receptors for protein surfaces, as the core size can be prepared commensurate to the size of the proteins

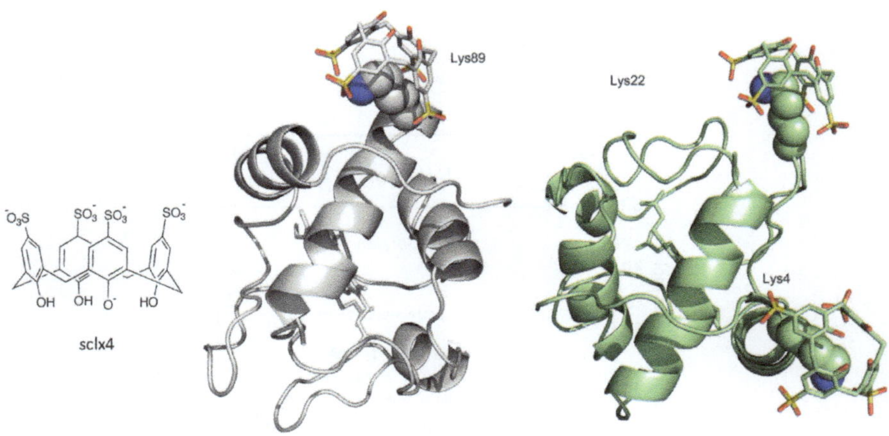

Figure 9.18 Structure of *p*-sulfonatocalix[4]arene ($sclx_4$) and crystal structures of the cytochrome c-$sclx_4$ complexes. The asymmetric unit comprises two molecules of cytochrome c (chain A in grey and chain B in green) and three molecules of $sclx_4$.
Reprinted by permission from Macmillan Publishers Ltd: *Nat. Chem.*, ref. 42. © 2012.

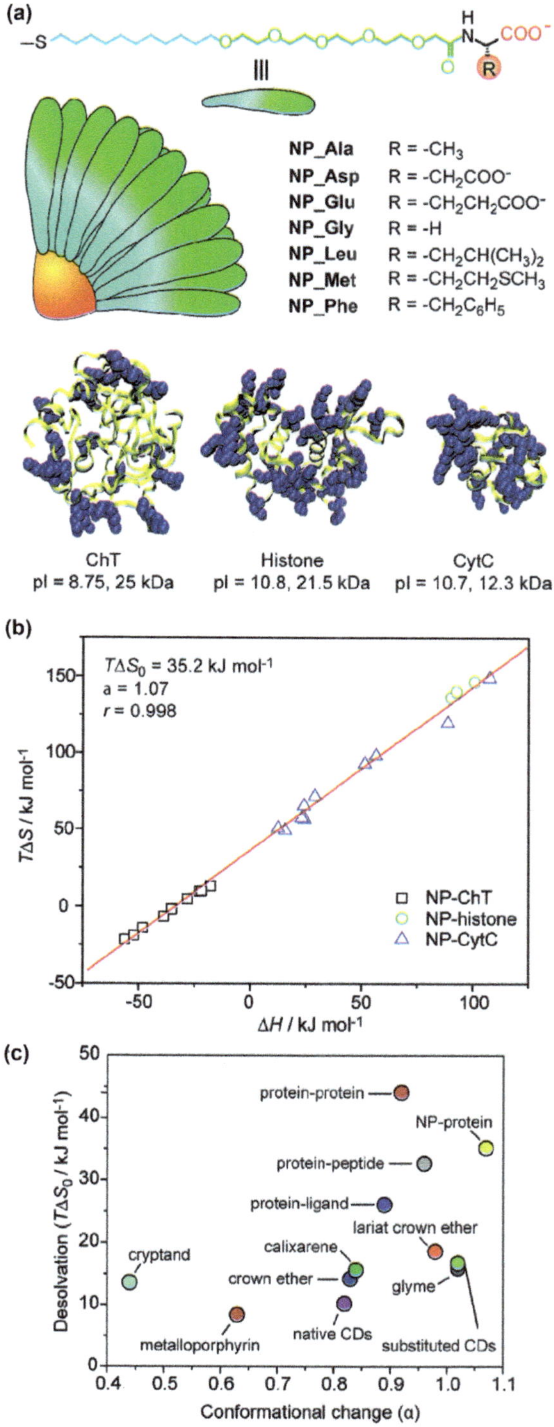
(a)
-S
H
N
COO⁻
O
R
NP_Ala R = -CH3
NP_Asp R = -CH2COO-
NP_Glu R = -CH2CH2COO-
NP_Gly R = -H
NP_Leu R = -CH2CH(CH3)2
NP_Met R = -CH2CH2SCH3
NP_Phe R = -CH2C6H5
ChT
pI = 8.75, 25 kDa
Histone
pI = 10.8, 21.5 kDa
CytC
pI = 10.7, 12.3 kDa
(b)
TΔS0 = 35.2 kJ mol-1
a = 1.07
r = 0.998
TΔS / kJ mol-1
ΔH / kJ mol-1
NP-ChT
NP-histone
NP-CytC
(c)
Desolvation (TΔS0 / kJ mol-1)
Conformational change (α)
protein-protein
NP-protein
protein-peptide
protein-ligand
lariat crown ether
cryptand
calixarene
crown ether
glyme
metalloporphyrin
native CDs
substituted CDs

and the particle interface can be tailored with desired functionality to selectively interact with different proteins.

A fundamental understanding of the thermodynamics of protein–NP interactions provides a foundation for biological applications.[46] The complexation process may be described by the following equations:

$$\text{Protein} + \text{NP} \rightleftharpoons \text{Protein–NP} \tag{9.2}$$

$$x\text{H}_2\text{Op} + y\text{H}_2\text{On} \rightleftharpoons (x+y-z)\text{H}_2\text{Op-n} + z\text{H}_2\text{O} \tag{9.3}$$

$$\text{Protein}\cdot x\text{H}_2\text{Op} + \text{NP}\cdot y\text{H}_2\text{On} \rightleftharpoons \text{Protein–NP}\cdot(x+y-z)\text{H}_2\text{Op-n} + z\text{H}_2\text{O} \tag{9.4}$$

In the above equations, H_2Op, H_2On and H_2Op-n are the water molecules associated with protein, NPs, and the protein–NP complex, respectively. Formation of the protein–NP complex described in eqn (9.2) is enthalpically favored ($\Delta H < 0$) and entropically opposed ($\Delta S < 0$). On the contrary, the solvent reorganization process involved in eqn (9.4) is entropically favored, where $\Delta H > 0$ and $\Delta S > 0$. The combined eqn (9.5), describing complex formation can be favored ($\Delta G < 0$) because it is either an enthalpy- or entropy-driven process, depending upon the contributions from eqns (9.3) and (9.4). With these thermodynamic considerations in mind, the Rotello group has demonstrated that the protein–NP interactions closely resemble PPIs.[47] Amino acid-terminated AuNPs were used as receptors for exploring their interactions with three positively charged proteins: ChT, histone, and cyt c, as shown in Figure 9.19a. The NP–protein complex formation was investigated using ITC. The study revealed a significant protein-dependent difference in the binding thermodynamics. Binding of NPs with ChT was enthalpy driven while the interaction with histone and cyt c was entropy driven. The slope (α) and entropy-derived value (TS) in Figure 9.19b provide a quantitative measure of the conformational change and the desolvation upon complex formation respectively. Enthalpy–entropy compensation analysis showed an excellent correlation between the two thermodynamic quantities. As shown in Figure 9.19c, the compensation coefficients of the NP–protein binding closely resembled that of the naturally occurring PPIs.

Figure 9.19 (a) Structural features and relative sizes of amino acid-functionalized AuNPs (NPs: nanoparticles) and proteins. The blue overlapping spheres in the proteins represent the positively charged residues on their surface. (b) Plots of entropy ($T\Delta S$) *versus* enthalpy (ΔH) for NP–protein interactions. (c) Slope (α) and intercept ($T\Delta S_0$) values for various host–guest systems. Protein–ligand interactions have been divided into protein–peptide and protein–other (protein–ligand) interactions. Reprinted with permission from ref. 47. © 2007 American Chemical Society.

9.3.4.1 Tuning the Catalytic Behavior of Enzymes using Nanoparticles

Mixed monolayer gold nanoparticles capped with negatively charged ligands have been shown to selectively bind and inhibit enzymes.[48] ChT is a serine protease with a cationic patch around its active site and the anionic AuNPs inhibit the activity of ChT, as shown in Figure 9.20a. The AuNPs induced a two-step inactivation mechanism featuring fast reversible inhibition followed by a slower irreversible process.[49] It was hypothesized that the initial inhibition was driven by electrostatic forces, whereas the slower

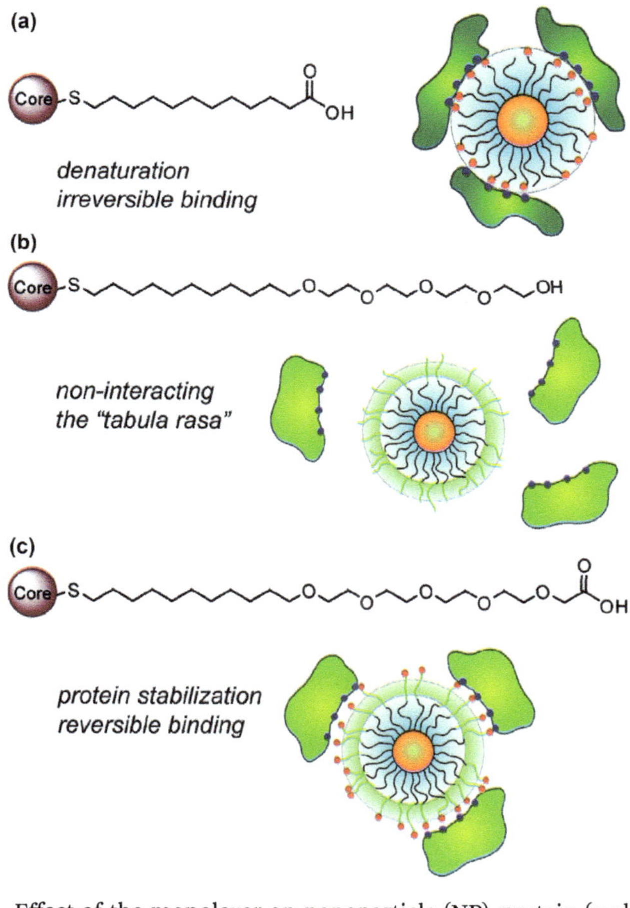

Figure 9.20 Effect of the monolayer on nanoparticle (NP)–protein (α-chymotrypsin, ChT) interactions. (a) Simple alkanethiol-based monolayer results in protein denaturation. (b) Triethyleneglycol (TEG)-functionalized particles are non-interacting. (c) Termination of the TEG layer with carboxylate groups results in reversible binding to ChT that stabilizes the protein against denaturation.
Reprinted with permission from ref. 49. © 2004 American Chemical Society.

denaturation process involved the interaction between the nonpolar interior of the NP monolayer with the hydrophobic residues on ChT (Figure 9.20b and c). This interaction was very strong, with $K_i = 10.4$ nM, with no inhibition observed with elastase, β-galactosidase (β-Gal), or cellular retinoic acid binding protein. In subsequent studies, AuNPs were functionalized with L-amino acids having oligoethylene glycol (OEG) linkers with different lengths.[50] Regulated denaturation of ChT could be obtained by varying the combination of the recognition elements (amino acids) and the OEG linkers, providing a tool for modulating the dynamic structure of proteins.

Non-covalent interactions between protein and NP can be an attractive strategy for altering enzyme selectivity. The association of ChT with a NP coated with a carboxylic acid monolayer (TCOOH) resulted in differential changes in rates of enzyme hydrolysis.[51] ChT with bound NP showed almost no activity change with cationic substrates, whereas enzyme activity decreased drastically with anionic substrates, as represented in Figure 9.21. Subsequent studies employed amino acid functionalized AuNPs (Figure 9.22a) to analyze the effect of receptor surface on protein surface binding.[52] AuNPs with a variety of amino acid side chain structures were fabricated to probe the effect of hydrophobic groups, additional anionic functionalities, cationic groups and H-bonding functionality on the protein–NP complex. The ability of ChT to hydrolyze its substrate *N*-succinyl-L-phenylaniline-*p*-nitroanilide (SPNA) was inhibited by complex formation with the amino acid-decorated AuNPs. The degree of inhibition depended on the nature of the side chains of the amino acid groups. NP–L-Asp, NP-L-Asn, NP–L-Glu with polar side chains showed 80% suppression of the ChT activity, whereas NP–L-Met, NP-L-Ala with hydrophobic side chain showed 60% inhibition of ChT activity. A plot of Gibbs free energy (ΔG) *versus* the hydrophobicity index of the side chains (Figure 9.22b) shows that hydrophobicity aids the ChT–NP interaction. Conformational changes in the ChT structure were induced by the polar amino side chains, whereas hydrophobic side chains had little effect on the secondary structure of ChT. Denaturation of ChT can be described by eqn (9.5):

$$\text{ChT}-N \xrightarrow{k} \text{ChT}-D \tag{9.5}$$

where ChT–*N* is the native form and ChT–*D* is the denatured form of ChT. A linear regression was established, confirming that the denaturation process follows a first-order reaction profile and its rate decreases with an increase in the hydrophobicity of the amino acid side chains (Figure 9.22c).

The enzymatic kinetics and substrate specificity of ChT changes when it associates with amino acid functionalized NPs.[53] The hydrolysis activity profile of different ChT–NP complexes was determined using different SPNA substrates with cationic, anionic, or neutral charge. With the cationic substrate, the activity of ChT–NP-L-Glu increased by approximtaely three-fold compared to ChT, whereas with the neutral and anionic SPNA substrates, the ChT–NP-L-Glu activity decreased by ~50% and ~95%, respectively

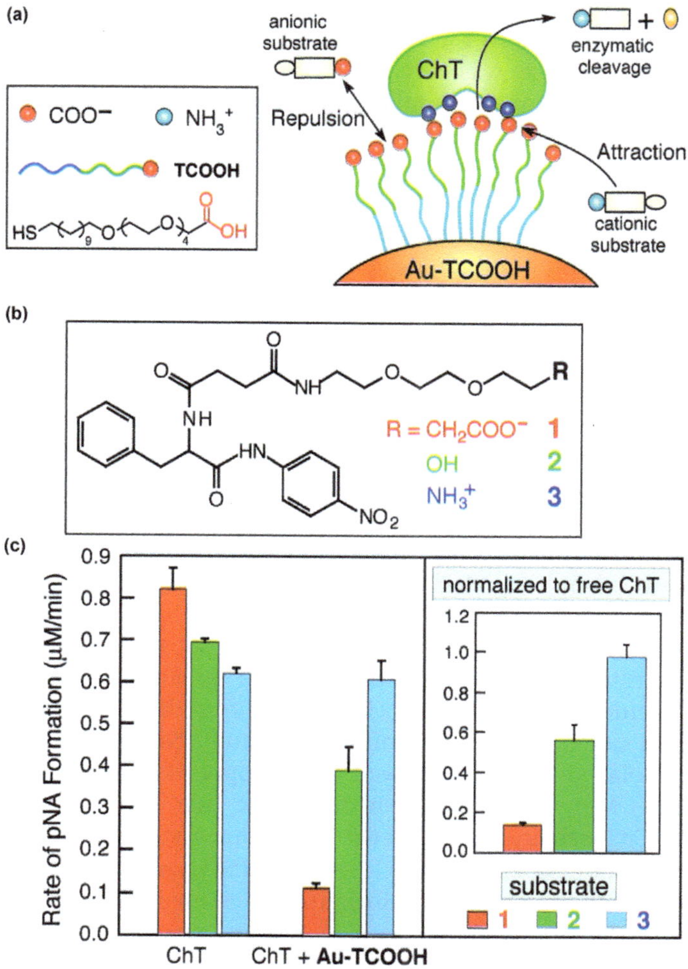

Figure 9.21 (a) Chemical structure of the TCOOH ligand and schematic depiction of substrate-monolayer-interaction-induced enzyme selectivity. (b) Structures of modified *N*-succinyl-L-phenylaniline-*p*-nitroanilide (SPNA) substrates. (c) Initial rates of α-chymotrypsin (ChT) hydrolysis of these modified substrates. Inset: normalized activity of Au-TCOOH-bound ChT toward substrates.

(Figure 9.23c). To further understand this phenomenon, appropriate features of the ChT–NP complexes were studied. The data suggest that the NP associates with the protein through surface binding, leaving the enzyme active pockets accessible for suitable substrates. Steric hindrance and electrostatic interaction with the anionic monolayer of the NPs are potential hurdles for the enzyme–substrate interaction (Figure 9.23d). The electrostatic interaction between the anionic monolayer and the cationic substrates

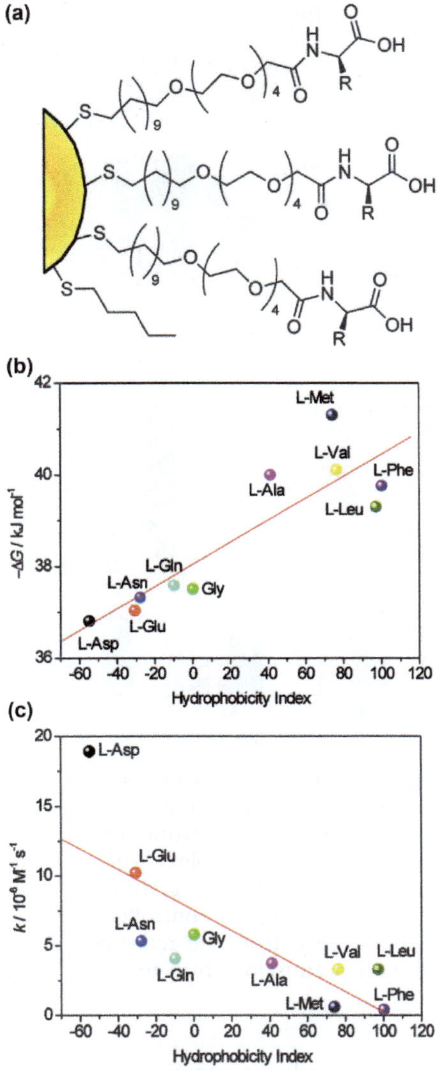

Figure 9.22 (a) Amino acid-decorated AuNPs (NPs: nanoparticles). (b) Correlation between the ΔG of NP–α-chymotrypsin (ChT) interaction and the hydrophobicity index of amino acid side chains. (c) Correlation between the denaturation rate constants (*k*) of ChT and the hydrophobicity index of amino acid side chains in nanoparticles.

enhances the accumulation of this substrate on the NP surface, increasing its local concentration and thereby increasing its catalytic activity. The activity of neutral substrate is moderately reduced by the steric hindrance, whereas the activity of the anionic substrate is greatly reduced by the electrostatic interaction between the enzyme and substrate. The catalytic

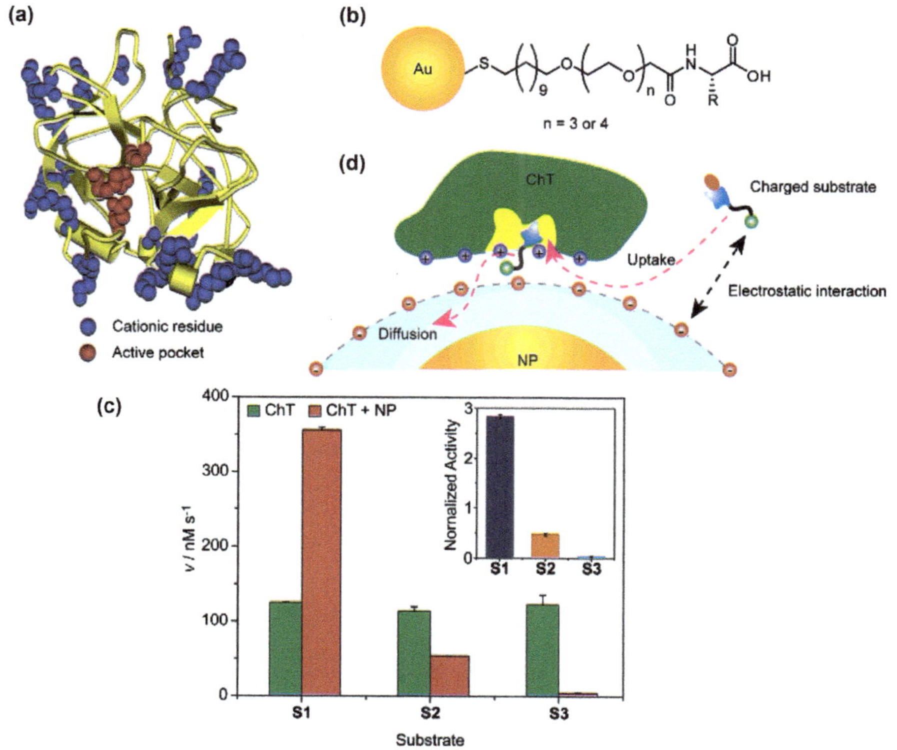

Figure 9.23 (a) Molecular structure of α-chymotrypsin (ChT). (b) Chemical structure of amino-acid-functionalized gold nanoparticles (AuNPs). (c) Generation rate (v) of 4-nitroaniline from different substrates under catalysis by ChT or ChT–NP-Glu complex). The inset shows the ratio of ChT-NP/ChT activities for the three different substrates. (d) Schematic representation of monolayer-controlled diffusion of the substrate into and the product away from the active pocket of NP-bound ChT. Reprinted with permission from ref. 53. © 2006 American Chemical Society.

behavior of ChT in these complexes depends on the properties of the amino acid side chains, as the k_{cat} and K_m values decrease smoothly according to the hydrophobicity index, whereas the k_{cat}/K_m is constant for the cationic substrate. The results show that the functionalized monolayers on NPs can be designed to fine-tune the substrate selectivity of surface-associated enzymes.

De and Rotello have examined the refolding of thermally denatured proteins by dicarboxylate functionalized gold nanoparticles (AuDA).[54] As shown in Figure 9.24, the enzymatic activity of thermally denatured ChT and papain was restored to ~93% and ~97%, respectively, by the presence of AuDA. The protein folding process mediated by AuDA can be rationalized as a two-step process: (1) binding of the AuDA with the unfolded proteins through electrostatic interactions: the protein–NP complex is negatively charged and

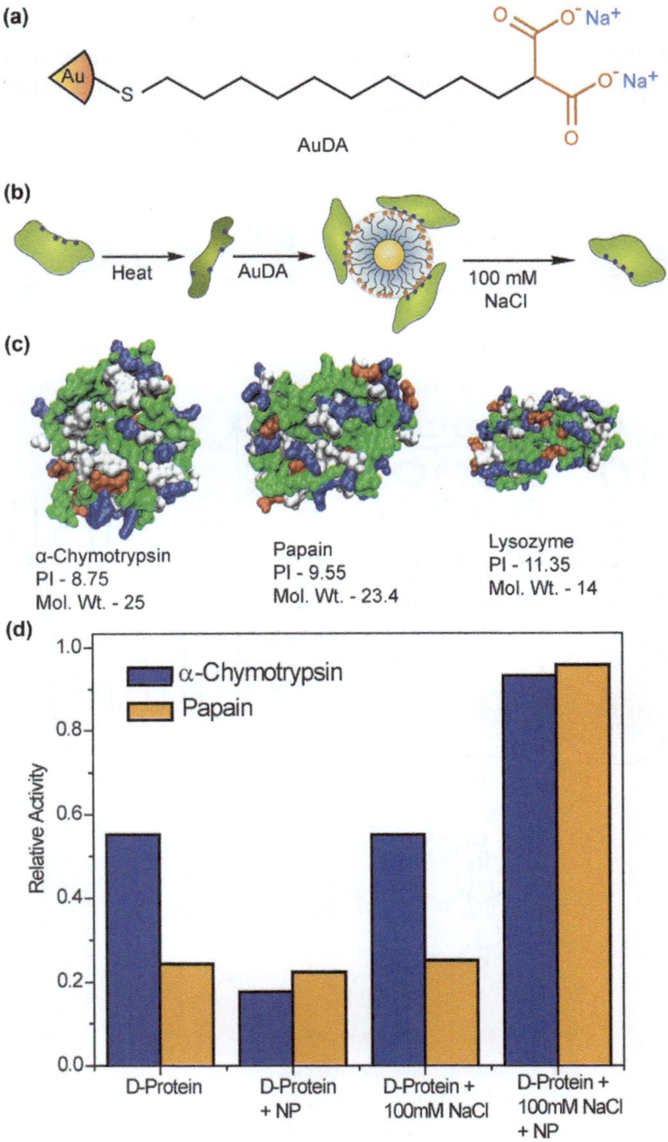

Figure 9.24 (a) Schematic representation of the structure of the gold dicarboxylate-functionalized nanoparticles (AuDA) (2 nm core) and (b) thermal denaturation followed by nanoparticle-mediated refolding of proteins. (c) Surface structural features of three positively charged proteins used in the refolding study: basic residues (blue); acidic residues (red); polar residues (green); and nonpolar residues (grey). (d) Enzymatic activity of thermally denatured α-chymotrypsin (ChT) and papain in the presence of AuDA and 100 mM NaCl solution in 5 mM sodium phosphate buffer (pH = 7.4).
Reproduced from ref. 54 with permission from the Royal Society of Chemistry.

does not aggregate but promotes correct refolding; (2) removal of the negatively charged NPs by increasing the salt concentration, thereby releasing the partially refolded proteins to fully assume a near-native structure.

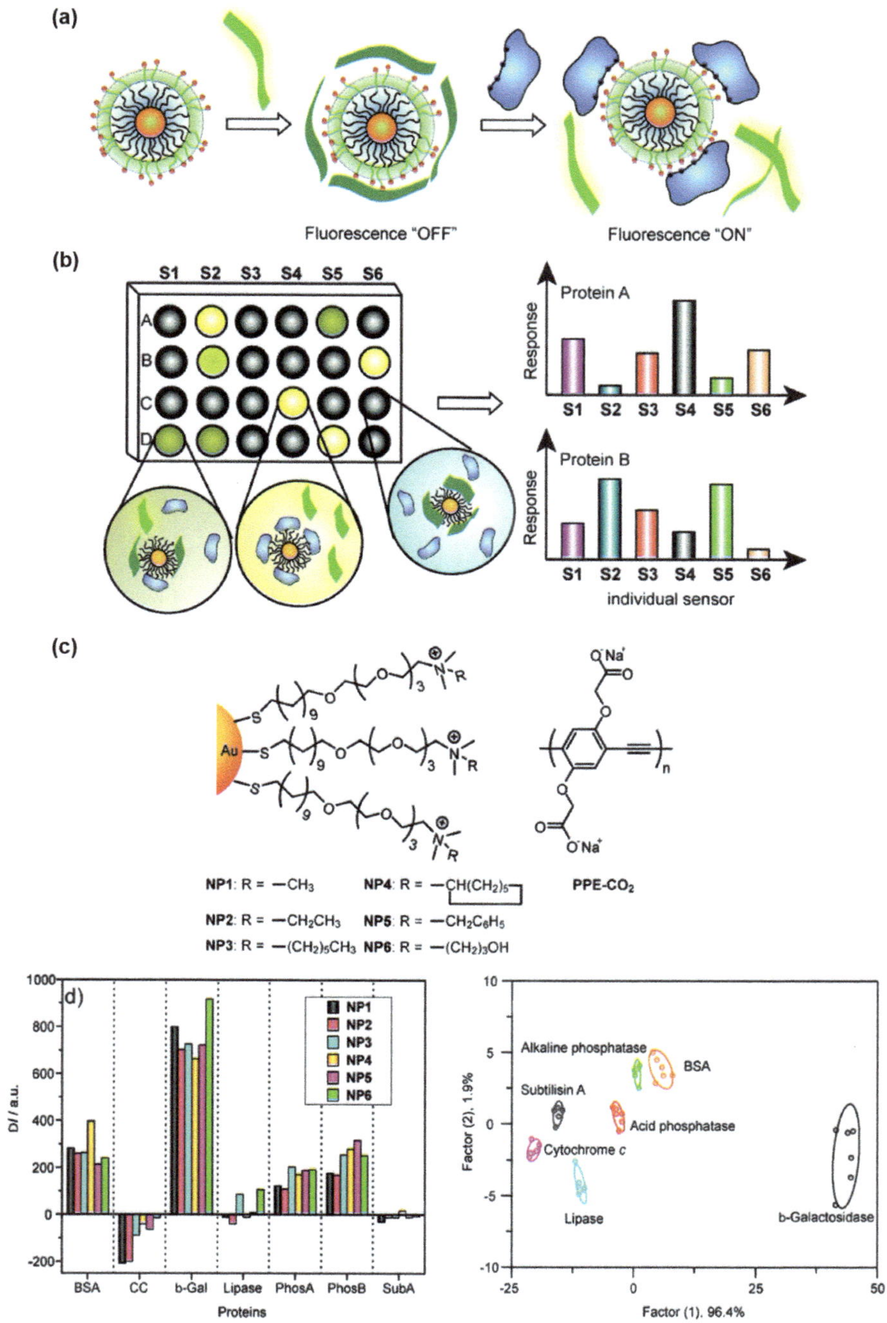

In addition to covalent modification, the NP surface functionality can be rapidly modified by hierarchal assembly methods. For example, the surface of cationic AuNPs with diaminohexane functionality was coated with the container molecule cucurbit[7]uril (CB[7]).[55] The CB[7] associated with the diaminohexane and created a modified NP-CB[7] surface, which was studied for association with model proteins like green fluorescent protein (GFP). NP-CB[7] was observed to form stronger protein interactions with GFP, and there was a change in NP–protein selectivity. The results show that hierarchal assembly methods can be used to rapidly construct novel hybrid NP–protein materials.

9.3.4.2 *Nanoparticle-based Protein Sensors*

There is extensive clinical evidence that changes in the concentration of biomarker proteins or the presence of irregular proteins in medical samples are indicative of cancer and other disease states. Thus, a sensitive, precise and convenient protein sensor array has many potential applications in medical diagnostics. An effective protein sensor requires a recognition agent with appropriate surface area to bind with a target protein and concomitant transduction of the binding event into a detectable signal.

Rotello and co-workers developed a sensor array containing six readily fabricated and structurally related cationic AuNPs with surface-associated fluorescent polymers for the detection, identification, and quantification of protein analytes.[56] In short, polymer fluorescence is quenched by AuNPs, and subsequent binding of the NPs with the protein analytes displaces the polymers, producing signaling patterns. A sensor array was developed based on the distinct fluorescence response from seven different proteins as target analytes. These patterns were quantitatively differentiated using linear discriminant analysis (LDA), with the sensor able to identify 52 out of 54 unknown protein samples (seven different proteins), exhibiting an accuracy of 94.2% (Figure 9.25). In further studies, the sensor system was modified to detect human serum proteins using arrays of NPs complexed with GFP (Figure 9.26).[57] Serum has an overall protein concentration of ~1 mM with >20 000 proteins at different concentrations. Fluorescence response was quantitatively

Figure 9.25 (a) Displacement of quenched fluorescent polymer by the protein analyte with the concomitant restoration of fluorescence. (b) Pattern generation through the differential release of fluorescent polymers from gold nanoparticles. (c) Chemical structure of cationic gold nanoparticles and anionic fluorescent polymer PPE-CO_2 ($n \approx 12$). (d) Fluorescence response (ΔI) patterns of the NP-PPE sensor array (NP1–NP6) against various proteins (CC: cytochrome c; β-Gal: β-galactosidase; PhosA: acid phosphatase; PhosB: alkaline phosphatase; SubA: subtilisin) and canonical score plot for the first two factors of simplified fluorescence response patterns. The 95% confidence ellipses for the individual proteins are also shown.

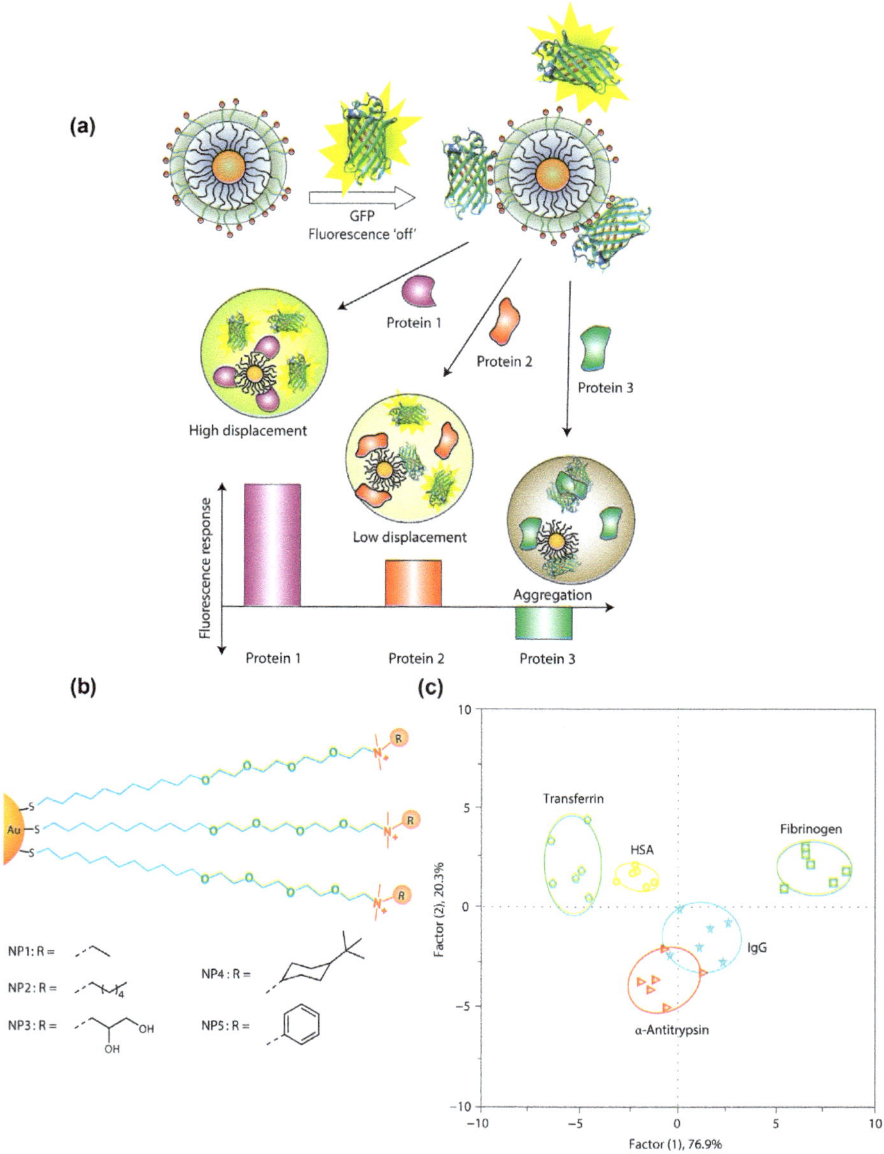

Figure 9.26 (a) Schematic illustration of the competitive binding between protein and quenched nanoparticle (NP)–green fluorescent protein (GFP) complexes and protein aggregation, leading to fluorescence light up or further quenching. (b) Chemical structure of cationic AuNPs. (c) Canonical score plot for the fluorescence patterns as obtained from linear discriminant analysis against five protein analytes at a fixed concentration (500 nM) with 95% confidence ellipses.
Reprinted by permission from Macmillan Publishers Ltd: *Nat. Nano.*, ref. 57, © 2009.

differentiated by LDA to identify minute (500 nM) changes in concentration of five major proteins in undiluted serum proteins, and an identification accuracy of 100% in buffer and 97% in human serum was achieved.

The sensitivity of the sensors for the detection of proteins was further enhanced by developing an approach termed enzyme amplified array sensing.[58] The enzyme–NP-based sensor array comprised (1) AuNPs as the receptors, (2) β-Gal as the enzyme, and (3) 4-methylumbelliferyl-β-D-galactopyranidose (MUG) as fluorogenic substrate. In this approach, cationic AuNPs electrostatically bind the anionic β-Gal, inhibiting the enzymatic activity without denaturation. Analyte proteins release β-Gal, restoring the enzymatic activity, which is reflected by the fluorescence readout signal (Figure 9.27). Target proteins (1 nM) were identified with 100% accuracy. Moreover, the sensor

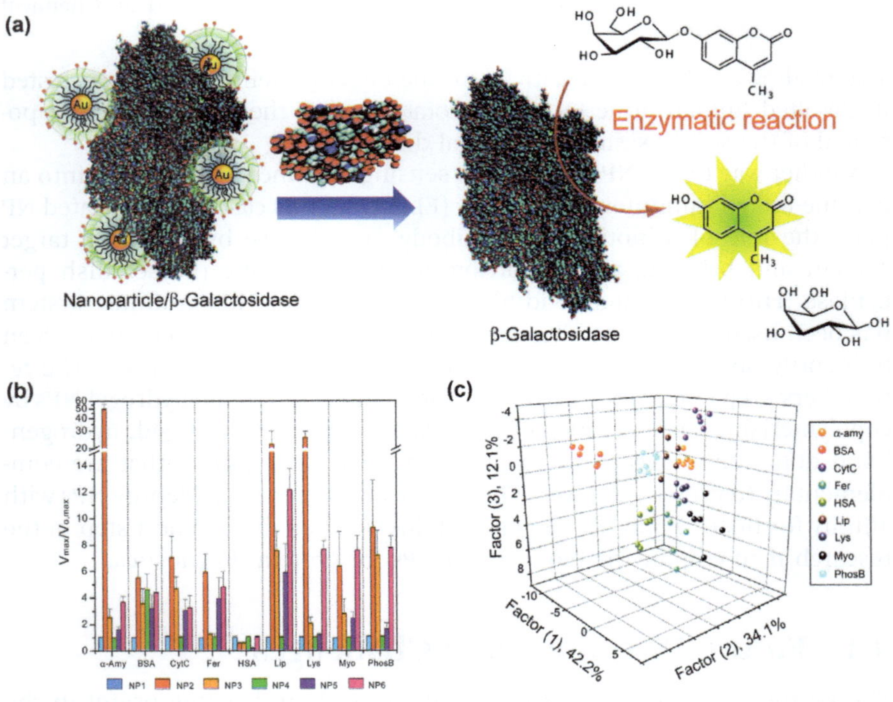

Figure 9.27 A schematic representation of a sensor element in the sensor comprised of β-galactoside (β-Gal) and cationic gold nanoparticles (AuNPs) and differentiation of proteins in 3D. (a) As shown, β-Gal is displaced from the β-Gal/AuNP complex by protein analytes, restoring the catalytic activity of β-Gal toward the fluorogenic substrate 4-methylumbelliferyl-β-D-galactopyranoside (MUG), resulting in an amplified signal for detection. (b) Differential protein pattern of the nine proteins at 1 nM. (c) Canonical score plot of the first three factors of fluorescence response patterns obtained through β-Gal/AuNP sensor array against nine target proteins in 1 nM concentration.
Reprinted with permission from ref. 58. © 2010 American Chemical Society.

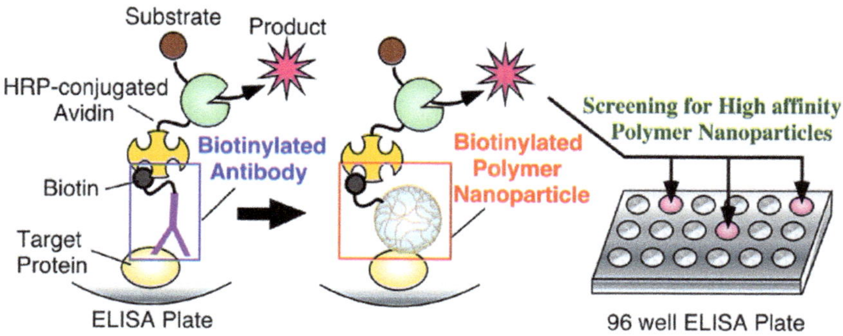

Figure 9.28 Concept of an enzyme-linked immunosorbent assay (ELISA)-mimic screen (middle and right). A standard ELISA using a biotinylated antibody is also illustrated (left).
Reprinted with permission from ref. 59. © 2012 American Chemical Society.

system showed identical sensitivity for the target proteins when it was tested in desalted human urine (~1.5 μM total protein), thereby showing the potential of the sensor system in medical diagnosis.

Another way to use NPs for protein sensing is to incorporate them into an enzyme-linked immunosorbent assay (ELISA). In this case, a biotinylated NP plays the role of a biotinylated antibody for selective binding to a target protein and bridging to a reporting enzyme conjugate (horseradish peroxidase (HRP)-conjugated avidin) (Figure 9.28). This ELISA-mimic system has been used recently by Shea and co-workers as a high-throughput screen to identify polymeric NPs with high and selective protein affinity.[59] The researchers prepared a library of biotinylated, multifunctional hydrogel NPs by copolymerizing various ratios of functional monomers (charged, hydrogen-bond donors and acceptors, hydrophobic, and hydrophilic) that are complementary to protein surfaces. The library was screened to identify NPs with affinity for histone and fibrinogen proteins. Presumably the next step in the research is to employ the biotinylated NPs for protein diagnostics.

9.4 Future Directions and Challenges

The synthetic receptors described here are likely to become useful in the near future as sensing arrays for various applications in medical diagnostics. In the longer term, they have great potential to become "conformational drugs",[60,61] that can alter protein structure and regulate protein function. Many key pharmaceutical issues have to be addressed before these molecules and nanoparticles can be used effectively for *in vivo* applications. The classic pharmaceutical challenges of ADMET (adsorption, distribution, metabolism, excretion, and toxicity) are usually discussed in terms of small-molecule drugs. With larger molecular aggregates or nanoparticles, a new set of ADMET problems arise. One is the difficulty in delivering nanoscale agents to a site of disease. In some cases, the enhanced permeation and retention effect can be used to our advantage, for example in tumor targeting. But in other cases, the rapid recognition of surface functionalized nanoparticles by the protective reticuloendothelial system is a major obstacle. A related concern is biofouling, that is, the undesired formation of a protein corona on the nanoparticle surface, which masks the targeting functionalization that produces the molecular recognition. Researchers who are entering the field of protein recognition should be encouraged by the remarkable advances that have been achieved over the last decade, with an expectation of accelerated progress in the years ahead.

Acknowledgements

The authors declare no conflict of interest. This work was supported by the National Instistutes of Health (GM077173).

References

1. S. Jones and J. M. Thornton, *Proc. Natl. Acad. Sci. U. S. A.*, 1996, **93**, 13.
2. L. Milroy, T. N. Grossmann, S. Hennig, L. Brunsveld and C. Ottmann, *Chem. Rev.*, 2014, **114**, 4695.
3. N. Horton and M. Lewis, *Protein Sci.*, 1992, **1**, 169.
4. L. Lo Conte, C. Chothia and J. Janin, *J. Mol. Biol.*, 1999, **285**, 2177.
5. J. Janin and C. Chothia, *J. Biol. Chem.*, 1990, **265**, 16027.
6. L. Young, R. L. Jernigan and D. G. Covell, *Protein Sci.*, 1994, **3**, 717.
7. P. Lijnzaad and P. Argos, *Proteins*, 1997, **28**, 333.
8. A. Fernandez and H. A. Scheraga, *Proc. Natl Acad. Sci. U. S. A.*, 2003, **100**, 113.
9. F. B. Sheinerman, R. Norel and B. Honig, *Curr. Opin. Struct. Biol.*, 2000, **10**, 153.
10. O. Keskin, B. Ma and R. Nussinov, *J. Mol. Biol.*, 2005, **345**, 1281.
11. J. A. Wells, *Biochemistry*, 1990, **29**, 8509.
12. B. C. Cunningham and J. A. Wells, *Science*, 1989, **244**, 1081.
13. I. S. Moreira, P. A. Fernandes and M. J. Ramos, *Proteins*, 2007, **68**, 803.
14. A. M. de Vos, M. Ultsch and A. A. Kossiakoff, *Science*, 1992, **255**, 306.
15. T. L. Nero, C. J. Morton, J. K. Holien, J. Wielens and M. W. Parker, *Nat. Rev. Cancer*, 2014, **14**, 248.
16. U. Samanta, D. Pal and P. Chakrabarti, *Proteins*, 2000, **38**, 288.
17. A. Fernandez, *FEBS Lett.*, 2002, **527**, 166.
18. A. A. Bogan and K. S. Thorn, *J. Mol. Biol.*, 1998, **280**, 1.
19. M. Sattler, H. Liang, D. Nettesheim, R. P. Meadows, J. E. Harlan, M. Eberstadt, H. S. Yoon, S. B. Shuker, B. S. Chang, A. J. Minn, C. B. Thompson and S. W. Fesik, *Science*, 1997, **275**, 983.
20. S. W. Muchmore, M. Sattler, H. Liang, R. P. Meadows, J. E. Harlan, H. S. Yoon, D. Nettesheim, B. S. Chang, C. B. Thompson, S. L. Wong, S. C. Ng and S. W. Fesik, *Nature*, 1996, **381**, 335.

21. I. S. Moreira, P. A. Fernandes and M. J. Ramos, *J. Phys. Chem. B*, 2007, **111**, 2697.
22. S. Fletcher and A. D. Hamilton, *J. R. Soc., Interface*, 2006, **3**, 215.
23. R. K. Jain and A. D. Hamilton, *Angew. Chem., Int. Ed.*, 2002, **41**, 641.
24. A. J. Wilson, K. Groves, R. K. Jain, H. S. Park and A. D. Hamilton, *J. Am. Chem. Soc.*, 2003, **125**, 4420.
25. T. Aya and A. D. Hamilton, *Bioorg. Med. Chem. Lett.*, 2003, **13**, 2651.
26. K. Groves, A. J. Wilson and A. D. Hamilton, *J. Am. Chem. Soc.*, 2004, **126**, 12833.
27. S. Fletcher and A. D. Hamilton, *New J. Chem.*, 2007, **31**, 623.
28. E. U. Weber-Ban, B. G. Reid, A. D. Miranker and A. L. Horwich, *Nature*, 1999, **401**, 90.
29. L. Baldini, A. J. Wilson, J. Hong and A. D. Hamilton, *J. Am. Chem. Soc.*, 2004, **126**, 5656.
30. H. Zhou, L. Baldini, J. Hong, A. J. Wilson and A. D. Hamilton, *J. Am. Chem. Soc.*, 2006, **128**, 2421.
31. F. G. Klärner and T. Schrader, *Acc. Chem. Res.*, 2012, **46**, 967.
32. P. Talbiersky, F. Bastkowski, F. G. Klärner and T. Schrader, *J. Am. Chem. Soc.*, 2008, **130**, 9824.
33. M. Kirsch, P. Talbiersky, J. Polkowska, F. Bastkowski, T. Schaller, H. de Groot, F. G. Klärner and T. Schrader, *Angew. Chem., Int. Ed.*, 2009, **48**, 2886.
34. D. Bier, R. Rose, K. Bravo-Rodriguez, M. Bartel, J. M. Ramirez-Anguita, S. Dutt, C. Wilch, F.G. Klärner, E. Sanchez-Garcia, T. Schrader and C. Ottmann, *Nat. Chem.*, 2013, **5**, 234.
35. S. Sinha, D. H. J. Lopes, Z. Du, E. S. Pang, A. Shanmugam, A. Lomakin, P. Talbiersky, A. Tennstaedt, K. McDaniel, R. Bakshi, P.-Y. Kuo, M. Ehrmann, G. B. Benedek, J. A. Loo, F.-G. Klärner, T. Schrader, C. Wang and G. Bitan, *J. Am. Chem. Soc.*, 2011, **133**, 16958.
36. S. B. Nimse and T. Kim, *Chem. Soc. Rev.*, 2013, **42**, 366.
37. Q. Lin, H. S. Park, Y. Hamuro, C. S. Lee and A. D. Hamilton, *Biopolymers*, 1998, **47**, 285.
38. H. S. Park, Q. Lin and A. D. Hamilton, *J. Am. Chem. Soc.*, 1999, **121**, 8.
39. H. S. Park, Q. Lin and A. D. Hamilton, *Proc. Natl. Acad. Sci. U. S. A.*, 2002, **99**, 5105.
40. S. Gordo, V. Martos, E. Santos, M. Menéndez, C. Bo, E. Giralt and J. de Mendoza, *Proc. Natl. Acad. Sci. U. S. A.*, 2008, **105**, 16426.
41. S. Dutt, C. Wilch and T. Schrader, *Chem. Commun.*, 2011, **47**, 5376.
42. R. E. McGovern, H. Fernandes, A. R. Khan, N. P. Power and P. B. Crowley, *Nat. Chem.*, 2012, **4**, 527.
43. T. Schrader, *Nat. Chem.*, 2012, **4**, 519.
44. R. Mout and V. M. Rotello, *Isr. J. Chem.*, 2013, **53**, 521.
45. K. Saha, S. S. Agasti, C. Kim, X. Li and V. M. Rotello, *Chem. Rev.*, 2012, **112**, 2739.
46. M. De, O. R. Miranda, S. Rana and V. M. Rotello, *Chem. Commun.*, 2009, 2157.

47. M. De, C. C. You, S. Srivastava and V. M. Rotello, *J. Am. Chem. Soc.*, 2007, **129**, 10747.
48. N. O. Fischer, C. M. McIntosh, J. M. Simard and V. M. Rotello, *Proc. Natl Acad. Sci. U. S. A.*, 2002, **99**, 5018.
49. A. Verma, J. M. Simard and V. M. Rotello, *Langmuir*, 2004, **20**, 4178.
50. C. C. You, M. De and V. M. Rotello, *Org. Lett.*, 2005, **7**, 5685.
51. R. Hong, T. Emrick and V. M. Rotello, *J. Am. Chem. Soc.*, 2004, **126**, 13572.
52. C. C. You, M. De, G. Han and V. M. Rotello, *J. Am. Chem. Soc.*, 2005, **127**, 12873.
53. C. C. You, S. S. Agasti, M. De, M. J. Knapp and V. M. Rotello, *J. Am. Chem. Soc.*, 2006, **128**, 14612.
54. M. De and V. M. Rotello, *Chem. Commun.*, 2008, 3504.
55. Y. C. Yeh, S. Rana, R. Mout, B. Yan, F. S. Alfonso and V. M. Rotello, *Chem. Commun.*, 2014, **50**, 5565.
56. C. C. You, O. R. Miranda, B. Gider, P. S. Ghosh, I. B. Kim, B. Erdogan, S. A. Krovi, U. H. F. Bunz and V. M. Rotello, *Nat. Nano.*, 2007, **2**, 318.
57. M. De, S. Rana, H. Akpinar, O. R. Miranda, R. R. Arvizo, U. H. F. Bunz and V. M. Rotello, *Nat. Chem.*, 2009, **1**, 461.
58. O. R. Miranda, H. T. Chen, C. C. You, D. E. Mortenson, X. C. Yang, U. H. F. Bunz and V. M. Rotello, *J. Am. Chem. Soc.*, 2010, **132**, 5285.
59. Y. Yonamine, Y. Hoshino and K. J. Shea, *Biomacromolecules*, 2012, **13**, 2952.
60. M. W. Peczuh and A. D. Hamilton, *Chem. Rev.*, 2000, **100**, 2479.
61. S. Fletcher and A. D. Hamilton, *Curr. Opin. Chem. Biol.*, 2005, **9**, 632.

CHAPTER 10

Synthetic Receptors for Polar Lipids

KASEY J. CLEAR AND BRADLEY D. SMITH*

Department of Chemistry and Biochemistry, 236 Nieuwland Science Hall, University of Notre Dame, Notre Dame, 46556 IN, USA
*Email: smith.115@nd.edu

10.1 Structures and Properties of Biological Lipids

Lipids are a diverse class of biomolecules with hydrophobic or amphiphilic structures. Lipid composition in living systems is very complex and >10 000 distinct lipid structures are possible from different combinations of building blocks.[1] The functional importance of lipids in biology can be divided into three broad categories.[2] First, as partially reduced hydrocarbon species, lipids serve as excellent energy storage molecules. Second, the self-assembly of polar lipids in aqueous environments leads to the formation of membranes, which serve as barriers for the compartmentalization of cellular components. Finally, lipids are active participants in cellular signaling and molecular recognition processes. Table 10.1 shows the major classes of biological lipids, a representative structure, and the typical physiological roles for each class.[3–13] The two most common carbon building blocks are fatty acids and isoprene. Condensation of a fatty acid with a long-chain alcohol leads to a wax, whereas esterification with glycerol produces acylglycerols and glycerophospholipids. Isoprene is a five-carbon building block which gives rise to the nonpolar terpenes, the weakly polar isoprenoids, and the polar eicosanoids and steroids.

Monographs in Supramolecular Chemistry No. 14
Synthetic Receptors for Biomolecules: Design Principles and Applications
Edited by Bradley D. Smith

Published by the Royal Society of Chemistry, www.rsc.org

Table 10.1 Major classes of lipids found in biological systems. R indicates a fatty acid tail of varying length and saturation.

Lipid class	Representative structure	Typical physiological roles	Refs
Fatty acid	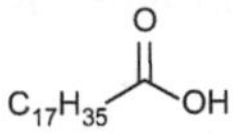 Oleic acid	Energy storage, membrane structure (as esters), and signaling	3
Waxes	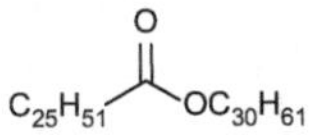 Beeswax component	Physical protection roles (*e.g.*, water proofing)	4
Acylglycerols	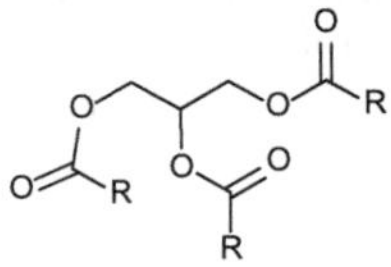 A triacylglycerol	Energy storage and signaling	5
Phospho-glycerolipids	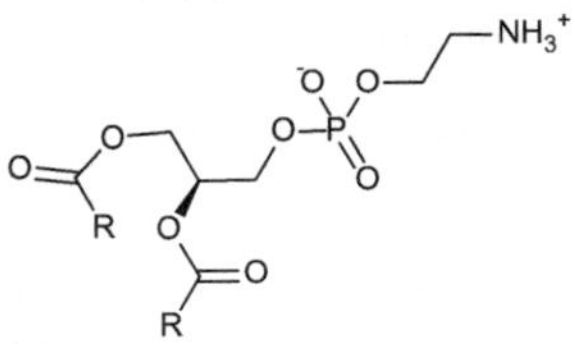 Phosphatidylethanolamine	Membrane lipids and signaling molecules and receptors	6
Sphingolipids	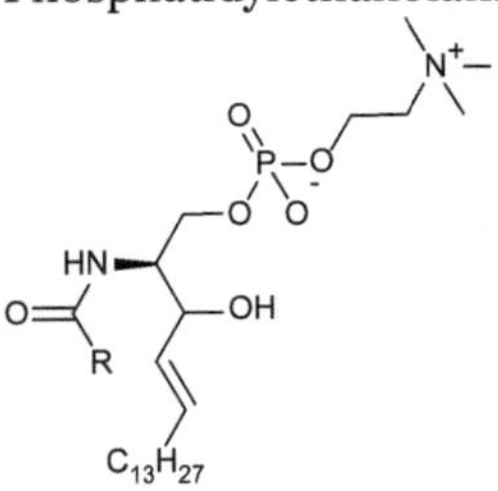 Sphingomyelin	Membrane lipids and signaling molecules	7
Terpenes	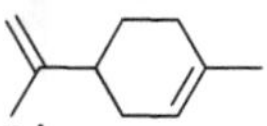 Limonene	Diverse structures and roles; many are useful as pharmaceutical agents	8
Isoprenoids	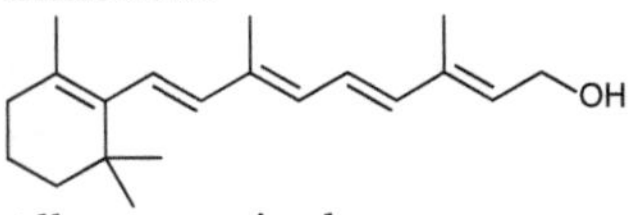 All-*trans* retinol	Primarily signaling and immune response	9,10
Eicosanoids	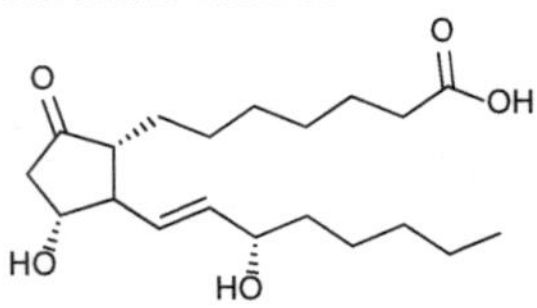 Prostaglandin E_1	Local hormones and signaling molecules; inflammation and immune response	11,12
Steroids	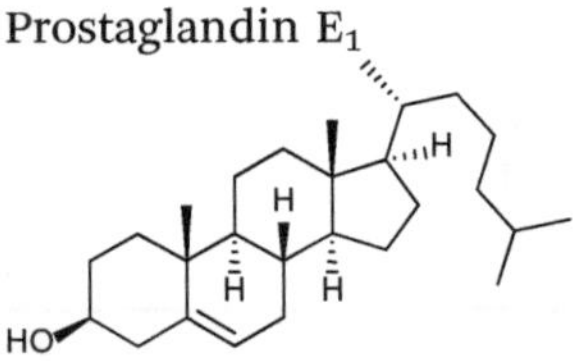 Cholesterol	Maintain membrane integrity and serve signaling roles	13

This chapter focuses on molecular recognition of polar lipids. To appreciate the recognition strategies for these molecules, it is important to first understand some key biophysical properties of polar lipids. As depicted in Table 10.2, the structure of a polar lipid molecule can be simplified as a polar headgroup linked to a hydrophobic tail.[14] In the presence of water, the amphiphilic molecules self-assemble to form aggregates that expose the polar headgroups to the aqueous solvent and shield the hydrophobic tails. The aggregate morphology depends on the relative size and shape of the

Table 10.2 Polar lipid self-assembly is controlled by the lipid packing parameter, P. Table adapted with permission from ref. 14.

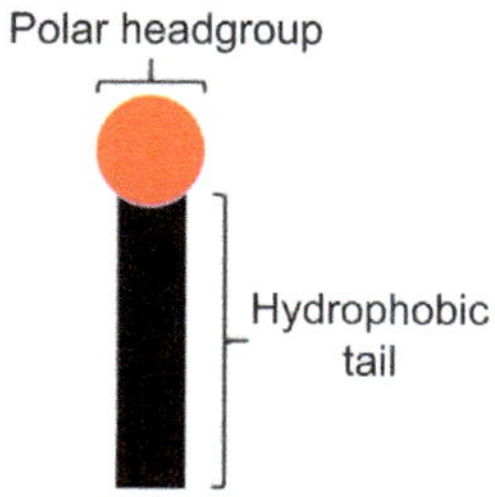

Lipid packing parameter:

$$P = \frac{V}{a \cdot l}$$

V = volume occupied by lipid

a = effective headgroup area

l = length of lipid

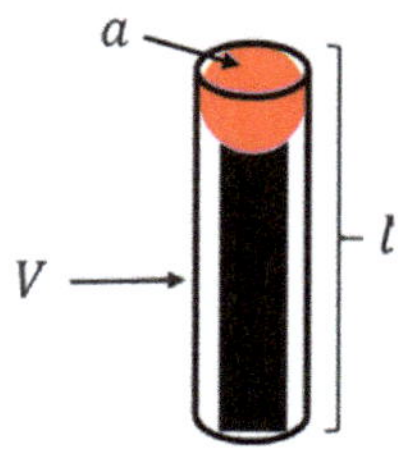

Lipids	Shape	Organization
Soaps Detergents Lysophospholipids	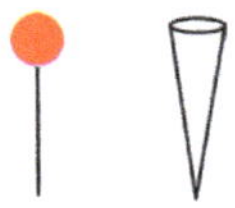 inverted cone $P < \frac{1}{3} - \frac{2}{3}$	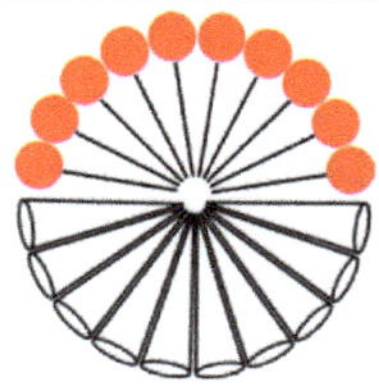 micelle (positive curvature)
Phosphatidyl - choline - serine - inositol Sphingomyelin	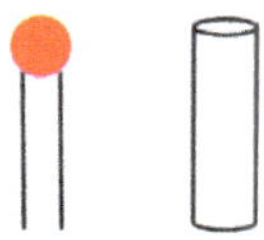 cylinder $P \sim 1$	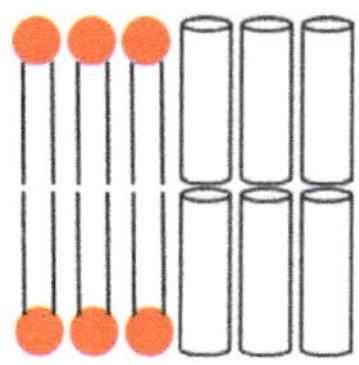 bilayer
Phosphatidyl - ethanolamine Phosphatidic acid Cholesterol Cardiolipin Lipid A	 cone $P > 1$	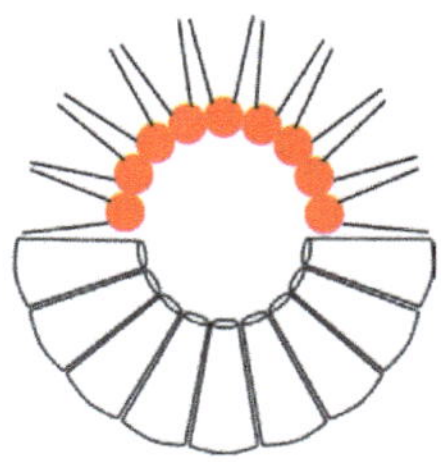 reverse micelle (negative curvature)

polar lipid. The lipid packing parameter, *P*, is useful for predicting morphology and surface curvature.[14] Polar lipids with a relatively large headgroup ($P<1$) tend to form micelles with positive curvature, while analogues with small headgroups ($P>1$) tend to form reverse micelles or cubic phases (negative curvature). The prevalent self-assembled structure found in biological systems is the unilamellar bilayer membrane ($P\sim1$) with two polar surfaces separated by a hydrophobic interior. In some large bilayer structures, such as the cell plasma membrane, the surface is relatively flat, but in other cases, such as smaller spherical vesicles (also known as liposomes), the surface can be highly curved.

Bilayer membrane structural properties, such as thickness and fluidity of the lipid core, are primarily determined by the length and saturation of the acyl chains. The fluid mosaic model of a cell membrane envisions the bilayer as a two-dimensional fluid in which the polar lipid molecules undergo rapid lateral diffusion. The lipid core of a chemically pure bilayer can transition from an ordered gel phase at low temperature to a disordered liquid phase at high temperatures. The phase transition temperature is lower when the acyl chains contain one or more *cis* double bonds. The presence of membrane additives, such as other types of polar lipids or transmembrane proteins, can alter these membrane properties in predictable ways that are reviewed eleswhere.[3]

The identity and relative abundance of polar lipids in biological membranes varies greatly across organisms and even depends on the cell type and intracellular location (Table 10.3). Lipid composition plays an important role in regulating membrane biophysical properties such as surface curvature and internal fluidity. The presence of sterols, such as cholesterol or ergosterol, greatly affect membrane packing and structural integrity.[15] Sterols are key players in the formation of local liquid-ordered membrane phases called 'lipid rafts', which also contain transmembrane proteins and are involved in biological signaling and trafficking processes.[16] Translocation of polar lipids across the membrane bilayer is called flip–flop. Unlike the rapid process of lateral membrane diffusion, uncatalyzed flip–flop is a slow process that can take hours to days.[17] Most mammalian cell membranes contain energy-consuming transport proteins (flippases) that maintain an

Table 10.3 Approximate polar lipid composition (by mass) in various biological membranes.[18,19]

	Red blood cell plasma membrane[18]	Mitochondrion membranes[18]	*Escherichia coli* bacterium[18]	*Candida albicans* plasma membrane[19]
Cholesterol	23	3	0	0
Ergosterol	0	0	0	40
Phosphatidylethanolamine	18	28	70	14
Phosphatidylserine	7	2	trace	10
Phosphatidylcholine	17	44	0	17
Sphingomyelin	18	0	0	0
Others	17	23	30	19

asymmetric transmembrane distribution of polar lipids. For example, mammalian plasma membranes typically have higher concentrations of phosphatidylserine (PS) and phosphatidylethanolamine (PE) in the interior monolayer and higher phosphatidylcholine (PC) and sphingomyelin (SM) concentrations in the outer monolayer.[18,19]

10.2 Biological Recognition of Polar Lipids

The structure and composition of biological membranes is maintained by the orchestrated action of various protein receptors that selectively associate with lipids and traffic them to specific membrane locations. In addition, there are receptors that monitor the membrane composition and packing.[20] Likewise, many cell signaling processes utilize a peripheral membrane protein to recognize one or more lipid headgroups within the context of a complicated multicomponent bilayer.[2]

Building on the structural concept that a polar lipid is an amphiphilic molecule comprised of a polar headgroup and hydrophobic tail, it is helpful to separate biological protein and peptide receptors into three supramolecular classes. The first class comprises 'container' receptors that encapsulate the hydrophobic tail, whereas the second class uses electrostatic and dipolar interactions to associate with the polar headgroup. The third class is ditopic receptors, which can simultaneously interact with both the polar headgroup and hydrophobic tail.

10.2.1 Biological Recognition of the Hydrophobic Tail

Many proteins act as biological receptors for the hydrophobic portion of polar lipids. Typically these receptors feature a deep pocket that can accommodate the lipophilic tail and shield it from the surrounding aqueous environment (Scheme 10.1). Selectivity and binding stoichiometry depends of the shape and dimensions of the binding pocket(s). For example, apolipoproteins associate with polar lipids and large numbers of weakly polar/nonpolar storage lipids, such as triacylglycerols and cholesteryl esters, to form soluble lipoprotein nanoparticles which serve to transport and circulate the lipids throughout the blood stream.[21] In addition, plants and animals both contain several classes of nonspecific lipid transport proteins, which can form bimolecular complexes with fatty acids, phospholipids and sterols.[22,23] In other cases, lipid transport proteins have the capacity to

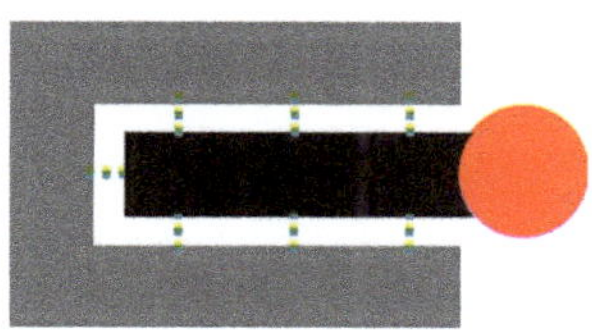

Scheme 10.1 Hydrophobic tail recognition of polar lipids.

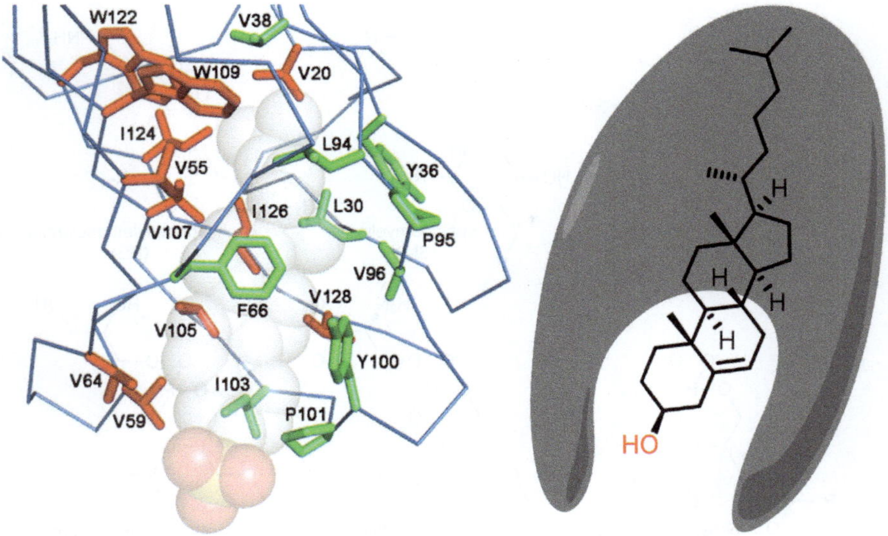

Figure 10.1 Bovine Niemann–Pick disease, type C2 (NPC2) cholesterol transport protein binds the hydrophobic tail of cholesterol. The crystal structure of the NPC2-cholesterol sulfate complex (left) and cartoon representation of the NPC2-cholesterol complex (right).
The crystal structure was originally published in *J. Biol. Chem.*, © 2007 The American Society for Biochemistry and Molecular Biology.[24]

selectively recognize the structure of the nonpolar tail. An example of this is NPC2, a sterol-selective transport protein which binds cholesterol with submicromolar dissociation constant.[24] In humans, the absence of NPC2 leads to Niemann–Pick type C disease, a fatal hereditary disorder characterized by abnormal lysosomal accumulation of cholesterol. The sterol-binding pocket of NPC2 selectively binds the nonpolar skeleton of cholesterol, and the associated complex is stabilized by interactions with a valine side chain and other hydrophobic residues in the binding pocket (Figure 10.1). There is no evidence for any interaction of the protein with the cholesterol hydroxyl.

10.2.2 Biological Recognition of the Polar Headgroup

Phospholipids are arguably the most important class of polar lipids, and they are classified by the headgroup structure. A summary of these names and structures are shown in Scheme 10.2. While there is great variation in the composition of the lipophilic tail in terms of length of acyl chains and degree of chain unsaturation, the variation in the acyl chain is largely ignored when considering the process of headgroup recognition. Inspection of the headgroup structures shows that the net charges are either zwitterionic or anionic. The sum of these charges within a patch of membrane determines the membrane's local electrostatic surface potential, which is sometimes a determining factor in attracting biological receptors to cell surfaces.

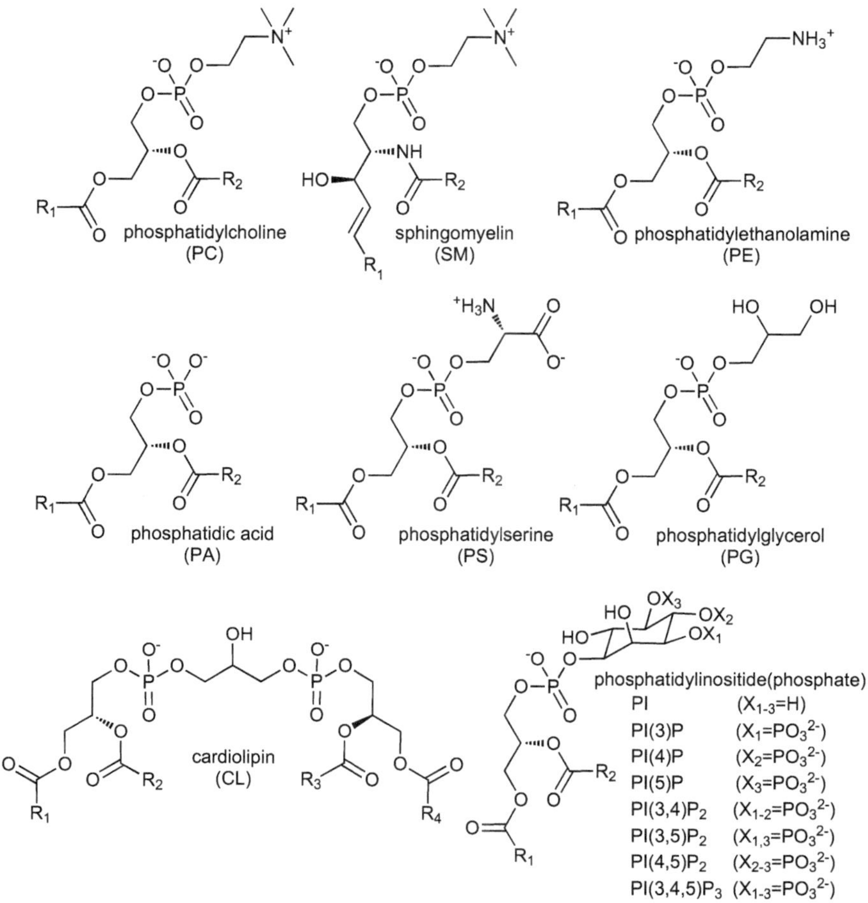

Scheme 10.2 Names and headgroup structures of phospholipids.

For example, membrane patches with high levels of negative charge attract positively charged proteins, as in the association of cationic cytochrome c to the anionic inner membrane of the mitochondria.[25]

10.2.2.1 *Headgroup Recognition by Peripheral Proteins*

Molecular recognition of specific polar lipids within a cell membrane is accomplished by peripheral membrane proteins, which use a combination of electrostatic and dipolar interactions to associate with the polar headgroup at the membrane surface. The peripheral proteins are typically high molecular weight structures containing smaller modules that are often referred to as lipid-binding domains.[26,27] The majority of currently known lipid binding domains and their polar lipid recognition targets are listed in Table 10.4. Most of the binding targets are phospholipids, and the supramolecular processes range from stereospecific recognition of lipid

Table 10.4 Currently known lipid binding domains found in peripheral membrane proteins. Table adapted by permission from Macmillan Publishers Ltd: *Nat. Rev. Mol. Cell. Biol.*, © 2008.[a,26]

Domain name	Domain size (aa)	Preferred target	Ca^{2+}-mediated
C1	~50	diacylglycerol, phorbol esters	No
PKC C2	~130	PS, primarily	Yes
PH	~125	PIs (diverse class, some specificity)	No
FYVE	60–70	PI(3)P	No
PX	~130	PI(3)P, primarily	No
PROPPIN	~500	$PI(3,5)P_2$, primarily	No
Gla	~45	PS	Yes
Annexin	~310	acidic phospholipids	Yes
Discoidin C2	~160	PS	No
ENTH	~150	$PI(4,5)P_2$, primarily	No
BAR	~240	acidic phospholipids	No
F-BAR	~320	acidic phospholipids	No
IMD	~250	acidic phospholipids, especially PIs	No

[a]aa: amino acids; PS: phosphatidylserine; PI: phosphatidylinositide.

headgroups (*e.g.*, FYVE domain) to selective recognition of membrane surface features such as electrostatic charge and curvature (*e.g.*, BAR domain). The remainder of this section will focus on the systems that achieve specific recognition of a polar headgroup, since this is the process that is usually mimicked by synthetic receptors. Lipid-binding domains differ in size from 50 to 500 amino acids and anionic (acidic) phospholipids are typically the preferred target. The phospholipid selectivity of the domains can vary with the parent protein. This variability is exemplified in two proteins containing the phosphatidylinositol $(PI)(P)_n$-specific pleckstrin homology (PH) domain. In phospholipase $C\delta_1$ (PLC δ_1), the PH domain binds selectivity to $PI(4,5)P_2$ whereas the PH domain of Bruton's tyrosine kinase is highly specific for phosphotidylinositides with phosphate groups at the 3-position, $PI(3,4)P_2$ and $PI(3,4,5)P_3$, despite 10-fold higher abundance of $PI(4,5)P_2$ in the membrane.[26] Some domains simply target phospholipids according to their charge. For example, annexin domains target phospholipids with anionic headgroups. Like the PKC C2 and Gla domains, annexin domains require Ca^{2+} to mediate the binding (see below).

Polar lipid headgroup recognition by a peripheral protein typically involves non-covalent interactions such as electrostatics, metal cation coordination, and hydrogen bonding. Shown in Scheme 10.3 is a schematic summary of four different biological strategies for improving polar lipid binding affinity or selectivity. The first is partial hydrophobic insertion of the protein receptor into the membrane (Scheme 10.3A), which increases binding affinity by decreasing the rate of dissociation. It is important to note that there is a class of peripheral proteins known as 'lipidated' proteins which fully insert into the membrane,[27] but here the focus is on reversible membrane binding proteins. A second strategy for enhancing lipid–protein

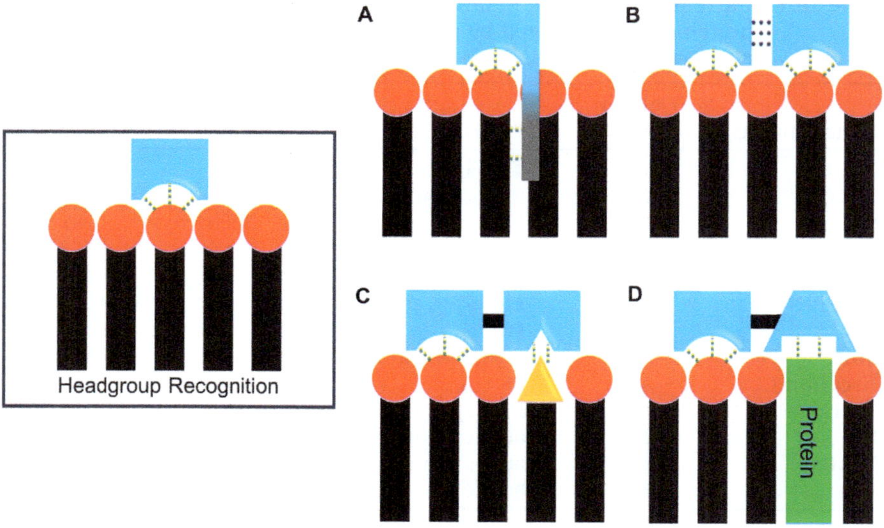

Scheme 10.3 (Left) Headgroup recognition of a polar lipid at a membrane surface using non-covalent interactions (*e.g.*, hydrogen bonding, electrostatics, or metal cation coordination). (Right) Biological strategies for improving polar lipid recognition at a membrane surface. (A) Partial hydrophobic insertion into membrane; (B) oligomerization of protein monomer upon lipid binding; (C) coincidence detection of two lipid types in a membrane; (D) coincidence detection of a lipid and transmembrane protein.

affinity is self-oligomerization of the protein on the membrane surface (Scheme 10.3B). This mechanism is exhibited by the endocytic protein dynamin, which binds PI(4,5)P_2 through its PH domain.

Homodimerization of dynamin at the membrane surface increases the affinity for PI(4,5)P_2 in the membrane by a factor of 1000.[26] Two strategies for improving binding selectivity involve 'coincidence detection', which requires the peripheral protein to simultaneously target two different lipids (Scheme 10.3C) or alternatively a lipid and a transmembrane protein (Scheme 10.3D). For example, the protein SNX1 contains adjacent PX and BAR domains and uses coincidence detection to selectively recognize highly curved PI(3)P-rich membranes.[26] Sometimes coincidence detection is employed within a single binding domain as in the case of the PKCα C2 domain, which simultaneously binds both PS and PIP_ns.[28]

In recent years, some of these peripheral protein recognition systems have been characterized at the atomic level. An example is the FYVE domain of early endosome antigen 1 (EEA1), which is recruited to early endosomal membranes by both PI(3)P and the GTPase Rab5 (coincidence detection of a lipid and protein) where it aids in endosomal fusion.[29] FYVE binding is a multistep process involving pH-induced activation of histidine residues, recruitment and binding to the membrane through nonspecific electrostatic interactions with PS and phosphatidic acid headgroups in the membrane, and partial hydrophobic insertion into the membrane.[30] In some cases, dimerization *via* coiled coil motifs occurs, allowing simultaneous binding of two PI(3)P headgroups. The crystal structure of the EEA1 FYVE domain bound to D-*myo*-inositol-1,3-bisphosphate reveals the origins of the specificity of PI(3)P and the importance of cationic arginine and histidine side chains (Figure 10.2). PI(3)P binds in a concave pocket containing centrally-located basic residues, with the 3-phosphate stabilized by Arg1375 and the 1-phosphate stabilized by Arg1370. Essential to PI(3)P-specific recognition over other PI isomers are contacts and hydrogen bonding between the 4,5,6 hydroxyl groups and Asp1352 and His1373 residues.

Another lipid-binding protein that has been characterized at the atomic level is annexin V (or annexin A5), a 35 kDa member of the annexin family of anionic phospholipid binding proteins. Annexins are sometimes labeled as PS-binding domains, but only because PS is the most abundant anionic phospholipid in mammalian cell membranes. Like most of the annexins, annexin V consists of four repeat domains, each coordinating approximately two Ca^{2+} ions. Annexin V binds very tightly to membranes containing PS, and the main driving force is coordination of protein-bound Ca^{2+} by the

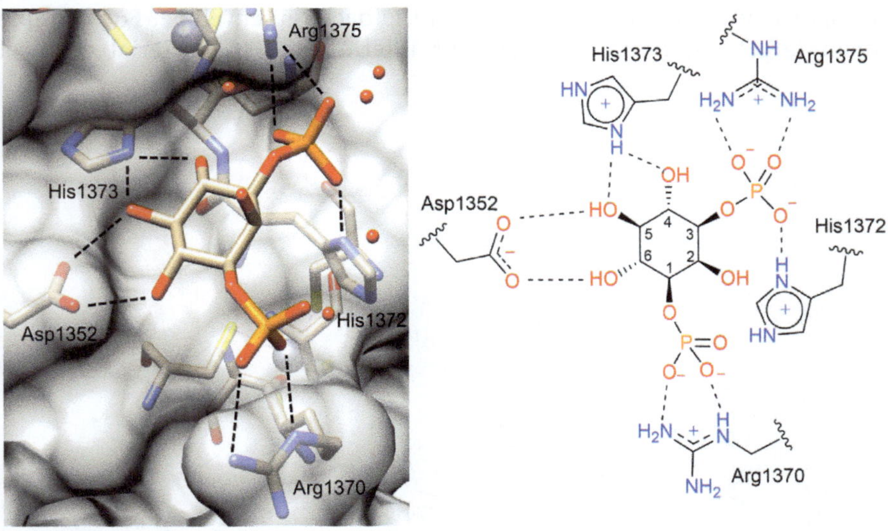

Figure 10.2 (Left) Crystal structure of the endosome antigen (EEA)1 FYVE domain bound to D-*myo*-inositol-1,3-bisphosphate, the headgroup of phosphataidylinositol-3-phosphate (PI(3)P) (Protein Data Bank code 1JOC). Molecular graphics image was produced using the Chimera package from the Computer Graphics Laboratory, University of California, San Francisco (CA, USA); supported by National Institutes of Health P41 RR-01081. (Right) Structural illustration of the binding interactions showing hydrogen bonding and salt bridges between protein side chains and PI(3)P.[31]

phosphate headgroups along with the entropic gain associated with dehydration of the large protein and membrane surface area.[32] Self-oligomerization of the protein on the membrane surface has also been suggested to occur under some conditions.[33] The crystal structure of annexin V bound to glycerophospho-L-serine illustrates the simultaneous coordination of Ca^{2+} by the protein and PS headgroup, along with secondary hydrogen-bonding interactions (Figure 10.3).[34] Synthetically modified versions of annexin V with attached reporter groups are also used widely as cell death imaging probes (see Section 10.3.2).

10.2.2.2 Headgroup Recognition by Peptide Natural Products

There are a number of relatively small peptide natural products that exhibit binding selectivity for membranes containing specific polar lipids, and the topic has been summarized recently in a review by Gao and Zheng.[35] Among the most well-characterized lipid receptors are the lantibiotic family of peptides, which are short (19–38 amino acids), highly crosslinked macrocyclic peptides produced by Gram-positive *Streptococcus* and *Streptomyces* bacterium. The term 'lantibiotic' is a portmanteau of the characteristic features of these peptides: the peptides contain a number of lanthionine (thioether) crosslinks and also possess antibiotic properties. They selectively recognize amphiphilic components in the bacterial envelope. For example, nisin targets lipid II, an important intermediate in bacterial cell well

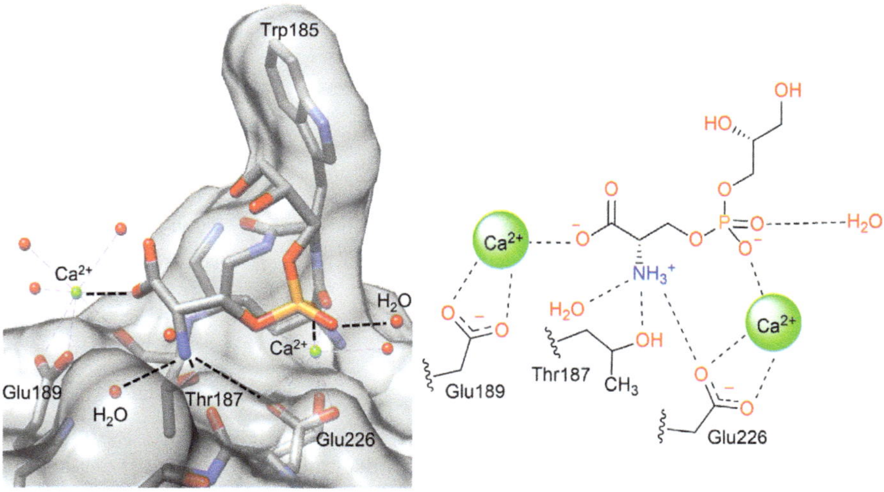

Figure 10.3 Phosphatidylserine (PS) headgroup recognition by annexin V, highlighting the requirement of two Ca^{2+} ions, which, interact with the PS phosphate and carboxylate groups. (Protein Data Bank code 1A8A). Molecular graphics image was produced using the Chimera package from the Computer Graphics Laboratory, University of California, San Francisco; supported by National Institutes of Health P41 RR-01081.

synthesis,[35] whereas the two related lantibiotics cinnamycin and duramycin have nanomolar affinity for membranes containing PE. These latter two structures contain three lanthionine linkages and they differ only at amino acid 2, containing either a lysine or arginine residue (Figure 10.4). The structure of cinnamycin bound to C_{12}-*lyso*-PE has been resolved by proton nuclear magnetic resonance (NMR).[36] As shown in the lower right of Figure 10.4, the PE ammonium cation associates within a cleft formed by the macrocycle and is stabilized by a salt bridge with an adjacent carboxylate. The size of the macrocycle cavity is too small to accommodate the larger trimethylammonium cation of the competing PC headgroup. Hydrophobic insertion into the membrane is thought to be important for strong

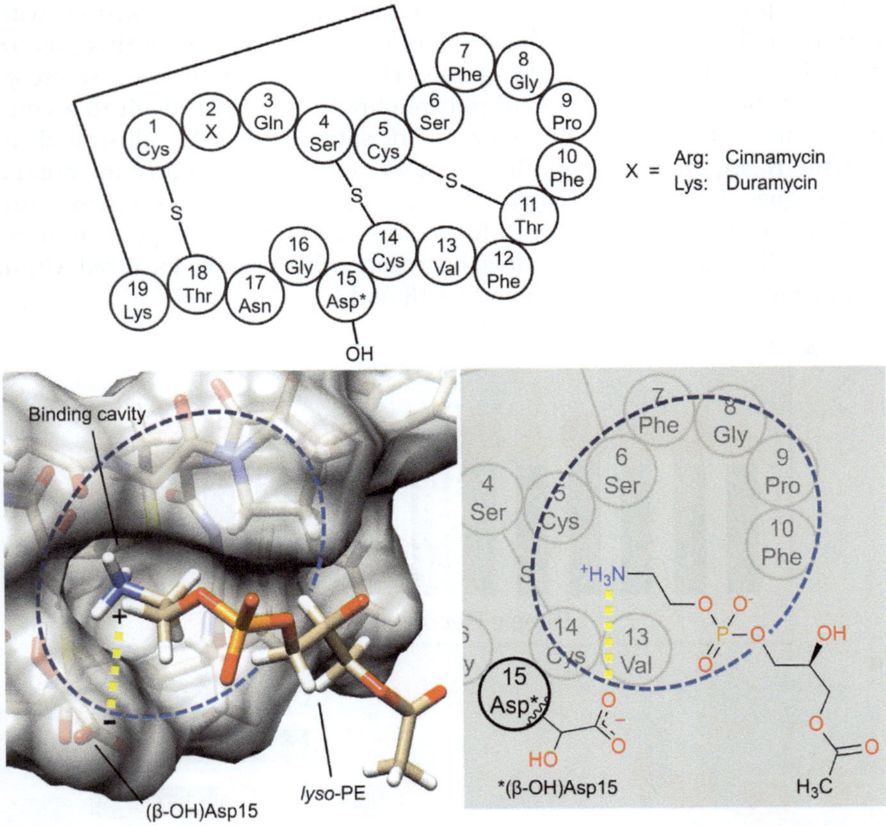

Figure 10.4 (upper) Peptide sequences of phosphatidylethanolomine (PE)-binding lantibiotics cinnamycin and duramycin. (lower left) Cinnamycin complexed with *lyso*-PE from nuclear magnetic resonance-resolved solution structure showing small binding cleft for ammonium of PE and adjacent carboxylate (Protein Data Bank code 2DDE).[36] (Lower right) Structural illustration of binding event. Molecular graphics image was produced using the Chimera package from the Computer Graphics Laboratory, University of California, San Francisco; supported by National Institutes of Health P41 RR-01081.

association and also contributes to the antibacterial activity of these peptides. The structures have been covalently modified to create lantibiotic-based molecular probes for diagnostic and therapeutic applications.[37]

Interestingly, plants also produce PE-binding compounds called cyclotides (cyclic peptides).[38] They are thought to use a PE recognition mechanism similar to that used by the lantibiotics, involving selective binding of the PE headgroup to a hydrophobic cavity formed by the cyclic peptide.

10.2.3 Biological Ditopic Recognition of Headgroup and Tail

There are several other classes of natural products that have biological activity due to an ability to recognize specific polar lipids. Moreover, the recognition processes seem to involve simultaneous ditopic interactions with both the polar headgroup and hydrophobic tail. The targets for this class of biological receptors are usually sterols. Presumably this is because recognition of the flexible acyl tails of phospholipids is more difficult to accomplish and less biologically relevant in some cases. As illustrated in Scheme 10.4A, these amphiphilic receptors can insert fully into a membrane and engage in ditopic recognition. The supramolecular complexes can undergo further self-assembly to form either a barrel stave-type membrane pore or, alternatively, a less defined aggregate, which can be localized within or outside the membrane (Scheme 10.4B and C).

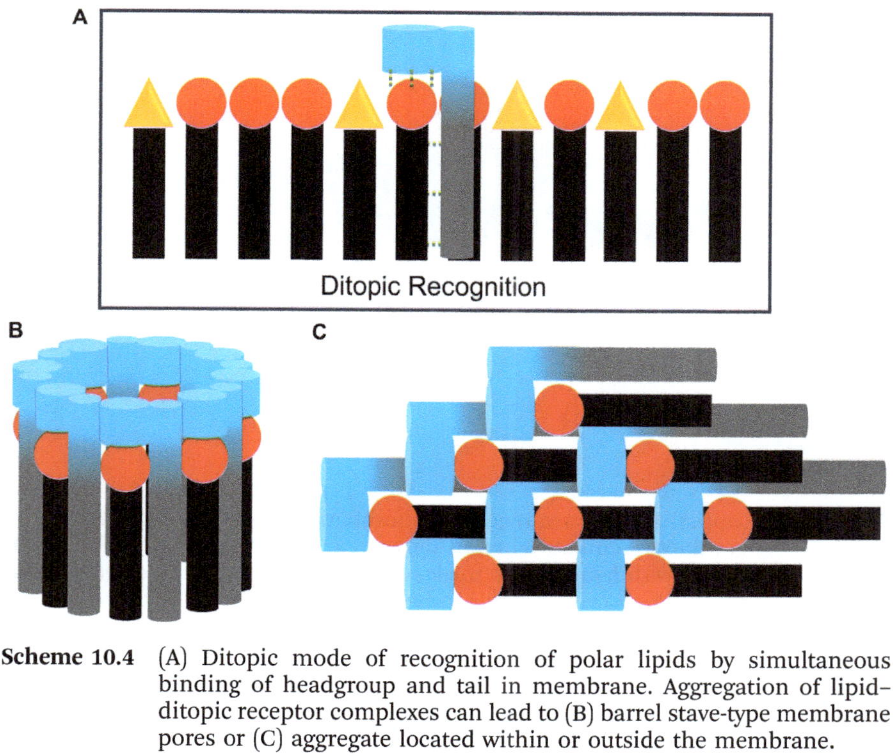

Scheme 10.4 (A) Ditopic mode of recognition of polar lipids by simultaneous binding of headgroup and tail in membrane. Aggregation of lipid–ditopic receptor complexes can lead to (B) barrel stave-type membrane pores or (C) aggregate located within or outside the membrane.

One class of protein-based ditopic receptors are the cholesterol-dependent cytolysins (CDCs). These proteins are produced by Gram-positive bacteria and assemble to form large (~250 Å diameter) pores, an important feature in the pathogenesis of these bacteria.[39] In most CDCs, the first step of the binding process involves recognition of cholesterol through a conserved cholesterol-binding motif, followed by oligomerization and pore formation. In contrast to the NPC2 cholesterol binding discussed previously, the cholesterol-binding pocket of CDCs is thought to also involve interactions between the sterol headgroup and adjacent polar residues.[40] An example of a ditopic phospholipid receptor is sticholysin, a pore-forming toxin that has been shown to recognize SM over PC by simultaneous interactions with both the choline headgroup and ceramide tail on SM.[41]

Another class of biological ditopic receptors for polar lipids are the pharmaceutically active macrocyclic polyenes isolated from *Streptomyces*. The best known compounds of this class are amphotericin B (AmB) and nystatin, both of which have similar structural features. They are both used clinically to treat fungal infections, but their pharmaceutical efficacies can be limited by undesired systemic toxicity.[42] The common mode of action for these compounds is thought to involve ditopic recognition of ergosterol, a steroid that is only present in the fungal membranes. The two polyenes appear to translate this recognition event into pharmaceutical activity by different mechanisms, although there is not yet a consensus on the fine details. It appears that nystatin does not cause membrane permeabilization and recent work suggests that it operates by sequestering the ergosterol, leading to inhibition of ergosterol-dependent proteins.[43] AmB, in contrast to nystatin, has been shown to form membrane pores in the manner illustrated in Scheme 10.4B, with the hydroxyl face of AmB lining the pore interior, and has been proposed to act by membrane permeabilization.[44] However, recent work indicates that pore formation is not required for antifungal activity[45] and that AmB forms extramebranous aggregates of AmB–ergosterol complexes (Scheme 10.4C), thereby removing essential ergosterol from the fungal membrane.[46]

Filipin is a sterol-binding polyene that does not possess antifungal activity, but it does bind to cholesterol. The naturally occurring substance is a mixture of isomers, and the major species, filipin III (Scheme 10.5), is regularly used as a fluorescent marker for cholesterol in histology and live cell imaging.[47,48] The details of its molecular recognition are unknown, but it appears to bind to cholesterol in a ditopic fashion and form an aggregate within the membrane.[49,50]

10.3 Synthetic Receptors for Polar Lipids

As first glance, polar lipids seem to be straightforward targets for synthetic receptors. They have long hydrophobic tails that should be readily encapsulated by synthetic container molecules and there is a wide array of

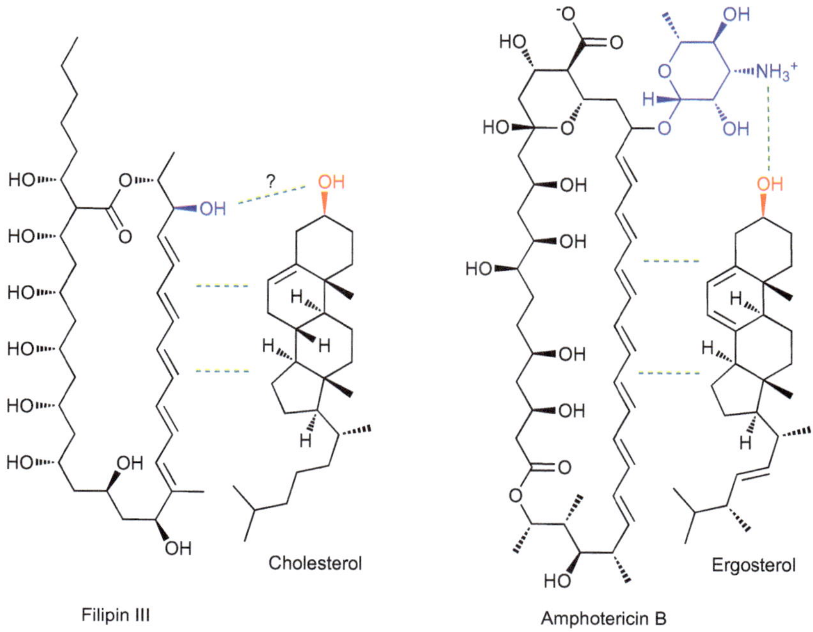

Scheme 10.5 Chemical structures of polyene macrocyclic natural products, filipin and amphotericin B (AmB) which act as ditopic receptors for cholesterol and ergosterol, respectively.

synthetic receptors for the phosphate, carboxylate and ammonium groups that are within the polar headgroups. However, the amphiphilic structure of polar lipids introduces several technical complications that make it challenging to quickly and accurately evaluate receptor binding. These complications include the propensity of polar lipids to self-assemble and form aggregates that cannot be readily monitored by standard spectrosopic techniques such as NMR. This problem is often circumvented by conducting binding studies in homogenous solution using molecules that represent the headgroup portion of the polar lipid structures. While the binding information may be useful for comparing different recognition motifs, it lacks the context of an assembled membrane. The most relevant way to conduct lipid binding studies is to incorporate the target structure into a model bilayer and conduct membrane binding studies. These types of studies have been summarized in reviews[28,51] and will not be repeated in detail here. Typically, membrane binding experiments include techniques such as ultracentrifugation, surface plasmon resonance, and fluorescence titrations using vesicles. Even in these cases, the dynamic and malleable structures of bilayer membranes can introduce experimental artifacts and systematic errors, and great care is often needed to collect reproducible binding data.

10.3.1 Synthetic Receptors for the Hydrophobic Tail

The hydrophobic nature of the tail of a polar lipid makes it an excellent target for supramolecular inclusion by water-soluble container molecules such as those listed in Table 10.5. Most studies have focused on recognition of steroids and other lipid molecules with well-defined structures. The receptors have hydrophilic exteriors and hydrophobic cavities with slightly different dimensions to accommodate lipid tail guests of varying sizes. Reviews outlining the fundamental strategies used in steroid recognition have been published.[52,53]

Members of the cyclodextrin family have been used as synthetic receptors for many polar lipids. Cyclodextrins (CDs) are macrocyclic oligosaccharides derived from starch, with ring sizes of six, seven, and eight glucopyranose units corresponding to the names α-, β-, and γCD.[54] The internal cavities are hydrophobic and lined with methine hydrogens, allowing for efficient inclusion of lipophilic guests (Scheme 10.6). βCD and methyl (MβCD) and 2-hydroxypropyl (HPβCD) variants have shown great utility for pharmaceutical applications related to the extraction and delivery of steroids such as cholesterol[62] and other hydrophobic guests.[54] As shown in Scheme 10.6, the cavity

Table 10.5 Common container molecules used for encapsulation of lipids and their typical guests.

Container molecule	Lipid guests	Refs
Cyclodextrin	Steroids, steroidal drugs, isoprenoids	52–57
Cucurbituril	Isoprenoids, steroidal drugs	58–59
Calixarene	Steroids	53, 60
Cyclophane	Steroids	52, 53
Naphthalene 'molecular tube'	Fatty acids	61

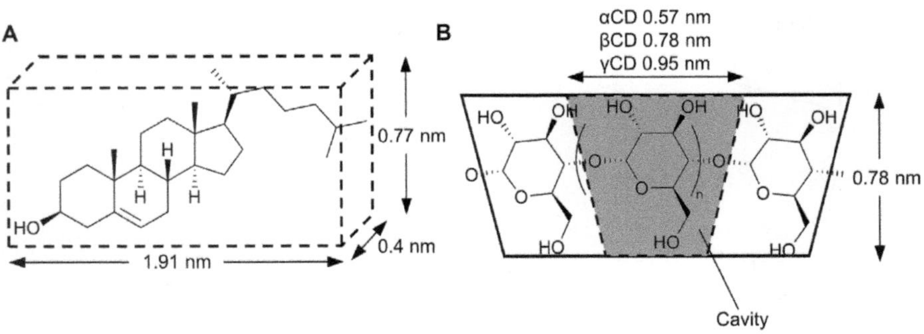

Scheme 10.6 Dimensions of (A) cholesterol and (B) cone-shaped cyclodextrin (CD) rings composed of 6, 7 or 8 D-glucose monomers (α, β, or γ-CD). Scheme adapted from ref. 62, © 2011, with permission from Elsevier.

diameter of βCD can accommodate cholesterol, with an association constant on the order of 10^4 M^{-1} in water (Table 10.6). It is notable that the more water soluble HPβCD binds cholesterol with a similar affinity as βCD. Since a cholesterol molecule is much longer than a single βCD, many dimeric CD systems have been designed to increase binding affinity.[57] Breslow and Zhang prepared a thioether-linked βCD dimer which showed a greater than 100-fold increase in association with cholesterol.[63] Interestingly, HPβCD did not exhibit the same cooperativity (only a five-fold increase), presumably because the 2-hydroxypropyl groups interfere with the formation of the 1:1 complex. A number of effective supramolecular strategies have been developed using βCDs as general modulators of cholesterol levels in the cell plasma membrane. A notable example is the use of HPβCD as a possible treatment for Niemann–Pick C disease, a cholesterol storage disorder.[64,65] Other lipid tail guests for βCD include the retinol derivative, A2E (Scheme 10.7), which can

Table 10.6 β-Cyclodextrin (CD) bimolecular association constants for monomeric and dimeric systems.[a]

Cyclodextrin	Guest	$K_{a\,(1:1)}$ (M^{-1}) in aqueous solution (25 °C)	Refs
βCD	Cholesterol	2×10^4	63
βCD	A2E bisretinoid	3×10^2	55
HPβCD	Cholesterol	2×10^4	63
βCD dimer (linker A)	Cholesterol	3×10^6	63
HPβCD dimer (linker A)	Cholesterol	1×10^5	63
βCD dimer (linker B)	A2E bisretinoid	1×10^3	55

[a]HP: 2-hydroxypropyl.

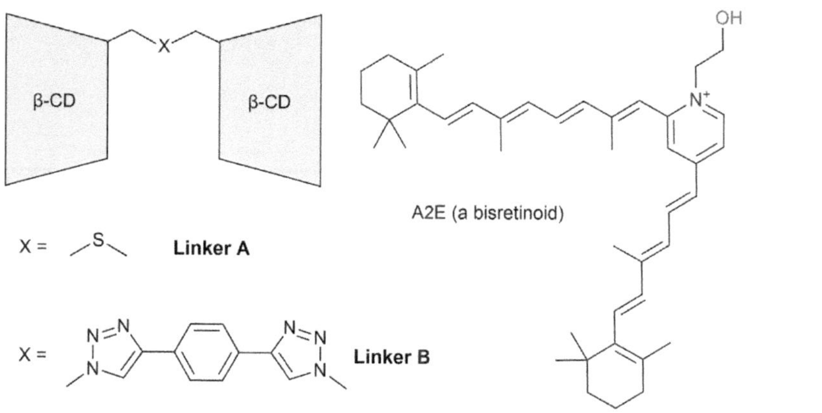

Scheme 10.7 (Left) Schematic of βCD dimers and the corresponding linker structures listed in Table 10.6. (Right) Structure of *N*-retinylidine-*N*-ethanolamine (A2E), a bisretinoid that accumulates in macular degenerative disease.

accumulate as lipofuscin in the retinal pigment epithelium and is associated with vision loss due to macular degenerative disease. A2E binds to βCD with $K_a \sim 10^2\ M^{-1}$, and association was improved by greater than three-fold in a dimeric βCD system that was able to bind both retinoid arms of A2E.[55] In addition to encapsulating A2E, the receptors were shown to protect the bis-retinoid from oxidation and decrease the amount lipofuscin in cell culture and *in vivo*.

Much less progress has been made in the area of synthetic receptors for fatty acid tails, but selective receptors are needed because there is a growing body of evidence suggesting that the length and saturation of fatty acids have major roles in disease[3,66] and fundamental biochemical processes.[67] Among the more recent examples of fatty acid receptors are the naphthalene-based molecular tube lipid sensors reported by Glass and co-workers (Figure 10.5).[61] These fluorescent receptors are formed from two naphthol dimers that are linked by either an amide bond **10.1** or a *trans* olefin **10.2**. Due to the amide linkage, **10.1** can adopt two conformations, an 'open' conformation in which both amide bonds are *trans*, and a 'closed' conformation where one amide switches to the *cis* rotamer, leading to the collapsed structure depicted in Figure 10.5, a structure that is not expected to effectively accommodate an alkyl chain. Since **10.2** contains the rigid alkene linker, it is expected to always adopt the open conformation. As anticipated, both receptors bind hydrophobic guests with higher affinity and selectivity for saturated, straight-chain alkanes. Furthermore, the more preorganized molecular tube **10.2** binds lipids ~100 more strongly than **10.1** (Table 10.7).

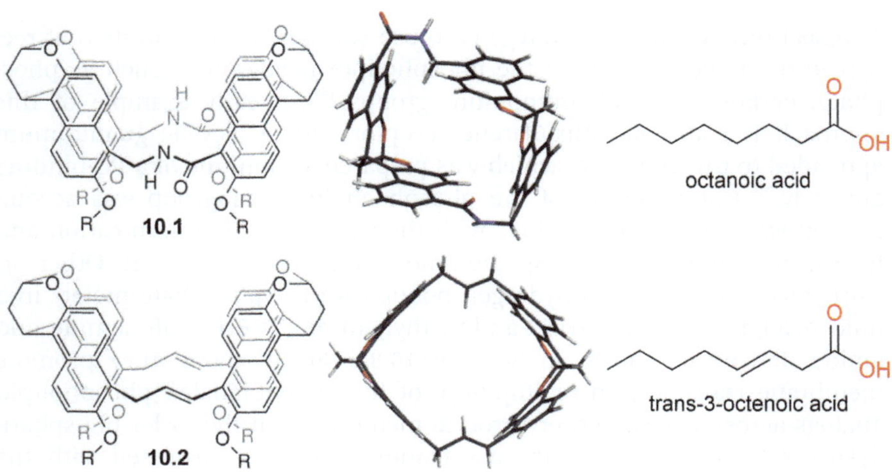

Figure 10.5 The chemical structures and structural models of molecular tubes **10.1** and **10.2**, highlighting the collapsed conformation of **10.1** and open conformation of **10.2** ($R = CH_2CO_2H$ in both structures).
Adapted with permission from ref. 61, © 2012 American Chemical Society. Shown on the right are the structures of two fatty acid guests.

Table 10.7 Binding constants and fluorescent response for the molecular tube hosts and fatty acid guests shown in Figure 10.5.[61]

Molecular tube	Guest	K_a (M^{-1}) in buffer (25 °C)	Fluorescent response (I_{sat}/I_0)
10.1	Octanoic acid	2.5×10^2	0.39
10.1	*Trans*-3-octenoic acid	0.92×10^2	0.61
10.2	Octanoic acid	1.7×10^4	1.78
10.2	*Trans*-3-octenoic acid	0.13×10^4	0.61

10.3.2 Synthetic Receptors for the Polar Headgroup

Small-molecule synthetic receptors for polar lipid headgroups have the potential to become diagnostic agents, chemical tools for cell biology and lipidomics research, and perhaps eventually pharmaceutical agents.[68] In a sense, the task of designing a synthetic receptor for the headgroup of a polar lipid is an anion or zwitterion recognition problem at the membrane surface. Thus, an excellent starting point is to consider the pool of known synthetic receptors for anions and zwitterions (Chapters 4 and 8). Another source of inspiration is the polar lipid recognition mechanisms used by the naturally occurring peripheral membrane proteins and polypeptides. In this section, the different types of synthetic receptors for polar lipid headgroups are classified according to the weak interactions that produce the recognition and they are hydrogen bonding and electrostatics, metal ion coordination, and reversible covalent bonding.

10.3.2.1 Hydrogen-bonding/Electrostatic Receptors

A logical receptor design strategy is to use supramolecular motifs that recognize the ionic residues in the phospholipid headgroups, such as phosphate, carboxylate, and ammonium groups.[68] An early example of this approach is a ditopic calix[6]arene receptor with a bicyclic guanidinium appended to the upper rim which was prepared as a mimic of a PC-binding antibody.[69] Complexation of the phosphocholine headgroup was accomplished *via* cation–π interactions with the trimethylammonium cation and hydrogen bonding with phosphate anion in an organic solvent. Other organic receptors that form hydrogen bonds with the phosphate moiety in a lipid headgroup are the tris(2-aminoethyl)amine (TREN) sulfonamide and amide derivatives shown in Scheme 10.8A. Furthermore, they promote membrane translocation or 'flip-flop' of fluorescent-labeled phospholipid analogs across vesicle and erythrocyte membranes in the order phosphatidylglycerol (PG) > PC > PE > PS. Association constants correlated with the acidity of the receptor NH groups, and all of the sulfonamides exhibited higher complexation than the amide derivatives.[68] In one case, a trifluoromethylsulfonamide derivative exhibited unusual binding and selective membrane translocation of PE.[70] A collaboration between the groups of Davis and Smith showed that a number of steroid-based receptors are able to

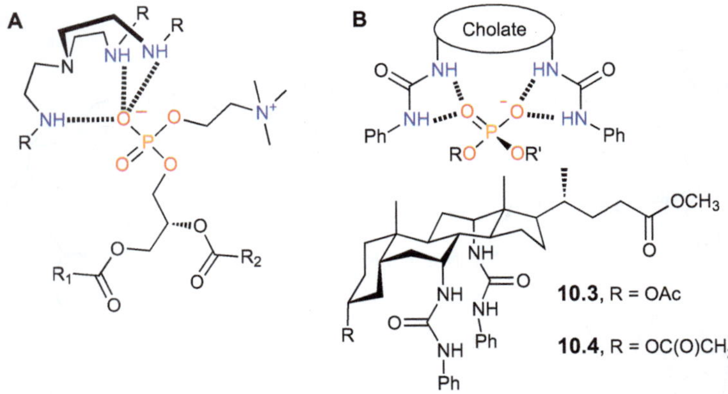

Scheme 10.8 (A) General structures of tris(2-aminoethylamine) (TREN) amide and sulfonamide hydrogen-bonding receptors for phospholipid headgroups, shown here interacting with the phosphate of phosphatidylcholine. (B) Structures of cholate bis-urea receptors and proposed model of phosphate headgroup recognition.

bind phospholipids and promote flip–flop (Scheme 10.8B). Lipophilic cholate bis-urea esters such as **10.3** were found to be more effective at PC binding and transport than the TREN-derived sulfonamides. Also, the cationic trimethylammonium-containing bis-urea receptor **10.4** facilitated PS flip–flop and induced PS exposure on the surface of erythrocytes, leading to thrombin activation.[71]

Burns and co-workers prepared picket-fence porphyrin receptors with appended hydrogen-bonding groups and found that they associate in organic media with the headgroup of PG, a major anionic component of bacterial cell membranes.[72] The bacterial envelope contains several saccharide-containing polar lipids that are potential targets for synthetic receptors. One of the most important is lipid A (Scheme 10.9) which is found in the outer membrane of Gram-negative bacteria. It contains two phosphate monoesters and is able to bind to bivalent cations such as Ca^{2+} and Mg^{2+}, forming complexes that stabilize the bacterial membrane. Lipid A also acts as the membrane anchoring portion of lipopolysaccharides and it is the causative agent in cases of septic shock. Synthetic receptors that sequester lipid A may help prevent septic shock, possess antibiotic properties on their own, or enhance the activity of existing antibiotics in combination therapies.

Polymyxin B is a naturally occurring polypeptide antibiotic that is selective for lipid A. Electrostatic interactions between the cationic ammonium groups of polymyxin B and the anion lipid A headgroup are thought to drive association and subsequent membrane insertion, leading to membrane disruption and antibiotic activity. Inspired by the structure and mechanism of polymyxin B, Savage and co-workers prepared a class of cationic bile salt derivatives called the ceragenins and tested them as antimicrobial agents (Scheme 10.9A).[73] The non-peptidic cholate derivatives (**10.5, 10.6, 10.7**) are

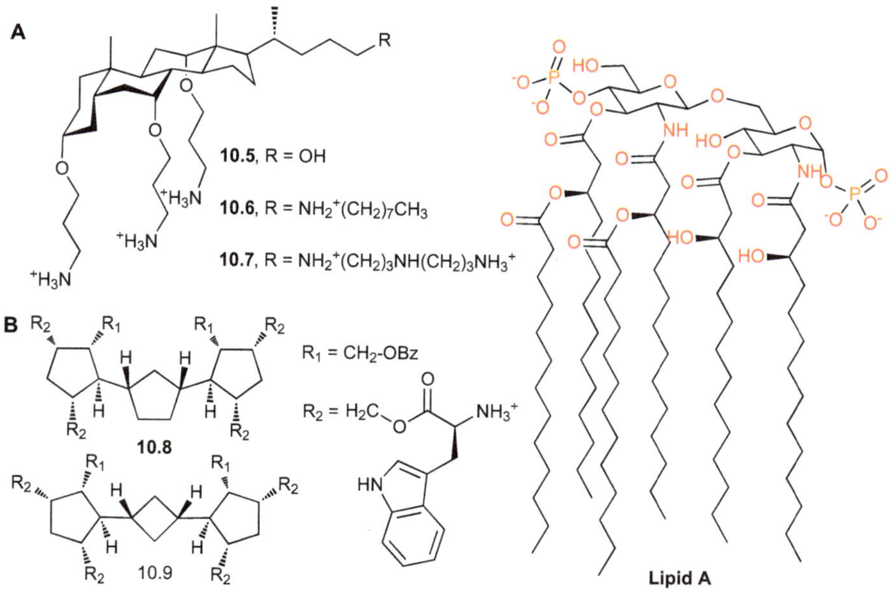

Scheme 10.9 Synthetic receptors for bacterial lipid A. (A) Cationic cholate derivatives and (B) cycloalkane compounds with appended L-Trp (C-terminal linkage).

facial amphiphiles that associate with membranes containing lipid A and produce bacterial membrane disruption and depolarization.[74] The extent of antibiotic activity depends on both the nature of the hydrophobic tail and the phospholipid content of the target membrane. Not only are the ceragenins active against some drug-resistant Gram-negative strains, they promote the diffusion of relative impermeable antibiotics across the bacterial outer membrane.

Substituted cycloalkanes **10.8** and **10.9** have also been studied as lipid A receptors.[75,76] The rigid scaffold supports a set of extended arms with ammonium and lipophilic aromatic groups that promote membrane association with lipid A. The binding stoichiometry was determined to be 1 : 1 with micromolar-range dissociation constants in aqueous buffer at low lipid concentrations. Variants without some of the hydrophobic side chains interacted weakly with lipid A, demonstrating the importance of these groups.[76]

In the case of mammalian cells, the most abundant anionic phospholipid is PS. Due to the concerted action of lipid transport proteins, the PS in the plasma membrane of healthy cells is sequestered within the interior monolayer. During cell death processes the PS transmembrane distribution is scrambled and its appearance on the cell surface is a signal for cell clearance by the innate immune system.[77] PS is also exposed on the surface of other cell types, including the blood vessels of many tumors.[78] Thus, synthetic receptors for PS are expected to have several uses as imaging or therapeutic agents. In the past two decades, several peptidic synthetic receptors for PS headgroups have been prepared as mimics of biological receptors in which hydrogen bonding and electrostatic interactions are utilized.[35] Perhaps the best characterized structures are the cyclic peptides prepared as mimics of the PS-binding protein, lactadherin.[79,80] Lactadherin is a protein secreted by macrophages that exhibits extraordinary selectivity for PS over other lipids, including the PS epimer, phosphatidyl-D-serine. Specific recognition of PS occurs through the Discoidin C2 domain (Table 10.4) of lactadherin. While no crystallization data on the lactadherin–PS complex are available, molecular docking experiments with diacetyl-PS suggest some key interactions in the energy-minimized complex, as illustrated in Figure 10.6.[79] These include salt bridges between Arg148 and the PS phosphate group and Asp80 and the PS ammonium group, and there also appears to be a hydrogen bond between His83 and the carboxylate group on PS. The hydrophobic residues Trp26 and Phe81 appear to engage in membrane anchoring by inserting into the interfacial region. It was hypothesized that the lactadherin-binding pocket could be mimicked by the cyclic peptides **10.10** and **10.11**, with the key residues linked by appropriate numbers of glycine linkers. The biphenyl group of **10.11** was shown to be important for membrane insertion of the peptide, and this receptor associated with lipid vesicles containing 20% PS with an 8 μM dissociation constant. The linear version of the peptide did not bind PS containing membranes at all, confirming the role of receptor preorganization in the association event. A fluorescently labeled version of **10.11** was able to differentiate between

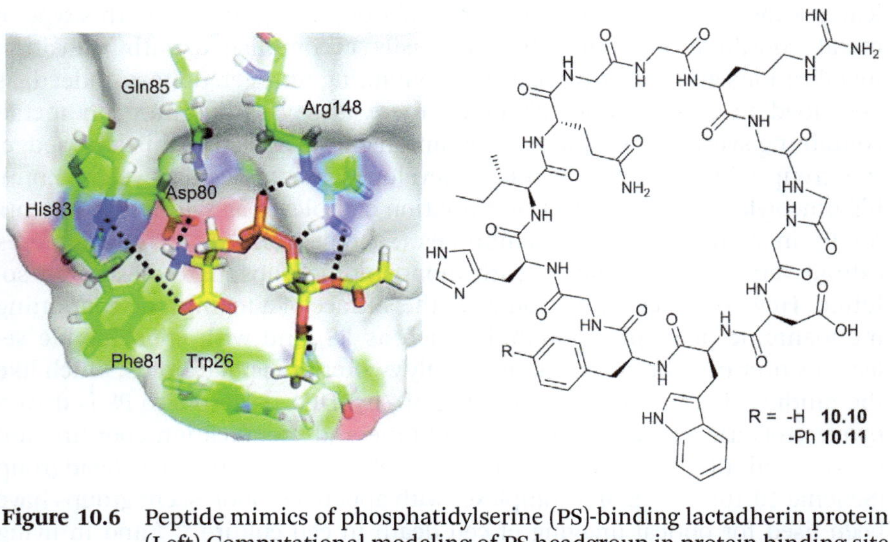

Figure 10.6 Peptide mimics of phosphatidylserine (PS)-binding lactadherin protein. (Left) Computational modeling of PS headgroup in protein binding site. Reprinted with permission from ref. 79, © 2011 American Chemical Society. (Right) Structure of cyclic peptides designed as small-molecule lactadherin mimics.

apoptotic cells and normal cells just like a much larger protein probe (fluorescent annexin V).

Another peptide approach used a consensus sequence for PS recognition that is found in both PKC and phosphatidylserine decarboxylase.[81] A synthetic peptide with the sequence FNFRLKAGQKIRFG based on the consensus sequence was shown to bind selectively but very weakly (low-μM K_d) to PS. More recently, a group took on the task of optimizing this sequence using alanine scanning for higher affinity binding to PS and prepared an apoptosis-imaging probe by attaching a radiometal chelator to the optimized peptide.[82] Combinatorial methods utilizing phage display techniques have also identified the peptide sequences SVSVGMKPSPRP[83] and LIKKPF and PGDLSR[84] as PS-selective with greater than two-fold selectivity for PS over PC. A cyclic peptide with the sequence CLSYYPSYC was also discovered by phage display as a selective PS-binding peptide.[85] While all of these peptides were explored as probes for apoptosis detection, very few mechanistic studies have been published at this point. There is also an example of a DNA aptamer template for PS, identified using a computational method called an entropic fragment-based approach.[86]

10.3.2.2 Metal Coordination-based Receptors

Since metal coordination is a key feature of several protein–lipid headgroup binding interactions (Section 10.2.2), many synthetic receptors have aimed to mimic this binding strategy. Most of the metal-assisted peripheral proteins use Ca^{2+} as the metal cation cofactor. However, it very hard to create a small-molecule chelating scaffold that strongly associates with Ca^{2+} yet leaves a vacant coordination site for headgroup recognition. Yet this type of metal coordination architecture is easily accomplished with transition metal cations and chelating scaffolds containing nitrogen donors. Indeed, as described in several other chapters, there is a vast literature on metal-containing synthetic receptors for oxyanion guests. Utilizing this knowledge, the group of Smith and coworkers developed low molecular weight Zn(II)-bisdipicolylamine (Zn_2BDPA) coordination complexes as synthetic receptors for PS in mammalian cell membranes (Scheme 10.10A).[87] The complexes exhibit very weak association with monoanionic phosphate diesters in solution. However, they bind strongly to the surface of a membrane containing monoanionic polar phospholipids, such as PS, and with considerable selectivity over membranes containing only zwitterionic polar lipids. Much like the binding of annexin V, the binding of Zn_2BDPA receptors to PS is driven by electrostatic attraction to the membrane and metal cation coordination to the anionic carboxylates and phosphates with the PS headgroup (Scheme 10.10B). Zn_2BDPA complexes with appended fluorescent groups have been used for optical imaging of cell death in cell culture[88,89] and in living animal models.[90–92] Receptor **10.12** is a Zn_2BDPA linked to a near-infrared fluorescent dye and was shown to detect tumor cell death in a living animal model caused by focal beam radiation therapy (Figure 10.7). This result

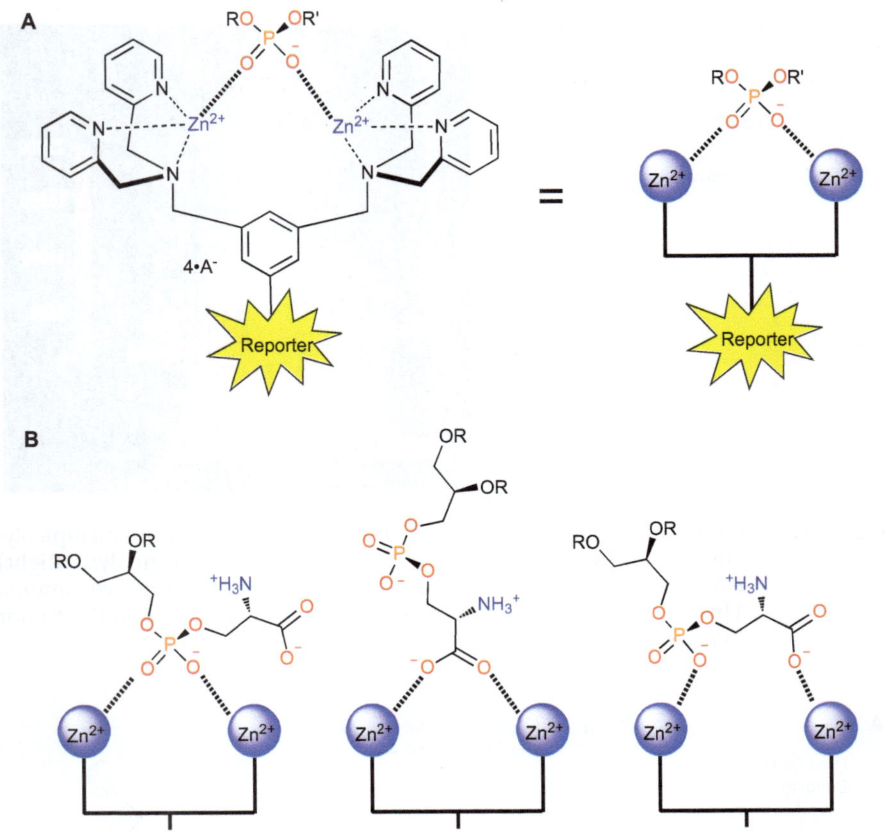

Scheme 10.10 Phosphate ester recognition using zinc(II)-bisdipicolylamine (Zn_2BDPA) coordination complexes. (A) Generic scheme for phosphate monoester binding. (B) Possible binding modes for phosphatidylserine headgroup recognition by Zn_2BDPA at the membrane surface.

suggests how *in vivo* cell death imaging technology could be used to assess the efficacy of tumor treatment within a few days, rather than the current waiting period of several months for patients undergoing standard care.

Effective clinical translation will require Zn_2BDPA imaging probes with very high PS affinity. Two strategies have been explored to achieve this outcome, covalent modification of the Zn_2BDPA structure and multivalent Zn_2BDPA probes (Scheme 10.11). In the former case, a library of compounds was prepared in which the Zn_2BDPA scaffold was functionalized with various appendages that could provide secondary hydrogen bonding and hydrophobic insertion into a PS-rich membrane.[93] Library screening identified a modified Zn_2BDPA receptor with appended 4-fluorophenethyl urea groups, and it was converted into a fluorescent imaging probe with excellent cell death imaging capabilities in cell culture and animal models of cell death. Parallel studies of fluorescent multivalent probes with two or four Zn_2BDPA

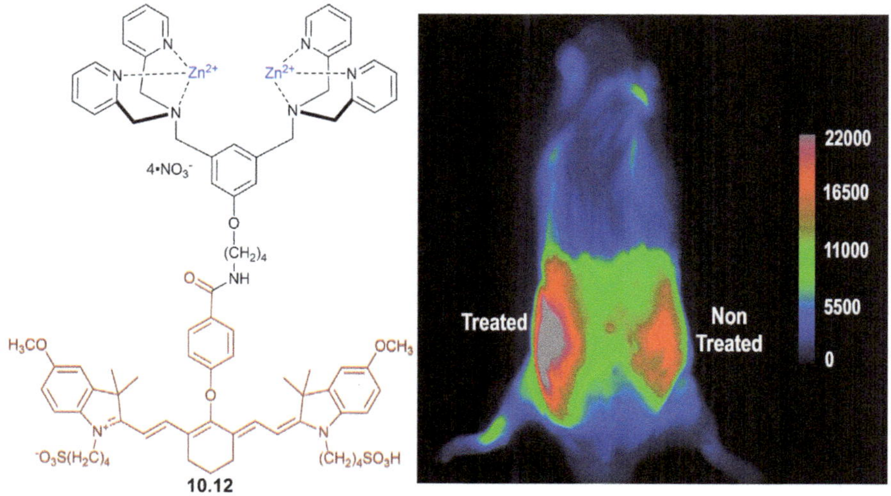

10.12

Figure 10.7 (Left) Cell death imaging probe **10.12** composed of zinc(II)-bisdipicolylamine (Zn_2BDPA) linked to a near-infrared fluorescent dye. (Right) False-colored fluorescence image of a living rat bearing two tumors. There is increased accumulation of the cell death probe in the tumor that has been treated by focal beam radiation.[90]

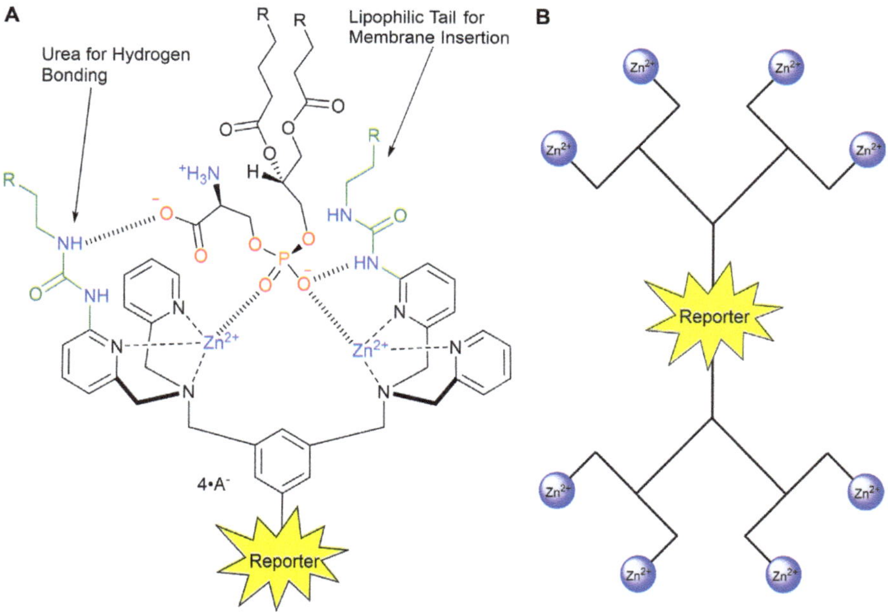

Scheme 10.11 Strategies for improved probe affinity for phosphatidylserine-rich membranes. (A) Modified Zn_2BDPA structure exhibits enhanced recognition of the PS headgroup by incorporation of secondary hydrogen bonding and hydrophobic interactions.[93] (B) Multivalent structure with four Zn_2BDPA groups.[94]

groups also demonstrated enhanced cell death targeting in several cell death models, and a Zn_2BDPA-coated liposomal system was effective at cell death recognition in a cancer cell chemotherapy model.[94,95]

Other synthetic receptors based on metal cation coordination have been developed. One such system also employs the coordination of two Zn^{2+} cations for the recognition of PS, but uses meta-substituted cyclens as the supporting scaffold.[96] Biological studies of this compound as a positron emission tomography imaging agent showed promise as a contrast agent for cell death *in vivo*. Do-Thanh *et al.* synthesized a bis-cyclen 'tweezer' compound as a sensor for PIP_n.[97] The sensor bound D-*myo*-inositol-1,4,5-triphosphate with or without zinc, but neither version was able to discriminate between the target inositide and inorganic phosphate.

Since the synthetic metal-containing receptors described above have affinity for cell membranes that are rich in anionic amphiphiles, it is not surprising that many of them exhibit selective affinity for the anionic surfaces of bacterial cells. A good example is the fluorescent near-infrared Zn_2BDPA probe **10.12**, which has been used to image bacterial infection in a living animal.[98] Furthermore, multivalent Zn_2BDPA probes have been shown to selectively bind and agglutinate bacteria;[99] thus, they have potential as therapeutic agents that can selectively remove bacteria from the blood stream.

10.3.2.3 Reversible Covalent Bonding With the Polar Headgroup

Reversible covalent bonding is not known to be a biological strategy for lipid headgroup recognition. However, the idea is attractive for synthetic receptors, since a common limitation is a high rate of receptor dissociation from the membrane surface. The best known example using this strategy is bivalent receptor **10.13** which was designed to bind to $PI(4,5)P_2$ utilizing two recognition elements (Figure 10.8).[100] The amino boronic acid recognizes the *cis*-diol on $PI(4,5)P_2$ through formation of a covalent boronic ester, and association is strengthened by simultaneous hydrogen bonding of the receptor ureas with lipid phosphates. The receptor was shown to serve as a small molecule mimic of the PH domain (Section 10.2.2), selectively binding with low micromolar affinity (K_d) to $PI(4,5)P_2$ over other PIP_n homologues. Receptor **10.13** performed remarkably well in biological studies, by entering cells and blocking access of the PLC-δ1 protein PH domain to the phospholipid headgroup. Other PIP_2-dependent processes were also altered by the receptor, including endocytic uptake of transferrin and formation of actin stress fibers. The study demonstrates the potential of synthetic receptors to have an intracellular effect by binding selectively to a signaling phospholipid. Other emerging work with boronic acid receptors shows that they have affinity for cell surface glycolipids, especially those containing sialic acid residues, which are often overexpressed on cancer cells;[101] interested readers are directed to a recent review on the use of boronic acids in biological chemistry[102] and Chapter 5 of this book.

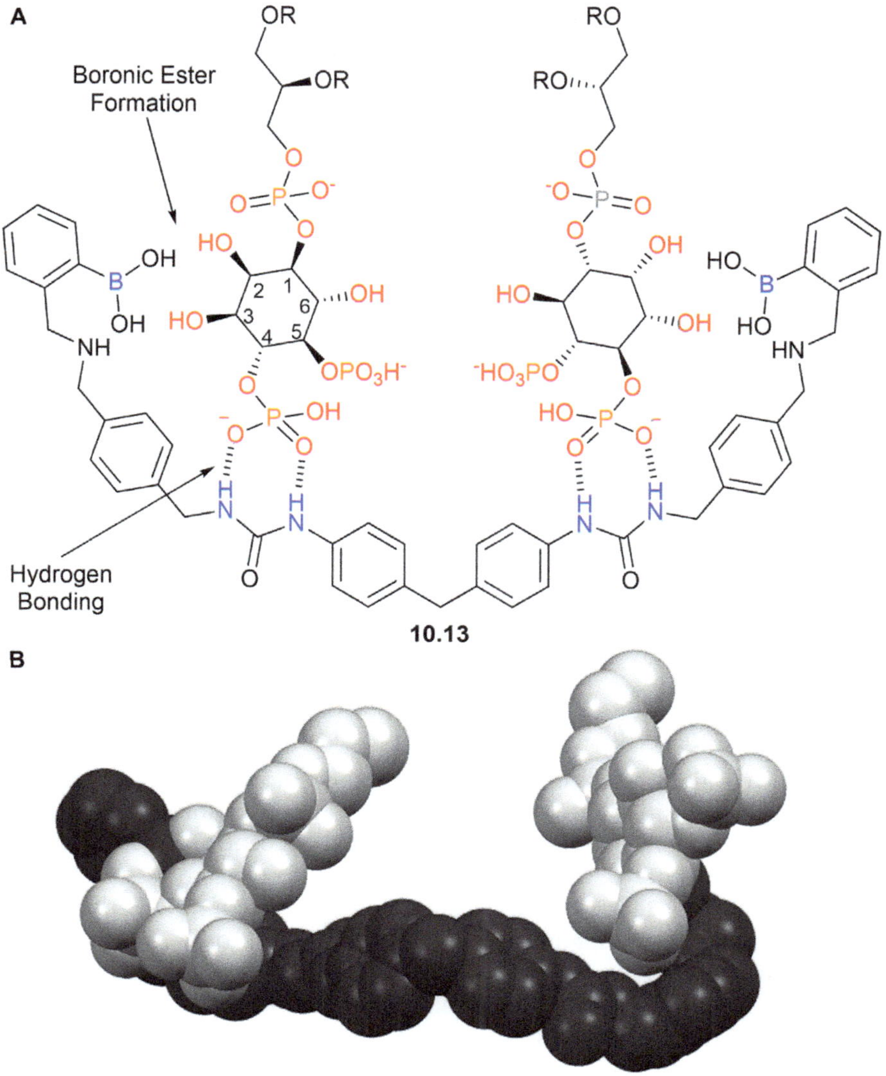

Figure 10.8 (A) Receptor **10.13** shown interacting with two phosphatidylinositol-4,5-bisphosphate ($PI(4,5)P_2$) headgroups through hydrogen bonding between the urea and phosphate and the site for boronic ester formation between the boronic acid and $PI(4,5)P_2$ *cis*-diol. (B) Modeling of **10.13** (dark) bound to two triphosphate headgroups (light).
Reprinted with permission from ref. 100, © 2011 American Chemical Society.

Glass and coworkers have explored synthetic receptors that form reversible covalent bonds with aminophospholipid headgroups. They designed a fluorescent ditopic indicator that contained an amine reactive aldehyde group and a coordinated Zn^{2+} cation for simultaneous recognition of the phosphate and amine moieties. The receptor was assessed for binding to the phosphate monoester headgroups of PS and PE.[103] Binding to either headgroup produced greater than 30-fold increase in fluorescence with two-fold stronger association for phosphoserine compared with phosphoethanolamine.

10.3.3 Synthetic Receptors for Ditopic Recognition of Headgroup and Tail

The sterol-selective polyene AmB (Section 10.2.3) is the prototype for ditopic recognition of the sterols. Indeed, many synthetic variants of AmB have been investigated as antifungal agents and this work has been reviewed.[104] However, most of these studies have evaluated biological activity and not directly measured specific lipid binding. Since nonselective binding of AmB to cholesterol is thought to play a major role in its toxicity to human cells, some studies have focused on sterol selectivity of synthetic variants of AmB. Recently, the Burke laboratory prepared synthetic versions of AmB and discovered that replacement of the 2′-hydroxyl group on the mycosamine headgroup with hydrogen leads to AmB derivative **10.14**, which does not bind cholesterol but retains much of its affinity for ergosterol (Scheme 10.12).[42] This remarkable result demonstrates that selective recognition of one sterol type is possible and may prove useful for designing improved, selective antifungal agents.

10.4 Future Directions and Challenges

Selective polar lipid recognition using a synthetic receptor is a relatively unexplored topic in supramolecular chemistry, with many potentially important biomedical applications. For example, a synthetic version of AmB that is capable of binding more selectively to ergosterol over cholesterol could be very valuable as an antifungal agent. Alternatively, improved synthetic receptors for cholesterol could be useful for sensitive cholesterol imaging and cell trafficking. Synthetic receptors for polar lipids are likely to play an important role in lipidome analysis by facilitating lipid separation

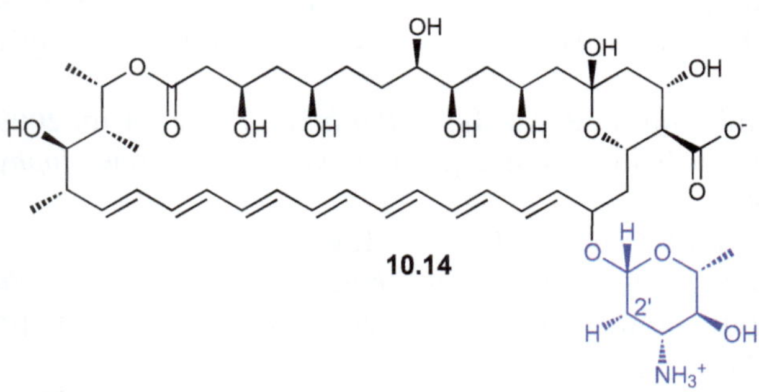

10.14

Scheme 10.12

and extraction processes. The emerging appreciation of the importance of the lipid tail structure suggests there is increasing need for polar lipid receptors that can distinguish fatty acid tails.[67,105] Binding specificity for phospholipid headgroups would also be very useful for a number of different applications, such as non-covalent labeling probes for mass spectrometry and molecular imaging applications. Despite several years of research, there remain few receptors that are selective for the mammalian lipid markers of cell death (PS and PE), the major bacterial lipids (lipid A, PG, PE), and the important PIP_n signaling lipids. In the more distant future, it is anticipated that synthetic receptors will be developed with catalytic properties, such as artificial lipases and phospholipases.

Acknowledgement

We are grateful for funding support from the National Institutes of Health (GM059078).

References

1. A. Shevchenko and K. Simons, *Nat. Rev. Mol. Cell Biol.*, 2010, **11**, 593.
2. G. van Meer, D. R. Voelker and G. W. Feigenson, *Nat. Rev. Mol. Cell Biol.*, 2008, **9**, 112.
3. V. Vangaveti, V. Shashidhar, G. Jarrod, B. T. Baune and R. L. Kennedy, *Ther. Adv. Endocrinol. Metab.*, 2010, **1**, 165.
4. L. Samuels, L. Kunst and R. Jetter, *Annu. Rev. Plant Biol.*, 2008, **59**, 683.
5. J. C. Gomez-Fernandez and S. Corbalan-Garcia, *Chem. Phys. Lipids*, 2007, **148**, 1.
6. T. Blom, P. Somerharju and E. Ikonen, *Cold Spring Harbor Perspect. Biol.*, 2011, **3**, a004713.
7. T. Hla and A. J. Dannenberg, *Cell Metab.*, 2012, **16**, 420.
8. E. Gonzalez-Burgos and M. P. Gomez-Serranillos, *Curr. Med. Chem.*, 2012, **19**, 5319.
9. F. Bouvier, A. Rahier and B. Camara, *Prog. Lipid Res.*, 2005, **44**, 357.
10. A. Eroglu and E. H. Harrison, *J. Lipid Res.*, 2013, **54**, 1719.
11. J. Z. Haeggström and C. D. Funk, *Chem. Rev.*, 2011, **111**, 5866.
12. M. W. Buczynski, D. S. Dumlao and E. A. Dennis, *J. Lipid Res.*, 2009, **50**, 1015.
13. B. Mesmin and F. R. Maxfield, *Biochim. Biophys. Acta*, 2009, **1791**, 636.
14. J. N. Israelachvili, S. Marčelja and R. G. Horn, *Q. Rev. Biophys.*, 1980, **13**, 121–200.
15. E. Dufourc, *J. Chem. Biol.*, 2008, **1**, 63.
16. *The Biology of Lipids: Trafficking, Regulation, and Function*, ed. K. Simons, Cold Spring Harbor Laboratory Press, Cold Spring Harbor, NY, 2011.
17. J. M. Boon and B. D. Smith, *Med. Res. Rev.*, 2002, **22**, 251.
18. B. Alberts, *Molecular Biology of the Cell*, Garland Science, 2008.

19. J. Löffler, H. Einsele, H. Hebart, U. Schumacher, C. Hrastnik and G. Daum, *FEMS Microbiol. Lett.*, 2000, **185**, 59.
20. J. C. Holthuis and A. K. Menon, *Nature*, 2014, **510**, 48.
21. A. Jonas, in *New Comprehensive Biochemistry*, ed. J. E. V. Dennis and E. Vance, Elsevier, 2002, vol. 36, p. 483.
22. T. H. Yeats and J. K. C. Rose, *Protein Sci.*, 2008, **17**, 191.
23. F. Schroeder, B. P. Atshaves, A. L. McIntosh, A. M. Gallegos, S. M. Storey, R. D. Parr, J. R. Jefferson, J. M. Ball and A. B. Kier, *Biochim. Biophys. Acta*, 2007, **1771**, 700.
24. S. Xu, B. Benoff, H. L. Liou, P. Lobel and A. M. Stock, *J. Biol. Chem.*, 2007, **282**, 23525.
25. M. Ott, J. D. Robertson, V. Gogvadze, B. Zhivotovsky and S. Orrenius, *Proc. Natl. Acad. Sci. U. S. A.*, 2002, **99**, 1259.
26. M. A. Lemmon, *Nat. Rev. Mol. Cell Biol.*, 2008, **9**, 99.
27. C. Wonhwa and R. V. Stahelin, *Annu. Rev. Biophys. Biomol. Struct.*, 2005, **34**, 119.
28. J. L. Scott, C. A. Musselman, E. Adu-Gyamfi, T. G. Kutateladze and R. V. Stahelin, *Integr. Biol.*, 2012, **4**, 247.
29. A. Simonsen, R. Lippe, S. Christoforidis, J. M. Gaullier, A. Brech, J. Callaghan, B. H. Toh, C. Murphy, M. Zerial and H. Stenmark, *Nature*, 1998, **394**, 494.
30. T. G. Kutateladze, *Biochim. Biophys. Acta*, 2006, **1761**, 868.
31. S. A. Lee, R. Eyeson, M. L. Cheever, J. Geng, V. V. Verkhusha, C. Burd, M. Overduin and T. G. Kutateladze, *Proc. Natl. Acad. Sci. U. S. A.*, 2005, **102**, 13052.
32. B. Jeppesen, C. Smith, D. F. Gibson and J. F. Tait, *J. Biol. Chem.*, 2008, **283**, 6126.
33. L. Ungethüm, H. Kenis, G. A. Nicolaes, L. Autin, S. Stoilova-McPhie and C. P. M. Reutelingsperger, *J. Biol. Chem.*, 2011, **286**, 1903.
34. M. Swairjo, N. Concha, M. Kaetzel, J. Dedman and B. Seaton, *Nat. Struct. Mol. Biol.*, 1995, **2**, 968.
35. J. M. Gao and H. Zheng, *Future Med. Chem.*, 2013, **5**, 947.
36. K. Hosoda, M. Ohya, T. Kohno, T. Maeda, S. Endo and K. Wakamatsu, *J. Biochem.*, 1996, **119**, 226.
37. M. Zhao, *Amino Acids*, 2011, **41**, 1071.
38. S. T. Henriques, Y.-H. Huang, M. A. R. B. Castanho, L. A. Bagatolli, S. Sonza, G. Tachedjian, N. L. Daly and D. J. Craik, *J. Biol. Chem.*, 2012, **287**, 33629.
39. M. A. Dunstone and R. K. Tweten, *Curr. Opin. Struct. Biol.*, 2012, **22**, 342.
40. J. Rossjohn, S. C. Feil, W. J. McKinstry, R. K. Tweten and M. W. Parker, *Cell*, 1997, **89**, 685.
41. P. S. Garcia, G. Chieppa, A. Desideri, S. Cannata, E. Romano, P. Luly and S. Rufini, *Toxicon*, 2012, **60**, 724.
42. B. C. Wilcock, M. M. Endo, B. E. Uno and M. D. Burke, *J. Am. Chem. Soc.*, 2013, **135**, 8488.

43. Y. M. te Welscher, M. R. van Leeuwen, B. de Kruijff, J. Dijksterhuis and E. Breukink, *Proc. Natl Acad. Sci. U. S. A.*, 2012, **109**, 11156.
44. T. E. Andreoli, *Ann. N.Y. Acad. Sci.*, 1974, **235**, 448.
45. K. C. Gray, D. S. Palacios, I. Dailey, M. M. Endo, B. E. Uno, B. C. Wilcock and M. D. Burke, *Proc. Natl Acad. Sci. U. S. A.*, 2012, **109**, 2234.
46. T. M. Anderson, M. C. Clay, A. G. Cioffi, K. A. Diaz, G. S. Hisao, M. D. Tuttle, A. J. Nieuwkoop, G. Comellas, N. Maryum, S. Wang, B. E. Uno, E. L. Wildeman, T. Gonen, C. M. Rienstra and M. D. Burke, *Nat. Chem. Biol.*, 2014, **10**, 400.
47. R. Bittman and S. A. Fischkoff, *Proc. Natl Acad. Sci. U. S. A.*, 1972, **69**, 3795.
48. G. Gimpl, *Subcell. Biochem.*, 2010, **51**, 1.
49. S. C. D. N. Lopes, E. Goormaghtigh, B. J. Costa Cabral and M. A. R. B. Castanho, *J. Am. Chem. Soc.*, 2004, **126**, 5396.
50. B. De Kruijff and R. A. Demel, *Biochim. Biophys. Acta*, 1974, **339**, 57.
51. H. Zhao and P. Lappalainen, *Mol. Biol. Cell*, 2012, **23**, 2823.
52. H.-J. Schneider, P. Agrawal and A. K. Yatsimirsky, *Chem. Soc. Rev.*, 2013, **42**, 6777.
53. P. Wallimann, T. Marti, A. Fürer and F. Diederich, *Chem. Rev.*, 1997, **97**, 1567.
54. C. Ortiz Mellet, J. M. Garcia Fernandez and J. M. Benito, *Supramolecular Systems in Biomedical Fields*, Royal Society of Chemistry, 2013, pp. 94–139.
55. M. M. Nociari, G. L. Lehmann, A. E. Perez Bay, R. A. Radu, Z. Jiang, S. Goicochea, R. Schreiner, J. D. Warren, J. Shan, S. Adam de Beaumais, M. Ménand, M. Sollogoub, F. R. Maxfield and E. Rodriguez-Boulan, *Proc. Natl Acad. Sci. U. S. A.*, 2014, **111**, E1402.
56. H. Jiang, S. Zhang and Q. Shi, *Asian J. Chem.*, 2011, **23**, 3783.
57. Y. Liu and Y. Chen, *Acc. Chem. Res.*, 2006, **39**, 681.
58. T. Minami, N. A. Esipenko, B. Zhang, L. Isaacs and P. Anzenbacher, *Chem. Commun.*, 2014, **50**, 61.
59. L. Isaacs, *Acc. Chem. Res.*, 2014, **47**, 2052.
60. E. V. Ukhatskaya, S. V. Kurkov, S. E. Matthews and T. Loftsson, *J. Pharm. Sci.*, 2013, **102**, 3485.
61. C. T. Avetta, B. J. Shorthill, C. Ren and T. E. Glass, *J. Org. Chem.*, 2012, **77**, 851.
62. G. Gimpl and K. Gehrig-Burger, *Steroids*, 2011, **76**, 216.
63. R. Breslow and B. Zhang, *J. Am. Chem. Soc.*, 1996, **118**, 8495.
64. A. I. Rosenbaum and F. R. Maxfield, *J. Neurochem.*, 2011, **116**, 789.
65. A. Tamura and N. Yui, *Sci. Rep.*, 2014, **4**, 4356.
66. A. Adan-Gokbulut, M. Kartal-Yandim, G. Iskender and Y. Baran, *Curr. Med. Chem.*, 2013, **20**, 108.
67. S. R. Clark, C. P. Thomas, V. J. Hammond, M. Aldrovandi, G. W. Wilkinson, K. W. Hart, R. C. Murphy, P. W. Collins and V. B. O'Donnell, *Proc. Natl. Acad. Sci. U. S. A.*, 2013, **110**, 5875.
68. T. N. Lambert and B. D. Smith, *Coord. Chem. Rev.*, 2003, **240**, 129.

69. J. O. Magrans, A. R. Ortiz, M. A. Molins, P. H. P. Lebouille, J. Sánchez-Quesada, P. Prados, M. Pons, F. Gago and J. de Mendoza, *Angew. Chem., Int. Ed. Engl.*, 1996, **35**, 1712.
70. J. M. Boon, R. Shukla, B. D. Smith, G. Licini and P. Scrimin, *Chem. Commun.*, 2002, 260.
71. J. M. Boon, T. N. Lambert, A. L. Sisson, A. P. Davis and B. D. Smith, *J. Am. Chem. Soc.*, 2003, **125**, 8195.
72. A. Alliband, F. A. Meece, C. Jayasinghe and D. H. Burns, *J. Org. Chem.*, 2013, **78**, 356.
73. X.-Z. Lai, Y. Feng, J. Pollard, J. N. Chin, M. J. Rybak, R. Bucki, R. F. Epand, R. M. Epand and P. B. Savage, *Acc. Chem. Res.*, 2008, **41**, 1233.
74. B. Ding, N. Yin, Y. Liu, J. Cardenas-Garcia, R. Evanson, T. Orsak, M. Fan, G. Turin and P. B. Savage, *J. Am. Chem. Soc.*, 2004, **126**, 13642.
75. K. M. Bucholtz, P. C. Gareiss, S. G. Tajc and B. L. Miller, *Org. Biomol. Chem.*, 2006, **4**, 3973.
76. R. D. Hubbard, S. R. Horner and B. L. Miller, *J. Am. Chem. Soc.*, 2001, **123**, 5810.
77. J. E. Vance and G. Tasseva, *Biochim. Biophys. Acta*, 2013, **1831**, 543.
78. S. Ran, A. Downes and P. E. Thorpe, *Cancer Res.*, 2002, **62**, 6132.
79. H. Zheng, F. Wang, Q. Wang and J. Gao, *J. Am. Chem. Soc.*, 2011, **133**, 15280.
80. A. S. Hosseini, H. Zheng and J. Gao, *Tetrahedron*, 2014, **70**, 7632.
81. K. Igarashi, M. Kaneda, A. Yamaji, T. C. Saido, U. Kikkawa, Y. Ono, K. Inoue and M. Umeda, *J. Biol. Chem.*, 1995, **270**, 29075.
82. C. Xiong, K. Brewer, S. Song, R. Zhang, W. Lu, X. Wen and C. Li, *J. Med. Chem.*, 2011, **54**, 1825.
83. R. Shao, C. Xiong, X. Wen, J. G. Gelovani and C. Li, *Mol. Imaging*, 2007, **6**, 417.
84. C. Burtea, S. Laurent, E. Lancelot, S. Ballet, O. Murariu, O. Rousseaux, M. Port, L. Vander Elst, C. Corot and R. N. Muller, *Mol. Pharm.*, 2009, **6**, 1903.
85. N. Thapa, S. Kim, I.-S. So, B.-H. Lee, I.-C. Kwon, K. Choi and I.-S. Kim, *J. Cell. Mol. Med.*, 2008, **12**, 1649.
86. M. Ashrafuzzaman, C. Y. Tseng, J. Kapty, J. R. Mercer and J. A. Tuszynski, *Nucleic Acid Ther.*, 2013, **23**, 418.
87. H. T. Ngo, X. Liu and K. A. Jolliffe, *Chem. Soc. Rev.*, 2012, **41**, 4928.
88. R. G. Hanshaw, C. Lakshmi, T. N. Lambert, J. R. Johnson and B. D. Smith, *ChemBioChem*, 2005, **6**, 2214.
89. C. Lakshmi, R. G. Hanshaw and B. D. Smith, *Tetrahedron*, 2004, **60**, 11307.
90. B. A. Smith, S. Xiao, W. Wolter, J. Wheeler, M. A. Suckow and B. D. Smith, *Apoptosis*, 2011, **16**, 722.
91. B. A. Smith, S. T. Gammon, S. Xiao, W. Wang, S. Chapman, R. McDermott, M. A. Suckow, J. R. Johnson, D. Piwnica-Worms and G. W. Gokel, *Mol. Pharm.*, 2011, **8**, 583.

92. B. A. Smith, W. J. Akers, W. M. Leevy, A. J. Lampkins, S. Xiao, W. Wolter, M. A. Suckow, S. Achilefu and B. D. Smith, *J. Am. Chem. Soc.*, 2009, **132**, 67.
93. A. J. Plaunt, K. M. Harmatys, W. R. Wolter, M. A. Suckow and B. D. Smith, *Bioconjugate Chem.*, 2014, **25**, 724.
94. B. A. Smith, K. M. Harmatys, S. Xiao, E. L. Cole, A. J. Plaunt, W. Wolter, M. A. Suckow and B. D. Smith, *Mol. Pharm.*, 2013, **10**, 3296.
95. S. Turkyilmaz, D. R. Rice, R. Palumbo and B. D. Smith, *Org. Biomol. Chem.*, 2014, **12**, 5645.
96. D. Oltmanns, S. Zitzmann-Kolbe, A. Mueller, U. Bauder-Wuest, M. Schaefer, M. Eder, U. Haberkorn and M. Eisenhut, *Bioconjugate Chem.*, 2011, **22**, 2611.
97. C.-L. Do-Thanh, M. M. Rowland and M. D. Best, *Tetrahedron*, 2011, **67**, 3803.
98. W. M. Leevy, S. T. Gammon, H. Jiang, J. R. Johnson, D. J. Maxwell, E. N. Jackson, M. Marquez, D. Piwnica-Worms and B. D. Smith, *J. Am. Chem. Soc.*, 2006, **128**, 16476.
99. S. Xiao, L. Abu-Esba, S. Turkyilmaz, A. G. White and B. D. Smith, *Theranostics*, 2013, **3**, 658.
100. L. H. Mak, S. N. Georgiades, E. Rosivatz, G. F. Whyte, M. Mirabelli, R. Vilar and R. Woscholski, *ACS Chem. Biol.*, 2011, **6**, 1382.
101. S. Geninatti Crich, D. Alberti, I. Szabo, S. Aime and K. Djanashvili, *Angew. Chem. Int. Ed.*, 2013, **52**, 1161.
102. G. Whyte, R. Vilar and R. Woscholski, *J. Chem. Biol.*, 2013, **6**, 161.
103. C. M. Cooley, K. S. Hettie, J. L. Klockow, S. Garrison and T. E. Glass, *Org. Biomol. Chem.*, 2013, **11**, 7387.
104. A. A. Volmer, A. M. Szpilman and E. M. Carreira, *Nat. Prod. Rep.*, 2010, **27**, 1329.
105. M. Pinot, S. Vanni, S. Pagnotta, S. Lacas-Gervais, L.-A. Payet, T. Ferreira, R. Gautier, B. Goud, B. Antonny and H. Barelli, *Science*, 2014, **345**, 693.

Subject Index

Page references to **figures, tables** and **text boxes** where not integral to text are shown in **bold**.